Principles and Clinical Diagnostic Applications of Surface-Enhanced Raman Spectroscopy

Principles and Clinical Diagnostic Applications of Surface-Enhanced Raman Spectroscopy

Edited by

YULING WANG

Department of Molecular Sciences, ARC Center of Excellence for Nanoscale BioPhotonics, Faculty of Science and Engineering, Macquarie University, Sydney, NSW, Australia

ELSEVIER

Elsevier
Radarweg 29, PO Box 211, 1000 AE Amsterdam, Netherlands
The Boulevard, Langford Lane, Kidlington, Oxford OX5 1GB, United Kingdom
50 Hampshire Street, 5th Floor, Cambridge, MA 02139, United States

Notices
Knowledge and best practice in this field are constantly changing. As new research and experience broaden our understanding, changes in research methods, professional practices, or medical treatment may become necessary.

Practitioners and researchers must always rely on their own experience and knowledge in evaluating and using any information, methods, compounds, or experiments described herein. In using such information or methods they should be mindful of their own safety and the safety of others, including parties for whom they have a professional responsibility.

To the fullest extent of the law, neither the Publisher nor the authors, contributors, or editors, assume any liability for any injury and/or damage to persons or property as a matter of products liability, negligence or otherwise, or from any use or operation of any methods, products, instructions, or ideas contained in the material herein.

British Library Cataloguing-in-Publication Data
A catalogue record for this book is available from the British Library

Library of Congress Cataloging-in-Publication Data
A catalog record for this book is available from the Library of Congress

ISBN: 978-0-12-821121-2

Publisher: Susan Dennis
Acquisitions Editor: Kathryn Eryilmaz
Editorial Project Manager: Andrea Dulberger
Production Project Manager: Debasish Ghosh
Cover Designer: Christian J. Bilbow

Typeset by MPS Limited, Chennai, India

Contents

List of contributors

Mujo Adanalic
Department of Chemistry, CENIDE and ZMB, University of Duisburg-Essen, Essen, Germany

Yifan Bao
State Key Laboratory of Physical Chemistry of Solid Surfaces, Collaborative Innovation Center of Chemistry for Energy Materials (*i*ChEM), College of Chemistry and Chemical Engineering, Xiamen University, Xiamen, P.R. China

Alois Bonifacio
Department of Engineering and Architecture (DIA), University of Trieste, Trieste, Italy

Maofeng Cao
State Key Laboratory of Physical Chemistry of Solid Surfaces, Collaborative Innovation Center of Chemistry for Energy Materials (*i*ChEM), College of Chemistry and Chemical Engineering, Xiamen University, Xiamen, P.R. China

Jaslyn Ru Ting Chen
Division of Chemistry and Biological Chemistry, School of Physical and Mathematical Sciences, Nanyang Technological University, Singapore

Yang Chen
Department of Laboratory Medicine, Fujian Medical University, Fuzhou, P.R. China

Jaebum Choo
Department of Chemistry, Chung-Ang University, Seoul, Republic of Korea

Anupam Das
Department of Chemistry, Chung-Ang University, Seoul, Republic of Korea

Shangyuan Feng
Key Laboratory of OptoElectronic Science and Technology for Medicine, Ministry of Education, Fujian Provincial Key Laboratory for Photonics Technology, Fujian Normal University, Fuzhou, P.R. China

Ren Hu
State Key Laboratory of Physical Chemistry of Solid Surfaces, Collaborative Innovation Center of Chemistry for Energy Materials (*i*ChEM), College of Chemistry and Chemical Engineering, Xiamen University, Xiamen, P.R. China

Joseph Irudayaraj
Department of Bioengineering, Cancer Center at Illinois, Nick Holonyak Micro and Nanotechnology Laboratory, University of Illinois at Urbana-Champaign, Urbana, IL, United States

Janina Kneipp
Department of Chemistry, Humboldt-Universität zu Berlin, Berlin, Germany

Charlynn Sher Lin Koh
Division of Chemistry and Biological Chemistry, School of Physical and Mathematical Sciences, Nanyang Technological University, Singapore

Shi Xuan Leong
Division of Chemistry and Biological Chemistry, School of Physical and Mathematical Sciences, Nanyang Technological University, Singapore

Yong Xiang Leong
Division of Chemistry and Biological Chemistry, School of Physical and Mathematical Sciences, Nanyang Technological University, Singapore

Xue Li
College of Chemistry and Molecular Sciences, Wuhan University, Wuhan, P.R. China

Duo Lin
Key Laboratory of OptoElectronic Science and Technology for Medicine, Ministry of Education, Fujian Provincial Key Laboratory for Photonics Technology, Fujian Normal University, Fuzhou, P.R. China

Xing Yi Ling
Division of Chemistry and Biological Chemistry, School of Physical and Mathematical Sciences, Nanyang Technological University, Singapore

Guokun Liu
State Key Laboratory of Marine Environmental Science, Fujian Provincial Key Laboratory for Coastal Ecology and Environmental Studies, Center for Marine Environmental Chemistry & Toxicology, College of the Environment and Ecology, Xiamen University, Xiamen, P.R. China

Yuan Liu
Department of Molecular Sciences, ARC Center of Excellence for Nanoscale BioPhotonics, Faculty of Science and Engineering, Macquarie University, Sydney, NSW, Australia

Nana Lyu
Department of Molecular Sciences, ARC Center of Excellence for Nanoscale BioPhotonics, Faculty of Science and Engineering, Macquarie University, Sydney, NSW, Australia

Sufang Qiu
Fujian Medical University Cancer Hospital, Fujian Cancer Hospital, Fuzhou, P.R. China

Bin Ren
State Key Laboratory of Physical Chemistry of Solid Surfaces, Collaborative Innovation Center of Chemistry for Energy Materials (*i*ChEM), College of Chemistry and Chemical Engineering, Xiamen University, Xiamen, P.R. China

Wen Ren
Department of Bioengineering, Cancer Center at Illinois, Nick Holonyak Micro and Nanotechnology Laboratory, University of Illinois at Urbana-Champaign, Urbana, IL, United States

Alison Rodger
Department of Molecular Sciences, ARC Center of Excellence for Nanoscale BioPhotonics, Faculty of Science and Engineering, Macquarie University, Sydney, NSW, Australia

Sebastian Schlücker
Department of Chemistry, CENIDE and ZMB, University of Duisburg-Essen, Essen, Germany

Ai-Guo Shen
School of Printing and Packaging, Wuhan University, Wuhan, P.R. China

Vi. Tran
Department of Chemistry, CENIDE and ZMB, University of Duisburg-Essen, Essen, Germany

Xiang Wang
State Key Laboratory of Physical Chemistry of Solid Surfaces, Collaborative Innovation Center of Chemistry for Energy Materials (*i*ChEM), College of Chemistry and Chemical Engineering, Xiamen University, Xiamen, P.R. China

Yuling Wang
Department of Molecular Sciences, ARC Center of Excellence for Nanoscale BioPhotonics, Faculty of Science and Engineering, Macquarie University, Sydney, NSW, Australia

Shuping Xu
State Key Laboratory of Supramolecular Structure and Materials, Institute of Theoretical Chemistry, College of Chemistry, Jilin University, Changchun, P.R. China

Sen Yan
State Key Laboratory of Physical Chemistry of Solid Surfaces, Collaborative Innovation Center of Chemistry for Energy Materials (*i*ChEM), College of Chemistry and Chemical Engineering, Xiamen University, Xiamen, P.R. China

Haishan Zeng
Imaging Unit—Integrative Oncology Department, BC Cancer Research Centre, Vancouver, BC, Canada; Department of Dermatology and Skin Science, Photomedicine Institute, University of British Columbia, Vancouver, BC, Canada

Yuying Zhang
Medical School of Nankai University, Tianjin, P.R. China

Xiao-Dong Zhou
The Centre of Analysis and Measurement of Wuhan University, Wuhan University, Wuhan, P.R. China

CHAPTER 1

Principles of surface-enhanced Raman spectroscopy

Xiang Wang[1], Guokun Liu[2], Ren Hu[1], Maofeng Cao[1], Sen Yan[1], Yifan Bao[1] and Bin Ren[1]
[1]State Key Laboratory of Physical Chemistry of Solid Surfaces, Collaborative Innovation Center of Chemistry for Energy Materials (*i*ChEM), College of Chemistry and Chemical Engineering, Xiamen University, Xiamen, P.R. China
[2]State Key Laboratory of Marine Environmental Science, Fujian Provincial Key Laboratory for Coastal Ecology and Environmental Studies, Center for Marine Environmental Chemistry & Toxicology, College of the Environment and Ecology, Xiamen University, Xiamen, P.R. China

1.1 Introduction

Raman spectroscopy, originated from the vibration of atomic nucleus, can provide the chemical fingerprint information of molecules. The unique Raman shifts and relative intensities of vibrational peaks reflect the bond strength and atoms forming the bond in a molecule, which enables us to determine the molecular structure and its interaction with environment. This feature endows Raman spectroscopy the power of qualitative analysis in identifying materials and molecular species and characterizing molecular behaviors. In addition, since the intensity of Raman peaks is proportional to the number and concentration of molecules, Raman spectroscopy can be a powerful tool for quantitative analysis [1−3].

Raman spectroscopy, similar to infrared (IR) spectroscopy, reflects the vibrational information of a molecule but follows different selection rules. A change in the polarizability during the vibration is required for a vibrational mode to be Raman active, whereas a change in the dipole moment is required to be IR active, which is highly dependent on the symmetry of the molecule. For a molecule with a center of symmetry, Raman active vibrations are normally IR inactive and vice versa, which is termed the principle of mutual exclusion. However, with the decreased symmetry, more and more vibrational modes will have both Raman and IR activity. Therefore these two techniques are highly complementary to each other. Both techniques have been developed into powerful tools in qualitative and quantitative chemical analysis. Especially, Raman spectroscopy has

Principles and Clinical Diagnostic Applications of Surface-Enhanced Raman Spectroscopy
DOI: https://doi.org/10.1016/B978-0-12-821121-2.00004-4

distinguished itself over IR spectroscopy with the following four merits. (1) Higher spectral (energy) resolution. Typically, the same vibrational mode shows a sharper peak in Raman spectrum than that in IR spectrum, which makes Raman spectroscopy better for multiplex qualitative analysis toward mixed compounds with similar structures. (2) Higher spatial resolution. Theoretically, Raman measurement could be carried out with any excitation wavelength from ultraviolet (UV) to visible and near-IR. The shorter the excitation wavelength, the higher the spatial resolution. (3) Nondestructive analysis. Any sample in solid, liquid, or gas states could be directly analyzed, as far as the laser power can be controlled below the damage threshold value. (4) Weak interference from water. As a bonus of the low polarizability change of the highly polar O−H bond during vibration, Raman spectroscopy is not sensitive to water. Thereby, Raman spectroscopy is favored over IR spectroscopy for the in situ clinical diagnosis of aqueous and wet samples.

However, the sensitivity of Raman spectroscopy is much lower than that of IR spectroscopy due to the too small Raman cross-section of target molecules, in which approximately one Raman photon is produced with 10^6 to 10^{10} incident photons. Therefore the development of Raman spectroscopy is much slower than IR spectroscopy until the invention of laser as the monochromatic and high-power light source in 1960s. Together with the resonance Raman effect, which is able to enhance the Raman scattering up to six orders of magnitude [4], Raman spectroscopy has been successfully applied to study bio-molecular structures [5]. Even with the resonance-enhanced Raman effect, the Raman cross-section for a molecule may still be about eight orders of magnitude smaller than its absorption cross-section at the same wavelength [6]. Therefore the low sensitivity of Raman spectroscopy prevented it to be a routine analytical tool towards trace analysis for over decades. In the mid of 1970s, the discovery of surface-enhanced Raman spectroscopy (SERS) completely changed the situation. With the aid of the strongly enhanced electric field on the surface of metal nanostructures as a result of the surface plasmon resonance (SPR) effect, the Raman signal of adsorbed molecules can be enhanced by over million times, providing a high sensitivity even down to the single-molecule level. Up to now, it is the only vibrational spectroscopy showing the single-molecule sensitivity under ambient condition. Yet, it still maintains the feature of multiplexing analysis thanks to the narrow Raman peaks. These two features are of highly importance and have boosted the use of SERS in biosensing and biomedical applications.

SERS relies on the use of the enhancement provided by Au or Ag nanostructures. The application of SERS in clinical diagnosis involves the design and fabrication of an active nanostructure, binding the analyte onto the nanostructure, and excitation/collection of SERS signals. The design and fabrication of a SERS-active surface is the foremost step and crucial to a successful application. In this chapter, we will guide you through the principles of SERS, including electromagnetic field enhancement and chemical enhancement, with emphasis on the SPR effect and its dependence on materials, size, shape, arrangement, and so on. Relevant practical issues will also be briefed from the mechanistic understanding. We hope this chapter will provide you the necessary background to understand the literatures and start your own journey of applying SERS for clinical diagnosis.

1.2 Surface plasmon resonance

SPR is the key to make SERS possible. The field of plasmonics is mainly developed on the basis of the resonant interaction between the light and free electrons of a metal. Back to the time of Roman Empire, this specific light-matter interaction was used to present colors of the Lycurgus cup [7]. The ability to strongly absorb light and enhance the electromagnetic field in the vicinity of metal surface attracts more and more interests to plasmonics, which results in a wide range of applications. In turn, SERS is still one of the most important tools to study plasmonic properties of nanostructures. Benefited from the rapid development of nanoscience and nanotechnology, the fields of SERS and plasmonics have experienced mutual beneficial development in the past four decades [8–10].

The term of surface plasmon is used to describe the collective oscillation of free electrons inside metals under the excitation of light. For simplicity the free electrons are assumed to be independent of each other and the nucleus is considered as a uniform positively charged background. Taking a thin metal film for example (shown in Fig. 1.1), when an

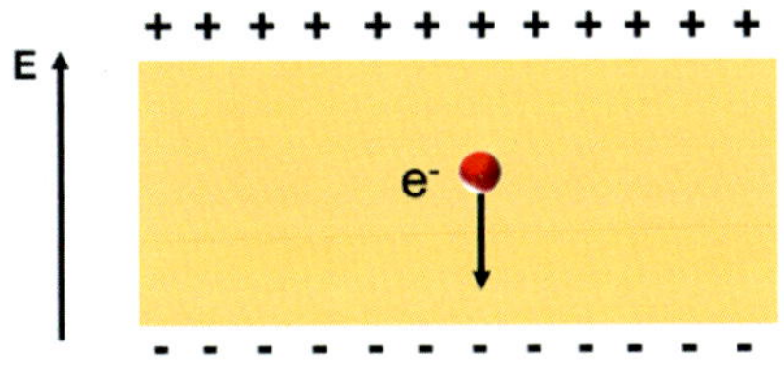

Figure 1.1 Oscillation of free electrons in a thin metal film.

external electric field is applied to the film, the free electrons will "feel" an electric force and be displaced from their initial position. As a result, negative static surface charges appear on one side and positive static surface charges appear on the other side of the film. The electrons can be accelerated by the electric field generated by the surface charges. When they arrive at their initial position, the accumulated momentum will enable the electrons to keep moving. Consequently, an electric field with an opposite direction will be formed. Assuming no damping in Jellium model, this process will repeat and generate the oscillation of free electrons [11]. This electric-field-driven oscillation involves all the electrons, which explicitly illustrates that it is a collective oscillation. The coupled electric field and charges involved during the collective oscillation is called a polariton.

Basically, there are two types of surface plasmon polaritons. One can propagate along the metal–dielectric interface and decay exponentially away from the interface, which is called propagated surface plasmon polaritons. The other one is the localized oscillation of the surface charge density at optical frequencies of a metal nanoparticle sustained by the closed geometrical boundaries, which is commonly referred to as localized surface plasmons (LSPs) [11]. In this case, the electrons with an effective mass respond to the applied external electric field but with an increasing phase lag as the frequency increases. When the phase lag approaches 90 degrees, the amplitude of the electron oscillation reaches a maximum, determined by the internal (Ohmic and radiation) damping of the system. This state corresponds to the localized surface plasmon resonance (LSPR). In this resonant state, free electrons transiently accumulate on the nanostructure surface and greatly increase the density of surface charges, leading to a tightly confined and strongly enhanced electric field on the surface [12]. These particular optical properties are governed by the optical response of free conduction electrons in metals, which will be discussed in the following section.

1.3 Optical properties of metals

To determine the optical properties of plasmonic materials, the electromagnetic response of free conduction electrons in the metal should be explicitly modeled and calculated. However, it is still a great challenge due to lots of complicated factors involved, such as the interaction of electrons with ions, the electron–electron correlations, the interaction between electrons and impurities, and the presence of surface. Therefore many approximations should be considered to describe this optical response. The Drude model [13,14] is widely used to describe the optical response of metals, where free electrons are uniformly and randomly distributed throughout the periodic structure of ions without restoring force. The total optical susceptibility is simply the sum of the individual polarizabilities of free electrons. The dielectric function of metals can be described with Drude model as [15]:

$$\varepsilon(\omega) = 1 - \frac{ne^2}{m\varepsilon_0}\frac{1}{\omega^2 + i\gamma\omega} \tag{1.1}$$

where n is the number of free electrons per unit volume, m is their effective mass, γ is the damping term which corresponds to the collision rate of free electrons with ions or impurities, ε_0 is the dielectric constant in vacuum, and ω is the incident frequency. The residual polarization from the positively charged ions and the influence of intraband and interband transitions can also be included, which is described in many textbooks [15]. Wherein, ω_p is defined as

$$\omega_p^2 = \frac{ne^2}{m\varepsilon_0} \tag{1.2}$$

and corresponds to the native oscillation frequency of the bulk plasmon.

Then we can obtain the real and imaginary parts of the dielectric function:

$$\mathrm{Re}(\varepsilon(\omega)) = 1 - \frac{\omega_p^2}{\omega^2 + \gamma^2} \tag{1.3}$$

$$\mathrm{Im}(\varepsilon(\omega)) = \frac{\omega_p^2\gamma}{\omega(\omega^2 + \gamma^2)} \tag{1.4}$$

The collision rate γ is inversely proportional to the relaxation time of free electrons, which is at the scale of $10^{14}\,\mathrm{s}^{-1}$ at room temperature. Because e, m, and ε_0 are constant, we can obtain $\omega_p^2 \propto n$ from Eq. (1.2), which means that the bulk frequency is governed by the density of free electrons. For dielectric materials, such as glass, due to the extremely low density of free electrons, $\omega_p \ll \omega$ at optical frequencies (at the scale of $10^{14} \sim 10^{15}\,\mathrm{s}^{-1}$); therefore, we can obtain $\omega_p^2/\omega^2 + \gamma^2 \ll 1$. As a result, the real part for dielectric materials is always positive. But for metals like Au and Ag with high conductivity, the density of conduction electrons is on the order of $6 \times 10^{28}\,\mathrm{m}^{-3}$, corresponding to a large ω_p of over

10^{16} s^{-1} [16]. In this case, $\omega_p^2/\omega^2 + \gamma^2 > 1$ and we have $\mathrm{Re}(\varepsilon(\omega)) < 0$. In addition, because optical frequency is relatively large, the imaginary part that determines the absorption will be small from Eq. (1.4). As it will be introduced in the next section, negative $\mathrm{Re}(\varepsilon(\omega))$ and small $\mathrm{Im}(\varepsilon(\omega))$ are the key to realize many interesting optical effects of metals, including LSPR, which is also in strong contrast with normal dielectrics with $\mathrm{Re}(\varepsilon(\omega))$ between 1 and 10 [15].

1.4 Local field enhancement

In this section, we will show that the negative real part of the dielectric function for metals enables the phase lag between the applied electric field and the collectively oscillated electrons and thus allows the existence of SPs and LSPR. Eventually the dramatically enhanced electric field in the vicinity of the metal surface due to the LSPR effect makes SERS possible.

Generally, the frequency-dependent polarizability of a metallic object is determined by the dielectric function of the metal and the surrounding environment. For simplicity, we consider the problem of a small metallic nanosphere excited by the electromagnetic field of light. The problem can be solved within the electrostatic approximation if the size of the sphere is much smaller than the wavelength of the light. Then the polarizability of the metallic sphere can be described as [11]:

$$\alpha = 4\pi\varepsilon_0 a^3 \frac{\varepsilon_{\mathrm{m}} - \varepsilon_{\mathrm{env}}}{\varepsilon_{\mathrm{m}} + 2\varepsilon_{\mathrm{env}}} \tag{1.5}$$

where a is the radius of the sphere, and ε_{m} and $\varepsilon_{\mathrm{env}}$ are the dielectric functions of the metal and environment, respectively. A very large polarizability can be achieved if the denominator is close to zero, when $\varepsilon_{\mathrm{m}} = -2\varepsilon_{\mathrm{env}}$. This is impossible for normal dielectrics because their ε_{m} are always positive. But for metals, such as Au and Ag, this condition can be met at the frequency when $\mathrm{Re}(\varepsilon(\omega)) = -2\varepsilon_{\mathrm{env}}$ and $\mathrm{Im}(\varepsilon(\omega))$ is small. Within the electrostatic approximation, for the metallic sphere, we can write [11]:

$$C_{\mathrm{scat}} = \frac{8\pi}{3} k^4 a^6 \left| \frac{\varepsilon_{\mathrm{m}} - \varepsilon_{\mathrm{env}}}{\varepsilon_{\mathrm{m}} + 2\varepsilon_{\mathrm{env}}} \right|^2 = \frac{k^4}{6\pi} |\alpha|^2 \tag{1.6}$$

$$C_{\mathrm{abs}} = 4\pi k a^3 \left\{ \frac{\varepsilon_{\mathrm{m}} - \varepsilon_{\mathrm{env}}}{\varepsilon_{\mathrm{m}} + 2\varepsilon_{\mathrm{env}}} \right\} = k\mathrm{Im}\{\alpha\} \tag{1.7}$$

where C_{scat} and C_{abs} are the scattering and absorption cross-sections, respectively, and k is the wavevector. We can see that the scattering and absorption can become very large under the same condition to achieve the maximal polarizability. We call such a state a resonance condition. The electric field inside the metallic sphere can be described as [11]:

$$E_{In} = \frac{3\varepsilon_{env}}{\varepsilon_m + 2\varepsilon_{env}} E_0 \tag{1.8}$$

where E_0 is the incident field. Therefore the electric field can be dramatically enhanced in the resonant state. The significance of the LSPR of metal nanoparticles lies in two aspects. On the one hand, the local fields around the nanoparticle as well as the absorption of light can be enhanced dramatically (even hundreds of times) in the resonant state. On the other hand, the electric field is strongly localized near the metal surface within a range of only 1/10 wavelength as shown in Fig. 1.2 due to the excitation of surface plasmon, which has been utilized to manipulate light at the nanoscale. These effects form the basis of a range of applications including plasmon-enhanced spectroscopies, medical diagnosis and photovoltaics.

The plasmonic effects described above can occur for any metallic nanoparticles, not restricted to spheres. Phenomenologically the metal nanoparticles act as an effective optical antenna to convert the propagating electromagnetic field into strongly localized near fields, giving rise to field enhancement. In this case, the optical response of metal nanoparticles is mainly controlled by the dimension of their axis along the incident

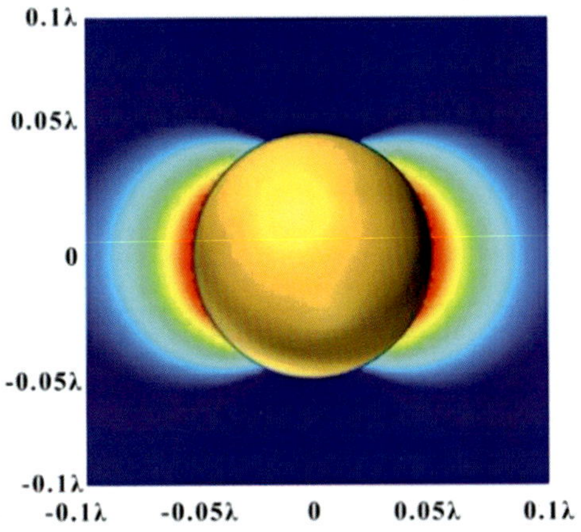

Figure 1.2 The near-field electric field distribution around a single Au sphere with a diameter of 50 nm excited at $\lambda = 633$ nm.

electric field, which is similar to the antenna at radio and telecom frequencies. At radio and telecom frequencies, the high permittivity of metals prevents the external electromagnetic field from penetrating into them, and therefore the optical response is completely determined by geometry. However, at higher frequencies like the visible region, the dielectric function of metals becomes lower in the absolute value. As a result, the electromagnetic field is able to penetrate deeper into metals, leading to the excitation of surface plasmons. Therefore it is interesting to see that the optical response of metallic nanoparticles is governed not only by the dielectric functions but also by the shape, size, and geometry. We will discuss these effects in Section 1.7.

1.5 Surface-enhanced Raman spectroscopy enhancement and $|E|^4$ approximation

Raman scattering is an inelastic photon scattering process involving two photons due to the excitation of vibrational modes. As shown in Fig. 1.3A, an electron at the ground state (S_0) absorbs the energy of the incident photon (ω_0) and transits to a "virtual state." This virtual state is a transient state originated from the perturbation of the electron cloud by light. For the Stokes process the electron decays back to an excited vibrational state (ν_1) in the electronic ground state (S_0) by scattering a photon with a lower frequency ω_{sc}, where $\omega_{sc} = \omega_0 - \omega_v$. ω_v is the energy difference between two vibrational states. Note that Raman scattering is instantaneous. Therefore the incident photon and scattering photon are directly

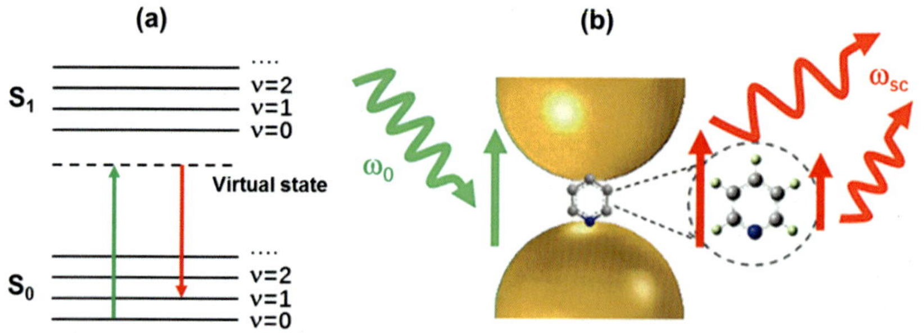

Figure 1.3 (A) Schematic Jablonski diagram for the electronic structure of the molecule. (B) Electromagnetic enhancement process of SERS where both excitation and scattering processes are enhanced. The green and red arrows represent incident and scattering photons, respectively [17]. *SERS*, surface-enhanced Raman spectroscopy.

connected in a coherent way through the scattering process. If the rate for the generation of the scattered photons is enhanced, there will be an increased rate for the incoming photons and vice versa [18]

A molecule near the metal surface experiences an enhanced local electric field due to the excitation of the surface plasmon, $E_{local}(\omega_0)$, which is much stronger than the incident electric field, $E_0(\omega_0)$. Alternatively, we can consider that the laser field is enhanced, which can produce more Raman processes. Here the enhancement factor (EF) is $M_{ex} = \left|E_{local}(\omega_0)/E_0(\omega_0)\right|^2$. From the Fermi golden rule the lifetime of a two-level system is proportional to the number of final states. Regarding the rate of radiative relaxation, the radiative decay rate is proportional to the number of electromagnetic states at the corresponding frequency [11]. The strongly enhanced electric field increases the local density of electromagnetic states, which in turn increases the radiation rate of the scattering process of the molecules and thus enhances the Raman process. Here the EF is $M_{sc} = \left|E_{local}(\omega_{sc})/E_0(\omega_{sc})\right|^2$ (where ω_{sc} is the frequency of Raman scattering). The above process can be understood in the following scenario: when a molecule is placed on the nanoparticle surface where the electric field at ω_{sc} is enhanced, the radiation rate of the Raman scattering process can be increased. Since the scattering is directly coupled to the excitation process, more photons can be drawn from the excitation laser to compensate the increased radiation rate of scattering. Therefore Raman process can benefit from both the excitation and scattering enhancements, as shown in Fig. 1.3B. Then the EF M for the whole process should be:

$$M = M_{ex} \times M_{sc} \tag{1.9}$$

The difference between EF at ω_0 and ω_{sc} sometimes can be ignored because Raman shift is small compared with the frequency range where the LSPR-induced local electric field enhancement presents an obvious change. As a result the EF for SERS can be simply written as

$$\mathrm{M} \approx M_{ex}^2 = \left|E_{local}(\omega_0)/E_0(\omega_0)\right|^4 \tag{1.10}$$

This is so-called $|E|^4$ approximation for SERS. Note that this expression includes a series of approximation, such as ignoring polarization issues between the incident and scattered field, molecular orientation with the surface field, and the validity of the optical reciprocal theorem [19–21]. However, it provides a useful way to roughly evaluate the enhancement in practical SERS experiments.

1.6 Choice of metals

From the description in Section 1.4, we have a clear idea that the resonance condition is determined by the real part ($Re(\varepsilon(\omega))$) of the dielectric function of metal and a negative real part is the key to realize LSPR. However, it is the imaginary part ($Im(\varepsilon(\omega))$) that limits how large the resonance can achieve. For a metal to be good for plasmonics the $Im(\varepsilon(\omega))$ should be small. Fig. 1.4 shows the dielectric function for various metals, and it can be found that Ag is the most ideal metal for plasmonics since $Im(\varepsilon(\omega))$ keeps small at wavelengths longer than 320 nm, almost over the whole UV-vis-near-IR range. It shows a very high enhancement and can be used in ultrasensitive detection, such as single-molecule detection. As shown in Fig. 1.4, Au and Cu can also be good plasmonic metals but only in the long wavelength range (typically longer that 600 nm), and Al is good for the UV range. Pd is not a good plasmonic metal in the visible range because its $Im(\varepsilon(\omega))$ is too large, indicating a large absorption.

Besides the dielectric function, many other factors should also be carefully considered in practical experiments, such as chemical stability. For example, although Li has almost ideal dielectric function as a plasmonic material across the whole visible range, it is not used in practical applications due to its low chemical stability. Cu is also easily oxidized in air, which prevents it from wide applications. Ag shows a very high enhancement. However, it has only moderate chemical stability, and it can be easily oxidized when exposed to air and can be further dissolved to form toxic Ag^+. Therefore special attention should be paid to the possible biotoxicity for in vivo applications. However, Ag is a good candidate for applications where the sensitivity is of top priority but toxicity is not of

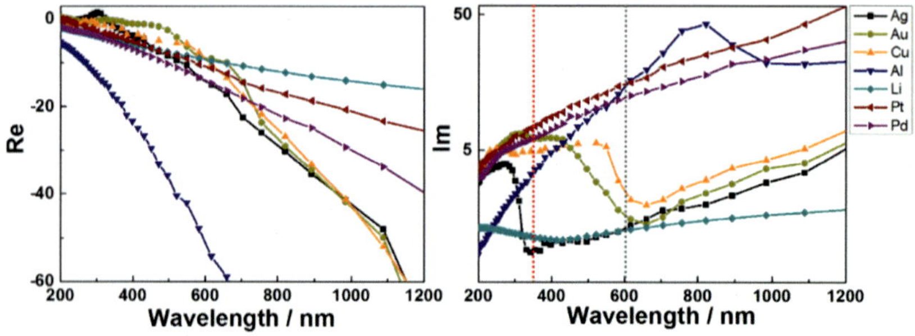

Figure 1.4 The real and imaginary part of dielectric functions for various metals. The red and gray dot lines are displayed to divide the UV-vis-NIR range into three part for readers' convenience.

central concern. Au is the most widely used plasmonic metal because of its high stability and good biocompatibility. It is most advantageous for biological and clinical applications. Furthermore, in many biological applications, where NIR lasers are preferred to use to avoid the photodamage of biomolecules or living cells, Au shows comparable SERS enhancement to Ag.

1.7 The effect of size and shape on the field enhancement

It has been stated above that it is the LSPR effect that dramatically enhances the electric field in the vicinity of the metal nanostructures and dominantly contributes to the SERS enhancement. In addition to materials the LSPR effect is further determined by the size and shape of metal nanostructures, which will be discussed in the following part.

The first is the size effect. Among small metal nanoparticles, spherical nanoparticles are the most commonly used objects for experimental and theoretical researches. For spherical nanoparticles with sizes (<10 nm) far smaller than the wavelength of the incident light and under the illumination of light, the entire nanoparticles can be considered to be in a uniformly distributed electric field. Therefore within this size range, the change in the size of nanoparticles has little influence on the local field EF and their enhancement is very small due to the strong damping from surface scattering [22]. With the increase of the size (>30 nm) of nanoparticles approaching that of the electric field, the electrostatic field approximation gradually fails, leading to an obvious size effect. Spherical nanoparticles in this range can often provide strong local field enhancement at the resonance wavelength. Because the scattering cross-section of the particle is in the sixth power relation with the size of the nanoparticle from Eq. (1.6), the scattering loss of the particle becomes serious when the particle is large. A systematic study of the size effect on the SERS enhancement by dispersing monodispersed Au nanoparticles (in the range of 16–160 nm) uniformly on a substrate reveals that the maximum enhancement can be obtained from the 130 nm Au nanoparticles with 632.8 nm excitation [23]. This result agrees well with the description above that there is an optimal size to achieve the strongest enhancement because small nanoparticles have a severe surface damping and large nanoparticles have a large radiation damping. Note that the optimal size may be different for a different excitation wavelength. In this study the nanoparticles were in a coupled state and the interparticle coupling may significantly affect the LSPR properties of nanostructures, which may result in a different nanoparticle size to achieve the highest SERS enhancement. A recent correlated single-nanoparticle LSPR-SEM-SERS study of Au nanorods with different sizes but supporting a same LSPR wavelength allows the systematical investigation of the size effect on SERS enhancement without bothering the interparticle coupling and different excitation efficiency. As shown in Fig. 1.5, larger nanorods show stronger scattering intensity but much weaker SERS intensity than smaller ones. Nanorods with a smaller size show stronger SERS signals as a result of the synergistic effect of stronger lightning-rod effect and weaker

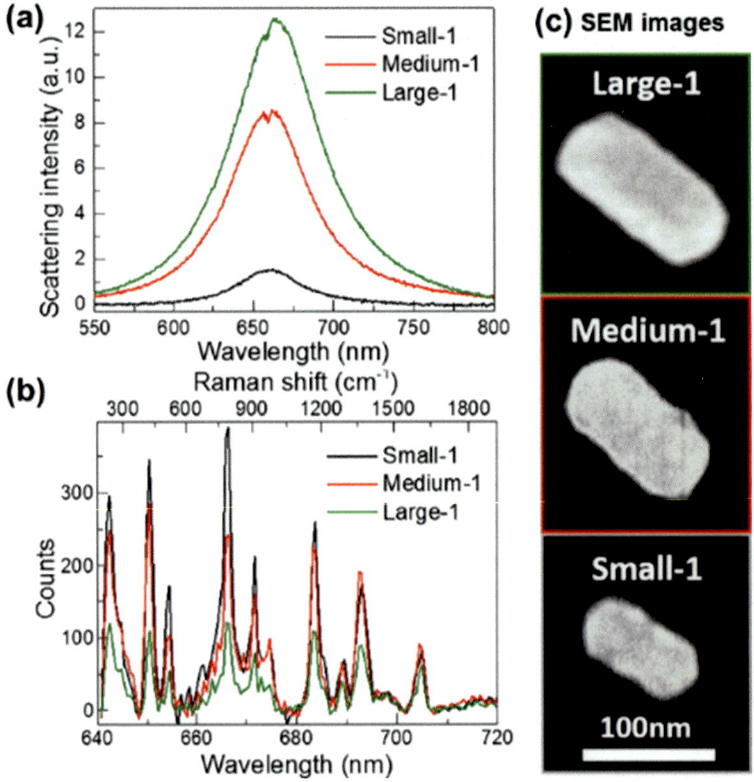

Figure 1.5 (A) The scattering intensity of three Au nanorods with different sizes measured by dark-field microscope. (B) Background-subtracted SERS spectra of the three Au nanorods. (C) SEM images of three Au nanorods with similar LSPR wavelength. The length and the diameter characterized through SEM for the small, medium, and large nanorods are 85.6 $\pm$ 5.5 nm/22.4 $\pm$ 2.1 nm, 116.3 $\pm$ 6.5 nm/34.6 $\pm$ 1.7 nm, and 129.9 $\pm$ 5.4 nm/45.8 $\pm$ 2.3 nm, respectively [24].

radiation damping. Such a size effect has been generalized in the coupling system of Au nanorod dimers with a same LSPR wavelength but different sizes [24].

The second is the shape effect. Surface plasmons are strongly influenced by the shape of nanoparticles. For example, a Au nanorod has two resonance modes at different wavelengths corresponding to the longitudinal and transverse directions. By changing the aspect ratio of nanorods, the overlap between the longitudinal plasmon band and excitation source can be tuned, resulting in different EFs [25]. Furthermore the local electric field distribution is different on the nanoparticles with different shapes. For example, the local electric field tends to be concentrated at the tip for nanoparticles with sharp features because of the lightning-rod effect. Various methods have been used to prepare nanoparticles of different shapes, such as spheres, rods, cubes, prisms, polyhedrons, and nanoflowers to tune the resonance mode of surface plasmon. Nanoflowers with sharp structures have been shown to have a strong surface enhancement [26].

In addition to above-mentioned effects of size and shape, LSPR and hence SERS also depend on surrounding environments. Therefore in order to realize a strong enhancement, it is crucial to find the LSPR position in the real measuring environment to select the right excitation wavelength before SERS measurements. A common practice to obtain the LSPR position of metal nanoparticles is to measure the extinction spectrum on an UV-vis spectrometer. However, the experimentally obtained extinction spectrum is contributed by both the absorption and scattering of nanostructures. Among them, the absorption results in the enhancement of the local field as well as the generation of heat, and the scattering dominates the radiation efficiency of the Raman processes. Therefore it will be erroneous or misleading for SERS measurements if the absorption process dominates. A dual-channel optical fiber-based UV-vis spectrometer can be used to extract the scattering and absorption contribution from the experimentally measured extinction spectra (collected in the transmission mode) and scattering spectra (collected at the direction perpendicular to the incident direction) [27]. Fig. 1.6 shows the normalized extinction spectra and the extracted absorption and scattering spectra for nanoparticles of different materials and shapes. The 60 nm Ag sphere shows the strongest scattering (Fig. 1.6B), and the scattering dominates in LSPR spectrum. Therefore Ag nanoparticles can be the best candidate for dark-field spectroscopy and a good probe for bioimaging [28,29]. As shown in Fig. 1.6C–E, Au nanorods, Au nanocages, and Pd nanosheets

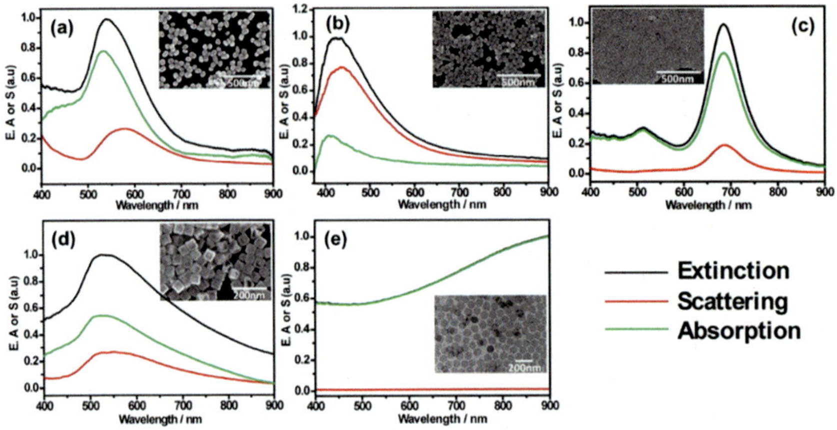

Figure 1.6 Extinction, scattering, and absorption spectra of various nanoparticles. (A) 60 nm Au spheres. (B) 60 nm Ag spheres. (C) Au nanorods. (D) Au nanocages. (E) Pd nanosheets [27].

present strong absorption contribution to the extinction, making them promising candidate for photothermal therapy [30–34].

1.8 Hot spots and various configurations for SERS

The local electric field enhancement provided by a single nanoparticle is too low to be used in practical applications. An effective strategy to boost the enhancement is to take advantage of the coupling between nanoparticles. When two nanoparticles are close enough (<5 nm) and the polarization direction of incident light is parallel to the center axis between them, a strong electric field will form in the gap. From a simple scenario, the two metal nanoparticles with a diameter of D separated by a distance of d can be regarded as a parallel plate capacitor placed in a uniform electrostatic field E_0, as shown in Fig. 1.7. If the particles are perfectly conductive, the electrostatic potential drop will be concentrated in the interparticle region, leading to a local field of about $E_{local} = E_0(D + d)/d$ [35]. The molecule sitting in the gap can experience a Raman enhancement of about $[(D+d)/d]^4$. With $D = 50$ nm and $d = 1$ nm, the EF can be as high as 6.7×10^6, which is much higher than that from a single particle (usually of 10^3) [17]. Considering the accumulated surface charges due to LSPR effect, the enhancement will be much stronger. Theoretical calculation shows that the EF increases exponentially with the decreased distance between two particles. For

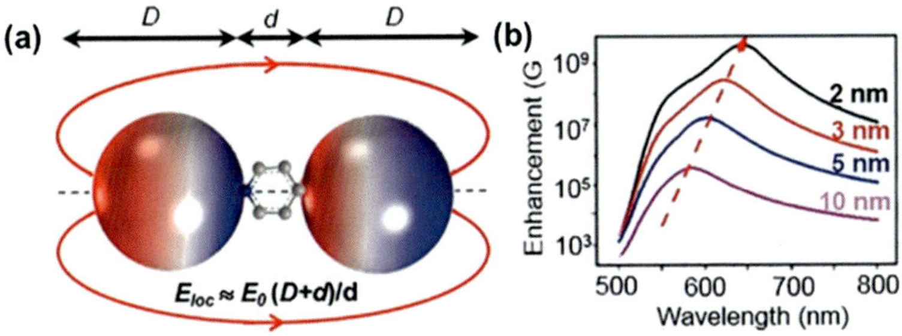

Figure 1.7 (A) A simplified model to understand the electromagnetic field enhancement inside the interparticle region. (B) Gap-dependent SERS enhancement of a Au nanoparticle dimer [17].

example, EF of an Ag dimer with 2 nm gap ($\sim 10^9$) is about 10^4 larger than that of a Ag sphere [8]. Therefore a single dimer can produce a SERS signal equal to that of tens of thousands of single nanoparticles, and the SERS signals in practical SERS enhancements are contributed by the small amount of aggregates of nanoparticles in solution or on surface. Therefore it is the LSPR response of the aggregates other than the isolated nanoparticles that contributes most to the overall SERS signal. However, their contribution to the overall extinction spectra is insignificant. Therefore it is important to measure the optical properties of aggregates with the addition of target molecules rather than in the as-synthesized sol of isolated nanoparticles to find the optimal excitation wavelength for practical SERS measurements.

According to above description about the importance of coupling between nanoparticles for SERS, it is necessary to create more sites with high enhancements (termed hot spots) to optimize the performance of SERS substrates. The field enhancement in the hot spot decreases exponentially not only with the increased distance of d between nanostructures as described in the previous section but also along the surface (along s direction) away from the core of the hot spot (Fig. 1.8A), representing a long-tail distribution around the hot spot. Taking a 30-nm Ag nanoparticle dimer with a gap distance of 2 nm as an example, the area on the surface where the EF is larger than 1/10 of the maximum value is only 0.68% of the total surface area. Consequently, only 0.6% of the total surface area contributes 80% of the total signals. As a result, subtle perturbation like molecular motion, surface diffusion, and laser heating will lead to fluctuation of signals, which is still a critical issue for reproducible and quantitative SERS detection [18,36].

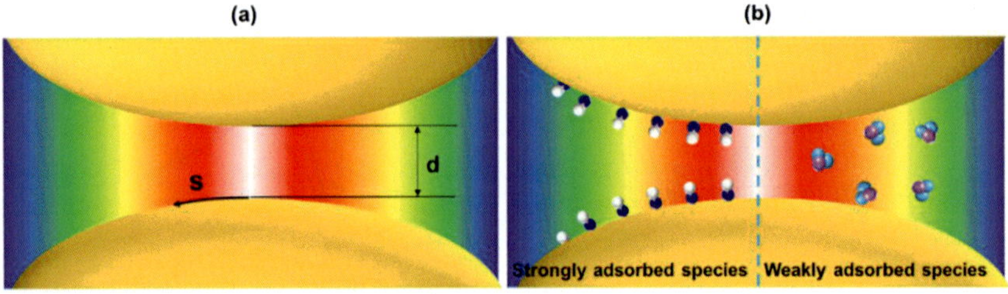

Figure 1.8 (A) Schematic of the gap between two nanoparticles. (B) The cases for strongly (left) and weakly (right) adsorbed molecules [10].

When the probe molecule approaches the surface, it will be either trapped by the surface if the incoming molecule possesses a kinetic energy smaller than the attractive surface potential, or desorbed from the surface if the kinetic energy is high enough. The SERS signals can be detected when the residence time of the probe molecule on surface matches the scale of the collection time. The residence time τ is given by $\tau = \tau_0 \exp(\Delta H_{ads}/RT)$, where τ_0 is the surface atom vibrational time (on the order of 10^{-12} s), ΔH_{ads} is the heat of adsorption, T is the temperature, and R is the gas constant. Consequently, the residence time is determined by the binding affinity of the molecule to surface. τ can be 1 s or longer at 300 K for $\Delta H_{ads} > 15\text{kcal/mol}$, which matches the typical collection time in SERS measurements [37,38]. For molecules containing functional groups that can chemisorb on the Au or Ag surface including thiol, pyridine, amine, nitrile, and carboxylic acid, the binding energy is so large that the molecules can stick on the surface to continuously "feel" the enhanced electric field and facilitate the Raman process (as shown in the left of Fig. 1.8B). However, for the weakly adsorbed molecules, they can only transiently stay on the surface and quickly move out of the hot spot (right of Fig. 1.8B). In addition, the impurity and other solvent molecules also prevent the probe molecules from approaching the surface. Therefore it is more difficult to capture and detect the weakly adsorbed molecules, which constitutes the key challenge for the practical applications of SERS. Ideally, an efficient SERS measurement requires the metal nanostructure with very narrow gap, hot spots easily accessible by the target molecules, and trapping of the target molecules inside the hot spot [10].

In SERS the substrate serves for both adsorption of molecules and enhancement of the Raman signal of the sample. The plasmonic colloidal solution (Fig. 1.9A) or solid plasmonic substrate (Fig. 1.9B) are commonly used as SERS substrates, and in this case the molecules are directly

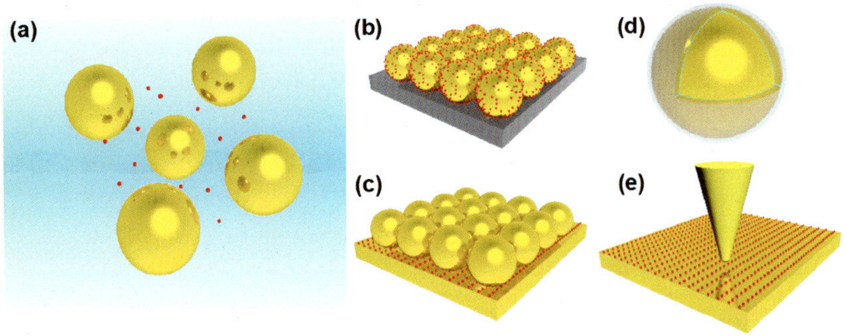

Figure 1.9 Different detection modes for plasmon-enhanced Raman spectroscopy. (A) in the colloidal solution; (B) on the solid substrate; (C) in the gap mode; (D) in shell-isolated nanoparticle-enhanced Raman spectroscopy (SHINERS); (E) in tip-enhanced Raman spectroscopy (TERS) [10].

adsorbed on the substrate. However, if the molecules of interest are on the support not on the nanoparticle surface, a gap-mode SERS can be used (Fig. 1.9C) [39–41]. Special attention should be paid to avoid the interference of the signal from nanoparticles surface or the charge transfer and chemical reactions due to the direct contact of sample with nanoparticles to that of target molecules. Shell-isolated nanoparticle-enhanced Raman spectroscopy (SHINERS) can be utilized to overcome the problem associated with the gap-mode SERS [42]. SHINERS utilizes Au or Ag nanoparticles coated with an ultrathin (1 ~ 5 nm) chemically and electronically inert shell (Fig. 1.9D). The shell can well separate the interaction of metal nanoparticles with the target molecules to minimize interference from nanoparticles and obtain the native chemical interaction of molecules with the support. However, the shell increases the distance between nanoparticles and the target molecules, usually resulting in a lower enhancement compared with that of bare nanoparticles. This problem may be overcome by using a shell of high dielectric index [43].

The difficulty to control the hot spot and the low-spatial resolution associated with SERS and SHINERS can be overcome by using tip-enhanced Raman spectroscopy (TERS) [44–46]. TERS combines scanning probe microscope (SPM) with Raman microscope by replacing the normal tip of SPM with a plasmonic Ag or Au tip. The Raman signal of the sample under the tip can be strongly enhanced due to the strongly confined electric field in the vicinity of the tip apex due to the LSPR and lightning-rod effect (Fig. 1.9E). By scanning the tip over a sample, TERS is able to obtain chemical fingerprint image with a spatial resolution of a few nanometers under the ambient condition or Ångstrom in ultra-high vacuum and at ultralow temperature [47–49]. In contrast to numerous hot spots in the illumination spot in SERS and SHINERS, there is only one hot spot in TERS. Therefore typically the sensitivity of TERS is much lower than that of SERS and SHINERS. However, for single-molecule study, TERS can provide as strong signal as the other two, as the single-molecule signal comes only from a single hot spot.

1.9 Estimation of the surface enhancement factor

The SERS EF is defined as the ratio of the SERS signal of a molecule on the SERS substrate and the Raman signal of the same molecule without SERS substrate when other conditions are consistent (such as the size of the laser spot). To obtain EF, the first step is to acquire the normal Raman signal of molecules. This is usually measured in a concentrated solution because Raman signal is so weak that a large number of molecules are needed to produce detectable signal. On the other hand, strong SERS signals can be detected even though only a few molecules are adsorbed on the SERS substrate. EF can be obtained by the following widely used formula:

$$\mathrm{EF} = \frac{I_{\mathrm{SERS}}/N_{\mathrm{surf}}}{I_{\mathrm{RS}}/N_{\mathrm{vol}}} \tag{1.11}$$

where I_{SERS} and I_{RS} are intensities of SERS and normal Raman signals, respectively, N_{surf} is the number of molecules adsorbed on the SERS substrate, and N_{vol} is the number of molecules measured in the solution. Usually, a molecule that can achieve full monolayer adsorption is used so that N_{surf} can be conveniently calculated. For the system that the molecule can be adsorbed as multiple layers or interacts weakly with the surface, Eq. (1.11) cannot be applied. Then, the analytical EF can be obtained by [15]:

$$\mathrm{EF} = \frac{I_{\mathrm{SERS}}/C_{\mathrm{SERS}}}{I_{\mathrm{RS}}/C_{\mathrm{RS}}} \tag{1.12}$$

where C_{SERS} is the concentration of molecule in solution in SERS measurement, and C_{RS} is the concentration of molecule in solution for taking the normal Raman spectrum. This calculation is quick and easy. However, the adsorption capacity of different molecules on a SERS substrate is different, which will result in different EF for the same SERS substrate. Therefore it cannot represent the performance of SERS substrate well and is more suitable for characterization of metal nanoparticle colloidal solution.

In fact, the enhancement at different points of the SERS substrate and even on the same nanoparticles is not uniform, and the EF calculated above is an average factor. Furthermore, when calculating and comparing the EF, the experimental conditions should be as consistent as possible and a same molecule should be used.

1.10 Chemical mechanism

Electromagnetic mechanism (EM) in the form of the enhanced electric field as a result of SPR dominates in the overall enhancement for SERS. But there are still experimental observations that cannot be explained with EM alone. For example, in EC-SERS measurements, it was found that the potential where the Raman intensity reaches the maximum depends on the energy of excitation lasers. It was proposed that a special interaction between the molecule, the plasmonic substrate, and the light that contributes to the enhancement, which is of chemical nature and called chemical mechanism (CM). CM can be categorized into three kinds [50—54]: (1) when the molecule is adsorbed on metal surface, the presence of metal leads to a redistribution of the electron cloud of molecule, which may result in an increase of the polarizability and thus the Raman intensity. (2) The molecule has such a strong chemical interaction with the metal surface that a surface complex can be formed, which in turn increase the polarizability of molecules or even shows resonance Raman effect in the laser wavelength. As a result, the Raman intensity can be enhanced. (3) When the excitation wavelength matches the energy difference between molecular orbitals and the Fermi level or a surface state of metals, there will be charge transfer between them. This photon-driven charge transfer (PICT) can increase the polarizability of molecules and thus enhance the Raman intensity. The chemical interaction will lead to the obvious change in the relative intensity of different Raman peaks of a same molecule with CM. Generally, the enhancement due to CM is relatively low (up to 100 times) compared with EM. Its influence on the absolute intensity of a SERS spectrum is small. However, such a small CM enhancement can considerably change the spectral pattern because the total SERS enhancement is a result of the EM enhancement times the CM enhancement.

CM is not a general process and is not necessarily an enhanced process for all molecules, because the polarizability of the probe molecules can be either increased or decreased after interaction with the substrate. The key experimental evidence of CM is to carry out laser wavelength-dependent electrochemical SERS study, which may help tune in and out of resonance of the molecule−substrate system with the change of surface potential. Especially when PICT is present, there will be a shift of the potential where the maximum SERS intensity is observed when the laser wavelength is changed [55]. Alternatively, CM is usually studied or verified by modeling the molecule interacting with metal clusters or periodic slabs using electronic structure methods [56−58]. However, the interpretations based on electronic structures are restrained to certain resonance condition or types of molecule−metal bonds and usually give semiquantitative analysis [59]. Molecular resonance and charge transfer resonance have been proposed to contribute to the chemical enhancements by coupling to the surface plasmon [55,60]. However, calculation of the charge-transfer state, which is strongly mixed with the plasmon excitation during SERS processes, is still challenging [54,61].

1.11 Spectral analysis

When a spectrum is experimentally obtained, further spectral analysis will be followed. For quantitative analysis of a certain species, a quantitative relationship between the peak intensity and the concentration should be established. However, for analysis of the interaction of molecule with the surface, usually the peak position and relative intensity of different peaks should be analyzed. All these parameters depend on how the molecule is interacting with the surface. For example, after the molecule is adsorbed on the surface, its symmetry may be lowered in the presence of the surface. In addition to the generation of new surface bond, some forbidden Raman modes may become Raman active. Even for a same molecule, the SER spectra usually show different features compared with that of the free molecule.

Furthermore, for a molecule adsorbed on a metal surface, the surface selection rule of SERS works: the vibrational modes with polarizability component along the enhanced surface electric field (usually perpendicular to the local surface) can be effectively enhanced. Consequently, the in-plane modes will be enhanced when the molecule stands up on the surface whereas the out-of-plane modes will be strengthened for flat-flying molecules. As a result, both absolute and relative intensities of different peaks depend on the molecular orientation and configuration on the surface. This is very common in SERS experiments. For example, the molecule may lie on the surface when the molecular surface coverage is low, whereas the molecule may stand up due to the interaction between molecules at a high-surface coverage. Therefore the relative intensities of SER spectra can change with the variation of the molecular surface coverage.

Note that the above surface selection rule may not be always met. For example, the electric field in the nanogap is so tightly confined that a strong field gradient can be formed. In this case the electric field may vary over the dimension of a molecule or even a vibrational mode. Then Raman scattering is determined by the electric quadrupole–dipole and quadrupole–quadrupole transition polarizability, leading to an enhancement that scales with $|\nabla E_{\mathrm{local}}|^4$ [62]. As a result the selection rule may change notably. For example, the in-plane modes of a Co(II)-tetraphenylporphyrin molecule lying flat on the surface were enhanced by an increased field gradient [63].

Furthermore, we should also consider the influence of LSPR on the relative intensity. Raman peaks at different frequencies experience different radiation enhancement. As a result, the relative intensity of SER spectrum is the shaping result of the original surface Raman spectrum by the LSP response. For example, although the chemical interaction remains the same for malachite green isothiocyanate (MGITC) molecules adsorbed on different Au nanorods, the relative intensity of different Raman peaks changes obviously as a result of the LSP response of different Au nanorods (as shown in Fig. 1.10A). It will be risky to extract the native chemical information from the relative intensity if such a plasmon shaping effect is not corrected. We found that the background continuum of SER spectrum in the above system is contributed by the photoluminescence (PL) of Au nanorods (it does not matter whether such a PL is originated from emission or electron scattering of Au). The PL background is originated from the bulk Au PL amplified by the LSPR of nanorod, similar to that SERS is the amplified result of normal Raman. With this understanding, the native Raman spectrum (i.e., $\mathrm{SERS_{CORR}}$) can be obtained from $\mathrm{SERS_{CORR}} = \mathrm{SERS_{sub}} \times \mathrm{PL_{bulk}}/\mathrm{Bg}$, where Bg is the fitted background in the original SERS spectrum (PL from Au nanorod), and $\mathrm{SERS_{sub}}$ is the SER spectrum after background subtraction (i.e., the true SER spectrum from MGITC), $\mathrm{PL_{bulk}}$ is the PL of bulk Au [64]. Fig. 1.10B shows that the SERS spectra of MGITC adsorbed on different Au nanorods present

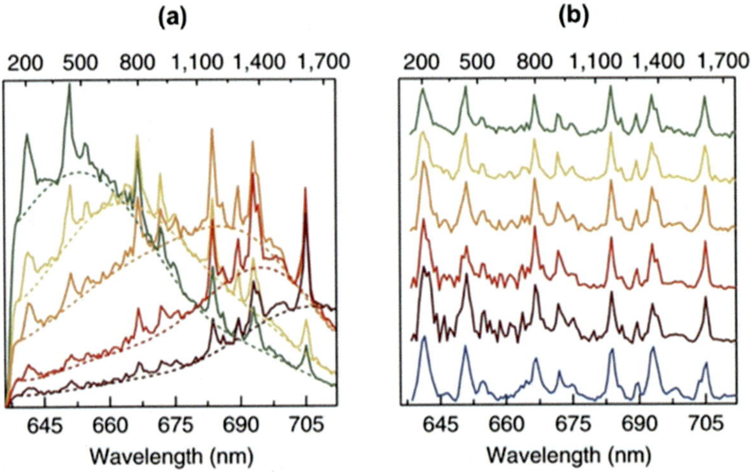

Figure 1.10 Original (A) and corrected (B) SER spectra for MGITC adsorbed on Au nanorods with different aspect ratio [64].

the same relative intensities after correction. This method is also found valid in more complicated systems, for example, Ag aggregates and TERS systems.

In addition to the background in the high-frequency region, the low-frequency continuum background in the SERS spectra has been ascribed to the electronic Raman scattering (ERS) signals from metal substrates. To extract both vibrational and electronic information from the measured SERS spectra in the low-frequency region (even down to 10 cm^{-1}), it is essential to obtain imaginary part of dynamical susceptibility $X''_{SERS}(\nu_m)$ by reducing the Bose-Einstein thermal factor $n(\nu_m)$, the scattering efficiency factor $(\nu_i - \nu_m)^3$, and plasmonic contribution f in Stokes branch:

$$X''_{SERS}(\nu_m) = \frac{I_{SERS}(\nu_i - \nu_m)}{f} \times \frac{1}{(\nu_i - \nu_m)^3} \times \frac{1}{[n(\nu_m) + 1]} \tag{1.13}$$

$$f = \frac{I_{ESERS}(\nu_i - \nu_m)}{I_{ERS}(\nu_i - \nu_m)}$$

where ν_i, ν_m are the photon energy and vibration energy of the molecule. I_{ESERS}, I_{ERS} are the intensity of surface-enhanced ERS and normal ERS. With such kind of correction, it allows the background free and intensity corrected SERS spectra to be obtained for more reliable analysis of molecular behavior on surface [65].

1.12 Selection of SERS substrates

SERS signal is highly dependent on the interfacial interaction between target and plasmonic nanostructures on both physical and chemical sides. Therefore building the maximal amounts of hottest spots [66–68] and enhancing the effective interaction of the target analyte with the hottest spots are the two crucial parameters determining the level of the selectivity and sensitivity of SERS for the target.

Theoretically, any material (metal, metal oxide, or semiconductor) with strong plasmon activity at the excitation wavelength can serve as a SERS substrate [69–76]. In practice, it is more common to use Ag and Au-based nanomaterials as SERS substrates to achieve a high sensitivity. Ag has a high plasmon quality in the whole visible to near-IR region, and Au is normally used for the red and near-IR regions (refer to Section 1.6 for more details). The much better chemical stability and biocompatibility of Au over Ag make Au a better choice for bioanalysis and clinical application. Chapter 2, *Nanoplasmonic materials for surface-enhanced Raman scattering*, will provide a detailed description about various nanoplasmonic materials for SERS.

The first SERS-active substrate was a roughened Ag electrode with a nanostructured surface morphology, which was prepared by the electrochemical oxidation and reduction cycles (ORC) [77]. Thereafter, various ORC methods have been developed for different metals to obtain clean and stable surfaces. But, ORC methods are hard to control the surface nanostructure. Soon later, the colloidal Ag and Au have been demonstrated to show high SERS enhancements [78,79]. With the rapid development of nanoscience and nanotechnology, it becomes routine to obtain Au and Ag nanostructures with highly controlled size, shape, and composition by wet chemical protocols [80]. These nanostructures can also be applied as basic building blocks to prepare solid SERS substrates, which is more convenient to use in comparison with colloidal ones [81–83]. More recently, SHINERS have been proposed to avoid the direct interaction and denature of biomolecules with the Au or Ag substrate with the protection of silica dielectric layer. Meanwhile, solid substrates prepared by vacuum or lithographic methods are usually clean with high uniformity and allows the batch preparation, which is very promising toward practical applications [84].

From the practical application point of view, colloidal and solid SERS substrates have their own unique advantages towards trace target detection from the following four aspects: (1) Scale-up fabrication: colloidal substrates are much easier to prepare and scale up than the solid ones, but normally displays less uniformity and wider size distribution. (2) Cleanliness of surface: the colloidal surfaces inevitably have surfactants (such as citrate) on the surface, which may produce interfering signal to that of target species. In comparison the impurities on solid substrate surfaces obtained during preparation and storage could be effectively removed by plasma treatment. (3) Reproducibility of SERS measurement: colloidal substrates usually show a better reproducibility than the solid ones due to the average effect of millions of hot spots in the three-dimensional detection volume. In comparison, the solid surfaces usually have only about hundreds to thousands of hot spots in the detection area, which has a higher requirement on the uniformity of fabrication processes. (4) Stability and shelf-life: both types of substrates have long-term stability under an inert storage environment.

1.13 SERS detection modes

Direct and indirect detection are two typical ways to perform SERS analysis, which will be detailed in Chapter 4, *Label-free surface-enhanced Raman scattering for clinical applications*, and Chapter 5, *Surface-enhanced Raman scattering nanotags design and synthesis*.

In brief, direct (label-free or intrinsic) detection presents the most unique advantage of SERS over fluorescence for clinic application, because it directly uses the molecular fingerprint information of the target molecule for detection without labeling (Fig. 1.11A). Therefore the conformation and orientation of adsorbed biomolecules can be extracted from the obtained Raman spectrum. Direct SERS detection has been successfully applied to the study of many common (macro-)biomolecules involved in living cells, such as amino acids, proteins, bases and nucleic acids, and some metabolites.

However, up to now, it is still an arduous task to extract the information of target species from SERS spectra for the following three reasons: (1) the Raman spectrum by direct detection is far too complex to be analyzed effectively, due to the complicated molecular structures of these biomolecules. (2) The SERS signals of interested molecules may be intrinsically weak and be easily overwhelmed by other irrelevant molecules. (3) It is still a big challenge to select an inherent biomolecule to directly reflect certain physical properties, like the local pH or the temperature inside a cell.

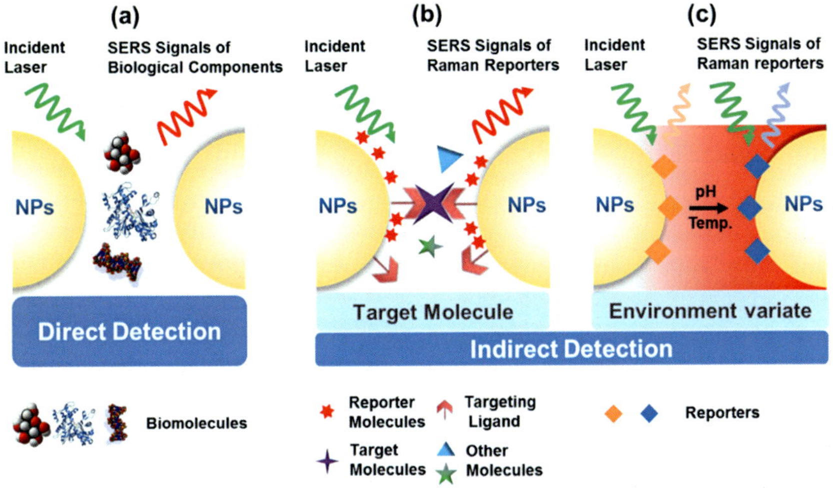

Figure. 1.11 (A) Scheme of the direct SERS detection, in which the SERS signal is from the biomolecules located on the SERS substrate. (B) Scheme of the indirect SERS detection of target molecules via specific recognition: the nanoparticles are first labeled with reporter molecules, followed by modification with targeting ligand to form SERS tag. In the presence of target, the target can specifically bind to the targeting ligand and bring the SERS tag to another SERS tag or a plasmonic metal substrate to form hot spot and give strong SERS signals of SERS tag. (C) Scheme of indirect SERS detection of the environment variate: the nanoparticles are modified with environment (pH, temperature, gas, etc.) sensitive molecules. Upon the change of environment, a change of SERS signals of the reporter molecule can be detected [17].

To overcome the above-mentioned problems of direct SERS detection, indirect (labeled or extrinsic) SERS detection has been developed, in which a reporter molecule with a large Raman cross-section is directly modified on the SERS substrate surface as a SERS tag, followed by functionalization with a layer of molecules that can specifically bind to a target species (Fig. 1.11B). Upon recognition with the target molecule, the SERS signal of reporter molecule (SERS tag) can be detected. Other molecules that cannot specifically bind to the targeting ligand will not interfere the detection. Alternatively, if the pH or temperature of the environment is more of concern, a reporter molecule that can respond to pH or temperature change can be used to report the local environment condition (Fig. 1.11C). In both cases, Au nanoparticle and tag molecules functionalized with strong anchoring groups like SH or NH_2 are preferred to use. The indirect detection enables the qualitative and quantitative bioanalysis of target molecules or the physical or chemical properties of the surroundings. Inevitably, it is unable to obtain the fingerprint information of target itself. In comparison with fluorescence spectroscopy, the indirect SERS possesses three merits: (1) a more stable signal and free of photobleaching, (2) a higher capability in multiplex detection with single-wavelength excitation attributing to the narrow peak width of Raman bands, and (3) a lower auto-fluorescence interference with near-IR excitation. Therefore the indirect SERS method has been widely applied in the clinical application for detection of protein and DNA, live-cell imaging, and cancer diagnosis [85–90].

No matter which detection mode will be used, the key point of successful measurement is to capture target on the SERS substrate surface throughout the detection time scale. It is necessary to strengthen the interaction between molecule and SERS substrate with the introduction of extra interfacial interaction, such as electrostatic modification, molecular steric effect, host–guest and biological recognitions, and chemical derivatization [91–95]. For the detection of target molecules having a weak affinity with the SERS substrate at low concentrations, it will be helpful to use dynamic SERS. This mode is able to enlighten the target signal by shortening the acquisition time down to the dwell time of the target in the detection area, which not only improves the SNR of the SERS spectrum by reducing the random contribution of noises or impurities but also could reveal the rare molecular characteristics at single-molecule level.

1.14 Key to the success of SERS measurements

Before starting SERS measurements, the SERS substrate should be carefully cleaned to remove the impurities inherited from the synthesis process and eliminate their interference to signal of target species. The residual citrate ions (the reducing and protecting agent during colloidal Au and Ag nanoparticles synthesis) have been found to interfere with the measurement of species weakly adsorbed on the surface or with low Raman cross-section. Iodide adsorption followed by electrochemical oxidation has been developed to obtain a relatively clean surface, which has been successfully applied for the SERS mapping of amino acids on the membrane of a living cell [96]. However, such an adsorption and oxidation process may sacrifice part of the SERS activity. Furthermore, if the impurities are trapped in the nanogap among nanoparticles, they will be hard to be removed experimentally. In these cases, it is necessary to use chemometric methods to treat the data, especially when they are assisted with the recently developed artificial intelligence methods. With chemmometrics, it will then become possible to isolate the signals of the target species from that of the impurities to improve the sensitivity and the imaging speed [97,98]. Readers who are interested in this topic are encouraged to follow Chapter 11, *Multivariate approaches for surface-enhanced Raman spectroscopy data analysis in clinics.*

Special attention should be paid to some key experimental parameters, including the SERS substrate, laser wavelength and power, and measurement environment and conditions (colloidal or solid states), to realize a successful SERS measurement with highly sensitive qualitative analysis and reliable quantitative analysis. A detailed guidelines for the qualitative and quantitative SERS analysis has been well discussed elsewhere [99] and see a more detailed experimental aspects of SERS for biological applications in Chapter 3, Experimental Aspects of Surface-Enhanced Raman Spectroscopy for Biological Applications.

During SERS measurement, it is necessary to well control the laser power and illumination time to exclude any potential interference from photo-induced chemical reactions [100,101], which can be checked if the SERS signal varies with the detection time. Then, it is important to ensure the SERS signal is indeed from the target by performing the target concentration titration to establish a correlation between the SERS intensity and concentration. For the quantitative analysis the higher quantification level demands more rigorous experimental design. In case of a rough estimation of target, the semiquantitative analysis is normally carried out by simply selecting one characteristic SERS peak and plotting the intensity versus the concentration. For a precise quantification the internal standard strategy, such as core-molecule-shell approach [102], is recommended to calibrate SERS signals.

References

[1] S. Ewen, G. Dent, Modern Raman Spectroscopy: A Practical Approach, 1st Ed, Wiley, 2005.
[2] G. Wu, Raman Spectroscopy: An Intensity Approach, World Scientific, 2016.
[3] G. Turrell, J. Corset, Raman Microscopy: Developments and Applications, 1st Ed, Academic Press, 1996.
[4] D.L. Rousseau, J.M. Friedman, P.F. Williams, The Resonance Raman Effect, 1st Ed, Springer Berlin Heidelberg, Berlin, 1979.
[5] A. Warshel, Interpretation of resonance Raman spectra of biological molecules, Annu. Rev. Biophysics Bioeng. 6 (1977) 273–300.
[6] R.H. Callender, A. Doukas, R. Crouch, K. Nakanishi, Molecular flow resonance Raman effect from retinal and rhodopsin, Biochemistry 15 (1976) 1621–1629.

[7] I. Freestone, N. Meeks, M. Sax, C. Higgitt, The Lycurgus Cup—a Roman nanotechnology, Gold. Bull. 40 (2007) 270–277.
[8] S.Y. Ding, J. Yi, J.F. Li, B. Ren, D.Y. Wu, R. Panneerselvam, et al., Nanostructure-based plasmon-enhanced Raman spectroscopy for surface analysis of materials, Nat. Rev. Mater. 1 (2016) 16021.
[9] J. Langer, D.J. de Aberasturi, J. Aizpurua, R.A. Alvarez-Puebla, B. Auguié, J.J. Baumberg, et al., Present and future of surface-enhanced Raman scattering, ACS Nano 14 (2020) 28–117.
[10] X. Wang, S.C. Huang, S. Hu, S. Yan, B. Ren, Fundamental understanding and applications of plasmon-enhanced Raman spectroscopy, Nat. Rev. Phys. 2 (2020) 253–271.
[11] S. Enoch, N. Bonod, Plasmonics-From Basics to Advanced Topics, 1st Ed, Springer-Verlag, Berlin Heidelberg, 2012.
[12] P. Biagioni, J.S. Huang, B. Hecht, Nanoantennas for visible and infrared radiation, Rep. Prog. Phys. 75 (2012) 024402.
[13] M. Dressel, G. Gruner, Electrodynamics of Solids: Optical Properties of Electrons in Matter, 1st Ed, Cambridge University Press, 2002.
[14] J.D. Jackson, Classical Electrodynamics, 3rd Ed, Wiley, 1998.
[15] E.C. Le Ru, P.G. Etchegoin, Principles of Surface-Enhanced Raman Spectroscopy and Related Plasmonic Effects, 1st Ed, Elsevier Science, 2009.
[16] L.J. Mendoza Herrera, D.M. Arboleda, D.C. Schinca, L.B. Scaffardi, Determination of plasma frequency, damping constant, and size distribution from the complex dielectric function of noble metal nanoparticles, J. Appl. Phys. 116 (2014) 233105.
[17] C. Zong, M. Xu, L.J. Xu, T. Wei, X. Ma, X.S. Zheng, et al., Surface-enhanced Raman spectroscopy for bioanalysis: reliability and challenges, Chem. Rev. 118 (2018) 4946–4980.
[18] S. Schlücker, W. Kiefer, Surface Enhanced Raman Spectroscopy: Analytical, Biophysical and Life Science Applications, 1st Ed, Wiley-VCH, 2010.
[19] S.Y. Ding, E.M. You, Z.Q. Tian, M. Moskovits, Electromagnetic theories of surface-enhanced Raman spectroscopy, Chem. Soc. Rev. 46 (2017) 4042–4076.
[20] E.C. Le Ru, P.G. Etchegoin, J. Grand, N. Félidj, J. Aubard, G. Lévi, et al., Surface enhanced Raman spectroscopy on nanolithography-prepared substrates, Curr. Appl. Phys. 8 (2008) 467–470.
[21] E.C. Le Ru, J. Grand, N. Félidj, J. Aubard, G. Lévi, A. Hohenau, et al., Experimental verification of the SERS electromagnetic model beyond the $|E|^4$ approximation: polarization effects, J. Phys. Chem. C. 112 (2008) 8117–8121.
[22] S.M. Jang, J.S. Park, S.M. Shin, C.J. Yoon, B.K. Choi, M.S. Gong, et al., Adsorption of 4-biphenylmethanethiolate on different-sized gold nanoparticle surfaces, Langmuir 20 (2004) 1922–1927.
[23] P.P. Fang, J.F. Li, Z.L. Yang, L.M. Li, B. Ren, Z.Q. Tian, Optimization of SERS activities of gold nanoparticles and gold-core–palladium-shell nanoparticles by controlling size and shell thickness, J. Raman Spectrosc. 39 (2008) 1679–1687.
[24] K.Q. Lin, J. Yi, S. Hu, B.J. Liu, J.Y. Liu, X. Wang, et al., Size effect on SERSs of gold nanorods demonstrated via single nanoparticle spectroscopy, J. Phys. Chem. C. 120 (2016) 20806–20813.
[25] C.J. Orendorff, L. Gearheart, N.R. Jana, C.J. Murphy, Aspect ratio dependence on surface enhanced Raman scattering using silver and gold nanorod substrates, Phys. Chem. Chem. Phys. 8 (2006) 165–170.
[26] J. Xie, Q. Zhang, J.Y. Lee, D.I.C. Wang, The synthesis of SERS-active gold nanoflower tags for in vivo applications, ACS Nano 2 (2008) 2473–2480.
[27] B.J. Liu, K.Q. Lin, S. Hu, X. Wang, Z.C. Lei, H.X. Lin, et al., Extraction of absorption and scattering contribution of metallic nanoparticles toward rational synthesis and application, Anal. Chem. 87 (2015) 1058–1065.

[28] K.J. Lee, P.D. Nallathamby, L.M. Browning, C.J. Osgood, X.H.N. Xu, In vivo imaging of transport and biocompatibility of single silver nanoparticles in early development of zebrafish embryos, ACS Nano 1 (2007) 133–143.
[29] X.H.N. Xu, W.J. Brownlow, S.V. Kyriacou, Q. Wan, J.J. Viola, Real-time probing of membrane transport in living microbial cells using single nanoparticle optics and living cell imaging, Biochemistry 43 (2004) 10400–10413.
[30] X. Huang, I.H. El-Sayed, W. Qian, M.A. El-Sayed, Cancer cell imaging and photothermal therapy in the near-infrared region by using gold nanorods, J. Am. Chem. Soc. 128 (2006) 2115–2120.
[31] E.B. Dickerson, E.C. Dreaden, X. Huang, I.H. El-Sayed, H. Chu, S. Pushpanketh, et al., Gold nanorod assisted near-infrared plasmonic photothermal therapy (PPTT) of squamous cell carcinoma in mice, Cancer Lett. 269 (2008) 57–66.
[32] S.E. Skrabalak, J. Chen, Y. Sun, X. Lu, L. Au, C.M. Cobley, et al., Gold nanocages: synthesis, properties, and applications, Acc. Chem. Res. 41 (2008) 1587–1595.
[33] X. Huang, S. Tang, X. Mu, Y. Dai, G. Chen, Z. Zhou, et al., Freestanding palladium nanosheets with plasmonic and catalytic properties, Nat. Nanotechnol. 6 (2011) 28–32.
[34] Y. Xia, W. Li, C.M. Cobley, J. Chen, X. Xia, Q. Zhang, et al., Gold nanocages: from synthesis to theranostic applications, Acc. Chem. Res. 44 (2011) 914–924.
[35] H. Xu, E. Bjerneld, J. Aizpurua, P. Apell, L. Gunnarsson, S. Petronis, et al., Interparticle coupling effects in surface-enhanced Raman scattering, Proc. SPIE 4258 (2001).
[36] E.C. Le Ru, P.G. Etchegoin, Single-molecule surface-enhanced Raman spectroscopy, Annu. Rev. Phys. Chem. 63 (2012) 65–87.
[37] W. Xi, B.K. Shrestha, A.J. Haes, Promoting intra- and intermolecular interactions in surface-enhanced Raman scattering, Anal. Chem. 90 (2018) 128–143.
[38] G.A. Somorjai, Y. Li, Introduction to Surface Chemistry and Catalysis, 2nd Ed, Wiley, 2010.
[39] K. Ikeda, N. Fujimoto, H. Uehara, K. Uosaki, Raman scattering of aryl isocyanide monolayers on atomically flat Au(111) single crystal surfaces enhanced by gap-mode plasmon excitation, Chem. Phys. Lett. 460 (2008) 205–208.
[40] K. Ikeda, J. Sato, N. Fujimoto, N. Hayazawa, S. Kawata, K. Uosaki, Plasmonic enhancement of raman scattering on non-SERS-active platinum substrates, J. Phys. Chem. C. 113 (2009) 11816–11821.
[41] S.Y. Chen, J.J. Mock, R.T. Hill, A. Chilkoti, D.R. Smith, A.A. Lazarides, Gold nanoparticles on polarizable surfaces as Raman scattering antennas, ACS Nano 4 (2010) 6535–6546.
[42] J.F. Li, Y.F. Huang, Y. Ding, Z.L. Yang, S.B. Li, X.S. Zhou, et al., Shell-isolated nanoparticle-enhanced Raman spectroscopy, Nature 464 (2010) 392–395.
[43] J.F. Li, Y.J. Zhang, S.Y. Ding, R. Panneerselvam, Z.Q. Tian, Core–shell nanoparticle-enhanced Raman spectroscopy, Chem. Rev. 117 (2017) 5002–5069.
[44] R.M. Stöckle, Y.D. Suh, V. Deckert, R. Zenobi, Nanoscale chemical analysis by tip-enhanced Raman spectroscopy, Chem. Phys. Lett. 318 (2000) 131–136.
[45] M.S. Anderson, Locally enhanced Raman spectroscopy with an atomic force microscope, Appl. Phys. Lett. 76 (2000) 3130–3132.
[46] N. Hayazawa, Y. Inouye, Z. Sekkat, S. Kawata, Metallized tip amplification of near-field Raman scattering, Opt. Commun. 183 (2000) 333–336.
[47] T.X. Huang, X. Cong, S.S. Wu, K.Q. Lin, X. Yao, Y.H. He, et al., Probing the edge-related properties of atomically thin MoS_2 at nanoscale, Nat. Commun. 10 (2019) 5544.
[48] R. Zhang, Y. Zhang, Z.C. Dong, S. Jiang, C. Zhang, L.G. Chen, et al., Chemical mapping of a single molecule by plasmon-enhanced Raman scattering, Nature 498 (2013) 82–86.

[49] J. Lee, K.T. Crampton, N. Tallarida, V.A. Apkarian, Visualizing vibrational normal modes of a single molecule with atomically confined light, Nature 568 (2019) 78–82.
[50] J.R. Lombardi, R.L. Birke, A unified approach to surface-enhanced Raman spectroscopy, J. Phys. Chem. C. 112 (2008) 5605–5617.
[51] P. Kambhampati, C.M. Child, M.C. Foster, A. Campion, On the chemical mechanism of surface enhanced Raman scattering: experiment and theory, J. Chem. Phys. 108 (1998) 5013–5026.
[52] General Discussion. Faraday Discuss. 2006, 132, 227–247.
[53] N. Valley, N. Greeneltch, R.P. Van Duyne, G.C. Schatz, A look at the origin and magnitude of the chemical contribution to the enhancement mechanism of surface-enhanced Raman spectroscopy (SERS): theory and experiment, J. Phys. Chem. Lett. 4 (2013) 2599–2604.
[54] J.R. Lombardi, R.L. Birke, The theory of surface-enhanced Raman scattering, J. Chem. Phys. 136 (2012) 144704.
[55] D.Y. Wu, J.F. Li, B. Ren, Z.Q. Tian, Electrochemical surface-enhanced Raman spectroscopy of nanostructures, Chem. Soc. Rev. 37 (2008) 1025–1041.
[56] S.M. Morton, D.W. Silverstein, L. Jensen, Theoretical studies of plasmonics using electronic structure methods, Chem. Rev. 111 (2011) 3962–3994.
[57] L. Jensen, C.M. Aikens, G.C. Schatz, Electronic structure methods for studying surface-enhanced Raman scattering, Chem. Soc. Rev. 37 (2008) 1061–1073.
[58] A.T. Zayak, Y.S. Hu, H. Choo, J. Bokor, S. Cabrini, P.J. Schuck, et al., Chemical Raman enhancement of organic adsorbates on metal surfaces, Phys. Rev. Lett. 106 (2011) 083003.
[59] R. Chen, L. Jensen, Interpreting the chemical mechanism in SERS using a Raman bond model, J. Chem. Phys. 152 (2020) 024126.
[60] J.R. Lombardi, R.L. Birke, A unified view of surface-enhanced Raman scattering, Acc. Chem. Res. 42 (2009) 734–742.
[61] R.L. Birke, J.R. Lombardi, W.A. Saidi, P. Norman, Surface-enhanced Raman scattering due to charge-transfer resonances: a time-dependent density functional theory study of Ag_{13}-4-mercaptopyridine, J. Phys. Chem. C. 120 (2016) 20721–20735.
[62] D.V. Chulhai, L. Jensen, Determining molecular orientation with surface-enhanced Raman scattering using inhomogenous electric fields, J. Phys. Chem. C. 117 (2013) 19622–19631.
[63] J. Lee, N. Tallarida, X. Chen, P. Liu, L. Jensen, V.A. Apkarian, Tip-enhanced Raman spectromicroscopy of Co(II)-tetraphenylporphyrin on Au(111): toward the chemists' microscope, ACS Nano 11 (2017) 11466–11474.
[64] K.Q. Lin, J. Yi, J.H. Zhong, S. Hu, B.J. Liu, J.Y. Liu, et al., Plasmonic photoluminescence for recovering native chemical information from surface-enhanced Raman scattering, Nat. Commun. 8 (2017) 14891.
[65] M. Inagaki, T. Isogai, K. Motobayashi, K.Q. Lin, B. Ren, K. Ikeda, Electronic and vibrational surface-enhanced Raman scattering: from atomically defined Au (111) and (100) to roughened Au, Chem. Sci. 11 (2020) 9807–9817.
[66] T.A. Laurence, G.B. Braun, N.O. Reich, M. Moskovits, Robust SERS enhancement factor statistics using rotational correlation spectroscopy, Nano Lett. 12 (2012) 2912–2917.
[67] W.H. Yang, J. Hulteen, G.C. Schatz, R.P. Van Duyne, A surface-enhanced hyper-Raman and surface-enhanced Raman scattering study of trans -1,2 - bis(4-pyridyl) ethylene adsorbed onto silver film over nanosphere electrodes. Vibrational assignments: experiment and theory, J. Chem. Phys. 104 (1996) 4313–4323.
[68] X. Zhang, R.P. Van Duyne, Optimized silver film over nanosphere surfaces for the biowarfare agent detection based on surface-enhanced Raman spectroscopy, MRS Proc. 876 (2011).

[69] Y. Yin, Z.Y. Li, Z. Zhong, B. Gates, Y. Xia, S. Venkateswaran, Synthesis and characterization of stable aqueous dispersions of silver nanoparticles through the Tollens process, J. Mater. Chem. 12 (2002) 522–527.
[70] Z.Q. Tian, B. Ren, D.Y. Wu, Surface-enhanced Raman scattering: from noble to transition metals and from rough surfaces to ordered nanostructures, J. Phys. Chem. B 106 (2002) 9463–9483.
[71] B. Ren, G.K. Liu, X.B. Lian, Z.L. Yang, Z.Q. Tian, Raman spectroscopy on transition metals, Anal. Bioanal. Chem. 388 (2007) 29–45.
[72] P.R. West, S. Ishii, G.V. Naik, N.K. Emani, V.M. Shalaev, A. Boltasseva, Searching for better plasmonic materials, Laser Photonics Rev. 4 (2010) 795–808.
[73] C. Rhodes, S. Franzen, J.-P. Maria, M. Losego, D.N. Leonard, B. Laughlin, et al., Surface plasmon resonance in conducting metal oxides, J. Appl. Phys. 100 (2006) 054905.
[74] S. Cong, Z. Wang, W. Gong, Z. Chen, W. Lu, J.R. Lombardi, et al., Electrochromic semiconductors as colorimetric SERS substrates with high reproducibility and renewability, Nat. Commun. 10 (2019) 678.
[75] M.D. Losego, A.Y. Efremenko, C.L. Rhodes, M.G. Cerruti, S. Franzen, J.-P. Maria, Conductive oxide thin films: model systems for understanding and controlling surface plasmon resonance, J. Appl. Phys. 106 (2009) 024903.
[76] L. Li, T. Hutter, A.S. Finnemore, F.M. Huang, J.J. Baumberg, S.R. Elliott, et al., Metal oxide nanoparticle mediated enhanced Raman scattering and its use in direct monitoring of interfacial chemical reactions, Nano Lett. 12 (2012) 4242–4246.
[77] M. Fleischmann, P.J. Hendra, A.J. McQuillan, Raman spectra of pyridine adsorbed at a silver electrode, Chem. Phys. Lett. 26 (1974) 163–166.
[78] M.E. Uppitsch, Observation of surface enhanced Raman spectra by adsorption to silver colloids, Chem. Phys. Lett. 74 (1980) 125–127.
[79] K.U. von Raben, R.K. Chang, B.L. Laube, Surface enhanced Raman scattering of $Au(CN)_2^-$ ions adsorbed on gold colloids, Chem. Phys. Lett. 79 (1981) 465–469.
[80] M.J. Banholzer, J.E. Millstone, L. Qin, C.A. Mirkin, Rationally designed nanostructures for surface-enhanced Raman spectroscopy, Chem. Soc. Rev. 37 (2008) 885–897.
[81] W. Lee, S.Y. Lee, R.M. Briber, O. Rabin, Self-assembled SERS substrates with tunable surface plasmon resonances, Adv. Funct. Mater. 21 (2011) 3424–3429.
[82] B. Mondal, S.K. Saha, Fabrication of SERS substrate using nanoporous anodic alumina template decorated by silver nanoparticles, Chem. Phys. Lett. 497 (2010) 89–93.
[83] S.Y. Lee, S.-H. Kim, M.P. Kim, H.C. Jeon, H. Kang, H.J. Kim, et al., Freestanding and arrayed nanoporous microcylinders for highly active 3D SERS substrate, Chem. Mater. 25 (2013) 2421–2426.
[84] J. Li, C. Chen, H. Jans, X. Xu, N. Verellen, I. Vos, et al., 300 mm wafer-level, ultra-dense arrays of Au-capped nanopillars with sub-10 nm gaps as reliable SERS substrates, Nanoscale 6 (2014) 12391–12396.
[85] X.X. Han, B. Zhao, Y. Ozaki, Surface-enhanced Raman scattering for protein detection, Anal. Bioanal. Chem. 394 (2009) 1719–1727.
[86] D. Graham, B.J. Mallinder, W.E. Smith, Detection and identification of labeled DNA by surface enhanced resonance Raman scattering, Biopolymers 57 (2000) 85–91.
[87] T. Vo-Dinh, L.R. Allain, D.L. Stokes, Cancer gene detection using surface-enhanced Raman scattering (SERS), J. Raman Spectrosc. 33 (2002) 511–516.
[88] J. Kneipp, H. Kneipp, A. Rajadurai, R.W. Redmond, K. Kneipp, Optical probing and imaging of live cells using SERS labels, J. Raman Spectrosc. 40 (2009) 1–5.
[89] J. Song, J. Zhou, H. Duan, Self-assembled plasmonic vesicles of SERS-encoded amphiphilic gold nanoparticles for cancer cell targeting and traceable intracellular drug delivery, J. Am. Chem. Soc. 134 (2012) 13458–13469.

[90] A. Samanta, K.K. Maiti, K.-S. Soh, X. Liao, M. Vendrell, U.S. Dinish, et al., Ultrasensitive near-infrared Raman reporters for SERS-based in vivo cancer detection, Angew. Chem. Int. Ed. 50 (2011) 6089–6092.
[91] L. He, Y. Liu, J. Liu, Y. Xiong, J. Zheng, Y. Liu, et al., Core–shell noble-metal@-metal-organic-framework nanoparticles with highly selective sensing property, Angew. Chem. Int. Ed. 52 (2013) 3741–3745.
[92] J. Mosquera, Y. Zhao, H.J. Jang, N. Xie, C. Xu, N.A. Kotov, et al., Plasmonic nanoparticles with supramolecular recognition, Adv. Funct. Mater. 30 (2020) 1902082.
[93] D.S. Grubisha, R.J. Lipert, H.-Y. Park, J. Driskell, M.D. Porter, Femtomolar detection of prostate-specific antigen: an immunoassay based on surface-enhanced Raman scattering and immunogold labels, Anal. Chem. 75 (2003) 5936–5943.
[94] S.S.R. Dasary, A.K. Singh, D. Senapati, H. Yu, P.C. Ray, Gold nanoparticle based label-free SERS probe for ultrasensitive and selective detection of trinitrotoluene, J. Am. Chem. Soc. 131 (2009) 13806–13812.
[95] S.E.J. Bell, N.M.S. Sirimuthu, Quantitative surface-enhanced Raman spectroscopy, Chem. Soc. Rev. 37 (2008) 1012–1024.
[96] M.D. Li, Y. Cui, M.X. Gao, J. Luo, B. Ren, Z.Q. Tian, Clean substrates prepared by chemical adsorption of iodide followed by electrochemical oxidation for surface-enhanced Raman spectroscopic study of cell membrane, Anal. Chem. 80 (2008) 5118–5125.
[97] C.K. Huang, M. Ando, H.O. Hamaguchi, S. Shigeto, Disentangling dynamic changes of multiple cellular components during the yeast cell cycle by in vivo multivariate Raman imaging, Anal. Chem. 84 (2012) 5661–5668.
[98] D. Feuerstein, K.H. Parker, M.G. Boutelle, Practical methods for noise removal: applications to spikes, nonstationary quasi-periodic noise, and baseline drift, Anal. Chem. 81 (2009) 4987–4994.
[99] A.I. Pérez-Jiménez, D. Lyu, Z. Lu, G. Liu, B. Ren, Surface-enhanced Raman spectroscopy: benefits, trade-offs and future developments, Chem. Sci. 11 (2020) 4563–4577.
[100] G.K. Liu, J. Hu, P.C. Zheng, G.L. Shen, J.H. Jiang, R.Q. Yu, et al., Laser-induced formation of metal – molecule – metal junctions between Au nanoparticles as probed by surface-enhanced Raman spectroscopy, J. Phys. Chem. C. 112 (2008) 6499–6508.
[101] Y.F. Huang, H.P. Zhu, G.K. Liu, D.Y. Wu, B. Ren, Z.Q. Tian, When the signal is not from the original molecule to be detected: chemical transformation of para-aminothiophenol on Ag during the SERS measurement, J. Am. Chem. Soc. 132 (2010) 9244–9246.
[102] W. Shen, X. Lin, C. Jiang, C. Li, H. Lin, J. Huang, et al., Reliable quantitative SERS analysis facilitated by core–shell nanoparticles with embedded internal standards, Angew. Chem. Int. Ed. 54 (2015) 7308–7312.

CHAPTER 2

Nanoplasmonic materials for surface-enhanced Raman scattering

Shi Xuan Leong, Yong Xiang Leong, Charlynn Sher Lin Koh, Jaslyn Ru Ting Chen and Xing Yi Ling
Division of Chemistry and Biological Chemistry, School of Physical and Mathematical Sciences, Nanyang Technological University, Singapore

2.1 The role of nanoplasmonic materials in surface-enhanced Raman scattering enhancement

Most surface-enhanced Raman scattering (SERS)-active substrates are based on nanoplasmonic materials, such as gold (Au), silver (Ag), and copper (Cu), because they can support surface plasmons in the visible light to near-infrared (NIR) wavelengths [1,2]. Upon light excitation, the metal's conduction band electrons collectively undergo resonant oscillation within the limited nanoscale surface area to give rise to localized surface plasmon resonances (LSPRs). This phenomenon creates a localized region of strong electromagnetic (EM) field at/near the metal surface (<10 nm) where both incident and scattered lights are amplified, resulting in a 10^3 to 10^8-fold SERS signal enhancement via an EM mechanism (Fig. 2.1A) [3–5]. Besides EM enhancement, SERS can also be enhanced via a chemical (CHEM) mechanism by up to 10^3-fold. This phenomenon typically occurs due to increased charge transfer between the nanoplasmonic surface and chemisorbed molecules [1]. Out of these two enhancement mechanisms, the former EM mechanism is often regarded as the main contributor to SERS signal enhancement. Consequently, most SERS-based research focuses on increasing EM field strengths of nanoplasmonic nanostructures. The detailed discussion on SERS enhancment mechanism can be found in Chapter 1, *Principles of surface-enhanced Raman spectroscopy*.

The most common way to improve EM enhancement is by engineering its hotspots, which are regions with exceptionally intense light confinement [6]. For zero- and one-dimensional (0D/1D) nanostructures, researchers strategically design nanoparticles with sharp tips and edges,

Principles and Clinical Diagnostic Applications of Surface-Enhanced Raman Spectroscopy
DOI: https://doi.org/10.1016/B978-0-12-821121-2.00005-6

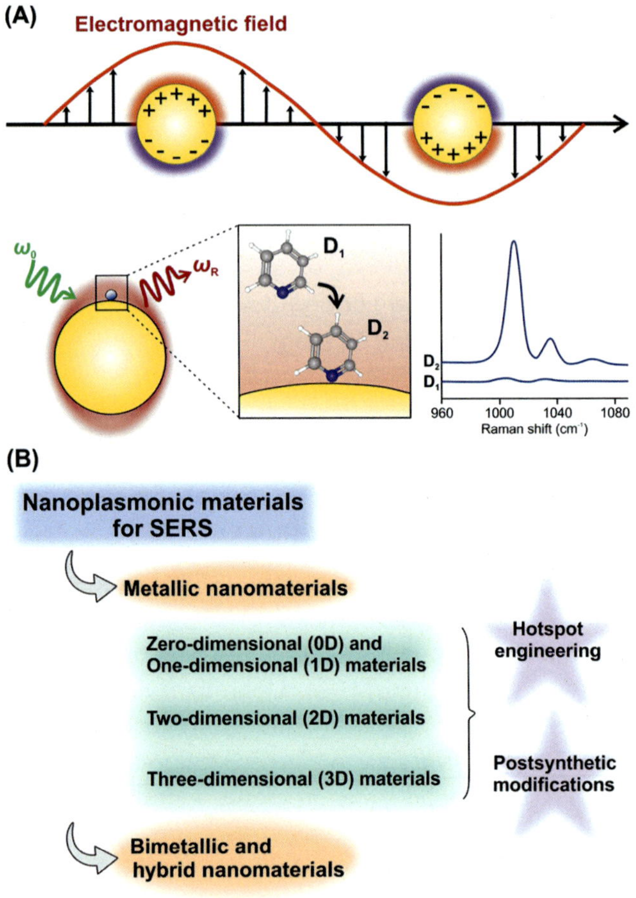

Figure 2.1 Nanoplasmonic materials for SERS enhancement. (A) Scheme depicting the resonant oscillation of conduction band electrons on the surface of nanoplasmonic materials and the formation of a localized region of strong electromagnetic field. (B) Summary of the various types of nanoplasmonic materials and common strategies employed to boost SERS signals. *SERS*, Surface-enhanced Raman scattering. *Reprinted and adapted with permission from C. Zhan, X.-J. Chen, J. Yi, J.-F. Li, D.-Y. Wu, Z.-Q. Tian, From plasmon-enhanced molecular spectroscopy to plasmon-mediated chemical reactions, Nat. Rev. Chem. 2 (9) (2018) 216–230. Copyright 2018 Springer Nature.*

such as nanoprisms and nanostars, because these features can concentrate the LSPRs and create EM hotspots via the "lightning-rod" effect [7]. Nanoparticles can also be assembled into a two-dimensional (2D) array to induce plasmonic coupling at the interparticle gap between neighboring nanoparticles and generate large areas of highly amplified EM fields.

Plasmonic coupling can also be extended over the *z*-axis in three-dimensional (3D) SERS substrates, where a hotspot volume of ≥ 0.5 μm enables full utilization of the focal volume of the excitation laser [5]. However, the field enhancements engendered by hotspot engineering are only effective if the target analyte can access these hotspots because of the surface-sensitive nature of the enhancement mechanisms. This is especially problematic for analytes that have low or no affinity to the plasmonic surface.

Therefore, to complement various hotspot engineering techniques, analyte manipulation strategies are also necessary to bring target analytes close to SERS-active surfaces. In particular, the SERS substrates can be physically modified to possess additional properties. For example, surface wetting properties of the platforms can be modified to confer superhydrophobicity to reduce solvent contact area and achieve analyte concentration within the area upon drying. Other emerging strategies include functionalization of highly porous metal-organic frameworks (MOFs) to selectively adsorb and concentrate guest analyte close to the plasmonic surface. These physical-based strategies are advantageous because they do not consider or modify the chemical nature of the analyte and require minimal amounts of sample for detection. To specifically capture analyte molecules, another analyte manipulation strategy is to chemically modify the nanoplasmonic surface with SERS nanotags or capturing probe molecules; these approaches will be explored in detail in subsequent chapters. Aside from analyte manipulation, novel nanostructure compositions such as bimetallic core–shell systems and hybrid nanostructures are also increasingly explored for better SERS enhancement capabilities. In addition to stronger SERS enhancements, these hybrid materials also combine the unique properties of both materials including tunable optical, catalytic, and magnetic properties to enable multifunctional applications.

In this chapter, we will discuss contemporary strategies to design nanoplasmonic SERS platforms to achieve heightened SERS performance with emphasis on applicability in the biomedical diagnosis (Fig. 2.1B). In terms of nanoplasmonic materials suitable for biomedical diagnosis, Ag is typically used for its strong plasmonic activity [8], while Au is preferred for its biocompatibility and stronger EM enhancement in the infrared to NIR region, which is the optimal spectral region for effective light absorption in most biological media [9]. We begin with a discussion on traditional hotspot engineering methods to fabricate Ag/Au platforms, which are categorized as 0D/1D, 2D, and 3D approaches. In each approach, we introduce various popular and/or emerging fabrication techniques and evaluate the optical properties and SERS performances of the as-discussed SERS platforms. We also discuss how analyte manipulation can be achieved via postsynthetic modifications of SERS platforms, focusing on physical alterations including conferring superhydrophobicity/omniphobicity and incorporation of MOFs. Finally, we explore unconventional nanoplasmonic materials including bimetallic and hybrid materials incorporating nonmetallic components before ending our discussion with a brief outlook in this research area.

2.2 Metallic nanoplasmonic materials

Metallic nanoparticles form the majority of SERS substrates due to their high plasmon efficiency and the range of established approaches available for morphological control and platform assembly. In this section, we will first look at different strategies for morphological control to increase hotspots in single plasmonic nanoparticles (0D/1D). Subsequently, we will discuss various methods for multidimensional (2D/3D) platform assembly to promote plasmonic coupling along *x*-, *y*-, and/or *z*-directions [10]. Finally, we will highlight emerging approaches for postsynthetic platform modifications using physical-based strategies.

2.2.1 Shape-controlled synthesis of individual nanoparticles (0D/1D)

Morphological manipulation of individual nanoparticles focuses on designing nanoparticles with specific structural features to enhance hotspot densities at the single particle level. In particular, anisotropic nanoparticles with regions of large curvatures and/or sharp tips and edges are of huge interest because these regions exhibit the strongest EM field enhancement resulting from the "lightning-rod" effect [1,4]. Common shape-controlled strategies can be broadly classified into two categories: (1) the use of molecular capping agents as shape-directing tools and (2) postsynthetic morphological modifications of as-synthesized nanoparticles (Fig. 2.2A).

2.2.1.1 Capping agent-directed shape-controlled nanoparticles

Molecular capping agents selectively passivate specific crystal planes against further growth through molecular interactions to enable seed particles to grow into anisotropic, shape-controlled nanoparticles. Common capping agents include polymers such as poly(vinylpyrrolidone) (PVP) and poly

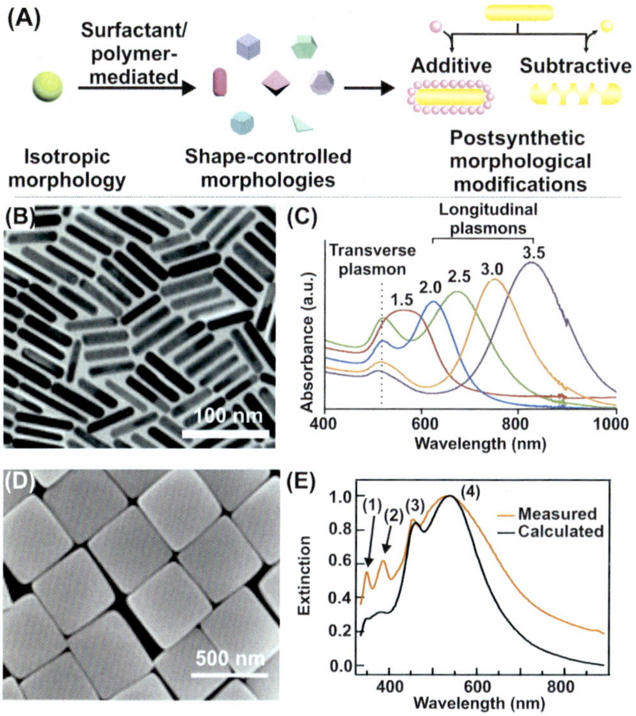

Figure 2.2 Classic 0D/1D plasmonic nanostructures. (A) Overview of advancements in strategies to fabricate 0D/1D plasmonic nanostructures. (B) TEM images of Au nanorods with average length/width ratio of (64 ± 6)/(12 ± 1) nm. (C) UV–vis–NIR absorbance spectra of Au nanorods with aspect ratios of 1.5 to 3.5, as indicated above the corresponding spectrum. The respective transverse and longitudinal plasmon resonance peaks characteristic of the nanorod are indicated as well. (D) SEM image of sharp Ag nanocubes with edge lengths = 400 ± 31 nm. (E) Measured and calculated extinction spectra of Ag nanocubes, where (1) denotes octupolar plasmon modes, (2) and (3) denote quadrupolar plasmon modes, and (4) denotes dipolar resonances. *NIR*, Near-infrared; *0D*, zero-dimensional; *1D*, one-dimensional; *Reprinted and adapted with permission from (A) H.K. Lee, Y.H. Lee, C.S.L. Koh, G.C. Phan-Quang, X. Han, C.L. Lay, et al., Designing surface-enhanced Raman scattering (SERS) platforms beyond hotspot engineering: emerging opportunities in analyte manipulations and hybrid materials, Chem. Soc. Rev. 48 (3) (2019) 731–756. (B) H. Chen, L. Shao, T. Ming, Z. Sun, C. Zhao, B. Yang, et al., Understanding the photothermal conversion efficiency of gold nanocrystals, Small 6 (20) (2010) 2272–2280. (C) S.E. Lohse, C.J. Murphy, The quest for shape control: a history of gold nanorod synthesis, Chem. Mater. 25 (8) (2013) 1250–1261. (D) Y. Lin, Y.-J. Zhang, W.-M. Yang, J.-C. Dong, F.-R. Fan, Y. Zhao, et al., Size and dimension dependent surface-enhanced Raman scattering properties of well-defined Ag nanocubes, Appl. Mater. Today 14 (2019) 224–232. (E) Y.H. Lee, H. Chen, Q.-H. Xu, J. Wang, Refractive index sensitivities of noble metal nanocrystals: the effects of multipolar plasmon resonances and the metal type, J. Phys. Chem. C 115 (16) (2011) 7997–8004. Copyright 2019 Royal Society of Chemistry. Copyright 2010 John Wiley & Sons, Inc. Copyright 2011 and 2013 American Chemical Society. Copyright 2018 Elsevier.*

(vinyl alcohol), as well as small molecules such as cetyltrimethylammonium bromide (CTAB), cetyltrimethylammonium chloride, and oleylamine.

One classic anisotropic nanoparticle morphology is the nanorod, typically synthesized via a seeded growth approach (Fig. 2.2B) [11,15–17]. In general, small preformed Au seeds are added to growth solutions containing Au(I) precursor, shape-directing CTAB molecules and weak reducing agents. It is postulated that CTAB molecules differentially block the long-axis crystal faces to induce growth along the short-axis faces to form nanorods. The nanorod aspect ratio (length over width) can be modulated through precise tuning of the growth conditions. For instance, a decrease in the seed concentration increases the nanorod aspect ratio from 1 to 7. Aspect ratios >7 can be achieved by applying more growth steps or by adding additional additives such as cosurfactants and organic additives [15,16,18]. For additional in-depth understanding of the developments in nanorod synthesis, interested readers can refer to relevant reviews cited here [12].

One of the most salient characteristics of plasmonic nanorods is that they possess multiple plasmon bands, namely, the longitudinal and transverse plasmon bands that correspond to light absorption and scattering along the nanorod's long and short axes, respectively. Unlike the single plasmon extinction of isotropic nanoparticles which are insensitive to the nanoparticle diameter, the energy of the nanorod's longitudinal plasmon can be tuned from visible ($\sim$600 nm) to infrared ($\sim$1500 nm) region by altering the nanorod aspect ratio (Fig. 2.2C). The LSPR λ_{max} generally increases linearly with an increase in Au nanorod aspect ratio from 2 to 5 [19,20]. This highly tunable property makes plasmonic nanorods excellent SERS substrates because we can readily manipulate the plasmon band position to match the excitation laser energy so as to achieve higher SERS enhancement via EM enhancement. Metallic nanorods also exhibit enhanced local EM field due to the "lightning-rod" effect, especially for nanorods with sharp tips. For instance, the calculated enhancement factor (EF) for 4-mercaptobenzoic acid self-assembled monolayers (4-MBA SAMs) sandwiched by Au nanorods on a planar Au substrate($\sim 10^8$) is an order of magnitude greater than that on Au spheres ($\sim 10^7$) [21].

Another classic shape-controlled morphology is the nanocube, whereby its sharp corners and edges provide greater localized EM field enhancement via the same "lightning-rod" effect (Fig. 2.2D) [13]. For instance, sharp nanocubes demonstrate approximately twofold increase in EF compared to the highly truncated, rounded counterparts using

1,4-benzenedithiol as the probe molecule [22]. In general, nanocubes are synthesized via the polyol reaction, using alcohols containing multiple hydroxyl groups such as ethylene glycol and diethylene glycol as the solvent and reductant. Shape control of perfect nanocubes bounded by {100} crystal planes can be achieved using PVP capping agents, which passivates growth along the <100> direction and/or induces growth along the <111> direction through selective crystal plane—PVP interactions [23,24]. Currently, Ag nanocubes with controllable edge lengths ranging from 13 to 400 nm have been synthesized by controlling the growth time, such as varying the amount of Ag seeds and/or amount of $AgNO_3$ precursor added [25,26]. Ag nanocubes also support multipolar LSPR behavior, whereby higher order resonance modes such as dipolar, quadrupolar, and octupolar resonances can be observed (Fig. 2.2E) [14,27]. This ability to support multiple LSPRs makes Ag nanocubes compatible with multiple common laser excitation wavelengths, for example, 532, 633, and 785 nm to generate intense plasmonic hotspots for SERS measurements even under different experimental conditions. Importantly, the narrow LSPRs of nanocubes due to their single crystalline nature give rise to high quality factors (Q-factor), which is desirable for highly sensitive SERS sensing.

Other than nanorods and nanocubes, there is also a growing interest in synthesizing a wide array of anisotropic morphologies including polyhedra such as tetrahedra, octahedra and icosahedra as well as nanostars and nanoflowers [23,28—30]. For instance, icosahedral Au nanoparticles are formed with high monodispersity and highly tunable size distribution ranging from 10 to 90 nm, using CTAB capping agents and citrate-capped Au seeds (Fig. 2.3A) [31]. Owing to the large number of hotspots from the presence of well-defined edges and corners, the as-synthesized icosahedral nanoparticles generate at least fourfold increase in SERS intensity compared to spherical nanoparticles, with various probe molecules including rhodamine 6G (R6G). In another example, highly symmetrical Au nanostars are grown from icosahedral Au seeds, using dimethylamine to stabilize the {321} high-index facets that make up the hexagonal pyramids at the nanostar tips (Fig. 2.3B) [32]. The symmetric Au nanostars show intense surface plasmon absorption bands in the NIR region, demonstrating higher single-particle SERS intensity (EF of 4.5×10^8) and increased signal reproducibility than irregularly shaped, asymmetric Au nanostars (EF of 1.2×10^8). In addition, thermodynamically unfavorable shapes with concave surfaces, high-index facets, and even branched structures can also be fabricated through the

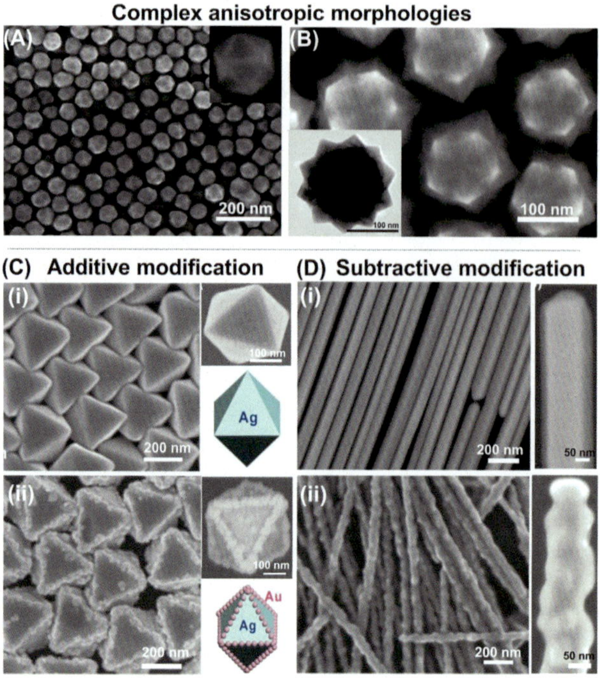

Figure 2.3 Complex anisotropic individual nanostructures. (A) SEM image of icosahedral Au nanoparticles. Inset shows (111) facets of a typical icosahedron. (B) SEM image of monodispersed Au nanostars. Inset shows TEM image of a single Au nanostar. (C) Additive postsynthetic morphological modification (i and ii): SEM images of as-fabricated Ag octahedra and modified selective edge Au-deposited Ag octahedra (SEGSO), respectively. (D) Subtractive postsynthetic morphological modification (i and ii): SEM images of as-fabricated and etched Ag nanowires, respectively. *Reprinted and adapted with permission from (A) K. Kwon, K.Y. Lee, Y.W. Lee, M. Kim, J. Heo, S.J. Ahn, et al., Controlled synthesis of icosahedral gold nanoparticles and their surface-enhanced Raman scattering property, J. Phys. Chem. C 111 (3) (2007) 1161–1165. (B) W. Niu, Y.A.A. Chua, W. Zhang, H. Huang, X. Lu, Highly symmetric gold nanostars: crystallographic control and surface-enhanced Raman scattering property, J. Am. Chem. Soc. 137 (33) (2015) 10460–10463. (C) Y. Liu, S. Pedireddy, Y.H. Lee, R.S. Hegde, W.W. Tjiu, Y. Cui, et al., Precision synthesis: designing hot spots over hot spots via selective gold deposition on silver octahedra edges, Small 10 (23) (2014) 4940–4950. (D) M.S. Goh, Y.H. Lee, S. Pedireddy, I.Y. Phang, W.W. Tjiu, J.M.R. Tan, et al., A chemical route to increase hot spots on silver nanowires for surface-enhanced Raman spectroscopy application, Langmuir 28 (40) (2012) 14441–14449. Copyright 2007, 2012 and 2015 American Chemical Society. Copyright 2014 John Wiley & Sons, Inc.*

co-operative influences of PVP as facet-stabilizing agents and controlled reaction kinetics. For example, tetrahexahedral Au nanoparticle containing high-index facets such as {210} are formed using a high PVP concentration and *N*,*N*-dimethylformamide, a weak reducing agent [35].

All in all, shape-controlled synthesis of nanoparticles can substantially engineer hotspots at the individual particle level by concentrating EM fields at sharp vertices and edges, leading to improved SERS performances. In addition, increased control over the nanoparticle morphology also facilitates better control over its LSPR absorption bands, allowing better overlap between the LSPR bands of nanoparticles and the excitation wavelength of Raman measurement for greater SERS activity. However, any slight deviation in the reaction conditions will easily result in a heterogeneous mixture of shapes and sizes. As such, greater mechanistic understanding of the exact influence exerted by different thermodynamic and kinetic parameters in controlling the shape evolution is necessary to ensure high reproducibility for large-scale mass production.

2.2.1.2 Postsynthetic morphological modifications

Besides morphological control during nanoparticle synthesis, the EM hotspot density of individual nanoparticles can also be manipulated and further enhanced via postsynthetic modifications, which can be categorized into (1) additive and (2) subtractive methods. The additive method increases EM hotspot density through selective deposition of additional plasmonic nanoparticles on specific regions or facets of as-synthesized nanoparticle templates [33]. By tuning the reaction conditions, additional particles can be directly deposited along the more reactive edges and vertices of the as-synthesized nanoparticle template to generate larger curvatures at these regions for even stronger field confinement. For instance, Au nanoparticles are deposited along the edges and vertices of Ag octahedra to form selective edge Au-deposited Ag octahedra (SEGSO) (Fig. 2.3C) [33]. Due to intraparticle plasmonic coupling between the Ag and Au surface plasmons, SEGSO demonstrates 15-fold stronger local EM field and 10-fold higher SERS EF along its Au-deposited edges and vertices than pure Ag octahedra. Such phenomenon is akin to the nanoparticle-on-mirror geometry which generates strong SERS due to interactions between the surface and propagating plasmons of nanoparticles and the thin metal film beneath [36]. Moreover, the precise decoration of original hotspot areas with additional hotspots, referred to as the "hotspot-over-hotspot" approach, further enables a threefold increase in single-particle SERS EF compared to nonselectively Au deposited-Ag octahedra (NSEGSO).

The subtractive method heightens SERS enhancement through controlled etching of surface atoms to increase nanoparticle surface roughness [34,37]. For instance, Ag octahedra nanoparticles are selectively transformed into truncated structures with etched edges and corners, followed by octapod structures with increasing etchant concentrations [37]. This produces intraparticle nanogaps, which concentrates local EM fields and also introduces intense LSPR absorption bands in the red and infrared regions. The unique structural features and broad resonance responses of the etched nanostructures thus contribute to a five to fifty-fold increase in SERS performance compared to the original octahedra at all excitation wavelengths (513, 633, and 785 nm). In a related study, smooth Ag nanowires are etched using a hydrogen peroxide/ammonia mixture to produce roughened nanowires resembling "beads-on-a-string" [34]. This similarly generates intraparticle nanogaps for more efficient EM field localization at these engineered hotspots. Consequently, SERS activity is observed along the entire etched nanowire length compared to the original smooth Ag nanowires that only possess SERS activity at the tips, thus increasing single-particle SERS EF (6.4×10^4) by approximately twofold (Fig. 2.3D). Importantly, these postsynthetic morphological modifications do not affect the intrinsic crystallinity of the as-synthesized nanoparticles, ensuring a high quality (Q)-factor of the nanoparticles' plasmon resonances [38].

To date, an impressive array of metallic nanoparticles with increasingly sophisticated morphologies has been reported, ranging from nanorods, nanocubes, to nanoflowers and nanostars. Thanks to advancements in nanoparticle synthesis and modification, we now have an extensive and continually expanding library of nanoparticle morphologies to achieve maximal EM field enhancement for better SERS detection at the individual particle level. However, the usefulness of these individual nanoparticles as ultrasensitive SERS platforms for macroscopic sensing is highly restricted by their nanometer scale SERS-active regions and insufficient hotspot density/strength. In addition, individual nanoparticles in colloidal dispersions are often susceptible to random, uncontrolled aggregation, which reduces the number of hotspots instead due to overaggregation [39]. Moreover, the hotspots are randomly distributed within colloidal dispersions, often resulting in inconsistent and irreproducible SERS signals over large areas [40]. Thus colloidal nanoparticles are still not ideal for practical sensing application.

2.2.2 Two-dimensional platforms for electromagnetic field enhancement

Organizing nanoparticles into 2D assemblies introduces plasmonic coupling along *x*- and *y*-axes to yield stronger and more consistent EM field enhancements, generating scalable SERS-active areas of up to centimeter scale [41]. In this section, we will discuss various bottom-up nanoparticle self-assembly processes, which are governed by various internanoparticle interactions, for example, van der Waals interactions, electrostatic forces, and hydrogen bonding. Emerging tools to drive and direct the formation of 2D plasmonic assemblies including (1) DNA-programmable self-assembly, (2) interfacial self-assembly, and (3) template-assisted self-assembly will be discussed here.

2.2.2.1 DNA-programmable self-assembly

DNA motifs are highly versatile surface ligands that enable nanoparticle assembly into various predesigned 2D patterns and nanoarchitectures with high structural integrity and nanoscale precision via highly specific complementary base pairing. In particular, DNA origami nanostructures, which are DNA strands of specific oligonucleotide sequences self-assembled into predetermined shapes, are rapidly gaining attention. The DNA origami nanostructures are encoded with specific sticky end oligonucleotides that are complementary to those on the nanoparticles to direct the two to connect together in a predetermined fashion to construct various low-dimensional 2D nanoparticle/DNA arrays. The interparticle distance, packing density, and final assembly pattern of the 2D arrays can be controlled by tailoring the lengths, sequences, and relative positions of the bridging oligonucleotides. This has significant influence over the overall SERS performance of the as-assembled 2D plasmonic arrays by modulating the density of EM hotspots. For instance, Au nanoparticles functionalized with internal-modified (A-B-Au-A-B) and complementary terminal-modified dithiol single-stranded (ss) DNA (A'-Au-A' and B'-Au-B') are used to form 2D nanoassemblies of three parallel rows, referred as 222 NA (Fig. 2.4A) [42]. Due to interparticle coupling conferred by the assembly, the LSPR peak of this DNA-directed Au nanoparticle array red shifts from 522 to 532 nm. Localized surface plasmon coupling at the interparticle junctions induces large EM field enhancement, resulting in ~69-fold increase in EF compared to that of isolated Au nanoparticles (Fig. 2.4A).

In addition, a wide array of assembly patterns can be customized using the DNA origami methodology. For instance, nanoflower-like structures formed from DNA origami bundles and Au nanoparticles can be assembled into 2D lattices of different symmetries including square and hexagonal lattices (Fig. 2.4B) [43]. The final lattice symmetry is controlled by the

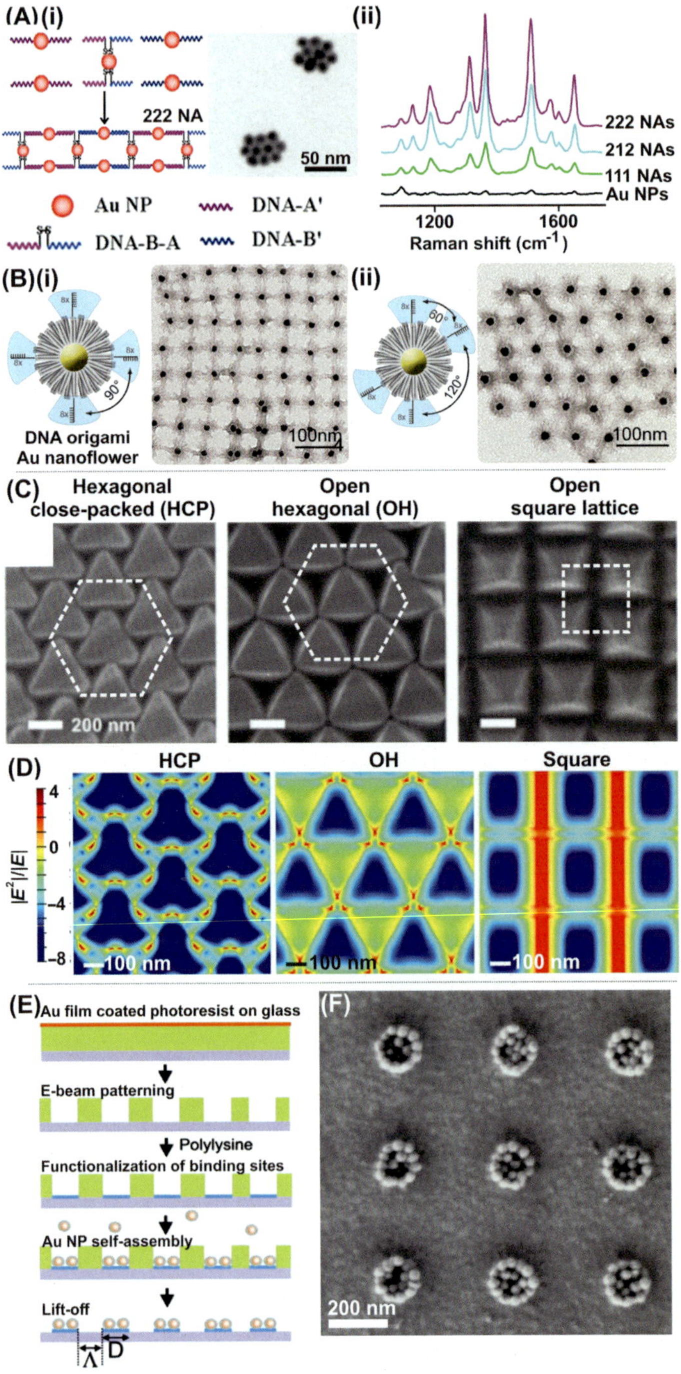
(A) (i)
222 NA
50 nm
Au NP
DNA-A'
DNA-B-A
DNA-B'
(ii)
222 NAs
212 NAs
111 NAs
Au NPs
1200
1600
Raman shift (cm-1)
(B) (i)
DNA origami
Au nanoflower
90°
100nm
(ii)
60°
120°
100nm
(C)
Hexagonal
close-packed (HCP)
Open
hexagonal (OH)
Open
square lattice
200 nm
(D)
HCP
OH
Square
|E²|/|E|
4
0
-4
-8
100 nm
100 nm
100 nm
(E)
Au film coated photoresist on glass
E-beam patterning
Polylysine
Functionalization of binding sites
Au NP self-assembly
Lift-off
Λ
D
(F)
200 nm

relative positions of attachment sites formed from ss DNA linkers, whereby attachment sites separated by 90 degrees result in a square lattice while separations of 120 and 60 degrees result in a hexagonal lattice (Fig. 2.3B). In another example, octahedral DNA origami frames with vertices encoded with Au nanoparticle attachment sites are used to control the precise position of these Au nanoparticles into discrete 2D arrays [45]. All in all, DNA motifs are versatile scaffolds that allow creation of various well-defined plasmonic 2D arrays. However, DNA-directed nanoassemblies for SERS-based application is still in its infancy, with few existing studies exploiting its full potential for SERS sensing.

2.2.2.2 Interfacial self-assembly

The interface of an immiscible two-fluid system (i.e., liquid−air or liquid−liquid) also provides one of the most effective techniques for organized self-assembly of 2D nanoparticle metacrystals. Nanoparticles spontaneously adsorb onto the interface and arrange into 2D metacrystals to minimize the interfacial free energy, most commonly facilitated by

◀ **Figure 2.4** 2D plasmonic platforms. (A) (i) Illustration of 2D Au nanoassemblies (222 NA) formation and corresponding TEM image, where 2 refers to the number of DNA strands conjugated to each NP. (ii) Comparison of SERS spectra of R6G adsorbed on Au NPs (black line) and as the nanoassemblies vary from 1D (111 NA, green line) to quasi-2D (212 NA, blue line) and 2D (222 NA, purple line). (B) DNA origami Au nanoflower assembly organized into (i) square lattices using hybridization sites at 90 degrees and (ii) hexagonal lattices using hybridization sites separated at 60 and 120 degrees. (C) SEM images of Ag octahedra-based metacrystals of hexagonal close-packed, open hexagonal, and square lattice superlattices. (D) Corresponding electric field distributions of various Ag-octahedra-based metacrystals at 532 nm. (E) Fabrication flowchart for nanocluster arrays (NCAs). (F) SEM image of NCA assembled with 40 nm Au NPs. *SERS*, Surface-enhanced Raman scattering; *2D*, two-dimensional. *Reprinted and adapted with permission from (A) Y. Yan, H. Shan, M. Li, S. Chen, J. Liu, Y. Cheng, et al., Internal-modified dithiol DNA−directed Au nanoassemblies: geometrically controlled self−assembly and quantitative surface−enhanced Raman scattering properties, Sci. Rep. 5 (1) (2015) 16715. (B) R. Schreiber, I. Santiago, A. Ardavan, A.J. Turberfield, Ordering gold nanoparticles with DNA origami nanoflowers, ACS Nano 10 (8) (2016) 7303−7306. (C and D) Y.H. Lee, W. Shi, H.K. Lee, R. Jiang, I.Y. Phang, Y. Cui, et al., Nanoscale surface chemistry directs the tunable assembly of silver octahedra into three two-dimensional plasmonic superlattices, Nat. Commun. 6 (2015) 6990−6990. (E and F) B. Yan, A. Thubagere, W.R. Premasiri, L.D. Ziegler, L. Dal Negro, B.M. Reinhard, Engineered SERS substrates with multiscale signal enhancement: nanoparticle cluster arrays, ACS Nano 3 (5) (2009) 1190−1202. Copyright 2015 Springer Nature. Copyright 2009 and 2016 American Chemical Society.*

Langmuir–Blodgett or Langmuir–Schaefer (LS) assembly techniques. For additional understanding of these film fabrication techniques, interested readers can refer to relevant publications and reviews cited here [46,47]. These 2D metacrystals can be transferred onto supporting substrates, or be directly used as substrate-less SERS platforms at the fluid–fluid interfacial region. In particular, substrate-less platforms are ideal for multiplex sensing across fluid interfaces and monitoring of biphasic reactions because these platforms have direct access to both immiscible fluids. For instance, assembly of Au nanorods at the oil–water interface can quantitatively detect both aqueous-soluble R6G and organic molecules such as oleic acid down to the nanomolar level [48]. However, detection using substrate-less interfacial SERS platforms is highly susceptible to signal fluctuations induced by external agitation including wind, natural ground vibrations and temperature changes. This reduces their robustness for practical sensing application. Hence, substrate-supported 2D metacrystals are still preferred for conventional single-phase detection and for fundamental investigations of metacrystal properties.

Such metacrystal engineering is essential to programme metacrystals of varying lattice structures, interparticle spacing and EM field enhancement strengths to achieve higher SERS efficiency. One notable example is the concept of "one nanoparticle, multiple superlattices" to achieve structural diversity using just one type of shape-controlled nanoparticle morphology [41,49,50]. This can be achieved by controlling the nanoparticle surface wettability via various techniques including using surface functionalization ligands of different hydrophobicity. For instance, as the nanoparticle hydrophobicity increases, Ag octahedral building blocks gradually change from a planar to a vertical configuration to form three distinct metacrystals of hexagonal close-packed (HCP), open hexagonal (OH) and open square arrays, respectively (Fig. 2.4C) [41]. Notably, the square array has the highest SERS enhancement among the three metacrystals when excited at 532 nm. This is despite the square array having the lowest packing density of ~33%, with a 7.5-fold increase in EF than both hexagonal arrays (packing densities of 67% and 89% for OH and HCP, respectively). This is because the square array has a larger hotspot region than the hexagonal lattices (unit hotspot area of ~22,650 vs 360 nm^2) and also more efficient SERS scattering due to its pyramidal structure (Fig. 2.4D). This highlights another important concept in nanoscale superlattice design that "more is not always better" for SERS, and precise control over the geometrical and spatial orientation of the building blocks in the metacrystals is essential.

These two key concepts are also demonstrated in a related study using Ag nanocubes functionalized with different ratios of hydrophilic/hydrophobic ligands to achieve different superlattices [49]. Another similar study capitalizes on solvent-dependent polymer conformational changes instead to manipulate the final configuration using Ag nanocubes functionalized with a PEG/C16 mixed monolayer [50]. Despite the advantages of using bulk interfaces to direct nanoparticle self-assembly, the as-assembled platforms are typically microscopically flat and thus vulnerable to laser misalignment during SERS measurements. Furthermore, the interfacial platforms typically require large volumes of sample amounts (ranging from milliliters to liters), which limits their use for miniaturized applications.

2.2.2.3 Template-assisted self-assembly

Another strategy to produce 2D plasmonic platforms is to use customized, prefabricated templates with geometrically constrained features, for example, pores or channels, to dictate the assembly of nanoparticle building blocks into larger, well-defined nanostructures. Various lithographic techniques including e-beam lithography, nanoimprint lithography, and photolithography can first be used to create suitable topographical templates and patterns [10]. Thereafter, nanoparticles can be self-assembled at the desired locations to create targeted nanoparticle clusters with uniform and spatially defined EM hotspots. For instance, spherical Au nanoparticles can self-assemble into nanoclusters of controlled sizes by using masks defined with varying binding site diameters, prefabricated by electron beam lithography (Fig. 2.4E and F) [44]. The as-fabricated nanocluster array is capable of detecting and differentiating three bacteria species, which cannot be achieved using nonpatterned substrates. In another example, SERS EF increases from 10^3-10^4 to 10^6-10^7 as the cluster size increases from monomers to teramers in nanocube- and octahedra-assembled nanoclusters [51]. Such integrated approaches are advantageous to attain better control over final hotspot locations on a substrate, thus achieving high signal reproducibility across the entire area. However, the self-assembly process is time-consuming and susceptible to defect formation, such as incomplete filling of nanoparticles at the predefined sites. Lack of precise control over the relative orientation of nanoparticles within a single nanocluster may also lead to inconsistent hotspot distribution within individual nanoclusters. Moreover, template-assisted self-assembly requires an additional preliminary step of creating the templates and may thus be not as time- or cost-efficient compared to other 2D strategies. Nonetheless, the template-based self-assembly method is still useful and advantageous for fundamental SERS investigations.

Without a question, critical developments in various bottom-up and integrated techniques have advanced the construction of low-dimensional 2D plasmonic platforms with better SERS enhancement and greater signal consistency for large-area scalability. However, the dimensionality of the engineered nanoplasmonic platforms is similarly crucial for optimal SERS performance. 2D SERS platforms in general tend to underutilize the 3D focal volume of the laser irradiation, which typically extends over micron-sized dimensions in all three Cartesian planes. This limits SERS sensitivity and also demands precise laser focusing due to a low tolerance of focal misalignment.

2.2.3 Three-dimensional platforms for electromagnetic field enhancement

3D SERS platforms overcome the limitations of 2D SERS platforms through efficient utilization of the entire 3D laser excitation volume. Here, we define 3D platforms as those with hotspot volumes ranging from submicrometer ($\geq$0.5 μm) to centimeter scale, complementary with majority of Raman spectrometers with laser focal depths of $\geq$0.5 μm [52]. 3D platforms possess enhanced SERS performance due to an increase in effective spatial hotspot density (or SERS-active volume in other words) within the laser excitation volume [53]. They also demonstrate improved detection sensitivities due to an increase in effective SERS-active surface area, which boost interactions between the plasmonic nanoparticles and target analyte. Moreover, 3D platforms also possess better tolerance to laser misfocus along the z-axis, where sufficient hotspots remain within the excitation volume to generate consistent SERS signals. This enables faster and more feasible onsite measurements without laborious laser alignment prior to SERS detection for practical application.

Here, we introduce various contemporary 3D SERS platforms and focus on their unique plasmonic properties that arise from their 3D configurations. The 3D SERS platforms are broadly discussed in terms of substrate-based and substrate-less platforms. Substrate-based platforms are formed from the assembly of nanostructures on a flat solid support, which provides mechanical stability and ensures better fabrication reproducibility. On the other hand, the solid support is absent from substrate-less platforms, which bestows greater mobility and enables greater accessibility to analyte and excitation laser from all spatial directions.

2.2.3.1 Multilayered 3D supercrystals

One common substrate-based approach to fabricate 3D platforms is to build multilayered nanoparticles on a solid substrate, forming highly ordered 3D supercrystals. In doing so, the particle height along the z-direction is increased to provide a high EM field. These 3D supercrystals are typically fabricated via sedimentation of a highly concentrated droplet of nanoparticle solution onto a flat surface, which dries to form close-packed supercrystals. For instance, a multilayered nanorod supercrystal of ~ 1 μm of 3D hotspot thickness constructed by the abovementioned droplet-based assembly yields a high SERS EF of $\sim 10^6$ due to increased interstitial hotspots on the supercrystal surface [54]. However, one challenge of droplet-based assembly of multilayered substrates/supercrystals is that there should be no perturbation to the system over a long period of time (hours to days) as the solution evaporates. In addition, to form high-quality supercrystals, the nanoparticles must be highly homogeneous and highly concentrated, which aggregates easily. More importantly, the close-packed nature of the as-fabricated supercrystals inhibits effective laser and analyte penetration into their internal bulk beyond ~ 2 μm, which forces the resulting EF to plateau rapidly after reaching the laser penetration threshold [55].

Consequently, open 3D multilayered structures are increasingly favored, because they allow deeper laser penetration and thus improve accessibility of both laser and analyte to the interior bulk. This advantage is clearly highlighted in the enhanced SERS EF of dual-structure 3D supercrystals fabricated from the self-assembly of Ag polyhedra in two distinct microenvironments within the droplet: one at the drying front and another at the air—water interface [56]. The resultant supercrystals compose of more open structures at the air—water interface and more densely packed structures at the drying front, demonstrating heightened SERS EF of 5.2×10^6 compared to that of uniform densely packed supercrystals at 1.6×10^6 (Fig. 2.5A and B). Besides the droplet-based method, open multilayered structures are also fabricated via the LS assembly technique, in which nanoparticle monolayers are successively stacked together at the air/water interface onto a substrate [59]. For instance, the SERS intensity-depth profile from the hyperspectral SERS imaging of an open 3D Ag nanowire mesh-like array exhibits 7 times deeper laser penetration depth than its close-packed woodpile-like counterpart [59]. In addition, the open 3D array demonstrates an eightfold increase in analytical enhancement factor (AEF) of up to 1.7×10^{11}, illustrating that more nanoparticles have access to the laser and are excited in the open structure.

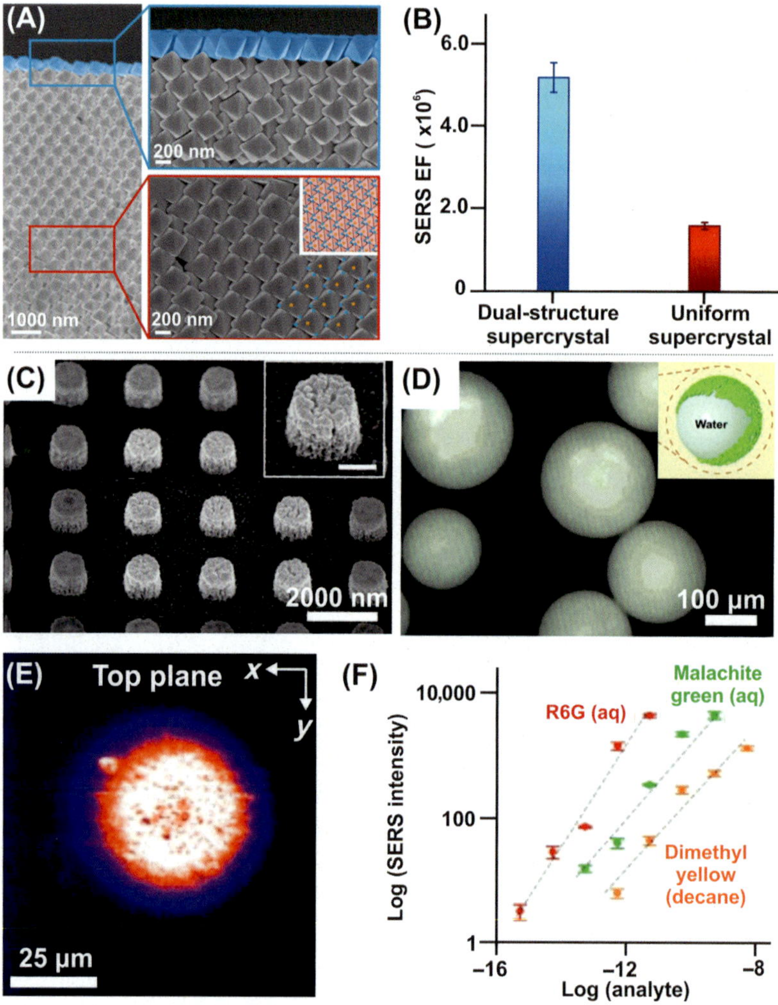

Figure 2.5 3D plasmonic platforms for EM field enhancement. (A) Cross-sectional SEM image of the dual-structure supercrystals with long range order. The supercrystal can be divided into two regions: the top layer is of an open structure while the bulk of the supercrystal is crystalline. (B) Comparison of SERS EFs of dual-structure and uniform supercrystals. (C) SEM image of nanoporous SiO_2 microcylinder arrays decorated with Au nanoparticles. (D) Microscopic image and scheme of plasmonic colloidosomes.(E) *x-y* SERS image when laser is focused on the top-plane of the colloidosome. (F) Linear intensity–concentration relationships of malachite green, rhodamine 6G, and dimethyl yellow spanning ~7 orders of magnitude from 10^{-15} to 10^{-8} mole. *EF*, Enhancement factor; *SERS*, surface-enhanced Raman scattering; *3D*, three-dimensional. *Reprinted and adapted with permission from (A and B) Y.H. Lee, C.L. Lay, W. Shi, H.K. Lee, Y. Yang, S. Li, et al., Creating two self-assembly micro-environments to achieve supercrystals with dual structures using polyhedral nanoparticles, Nat. Commun. 9 (1) (2018) 2769. (C) S.Y. Lee, S.-H.*

Overall, the assembly of nanoparticles into micrometer scale structures effectively exploits multidimensional plasmonic coupling for improved SERS EF. However, such multilayered nanoparticle assemblies still suffer from low particle usage efficiency, in which only the outermost nanoparticle layers (up to ~16 μm for open structures) can be accessed by the target analyte molecules.

2.2.3.2 Template-based 3D platforms

To reduce waste of plasmonic materials, one alternative is to deposit the plasmonic nanoparticles on cost-effective, nonplasmonic interior supporting templates. In this way, only the outermost layers, which are accessible to the incoming analytes, are SERS active. Various lithographic methods such as nanoimprint lithography are typically used to form the interior supports, which is then coated with plasmonic material via different physical and colloidal methods, including sputtering and self-assembly [60,61]. In one study, Au nanoparticles were assembled on porous microcylindrical templates to form SERS-active microcylinders [57]. Plasmonic coupling among Au nanoparticles embedded within the nanopores is indicated by a red shift in the reflectance spectrum from 520 nm (for isolated Au nanoparticles) to 650 nm (for Au nanoparticle-decorated microcylinders). The presence of 3D porosity also enables a higher loading of Au nanoparticles to enhance SERS signals by >10-fold compared to nonporous SERS-active microcylinders (Fig. 2.5C). A separate study uses photolithography to fabricate polymeric pyramidal structures, followed by external coating with Ag film and Ag nanocube assembly [60]. The 3D pyramidal nanostructures achieve high hotspot dimension from 5 to 15 μm in height, exhibiting hotspots in three orthogonal directions to give an EF of 2.6×10^6. In summary, depositing plasmonic materials on 3D interior supporting template utilizes nanoparticle in a more economic manner. Moreover, there is much flexibility to design a variety of supporting structures with

◀ *Kim, M.P. Kim, H.C. Jeon, H. Kang, H.J. Kim, et al., Freestanding and arrayed nanoporous microcylinders for highly active 3D SERS substrate, Chem. Mater. 25 (12) (2013) 2421–2426. (D–F) G.C. Phan-Quang, H.K. Lee, I.Y. Phang, X.Y. Ling, Plasmonic colloidosomes as three-dimensional SERS platforms with enhanced surface area for multiphase sub-microliter toxin sensing, Angew. Chem. Int. Ed. 54 (33) (2015) 9691–9695. Copyright 2018 Springer Nature. Copyright 2013 American Chemical Society. Copyright 2015 Wiley & Sons, Inc.*

well-controlled shapes and sizes using lithography, compared to supercrystal formation. However, similar to template-assisted fabrication of 2D SERS platforms, any template defects will be transferred to the platforms. An additional step is also required to produce the supporting templates.

2.2.3.3 Substrate-less 3D platforms

An alternative to substrate-based SERS platforms is substrate-less platforms, which are free-standing structures that do not require any solid support. Here, we will discuss emerging 3D substrate-less SERS platforms such as plasmonic colloidosomes and plasmonic liquid marbles (PLMs). In general, PLMs and plasmonic colloidosomes are spherical millimeter-/micrometer-sized SERS platforms formed via the spontaneous assembly of nanoparticles at the immiscible fluid–fluid interface (Fig. 2.5D). The thickness of the nanoparticle shell in these milli-/micro-droplets can be readily modulated by controlling the concentration of nanoparticles used during PLM/ colloidosome formation. Upon formation, both platforms possess multiple layers of closely packed metallic nanoparticles to give rise to uniform and intense EM field across the entire 3D shell, due to extensive plasmonic coupling in all three Cartesian planes (Fig. 2.5E). Coupled with greater access to analyte and laser penetration from all spatial directions, this enables better utilization of the 3D hotspot density to achieve high AEF of 10^6–10^9, which can be applied for ultratrace SERS detection [40,58,62]. For instance, 2 μL Ag PLMs can detect BPA, a toxic phenolic estrogen, at a detection limit down to 10 amol (or 1 pg mL^{-1}), which is $>10^7$-fold better than the UV–vis method [63]. Plasmonic Ag colloidosomes with diameter of 40 μm can also detect R6G and other aqueous- or organic-soluble analytes spanning ~7 orders of magnitude (10^{-15} to 10^{-8} mole), notably with a 0.5-fmole detection limit for R6G (Fig. 2.5F) [58]. In addition, these 3D substrate-less platforms can better tolerate laser misalignment compared to their 2D counterparts, because the curved surfaces and extended *z*-dimension provide multiple focal planes throughout the PLM/colloidosomes to ensure that the laser is always focused on the platform.

Another unique and important feature of these 3D substrate-less platforms is their ability to effectively isolate the encapsulated analyte samples within their aqueous cores to prevent sample cross-talk. For example, batches of colloidosomes can be flowed through the same microfluidic channel for rapid throughput detection of multiple samples sequentially without any channel contamination or cross-talk [63]. Moreover, SERS detection using these miniaturized platforms only require submicroliter to microliter amounts of analyte solution, which is highly appealing in the field of medical diagnosis where sample availability may be scarce. This spells immense promise for future developments of portable sensors based on plasmonic colloidosomes or PLMs to enable rapid, accurate onsite medical diagnosis. Hence, particle-assembled 3D substrate-less SERS platforms demonstrate superior potential for interfacial and/or small-volume applications compared to substrate-based counterparts. While most platforms are water-in-oil PLM/colloidosomes thus far, moving forward, we envision that the future development of other emulsion systems such as oil-in-water capsules can bring forth immense potential in multiphasic analysis.

Ultimately, the rational design of an appropriate SERS platform, for example, material, substrate type, and fabrication method, is dictated by the specific application in question. However, there are many molecules of interest, especially biomolecules, that have no special affinity to metal surfaces and/or poor Raman cross-sections and cannot be detected easily even using SERS platforms of strong SERS capabilities. Hence, relying solely on hotspot engineering to develop SERS platforms that can be utilized for real-world application in fields ranging from biomedical diagnosis [64,65] to environmental, food, and industrial monitoring [59,66,67] remains a huge challenge, even if possible. It is thus critical to complement hotspot engineering with additional analyte manipulation strategies to ensure that the target analytes can better access the optimally designed EM hotspots.

2.2.4 Analyte manipulation strategies

Analyte manipulation strategies aim to improve nanoparticle–analyte interactions by modifying the SERS substrates to concentrate analyte molecules close to designer plasmonic surfaces. For example, SERS substrates can be made superhydrophobic to decrease aqueous contact area and concentrate analyte within a confined area upon drying. These approaches are highly versatile because the methods are applicable to a large range of SERS substrates and analytes with different chemical functionalities. In addition, these modifications are advantageous for biomedical diagnosis because minimal sample quantities is required due to analyte concentrating effects.

In this section, we discuss two emerging types of analyte-concentrating strategies achieved via physical modification of the SERS platform: (1) altering the surface wetting properties of plasmonic surfaces to reduce solvent spread and (2) incorporating MOFs with SERS substrates to encapsulate gaseous analytes. We note that another major approach to manipulate analyte is by chemically modifying the plasmonic surface to capture the target analyte and this will be discussed in detail in subsequent chapters.

2.2.4.1 Modifying surface wetting properties of plasmonic surfaces

For trace analyte detection, solvent spread poses a huge problem because it significantly reduces the amount of analyte per unit area and hence decreases the SERS signal intensities. To address this issue, SERS substrates can be made superhydrophobic (contact angle of the droplet is >150 degrees) so that the solvent contact area is minimized (Fig. 2.6A) [68]. For example, an aqueous droplet containing R6G dries over an area of $50 \times 50\ \mu m^2$ when dropcasted on a superhydrophobic C_4F_8-functionalized Ag-coated silicon nanopillars platform compared to a large area of $5 \times 5\ mm^2$ on a smooth platform (Fig. 2.6B) [69]. The confinement of the drying spot results in a 10^4-fold analyte concentration between the two types of surfaces. In general, the hydrophobicity of the platform can be increased via two methods: (1) modifying the surface chemistry of SERS substrates and (2) increasing the nano-/microscale surface roughness so that the solid–liquid interfacial energy can be increased and air pockets between the substrate surface and the solvent droplet can be formed [70,71]. The former is realized via the functionalization of hydrophobic fluorocarbons such as perfluorodecanethiol (PFDT), while the latter via nanoparticle deposition/assembly or using roughened template supports such as nanopillars. Typically, to achieve superhydrophobicity, both the surface chemistry and surface roughness must be modified. For example, the contact angles for smooth PFDT-functionalized Ag film surface and nonfunctionalized Ag nanocube array only reaches up to ~77 and ~131 degrees, respectively while the PFDT-functionalized Ag nanocube array platform achieves ~169 degrees [68]. The resultant high analyte concentrating effect of the superhydrophobic PFDT-functionalized Ag nanocube array enables detection of R6G at 10^{-16} M, with an AEF of 10^{11}. Therefore altering the surface wetting properties is highly beneficial in obtaining amplified SERS signals due to analyte concentrating effects. This is particularly advantageous for trace analyte detection in biological matrices since the samples are often diluted [72].

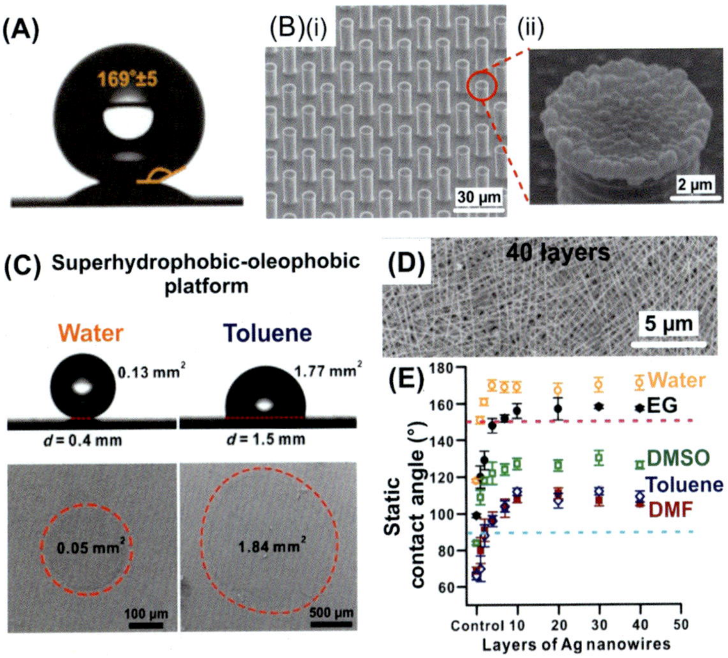

Figure 2.6 Postsynthetic modifications of surface wetting properties. (A) Contact angle of an aqueous droplet on a superhydrophobic surface. (B) (i) An array overview and (ii) close-up SEM images of silicon nanopillars with Ag nanoparticle deposition. (C–E) Superhydrophobic–oleophobic platform comprising nanowire mesh-like arrays. (C) Contact angle and SEM images of water and toluene droplets deposited on the surface (D) SEM image of the nanowire arrays. (E) Changes in static contact angles of solvents with various polarities as the number of nanowire layers increase. *Reprinted and adapted with permission from (A) H.K. Lee, Y.H. Lee, Q. Zhang, I.Y. Phang, J.M.R. Tan, Y. Cui, et al., Superhydrophobic surface-enhanced Raman scattering platform fabricated by assembly of Ag nanocubes for trace molecular sensing, ACS Appl. Mater. Interf. 5 (21) (2013) 11409–11418. (B) F. De Angelis, F. Gentile, F. Mecarini, G. Das, M. Moretti, P. Candeloro, et al., Breaking the diffusion limit with super-hydrophobic delivery of molecules to plasmonic nanofocusing SERS structures, Nat. Photonics 5 (11) (2011) 682–687. (C–E) X. Li, H.K. Lee, I.Y. Phang, C.K. Lee, X.Y. Ling, Superhydrophobic-oleophobic Ag nanowire platform: an analyte-concentrating and quantitative aqueous and organic toxin surface-enhanced Raman scattering sensor, Anal. Chem. 86 (20) (2014) 10437–10444. Copyright 2013 and 2014 American Chemical Society. Copyright 2011 Springer Nature.*

However, a limitation of purely superhydrophobic surfaces is the inability to exhibit similar analyte concentrating effect for less polar/organic solvents. To overcome this inadequacy, the surface can be imbued with oleophobic properties by simply increasing the surface roughness since organic solvents have smaller surface tensions compared to water (Fig. 2.6C) [59]. For instance, 3D Ag nanowire mesh-like arrays functionalized with PFDT exhibit analyte concentration factors of 100-fold and 8-fold for water and toluene, respectively (Fig. 2.6D and E) [59]. In another study, omniphobic stacked-disk nanotower arrays achieves a limit of detection of 10^{-15} M for R6G in water and 10^{-14} M for Nile red in toluene [73]. The surface wetting properties can thus be modified according to the nature of the solvent or sample matrix to attain analyte concentrating effects. Moreover, a combination of different surface wetting properties can be simultaneously utilized to manipulate behavior of the solvent droplet on the SERS substrate. A nanodendritic Au substrate that possesses both hydrophobic and hydrophilic elements on its surface enables polar solvent droplets to be confined within the small hydrophilic region surrounded by the hydrophobic region and achieves an EF of 10^8 for R6G in water [74].

Therefore modification of surface wetting properties presents a relatively straightforward way to concentrate analyte molecules without altering their chemical properties. It allows concentration of dilute samples in small volumes without the need for sample pretreatment or other concentration processes which are prone to sample loss. This is especially important in the trace analysis of biomolecules/biomarkers for biomedical analysis. However, this method is limited to analytes which are soluble in liquids and lacks selectivity in capturing desired analyte molecules in complex biological matrices. Further, it is unsuitable for direct detection of gaseous analytes such as volatile organic compounds (VOCs) and polycyclic aromatic hydrocarbons (PAHs).

2.2.4.2 Metal-organic frameworks

Gas-phase SERS detection is inherently more challenging compared to liquid-phase SERS detection due to the high mobility, low affinity to plasmonic surfaces, and much more diffuse nature of molecules in the gaseous state [5]. Currently available sample pretreatment procedures such as the use of thermoelectric cooling lack the ability to simultaneously address all the challenges involved [67]. An emerging approach is to incorporate MOFs with plasmonic substrates (Fig. 2.7A). MOFs are 3D, highly porous and crystalline networks of metal ions connected by organic polymers, where gaseous molecules may diffuse into and become temporarily confined within. The functionalization of MOFs with plasmonic substrates thus enables the adsorption and concentration of gaseous molecules close

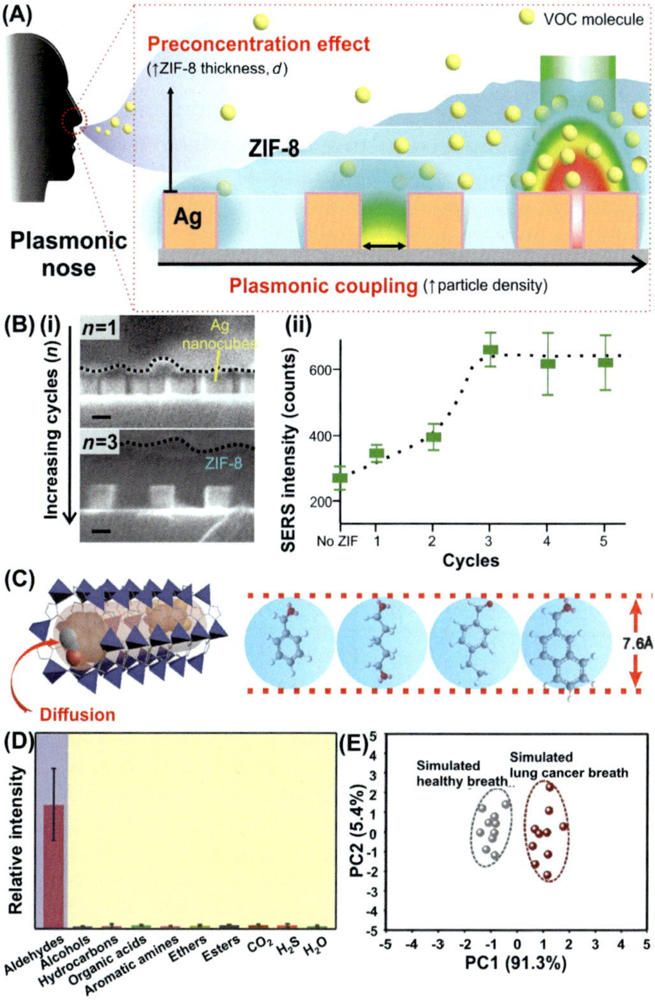

Figure 2.7 Metal-organic frameworks (MOFs). (A) Schematic depicting the "plasmonic nose," comprising Ag nanocube array with ZIF-8 (Ag@ZIF). (B) (i) Cross-sectional SEM image of Ag@ZIF-8 after *n*-cycles of ZIF growth and (ii) corresponding increased SERS intensity observed as thickness of ZIF increases. (C) Schematic illustrating the analyte size selectivity endowed by ZIF and (D) the corresponding selective SERS detection of aldehydes vapor compared to other interference. (E) Principal component analysis (PCA) of set of simulated breath samples of healthy and lung cancer. *SERS*, Surface-enhanced Raman scattering; *ZIF-8*, zeolitic imidazolate framework-8. *Reprinted and adapted with permission from (A and B) C.S.L. Koh, H.K. Lee, X. Han, H.Y.F. Sim, X.Y. Ling, Plasmonic nose: integrating the MOF-enabled molecular preconcentration effect with a plasmonic array for recognition of molecular-level volatile organic compounds, Chem. Commun. 54 (20) (2018) 2546–2549. (C–E) X. Qiao, B. Su, C. Liu, Q. Song, D. Luo, G. Mo, et al., Selective surface enhanced Raman scattering for quantitative detection of lung cancer biomarkers in superparticle@MOF structure, Adv. Mater. 30 (5) (2018) 1702275. Copyright 2018 The Royal Society of Chemistry. Copyright 2017 John Wiley & Sons, Inc.*

to the plasmonic surface. Notably, the pore size of MOFs can be controlled by varying the identities of the metal ion and/or organic linker, allowing flexibility in novel structure design and ability to customize the MOF for specific analyte capture [75–80]. In one study, zeolitic imidazolate framework-8 (ZIF-8) coated on a monolayer of Ag nanocubes is employed to adsorb and detect PAHs [e.g., 2-naphthalenethiol (2NT)] and VOCs (e.g., toluene) [67]. The fabricated platform, termed "plasmonic nose," possesses a detection limit of 50 ppb for 2NT and 200 ppm for toluene, respectively, and achieving a consistent approximately twofold improvement compared to a control platform in the absence of ZIF-8 (Fig. 2.7B). In the context of biomedical analysis, exhaled breath analysis using SERS presents a simple, noninvasive technique to obtain biomarkers correlating to diseases [81,82]. For instance, using a 3D Au core–shell coated with ZIF-8, lung cancer biomarker 4-ethylbenzaldehyde can be detected in exhaled breath down to the ppb level (Fig. 2.7C–E) [64]. The MOF shell with pore aperture of about 7.6 Å excludes larger molecules, allowing selective detection of small aldehydes in the complex matrix. Through these examples, the potential of utilizing MOFs in the selective encapsulation of gaseous analytes to promote gas-phase SERS sensing appears promising.

However, it is important to consider several factors when employing MOFs as part of the intended SERS substrate. First, the MOF itself should not have SERS signals in the region of interest as it would pose significant interference to signals resulting from the analyte(s). Next, the pore size of the MOF should be carefully selected to allow desired molecules to enter/exit the structure, while excluding potentially interfering molecules. Finally, it is important to consider that the efficacy of a selected MOF is generally limited by the rate of diffusion of gaseous analyte molecules as well as its relative rate of adsorption within the cavities.

In overall, there has been an expansion of various types and configuration of SERS platforms moving from shape-controlled individual nanoparticles to 2D and 3D plasmonic platforms over the past four decades. This rapid development in nanoplasmonic materials and platforms is driven by the need to maximize SERS sensing capability for various scenarios. Apart from traditional hotspot engineering, the incorporation of various analyte manipulation strategies is also critical to overcome existing performance bottlenecks in terms of sensitivity and practicality. A brief overview of the latest developments in hotspot engineering and postsynthetic platform modifications is summarized in a nonexhaustive list in Table 2.1.

Table 2.1 Current progress in development of novel nanoplasmonic platforms.

Approach	SERS platform	Enhancement factor (EF) or analytical enhancement factor (AEF)	Comments	Ref.
0D/1D nanoparticles				
Capping agent-directed growth	Ag nanocubes (0D)	NA	Supports multipolar LSPR behavior	[14,27]
	Au icosahedra (0D)	NA	Fourfold stronger intensity than spherical NPs	[31]
	Symmetric Au nanostars (0D)	4-Mercaptobenzoic acid: 4.5×10^8	Higher signal reproducibility than irregularly shaped, asymmetric Au nanostars	[32]
	Au tetrahexahedral nanoparticles (0D)	NA	Contains numerous high-index facets including {210}	[35]
	Au nanorods (1D)	NA	Possess multiple plasmon bands (longitudinal and transverse plasmon bonds)	[11,15–17]
Postsynthetic morphological modifications	Selective edge Au-deposited Ag octahedral (SEGSO, 0D)	4-Methylbenzenethiol: 1.1×10^5 *Single-particle level	15-fold and 10-fold stronger field and EF than bare Ag octahedral	[60]
	Etched Ag nanowires (1D)	6.4×10^4 *Single-particle level	Uniform SERS activity along entire length of etched nanowire; as-synthesized nanowires are SERS-active only at the tips	[34]

(Continued)

Table 2.1 (Continued)

Approach	SERS platform	Enhancement factor (EF) or analytical enhancement factor (AEF)	Comments	Ref.
2D plasmonic platforms				
DNA-programmable self-assembly	Lattices of DNA/Au nanoparticle nanoflowers	NA	NA	[43]
	Octahedral DNA origami frames with Au attachment sites at vertices	NA	NA	[45]
Interfacial self-assembly	Vertically aligned Au nanorods assembled at liquid–liquid interface	Rhodamine 6G: 10^6 (AEF)	10^4-fold better than random-oriented Au nanorods system	[48]
	Plasmonic hexagonal and open square Ag metacrystals	4-Methylbenzenethiol: 9.9×10^4	Square array has the highest EF despite having the lowest packing density (~33%)	[41]
Template-assisted self-assembly	Nanoclusters of spherical Au nanoparticles	NA	NA	[44]

3D plasmonic platforms

Multilayered 3D supercrystals	Open Ag nanowire mesh-like array	Melamine: 1.7×10^{11} (AEF)	SERS AEF is ~ 8-fold higher than that of the close-packed woodpile-like counterpart	[59]
	Dual-structure Ag polyhedral supercrystals	4-Methylbenzenethiol: 5.2×10^{6}	SERS EF is ~3.3-fold higher than that of uniform densely packed supercrystals	[56]
Template-based synthesis	Nanoporous microcylinder loaded with Au nanoparticles	Benzenethiol: 6.5×10^{5}	SERS EF is ~10-fold better than that of	[57]
	Photolithographically fabricated polymeric pyramidal structures decorated with Ag nanocubes	4-Methylbenzenthiol: 1.2×10^{6}	NA	[60]
Substrate-less platforms	Ag nanocube plasmonic colloidosome	Rhodamine 6G: 2.0×10^{6} (AEF)	NA	[58]

(*Continued*)

Table 2.1 (Continued)

Approach	SERS platform	Enhancement factor (EF) or analytical enhancement factor (AEF)	Comments	Ref.
Analyte manipulation strategies				
PFDT-functionalized Ag nanocube array	Superhydrophobic	Rhodamine 6G: 10^{11} (AEF)	Analyte-concentrating factor of 14-fold	[68]
PFDT-functionalized Ag nanowire mesh-like array formed by Langmuir–Schaefer technique		NA	Analyte concentrating factor of 100-fold for water and 8-fold for toluene	[59]
Stacked-disk nanowire arrays formed by phase-shift interference lithography	Superomniphobic	NA	Limit of detection of 10^{-15} M for R6G in water and 10^{-14} M for Nile red in toluene	[73]
Nanodendritic Au substrate fabricated via electrochemical deposition	Superwettable	Rhodamine 6G: 10^{8}	NA	[74]
MOF-encapsulated Ag nanocube assembly (plasmonic nose)	MOF	4-Methylbenzenthiol: 7.0×10^{6} (AEF) toluene: 2.0×10^{5} (AEF)	~2-fold improvement in detection limit compare to an Ag-only control platform	[67]

LSPR, Localized surface plasmon resonance; *MOF*, metal-organic framework; *PFDT*, perfluorodecanethiol; *SERS*, surface-enhanced Raman scattering; *0D*, zero-dimensional; *1D*, one-dimensional; *3D*, three-dimensional. *EF value is at single-particle level.

2.3 Nonconventional surface-enhanced Raman scattering platforms

Besides conventional noble metal materials, recent efforts have been devoted to identifying alternative plasmonic configurations such as bimetallic core–shell systems, as well as hybrid nanostructures incorporating nonmetallic materials including graphene and semiconductors. These burgeoning hybrid systems are reported to exhibit higher SERS enhancement while possessing better tunability of optical, catalytic, and magnetic properties, which increases their practicality for biomedical application. In this section, we focus on the abovementioned nonconventional plasmonic configurations and discuss their advantages and future work as potential nanoplasmonic materials for SERS.

2.3.1 Bimetallic systems

One common configuration of bimetallic nanoparticles is the core–shell system (core@shell), in which the core metal is surrounded by a second metallic shell layer (Fig. 2.8A). Due to the synergistic electronic coupling between the two metals, bimetallic nanoparticles exhibit highly tunable optical properties over a broad range of excitation wavelengths. The optical properties can be readily modulated by varying the thickness of the core and/or shell, with a general red-shift of the LSPR band with an increase in shell thickness due to the occurrence of plasmonic percolation (Fig. 2.8A) [83]. Notably, this red-shift toward the NIR region is highly appealing for biomedical application because NIR excitation is often employed for noninvasive detection to reduce tissue and cellular damage [83,88]. For instance, Au@Ag core–shell nanoparticles prepared via stepwise reduction of Ag^+ ions on Au nanoparticles by ascorbic acid demonstrate a larger extinction contribution above 600 nm compared to pure Ag nanoparticles of similar size [89]. Spiky bimetallic nanoparticles (Ag core and Au spikes) similarly showcase a red-shift of the plasmon resonance modes into the NIR region with an increase in spike length. Spike lengths >50 nm further enable strong EM field enhancement by up to 10^4 to serve as efficient light absorbers in the NIR regime [90].

The strong EM field enhancements enable core–shell nanoparticles to demonstrate higher SERS efficiencies than the individual monometallic nanoparticles. For instance, Au@Ag core–shell nanoparticles have been employed to detect DNA and RNA, with consistent SERS signals and a highly sensitive detection limit of 20 fM [91]. Without the additional Ag

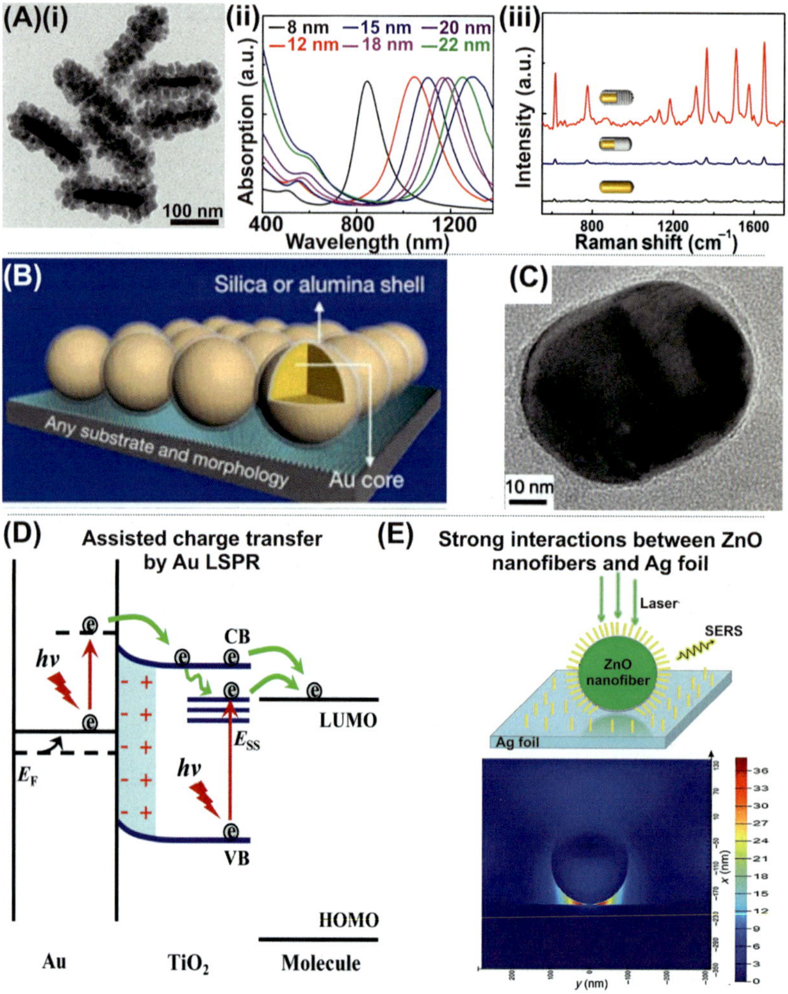

Figure 2.8 Unconventional nanoplasmonic platforms. (A) Bimetallic Au/Ag core–-shell superstructures. (i) TEM image of core–shell superstructures (ii) UV–vis–NIR absorption spectra of core–shell superstructures with different Ag shell thickness, (iii) SERS spectra of Au NRs (black), smooth Au/Ag core–shell NRs (blue), and rough Au/Ag core–shell NRs. (B) Schematic illustration of SHINERS; shell-isolated detection mode. (C) TEM image of Au@MnO_2 NPs synthesized at pH 9.5. (D) Schematic illustration of the enhancement mechanism for SERS of 4-MBA absorbed on Au-TiO_2. (E) Schematic of ZnO nanofibers deposited on the Ag foil surface functionalized with probe molecules (top) and corresponding FDTD simulated electric field distribution. Large EM field enhancement at the gap between ZnO nanofibers and Ag foil occurs due to strong exciton–plasmon interactions between the two (bottom). *EF*, Enhancement factor; *FDTD*, finite-difference time-domain; *NIR*, near-infrared; *SERS*, surface-enhanced Raman scattering; *SHINERS*, shell-isolated nanoparticle Raman

spectroscopy; *4-MBA*, 4-mercaptobenzoic acid. *Reprinted and adapted with permission from (A) L. Dai, L. Song, Y. Huang, L. Zhang, X. Lu, J. Zhang, et al., Bimetallic Au/Ag core–shell superstructures with tunable surface plasmon resonance in the near-infrared region and high performance surface-enhanced Raman scattering, Langmuir 33 (22) (2017) 5378–5384. (B) J.F. Li, Y.F. Huang, Y. Ding, Z.L. Yang, S.B. Li, X.S. Zhou, et al., Shell-isolated nanoparticle-enhanced Raman spectroscopy, Nature 464 (7287) (2010) 392–395. (C) X.-D. Lin, V. Uzayisenga, J.-F. Li, P.-P. Fang, D.-Y. Wu, B. Ren, et al., Synthesis of ultrathin and compact Au@MnO_2 nanoparticles for shell-isolated nanoparticle-enhanced Raman spectroscopy (SHINERS), J. Raman Spectrosc. 43 (1) (2012) 40–45. (D) X. Jiang, X. Sun, D. Yin, X. Li, M. Yang, X. Han, et al., Recyclable Au–TiO_2 nanocomposite SERS-active substrates contributed by synergistic charge-transfer effect, Phys. Chem. Chem. Phys. 19 (18) (2017) 11212–11219. (E) W. Song, W. Ji, S. Vantasin, I. Tanabe, B. Zhao, Y. Ozaki, Fabrication of a highly sensitive surface-enhanced Raman scattering substrate for monitoring the catalytic degradation of organic pollutants, J. Mater. Chem. A 3 (25) (2015) 13556–13562. Copyright 2017 American Chemical Society. Copyright 2010 Springer Nature. Copyright 2011 John Wiley & Sons, Inc. Copyright 2015 and 2017 Royal Society of Chemistry.*

shell, the Au nanoparticles (diameters of 13 nm) alone do not offer sufficient EM field enhancement for DNA detection at relevant target concentrations (<1 nM). In another study, smooth and roughened Au@Ag core–shell nanorods record a respective 12-fold and 50-fold increase in EF value (5.1×10^7 and 2.0×10^8, respectively) than that of monometallic Au nanorods (4.0×10^6) (Fig. 2.8A) [83]. In addition, tailoring the morphology or spatial atomic arrangement of the outer metallic layer can lead to better plasmonic coupling and denser SERS hotspots for higher detection sensitivity. For instance, sharp Ag spikes generates a fivefold increase in EM field enhancement compared to blunt Ag spikes, enabling a detection limit of 10^{-15} M using crystal violet and methylene blue as probes [92].

Bimetallic nanoparticles are thus highly appealing due to their enhanced functionalities by integrating the advantages of both metals to allow better SERS performance as well as highly tunable plasmonic responses. However, further work is still required to understand the underlying mechanisms such as the synergistic effects between different combination of metals and their resultant properties so as to develop better bimetallic systems to increase its ability to enhance SERS signals.

2.3.2 Hybrid nanoplasmonic platforms

Hybrid platforms are formed by modifying metallic nanoplasmonic materials with secondary functional materials such as oxides [93], graphene

[94], and semiconductors [95], in contrast to purely metallic nanoplasmonic platforms. Here, we will focus on three emerging platforms, namely, shell-isolated nanoparticles (SHIN), graphene-based, and semiconductor-based hybrid plasmonic substrates.

2.3.2.1 Shell-isolated nanoparticles

SHINs are made up of a plasmonic core surrounded by an ultrathin dielectric shell (usually silica or alumina) (Fig. 2.8B). Similar to metallic SERS platforms, the plasmonic cores of SHINs enhance local EM field to boost the Raman signals of probe molecules adsorbed on the shell, with an EF of up to 10^5 [96].

An alternative Raman technique using SHINs known as SHIN Raman spectroscopy (SHINERS) offers several advantages over the traditional contact mode of SERS [84]. First, direct contact between the plasmonic surface and target analytes may distort the SERS signals due to changes in the analyte's electronic structures arising from charge transfer. In comparison, the deposited shell in SHIN prevents this direct contact. This is especially crucial to maintain the structural integrity of biological samples for accurate SERS detection and analysis. Second, the inert protective shell also avoids the occurrence of potential photocatalytic reactions triggered by LSPR-induced hot electrons that will convert the target analytes into other species. The shell also increases the durability of the plasmonic metallic nanoparticles due to the inertness and high chemical stability of the outer shell. Next, the physical properties of SHIN can be easily tuned by controlling parameters such as core/dielectric shell material, shell thickness, core diameter, and overall nanoparticle shape [93,96]. For instance, Au nanoparticles coated with MnO_2 remain stable in strong alkaline systems, which is otherwise not possible for SHINs with SiO_2 and Al_2O_3 coatings (Fig. 2.8C) [85]. It is noteworthy that an ultrathin shell is necessary to maintain satisfactory SERS signals because the strength of EM field decreases with increasing distance to the particle surface [97]. This is well illustrated in a study conducted using a fixed Au core of 55 nm in diameter with varied silica shell thickness from 2 to 20 nm, in which an increase in shell thickness registers an exponential decrease in SHINERS intensity of pyridine [93]. 3D finite-difference time-domain (FDTD) simulation further indicates that the optical EM field enhancement for the 2-nm shell is approximately 2 times greater than that of the 4-nm shell. The SHINERS technique has been applied in several proof-of-concept diagnostic studies to detect circulating tumor cells in the blood, as well as identify human breast lesions and microcalcifications in frozen breast tissue [98,99].

However, one current challenge is to fabricate ultrathin shells that are pinhole-free to ensure signal consistency and accuracy. Analytes can access and adsorb strongly onto the plasmonic core via the pinholes, producing comparably stronger SERS signals which will mask other useful signals and result in signal misinterpretation. Besides resolving the pinhole issue, it is also necessary to develop robust functionalization strategies to anchor the shell with various capturing agents so as to selectively capture different target molecules and avoid matrix interference from complex biological media.

2.3.2.2 Graphene-based hybrid surface-enhanced Raman scattering substrates

Graphene consists of a thin monolayer of carbon atoms that are bound in a 2D honeycomb lattice. On its own, it exhibits a continuous energy band gap, large surface area with excellent molecular enrichment ability [5]. While graphene and graphene-based materials such as graphene oxide (GO) and reduced graphene oxide (rGO) exhibit inherently low Raman scattering cross-sections, their synergistic combination with metallic nanoparticles can increase SERS performance by 10–100-fold [100]. This is primarily achieved via the chemical enhancement (CHEM) mechanism, whereby ground-state charge transfer processes between the graphene π-electrons and the analytes enhances electron–phonon coupling and induces additional Raman enhancement. Strong π-coupling between graphene and the analytes further promotes the selective enhancement of vibrational modes involving π- or lone-pair electrons. The synergistic influence of incorporating graphene-based materials is clearly illustrated in this example, wherein the addition of GO to Au nanoparticles results in an increase in EF from 1.2×10^9 to 3.8×10^{11} [101]. The extent of CHEM enhancement contributed by graphene-based materials is reported to be related to the degree of GO chemical reduction [102]. For example, mildly reduced GO enhances CHEM enhancement by up to ~10-fold for rhodamine B (RhB) as compared to graphene [102].

Other than pure CHEM, graphene-based materials can also contribute to SERS via synergistic EM mechanism. Upon close contact with metallic nanostructures which are of higher work function, electron flow from graphene to the metallic surface through p-doping will result in higher EM enhancement from the metal. For instance, a hybrid R6G/Au nanoparticle/graphene platform is reported to have an 86-fold and 4-fold increase in R6G SERS signals than a R6G/graphene platform and a R6G/Au nanoparticle platform respectively [100]. However, such synergy is strongly influenced by the relative work function of the metal with respect to the graphene. Using metals of lower work function (e.g., Ag instead of Au) will result in n-doping instead of p-doping, and therefore weaken the CHEM enhancement.

Apart from synergistic EM/CHEM enhancement, the combination of graphene and plasmonic metal also presents other advantages. For one, its chemical inertness helps prevent the corrosion and oxidation of metallic nanoparticles. The use of graphene as a support matrix for the uniform deposition of metallic nanoparticles can also prevent aggregation of metallic nanoparticles to achieve higher signal stability and reproducibility [103]. Moreover, unlike metallic substrates in which intense fluorescence background can mask the analyte's vibrational fingerprints, graphene can quench this background via Forster resonance energy transfer and thus enable the detection of fluorescent analytes [104,105]. However, one distinct limitation of employing graphene/metal hybrid platforms is the intrinsic carbon peaks of graphene at 1580 and 2700 cm^{-1}, which may overshadow the analyte's SERS fingerprint and thus hinder sensing and differentiation of trace analyte compounds. In addition, chemical enhancement induced by graphene is largely limited to aromatic compounds with π-systems in order to have significant $\pi-\pi$ interactions. This limits the application scope of these hybrid platforms. The biocompatibility of graphene-based substrates as well as their interactions with biological systems also demands further in-depth investigations in order to realize their potential for biological applications [106].

2.3.2.3 Semiconductor-based hybrid platforms

Nanostructured semiconductors such as ZnO and CuO are another emerging alternative that can be incorporated into metallic SERS platforms to enhance SERS by up to 10^3-fold via various mechanisms such as charge transfer, exciton and molecular resonance. For instance, the large energy difference between a metal's photoexcited state and the analyte's LUMO hinders direct charge transfer for SERS chemical enhancement. The semiconductor thus serves as a bridge to facilitate electron transfer via the semiconductor's conduction band. This phenomenon is clearly demonstrated in Au-TiO_2 nanocomposites, which have an EF of 7×10^3 using MBA as probe molecule, more than threefold increase in SERS intensity compared to Au-only and TiO_2-only controls (Fig. 2.8D) [86]. Notably, it is theoretically feasible to modulate various properties of semiconductors including band gap, conduction and valence band edges and refractive index to match the energy levels of a target analyte for optimal SERS enhancement by maximizing the charge transfer effect. For instance, replacing Au with Ag in a sandwiched TiO_2/MBA/Au system induces charge transfer due to a lower Ag Fermi level relative to the surface state energy level of TiO_2, therefore leading to stronger SERS intensities [107].

Besides charge transfer, 1D high refractive index semiconductors, such as ZnO can also enhance the local EM field at the heterojunction [87]. Interactions between the semiconductor's electronic states and the dielectric-confined EM modes of the metal facilitate metal-to-semiconductor charge transfer and prevent the recombination of electron−hole pairs. Consequently, the EM field is highly localized at the semiconductor/metal gap, thereby contributing to a large EM enhancement. Such semiconductor-facilitated EM enhancement is similarly observed in a ZnO nanofibers/Ag foil hybrid platform, which demonstrates a detection limit of 10^{-12} M using *p*-aminothiophenol and with a >30-fold increase in EF compared to an Ag-only platform (Fig. 2.8E). In addition, photocatalytic semiconductors enable self-cleaning capabilities in the hybrid plasmonic platforms for greater platform stability and recyclability. For instance, by subjecting the hybrid SERS platform to UV light, enhanced photocatalytic degradation of analytes can be readily achieved due to improved charge separation in the platform, and thus allow the hybrid SERS platforms to be recycled [86].

Although semiconductor-based hybrid platforms are highly versatile, there is also a current lack of research for selective SERS detection of specific target analytes so as to minimize interfering signals from other molecules present in the complex biological media. Furthermore, existing semiconductor-based hybrid platforms typically rely on chemical enhancement via charge transfer to improve overall SERS enhancement, requires chemical or physical binding, with several but limited reports on boosting SERS via EM enhancement. The main limitation is that this requires chemical or physical bonding which severely restricts the analyte choice to thiolated or aromatic analytes.

In summary, while metallic materials are conventionally employed as SERS platform, hybrid materials exploiting the synergistic effects between nonmetallic and metallic materials are on the rise. Nonetheless, the type of

Table 2.2 Table of different hybrid SERS platforms with their advantages and biological applications.

SERS platform	Advantages	Demonstration of biological application	Ref.
Bimetallic nanomaterials			
Smooth and roughened Au@Ag core–shell nanorods	• High EF of 10^7 and 10^8 for rhodamine 6G, respectively • Red-shift of LSPR toward the NIR region	NA	[83]
Au–Ag bimetallic nanoparticles with sharp spikes	Single-molecule SERS detection limit of 10^{-15} M for crystal violet and methylene blue	NA	[92]
Au@Ag core–shell nanoparticles	Higher SERS efficiency in the NIR region than that of pure Ag nanoparticles	NA	[89]
Spiky Ag–Au bimetallic nanoparticles	Red-shift of LSPR toward the NIR region with an increase in spike length	NA	[90]
Shell-isolated nanoparticles			
Ag@MnO_2	Increased stability in highly alkaline systems compared to SiO_2 and Al_2O_3 coated Ag SHINs	NA	[85]
Ag@SiO_2	• High nanoparticle stability and signal reproducibility • High specificity, sensitivity, and accuracy	Label-free SERS differentiation of circulating tumor cells from normal cells	[98]
Silica-coated Au and Ag nanoparticles	Silica-coated Au nanoparticles are more suitable for *in situ* studies because biomolecules will irreversibly bind to the Ag surface	Study of protein–nanoparticle/cell–nanoparticle interactions	[108]

(Continued)

Table 2.2 (Continued)

SERS platform	Advantages	Demonstration of biological application	Ref.
Au@SiO_2	Label-free detection of the molecular fingerprint for multiplex detection of various biomolecules such as lipids, proteins, nucleic acids	Detection of human breast lesions and microcalcifications	[99]
Graphene-based substrates			
Graphene oxide-Au nanoparticles	Increase in EF from 1.2×10^9 to 3.8×10^{11}	NA	[101]
Reduced graphene oxide-gold nanostar (rGO-NS)	• Large surface area • Higher sensitivity toward aromatic molecules and increased stability than Au NS • rGO prevents aggregation of Au NS without needing capping agent	Anticancer drug (doxorubicin) loading and release for chemotherapeutic purposes	[109]
Ag@rGO@Au	• Large surface area • Promotion of rGO Raman scattering by over 70 times	Differentiation of tumor cells from normal cells	[110]
Semiconductor and semiconductor-based substrates			
Au-TiO_2	Threefold increase in SERS intensity compared to Au-only and TiO_2-only controls	NA	[86]
ZnO nanofibers deposited on Ag foil	• Detection limit of 10^{-12} M for *p*-aminothiophenol • >30-Fold increase in EF compared to a Ag-only platform	NA	[87]

EF, Enhancement factor; LSPR, localized surface plasmon resonance; NIR, near-infrared; SHIN, shell-isolated nanoparticle; SERS, surface-enhanced Raman scattering.

SERS substrate one should employ lies in the specific application that one wants to use it for. A nonexhaustive list of these different hybrid SERS platforms with their advantages and applications are compiled in Table 2.2.

2.4 Conclusion and outlook

In this chapter, we discussed the recent developments in nanoplasmonic materials and platforms from 0D/1D to 2D and 3D levels to increase SERS sensitivity and reproducibility through hotspot engineering. Advances in bottom-up and top-down approaches enable the precise engineering of plasmonic nanostructures of tailored morphologies, as well as their controlled assembly into multidimensional SERS platforms with strong plasmonic coupling and increased field enhancement. Beyond manipulating the SERS capabilities of SERS platforms, we also demonstrate how analyte modification strategies are integral to detect molecules with no affinity to metal surfaces. Two main physical analyte-capturing strategies, which exploit different working principles to concentrate these analytes close to the surface, are highlighted, namely, (1) modification of surface wetting properties and (2) use of MOFs as molecule sorbent layers. Furthermore, we provide an overview of nonconventional nanoplasmonic configurations with improved EM and chemical properties for better SERS performance. These include bimetallic core—shell systems and hybrid nanostructures that couple metallic substrates with other functional materials.

However, several major challenges need to be addressed before we can employ these nanoplasmonic materials and platforms for actual practical application in various fields ranging from biomedicine, pharmaceutics to environmental analysis. First, many existing SERS-active nanomaterials and substrates involve the use of toxic organic molecules and/or surfactants during their synthesis. Thus, there is a need to thoroughly investigate their toxicity and biocompatibility as well as a parallel development of new strategies to reduce the toxicity and/or develop alternative biocompatible materials. Next, it remains a challenge to construct large-area and homogeneous 3D plasmonic architectures to achieve uniform and high SERS signals without sacrificing light accessibility, which thus requires further progress in nanofabrication approaches. Finally, existing research on hybrid SERS platforms is still largely limited to enhancement on the single-particle level with few studies focused on large-area fabrication of hybrid SERS platforms or on the precise engineering of multidimensional hybrid SERS platforms. Hence, we anticipate future developments in these areas, as well as potential synergistic combination of these hybrid SERS platforms with analyte manipulation techniques, are crucial in optimizing these nanoplasmonic materials.

All in all, SERS performance of nanoplasmonic materials can be enhanced by (1) directly modulating EM field strength through hotspot engineering, (2) analyte manipulation for increased analyte—surface contact, and (3) incorporation with other functional materials. Collectively, multifaceted development of these strategies can lead to key breakthroughs in the design of next-generation SERS-active materials and platforms for real-world analytical applications.

References

[1] M.F. Cardinal, E. Vander Ende, R.A. Hackler, M.O. McAnally, P.C. Stair, G.C. Schatz, et al., Expanding applications of SERS through versatile nanomaterials engineering, Chem. Soc. Rev. 46 (13) (2017) 3886—3903.

[2] B. Sharma, R.R. Frontiera, A.-I. Henry, E. Ringe, R.P. Van Duyne, SERS: materials, applications, and the future, Mater. Today 15 (1) (2012) 16—25.

[3] C. Zhan, X.-J. Chen, J. Yi, J.-F. Li, D.-Y. Wu, Z.-Q. Tian, From plasmon-enhanced molecular spectroscopy to plasmon-mediated chemical reactions, Nat. Rev. Chem. 2 (9) (2018) 216—230.

[4] S.-Y. Ding, E.-M. You, Z.-Q. Tian, M. Moskovits, Electromagnetic theories of surface-enhanced Raman spectroscopy, Chem. Soc. Rev. 46 (13) (2017) 4042—4076.

[5] H.K. Lee, Y.H. Lee, C.S.L. Koh, G.C. Phan-Quang, X. Han, C.L. Lay, et al., Designing surface-enhanced Raman scattering (SERS) platforms beyond hotspot engineering: emerging opportunities in analyte manipulations and hybrid materials, Chem. Soc. Rev. 48 (3) (2019) 731—756.

[6] S.-Y. Ding, J. Yi, J.-F. Li, B. Ren, D.-Y. Wu, R. Panneerselvam, et al., Nanostructure-based plasmon-enhanced Raman spectroscopy for surface analysis of materials, Nat. Rev. Mater. 1 (6) (2016) 16021.

[7] J.I. Gersten, The effect of surface roughness on surface enhanced Raman scattering, J. Chem. Phys. 72 (10) (1980) 5779—5780.

[8] E.J. Zeman, G.C. Schatz, An accurate electromagnetic theory study of surface enhancement factors for silver, gold, copper, lithium, sodium, aluminum, gallium, indium, zinc, and cadmium, J. Phys. Chem. 91 (3) (1987) 634—643.

[9] T.K. Sau, A.L. Rogach, F. Jäckel, T.A. Klar, J. Feldmann, Properties and applications of colloidal nonspherical noble metal nanoparticles, Adv. Mater. 22 (16) (2010) 1805—1825.

[10] M. Jahn, S. Patze, I.J. Hidi, R. Knipper, A.I. Radu, A. Mühlig, et al., Plasmonic nanostructures for surface enhanced spectroscopic methods, Analyst 141 (3) (2016) 756—793.

[11] H. Chen, L. Shao, T. Ming, Z. Sun, C. Zhao, B. Yang, et al., Understanding the photothermal conversion efficiency of gold nanocrystals, Small 6 (20) (2010) 2272—2280.

[12] S.E. Lohse, C.J. Murphy, The quest for shape control: a history of gold nanorod synthesis, Chem. Mater. 25 (8) (2013) 1250–1261.
[13] Y. Lin, Y.-J. Zhang, W.-M. Yang, J.-C. Dong, F.-R. Fan, Y. Zhao, et al., Size and dimension dependent surface-enhanced Raman scattering properties of well-defined Ag nanocubes, Appl. Mater. Today 14 (2019) 224–232.
[14] Y.H. Lee, H. Chen, Q.-H. Xu, J. Wang, Refractive index sensitivities of noble metal nanocrystals: the effects of multipolar plasmon resonances and the metal type, J. Phys. Chem. C 115 (16) (2011) 7997–8004.
[15] B. Nikoobakht, M.A. El-Sayed, Preparation and growth mechanism of gold nanorods (NRs) using seed-mediated growth method, Chem. Mater. 15 (10) (2003) 1957–1962.
[16] N.R. Jana, L. Gearheart, C.J. Murphy, Wet chemical synthesis of high aspect ratio cylindrical gold nanorods, J. Phys. Chem. B 105 (19) (2001) 4065–4067.
[17] N.R. Jana, L. Gearheart, C.J. Murphy, Seed-mediated growth approach for shape-controlled synthesis of spheroidal and rod-like gold nanoparticles using a surfactant template, Adv. Mater. 13 (18) (2001) 1389–1393.
[18] A. Gole, C.J. Murphy, Seed-mediated synthesis of gold nanorods: role of the size and nature of the seed, Chem. Mater. 16 (19) (2004) 3633–3640.
[19] P.K. Jain, K.S. Lee, I.H. El-Sayed, M.A. El-Sayed, Calculated absorption and scattering properties of gold nanoparticles of different size, shape, and composition: applications in biological imaging and biomedicine, J. Phys. Chem. B 110 (14) (2006) 7238–7248.
[20] C.J. Orendorff, C.J. Murphy, Quantitation of metal content in the silver-assisted growth of gold nanorods, J. Phys. Chem. B 110 (9) (2006) 3990–3994.
[21] C.J. Orendorff, A. Gole, T.K. Sau, C.J. Murphy, Surface-enhanced Raman spectroscopy of self-assembled monolayers: sandwich architecture and nanoparticle shape dependence, Anal. Chem. 77 (10) (2005) 3261–3266.
[22] J.M. McLellan, A. Siekkinen, J. Chen, Y. Xia, Comparison of the surface-enhanced Raman scattering on sharp and truncated silver nanocubes, Chem. Phys. Lett. 427 (1) (2006) 122–126.
[23] Y. Sun, Y. Xia, Shape-controlled synthesis of gold and silver nanoparticles, Science 298 (5601) (2002) 2176.
[24] Z.L. Wang, Transmission electron microscopy of shape-controlled nanocrystals and their assemblies, J. Phys. Chem. B 104 (6) (2000) 1153–1175.
[25] Q. Zhang, W. Li, C. Moran, J. Zeng, J. Chen, L.-P. Wen, et al., Seed-mediated synthesis of Ag nanocubes with controllable edge lengths in the range of 30 – 200 nm and comparison of their optical properties, J. Am. Chem. Soc. 132 (32) (2010) 11372–11378.
[26] Y. Wang, Y. Zheng, C.Z. Huang, Y. Xia, Synthesis of Ag nanocubes 18–32 nm in edge length: the effects of polyol on reduction kinetics, size control, and reproducibility, J. Am. Chem. Soc. 135 (5) (2013) 1941–1951.
[27] L.J. Sherry, S.-H. Chang, G.C. Schatz, R.P. Van Duyne, B.J. Wiley, Y. Xia, Localized surface plasmon resonance spectroscopy of single silver nanocubes, Nano Lett. 5 (10) (2005) 2034–2038.
[28] B. Wiley, Y. Sun, Y. Xia, Synthesis of silver nanostructures with controlled shapes and properties, Acc. Chem. Res. 40 (10) (2007) 1067–1076.
[29] A. Tao, P. Sinsermsuksakul, P. Yang, Polyhedral silver nanocrystals with distinct scattering signatures, Angew. Chem. Int. Ed. 45 (28) (2006) 4597–4601.
[30] L.-F. Zhang, S.-L. Zhong, A.-W. Xu, Highly branched concave Au/Pd bimetallic nanocrystals with superior electrocatalytic activity and highly efficient SERS enhancement, Angew. Chem. Int. Ed. 52 (2) (2013) 645–649.

[31] K. Kwon, K.Y. Lee, Y.W. Lee, M. Kim, J. Heo, S.J. Ahn, et al., Controlled synthesis of icosahedral gold nanoparticles and their surface-enhanced Raman scattering property, J. Phys. Chem. C 111 (3) (2007) 1161–1165.
[32] W. Niu, Y.A.A. Chua, W. Zhang, H. Huang, X. Lu, Highly symmetric gold nanostars: crystallographic control and surface-enhanced Raman scattering property, J. Am. Chem. Soc. 137 (33) (2015) 10460–10463.
[33] Y. Liu, S. Pedireddy, Y.H. Lee, R.S. Hegde, W.W. Tjiu, Y. Cui, et al., Precision synthesis: designing hot spots over hot spots via selective gold deposition on silver octahedra edges, Small 10 (23) (2014) 4940–4950.
[34] M.S. Goh, Y.H. Lee, S. Pedireddy, I.Y. Phang, W.W. Tjiu, J.M.R. Tan, et al., A chemical route to increase hot spots on silver nanowires for surface-enhanced Raman spectroscopy application, Langmuir 28 (40) (2012) 14441–14449.
[35] D.Y. Kim, S.H. Im, O.O. Park, Synthesis of tetrahexahedral gold nanocrystals with high-index facets, Cryst. Growth Des. 10 (8) (2010) 3321–3323.
[36] Y.H. Lee, C.K. Lee, B. Tan, J.M. Rui Tan, I.Y. Phang, X.Y. Ling, Using the Langmuir–Schaefer technique to fabricate large-area dense SERS-active Au nanoprism monolayer films, Nanoscale 5 (14) (2013) 6404–6412.
[37] M.J. Mulvihill, X.Y. Ling, J. Henzie, P. Yang, Anisotropic etching of silver nanoparticles for plasmonic structures capable of single-particle SERS, J. Am. Chem. Soc. 132 (1) (2010) 268–274.
[38] H.T. Chorsi, Y. Lee, A. Alù, J.X.J. Zhang, Tunable plasmonic substrates with ultrahigh Q-factor resonances, Sci. Rep. 7 (1) (2017) 15985.
[39] S.S.B. Moram, C. Byram, S.N. Shibu, B.M. Chilukamarri, V.R. Soma, Ag/Au nanoparticle-loaded paper-based versatile surface-enhanced Raman spectroscopy substrates for multiple explosives detection, ACS Omega 3 (7) (2018) 8190–8201.
[40] X. Han, H.K. Lee, Y.H. Lee, W. Hao, Y. Liu, I.Y. Phang, et al., Identifying enclosed chemical reaction and dynamics at the molecular level using shell-isolated miniaturized plasmonic liquid marble, J. Phys. Chem. Lett. 7 (8) (2016) 1501–1506.
[41] Y.H. Lee, W. Shi, H.K. Lee, R. Jiang, I.Y. Phang, Y. Cui, et al., Nanoscale surface chemistry directs the tunable assembly of silver octahedra into three two-dimensional plasmonic superlattices, Nat. Commun. 6 (2015) 6990.
[42] Y. Yan, H. Shan, M. Li, S. Chen, J. Liu, Y. Cheng, et al., Internal-modified dithiol DNA–directed Au nanoassemblies: geometrically controlled self–assembly and quantitative surface–enhanced Raman scattering properties, Sci. Rep. 5 (1) (2015) 16715.
[43] R. Schreiber, I. Santiago, A. Ardavan, A.J. Turberfield, Ordering gold nanoparticles with DNA origami nanoflowers, ACS Nano 10 (8) (2016) 7303–7306.
[44] B. Yan, A. Thubagere, W.R. Premasiri, L.D. Ziegler, L. Dal Negro, B.M. Reinhard, Engineered SERS substrates with multiscale signal enhancement: nanoparticle cluster arrays, ACS Nano 3 (5) (2009) 1190–1202.
[45] Y. Tian, T. Wang, W. Liu, H.L. Xin, H. Li, Y. Ke, et al., Prescribed nanoparticle cluster architectures and low-dimensional arrays built using octahedral DNA origami frames, Nat. Nanotechnol. 10 (7) (2015) 637–644.
[46] A.R. Tao, J. Huang, P. Yang, Langmuir – Blodgettry of nanocrystals and nanowires, Acc. Chem. Res. 41 (12) (2008) 1662–1673.
[47] V. Santhanam, J. Liu, R. Agarwal, R.P. Andres, Self-assembly of uniform monolayer arrays of nanoparticles, Langmuir 19 (19) (2003) 7881–7887.
[48] K. Kim, H.S. Han, I. Choi, C. Lee, S. Hong, S.-H. Suh, et al., Interfacial liquid-state surface-enhanced Raman spectroscopy, Nat. Commun. 4 (1) (2013) 2182.
[49] Y. Yang, Y.H. Lee, I.Y. Phang, R. Jiang, H.Y.F. Sim, J. Wang, et al., A chemical approach to break the planar configuration of Ag nanocubes into tunable two-dimensional metasurfaces, Nano Lett. 16 (6) (2016) 3872–3878.

[50] Y. Yang, Y.H. Lee, C.L. Lay, X.Y. Ling, Tuning molecular-level polymer conformations enables dynamic control over both the interfacial behaviors of Ag nanocubes and their assembled metacrystals, Chem. Mater. 29 (14) (2017) 6137–6144.
[51] J. Henzie, S.C. Andrews, X.Y. Ling, Z. Li, P. Yang, Oriented assembly of polyhedral plasmonic nanoparticle clusters, Proc. Natl. Acad. Sci. 110 (17) (2013) 6640.
[52] K.P.J. Williams, G.D. Pitt, D.N. Batchelder, B.J. Kip, Confocal Raman microspectroscopy using a stigmatic spectrograph and CCD detector, Appl. Spectrosc. 48 (2) (1994) 232–235.
[53] C. Srichan, M. Ekpanyapong, M. Horprathum, P. Eiamchai, N. Nuntawong, D. Phokharatkul, et al., Highly-sensitive surface-enhanced Raman spectroscopy (SERS)-based chemical sensor using 3D graphene foam decorated with silver nanoparticles as SERS substrate, Sci. Rep. 6 (1) (2016) 23733.
[54] R.A. Alvarez-Puebla, A. Agarwal, P. Manna, B.P. Khanal, P. Aldeanueva-Potel, E. Carbó-Argibay, et al., Gold nanorods 3D-supercrystals as surface enhanced Raman scattering spectroscopy substrates for the rapid detection of scrambled prions, Proc. Natl. Acad. Sci. 108 (20) (2011) 8157.
[55] M. Chen, I.Y. Phang, M.R. Lee, J.K.W. Yang, X.Y. Ling, Layer-by-layer assembly of Ag nanowires into 3D woodpile-like structures to achieve high density "hot spots" for surface-enhanced Raman scattering, Langmuir 29 (23) (2013) 7061–7069.
[56] Y.H. Lee, C.L. Lay, W. Shi, H.K. Lee, Y. Yang, S. Li, et al., Creating two self-assembly micro-environments to achieve supercrystals with dual structures using polyhedral nanoparticles, Nat. Commun. 9 (1) (2018) 2769.
[57] S.Y. Lee, S.-H. Kim, M.P. Kim, H.C. Jeon, H. Kang, H.J. Kim, et al., Freestanding and arrayed nanoporous microcylinders for highly active 3D SERS substrate, Chem. Mater. 25 (12) (2013) 2421–2426.
[58] G.C. Phan-Quang, H.K. Lee, I.Y. Phang, X.Y. Ling, Plasmonic colloidosomes as three-dimensional SERS platforms with enhanced surface area for multiphase submicroliter toxin sensing, Angew. Chem. Int. Ed. 54 (33) (2015) 9691–9695.
[59] X. Li, H.K. Lee, I.Y. Phang, C.K. Lee, X.Y. Ling, Superhydrophobic-oleophobic Ag nanowire platform: an analyte-concentrating and quantitative aqueous and organic toxin surface-enhanced Raman scattering sensor, Anal. Chem. 86 (20) (2014) 10437–10444.
[60] Q. Zhang, Y.H. Lee, I.Y. Phang, C.K. Lee, X.Y. Ling, Hierarchical 3D SERS substrates fabricated by integrating photolithographic microstructures and self-assembly of silver nanoparticles, Small 10 (13) (2014) 2703–2711.
[61] R. Alvarez-Puebla, B. Cui, J.-P. Bravo-Vasquez, T. Veres, H. Fenniri, Nanoimprinted SERS-active substrates with tunable surface plasmon resonances, J. Phys. Chem. C 111 (18) (2007) 6720–6723.
[62] X. Han, C.S.L. Koh, H.K. Lee, W.S. Chew, X.Y. Ling, Microchemical plant in a liquid droplet: plasmonic liquid marble for sequential reactions and attomole detection of toxin at microliter scale, ACS Appl. Mater. Interf. 9 (45) (2017) 39635–39640.
[63] G.C. Phan-Quang, E.H.Z. Wee, F. Yang, H.K. Lee, I.Y. Phang, X. Feng, et al., Online flowing colloidosomes for sequential multi-analyte high-throughput SERS analysis, Angew. Chem. Int. Ed. 56 (20) (2017) 5565–5569.
[64] X. Qiao, B. Su, C. Liu, Q. Song, D. Luo, G. Mo, et al., Selective surface enhanced Raman scattering for quantitative detection of lung cancer biomarkers in superparticle@MOF structure, Adv. Mater. 30 (5) (2018) 1702275.
[65] X.X. Han, P. Pienpinijtham, B. Zhao, Y. Ozaki, Coupling reaction-based ultrasensitive detection of phenolic estrogens using surface-enhanced resonance Raman scattering, Anal. Chem. 83 (22) (2011) 8582–8588.
[66] S. Ben-Jaber, W.J. Peveler, R. Quesada-Cabrera, E. Cortés, C. Sotelo-Vazquez, N. Abdul-Karim, et al., Photo-induced enhanced Raman spectroscopy for universal

ultra-trace detection of explosives, pollutants and biomolecules, Nat. Commun. 7 (1) (2016) 12189.
[67] C.S.L. Koh, H.K. Lee, X. Han, H.Y.F. Sim, X.Y. Ling, Plasmonic nose: integrating the MOF-enabled molecular preconcentration effect with a plasmonic array for recognition of molecular-level volatile organic compounds, Chem. Commun. 54 (20) (2018) 2546–2549.
[68] H.K. Lee, Y.H. Lee, Q. Zhang, I.Y. Phang, J.M.R. Tan, Y. Cui, et al., Superhydrophobic surface-enhanced Raman scattering platform fabricated by assembly of Ag nanocubes for trace molecular sensing, ACS Appl. Mater. Interf. 5 (21) (2013) 11409–11418.
[69] F. De Angelis, F. Gentile, F. Mecarini, G. Das, M. Moretti, P. Candeloro, et al., Breaking the diffusion limit with super-hydrophobic delivery of molecules to plasmonic nanofocusing SERS structures, Nat. Photonics 5 (11) (2011) 682–687.
[70] A.B.D. Cassie, Contact angles, Discuss. Faraday Soc. 3 (0) (1948) 11–16.
[71] A.B.D. Cassie, S. Baxter, Wettability of porous surfaces, Trans. Faraday Soc. 40 (0) (1944) 546–551.
[72] Y.-C. Kao, X. Han, Y.H. Lee, H.K. Lee, G.C. Phan-Quang, C.L. Lay, et al., Multiplex surface-enhanced Raman scattering identification and quantification of urine metabolites in patient samples within 30 min, ACS Nano 14 (2) (2020) 2542–2552.
[73] T.Y. Jeon, J.H. Kim, S.-G. Park, J.-D. Kwon, D.-H. Kim, S.-H. Kim, Stacked-disk nanotower arrays for use as omniphobic surface-enhanced Raman scattering substrates, Adv. Opt. Mater. 4 (11) (2016) 1893–1900.
[74] Y. Song, T. Xu, L.-P. Xu, X. Zhang, Superwettable nanodendritic gold substrates for direct miRNA SERS detection, Nanoscale 10 (45) (2018) 20990–20994.
[75] K.-S. Lin, A.K. Adhikari, C.-N. Ku, C.-L. Chiang, H. Kuo, Synthesis and characterization of porous HKUST-1 metal organic frameworks for hydrogen storage, Int. J. Hydrog. Energy 37 (18) (2012) 13865–13871.
[76] B. Wang, A.P. Côté, H. Furukawa, M. O'Keeffe, O.M. Yaghi, Colossal cages in zeolitic imidazolate frameworks as selective carbon dioxide reservoirs, Nature 453 (7192) (2008) 207–211.
[77] M. Anbia, V. Hoseini, S. Sheykhi, Sorption of methane, hydrogen and carbon dioxide on metal-organic framework, iron terephthalate (MOF-235), J. Ind. Eng. Chem. 18 (3) (2012) 1149–1152.
[78] H. Duo, H. Tang, J. Ma, X. Lu, L. Wang, X. Liang, Iron-based metal–organic framework as an effective sorbent for the rapid and efficient removal of illegal dyes, N. J. Chem. 43 (38) (2019) 15351–15358.
[79] H.T.D. Nguyen, Y.B.N. Tran, H.N. Nguyen, T.C. Nguyen, F. Gándara, P.T.K. Nguyen, A series of metal–organic frameworks for selective CO_2 capture and catalytic oxidative carboxylation of olefins, Inorg. Chem. 57 (21) (2018) 13772–13782.
[80] G.C. Phan-Quang, N. Yang, H.K. Lee, H.Y.F. Sim, C.S.L. Koh, Y.-C. Kao, et al., Tracking airborne molecules from afar: three-dimensional metal–organic framework-surface-enhanced Raman scattering platform for stand-off and real-time atmospheric monitoring, ACS Nano 13 (10) (2019) 12090–12099.
[81] T.D.C. Minh, D.R. Blake, P.R. Galassetti, The clinical potential of exhaled breath analysis for diabetes mellitus, Diabetes Res. Clin. Pract. 97 (2) (2012) 195–205.
[82] C.L. Wong, U.S. Dinish, M.S. Schmidt, M. Olivo, Non-labeling multiplex surface enhanced Raman scattering (SERS) detection of volatile organic compounds (VOCs), Anal. Chim. Acta 844 (2014) 54–60.
[83] L. Dai, L. Song, Y. Huang, L. Zhang, X. Lu, J. Zhang, et al., Bimetallic Au/Ag core–shell superstructures with tunable surface plasmon resonance in the near-infrared

region and high performance surface-enhanced Raman scattering, Langmuir 33 (22) (2017) 5378–5384.
[84] J.F. Li, Y.F. Huang, Y. Ding, Z.L. Yang, S.B. Li, X.S. Zhou, et al., Shell-isolated nanoparticle-enhanced Raman spectroscopy, Nature 464 (7287) (2010) 392–395.
[85] X.-D. Lin, V. Uzayisenga, J.-F. Li, P.-P. Fang, D.-Y. Wu, B. Ren, et al., Synthesis of ultrathin and compact Au@MnO_2 nanoparticles for shell-isolated nanoparticle-enhanced Raman spectroscopy (SHINERS), J. Raman Spectrosc. 43 (1) (2012) 40–45.
[86] X. Jiang, X. Sun, D. Yin, X. Li, M. Yang, X. Han, et al., Recyclable Au–TiO_2 nanocomposite SERS-active substrates contributed by synergistic charge-transfer effect, Phys. Chem. Chem. Phys. 19 (18) (2017) 11212–11219.
[87] W. Song, W. Ji, S. Vantasin, I. Tanabe, B. Zhao, Y. Ozaki, Fabrication of a highly sensitive surface-enhanced Raman scattering substrate for monitoring the catalytic degradation of organic pollutants, J. Mater. Chem. A 3 (25) (2015) 13556–13562.
[88] A.J. McGrath, Y.-H. Chien, S. Cheong, D.A.J. Herman, J. Watt, A.M. Henning, et al., Gold over branched palladium nanostructures for photothermal cancer therapy, ACS Nano 9 (12) (2015) 12283–12291.
[89] A.K. Samal, L. Polavarapu, S. Rodal-Cedeira, L.M. Liz-Marzán, J. Pérez-Juste, I. Pastoriza-Santos, Size tunable Au@Ag core–shell nanoparticles: synthesis and surface-enhanced Raman scattering properties, Langmuir 29 (48) (2013) 15076–15082.
[90] S. Pedireddy, A. Li, M. Bosman, I.Y. Phang, S. Li, X.Y. Ling, Synthesis of spiky Ag–Au octahedral nanoparticles and their tunable optical properties, J. Phys. Chem. C 117 (32) (2013) 16640–16649.
[91] Y.C. Cao, R. Jin, C.A. Mirkin, Nanoparticles with Raman spectroscopic fingerprints for DNA and RNA detection, Science 297 (5586) (2002) 1536–1540.
[92] R.V. William, G.M. Das, V.R. Dantham, R. Laha, Enhancement of single molecule Raman scattering using sprouted potato shaped bimetallic nanoparticles, Sci. Rep. 9 (1) (2019) 10771.
[93] J.R. Anema, J.-F. Li, Z.-L. Yang, B. Ren, Z.-Q. Tian, Shell-isolated nanoparticle-enhanced Raman spectroscopy: expanding the versatility of surface-enhanced Raman scattering, Annu. Rev. Anal. Chem. 4 (1) (2011) 129–150.
[94] W. Fan, Y.H. Lee, S. Pedireddy, Q. Zhang, T. Liu, X.Y. Ling, Graphene oxide and shape-controlled silver nanoparticle hybrids for ultrasensitive single-particle surface-enhanced Raman scattering (SERS) sensing, Nanoscale 6 (9) (2014) 4843–4851.
[95] B. Yang, S. Jin, S. Guo, Y. Park, L. Chen, B. Zhao, et al., Recent development of SERS technology: semiconductor-based study, ACS Omega 4 (23) (2019) 20101–20108.
[96] J. Xu, Y.-J. Zhang, H. Yin, H.-L. Zhong, M. Su, Z.-Q. Tian, et al., Shell-isolated nanoparticle-enhanced Raman and fluorescence spectroscopies: synthesis and applications, Adv. Opt. Mater. 6 (4) (2018) 1701069.
[97] P.L. Stiles, J.A. Dieringer, N.C. Shah, R.P. Van Duyne, Surface-enhanced Raman spectroscopy, Annu. Rev. Anal. Chem. 1 (1) (2008) 601–626.
[98] K. Niciński, J. Krajczewski, A. Kudelski, E. Witkowska, J. Trzcińska-Danielewicz, A. Girstun, et al., Detection of circulating tumor cells in blood by shell-isolated nanoparticle-enhanced Raman spectroscopy (SHINERS) in microfluidic device, Sci. Rep. 9 (1) (2019) 9267.
[99] C. Zheng, W. Shao, S.K. Paidi, B. Han, T. Fu, D. Wu, et al., Pursuing shell-isolated nanoparticle-enhanced Raman spectroscopy (SHINERS) for concomitant detection of breast lesions and microcalcifications, Nanoscale 7 (40) (2015) 16960–16968.

[100] X.-k Kong, Q.-w Chen, Z.-y Sun, Enhanced SERS of the complex substrate using Au supported on graphene with pyridine and R6G as the probe molecules, Chem. Phys. Lett. 564 (2013) 54–59.
[101] Z. Fan, R. Kanchanapally, P.C. Ray, Hybrid graphene oxide based ultrasensitive SERS probe for label-free biosensing, J. Phys. Chem. Lett. 4 (21) (2013) 3813–3818.
[102] X. Yu, H. Cai, W. Zhang, X. Li, N. Pan, Y. Luo, et al., Tuning chemical enhancement of SERS by controlling the chemical reduction of graphene oxide nanosheets, ACS Nano 5 (2) (2011) 952–958.
[103] H. Lai, F. Xu, Y. Zhang, L. Wang, Recent progress on graphene-based substrates for surface-enhanced Raman scattering applications, J. Mater. Chem. B 6 (24) (2018) 4008–4028.
[104] Z. Wang, S. Wu, L. Colombi Ciacchi, G. Wei, Graphene-based nanoplatforms for surface-enhanced Raman scattering sensing, Analyst 143 (21) (2018) 5074–5089.
[105] C.-W. Huang, H.-Y. Lin, C.-H. Huang, K.-H. Lo, Y.-C. Chang, C.-Y. Liu, et al., Fluorescence quenching due to sliver nanoparticles covered by graphene and hydrogen-terminated graphene, Appl. Phys. Lett. 102 (5) (2013) 053113.
[106] T. Wang, S. Zhu, X. Jiang, Toxicity mechanism of graphene oxide and nitrogen-doped graphene quantum dots in RBCs revealed by surface-enhanced infrared absorption spectroscopy, Toxicol. Res. 4 (4) (2015) 885–894.
[107] X. Jiang, X. Li, X. Jia, G. Li, X. Wang, G. Wang, et al., Surface-enhanced Raman scattering from synergistic contribution of metal and semiconductor in TiO_2/MBA/Ag(Au) and Ag(Au)/MBA/TiO_2 assemblies, J. Phys. Chem. C 116 (27) (2012) 14650–14655.
[108] D. Drescher, I. Zeise, H. Traub, P. Guttmann, S. Seifert, T. Büchner, et al., In situ characterization of SiO_2 nanoparticle biointeractions using brightsilica, Adv. Funct. Mater. 24 (24) (2014) 3765–3775.
[109] Y. Wang, L. Polavarapu, L.M. Liz-Marzán, Reduced graphene oxide-supported gold nanostars for improved SERS sensing and drug delivery, ACS Appl. Mater. Interf. 6 (24) (2014) 21798–21805.
[110] N. Yi, C. Zhang, Q. Song, S. Xiao, A hybrid system with highly enhanced graphene SERS for rapid and tag-free tumor cells detection, Sci. Rep. 6 (1) (2016) 25134.

CHAPTER 3

Experimental aspects of surface-enhanced Raman scattering for biological applications

Shuping Xu
State Key Laboratory of Supramolecular Structure and Materials, Institute of Theoretical Chemistry, College of Chemistry, Jilin University, Changchun, P.R. China

3.1 Combination ways of surface-enhanced Raman scattering substrates with the analytical systems

The most significant difference between surface-enhanced Raman scattering (SERS) detections and Raman detections lies in the participation of the SERS substrates. By reviewing many SERS approaches applied for biosystems, how to combine the SERS substrate with the detected system is an inevitable but skillful point. The typical binding ways can be cataloged according to two types of SERS substrates: the colloidal plasmonic nanoparticles (NPs) and the solid-supported metal nanostructures. The detailed description about nanoplasmonic material has been discussed in Chapter 2, *Nanoplasmonic materials for surface-enhanced Raman scattering.*

The direct and indirect SERS detections demonstrate different combination ways. Direct SERS detections (see Chapter 4, *Label-free surface-enhanced Raman scattering for clinical application*) propose the "close touch" of analytes to the SERS substrates, but the indirect SERS strategies refer to the sensing process that the analytes join in one intermediate step. Many review papers, especially in 1999–2000, have summarized the direct SERS studies from the category aspects of analytes [1–3]. Here, the combination ways of the direct SERS methods that can avoid false-positive effects will be discussed in detail from the experimental perspective. The indirect/labeling SERS detection by using the colloidal metal NPs will be detailed in Chapter 5, *Surface-enhanced Raman scattering nanotags design and synthesis.*

Principles and Clinical Diagnostic Applications of Surface-Enhanced Raman Spectroscopy
DOI: https://doi.org/10.1016/B978-0-12-821121-2.00010-X

3.1.1 Colloidal metal nanoparticles

The colloidal metal NPs are restricted by a three-dimensional boundary of a surface. They are usually tens to about one hundred nanometers, showing the localized surface plasmon feature. They have been widely applicable for any size of bioanalytes. For the molecules with the size smaller or comparable to those of metal NPs, mixing the metal colloids directly with the analytes is usually adopted. For some large detection system as cell or larger ones, because the metal NPs are only partially attached to the interesting locations, the targeting or binding strategies will be employed to selectively seek objective locations in the complicated systems or living bodies.

3.1.1.1 The combination ways of metal nanoparticles to different sized biosystems

3.1.1.1.1 Electrostatic interaction

For many small sizes of biomolecules that are comparable to or smaller than the metal NPs, a widely used combination way is to mix them in a liquid phase simply. The most commonly used metal colloids are prepared by the citrate reduction reactions [4,5], in which the citrate is used as both a reducer and a surface ligand that also contributes SERS signals [6]. The size of the produced NPs becomes larger with the dose of citrate decreasing. And the citrate provides negative surface charges for NPs. Owing to the replacement of citrate by stronger binding molecules or the surface electrostatic adsorption effect, the analytes would be adjacent to the metal NPs. The fluorescence of some chromophores can be well quenched, which allows high sensitivity of SERS detections. The mixing ratio of metal colloids and analytes is trial-to-trial dependent. The adsorption affinity, the solvent of the analyte, the colloidal concentration and analyte concentration, etc., should all be considered. Usually, they first start with different volume ratios of the colloid and the analyte to achieve the largest dose amount of analytes before the occurrence of aggregation of colloids. For the bacteria measurement, the AgNPs:*Escherichia coli* volume ratios with 1:1, 2:1, and 3:1 were compared [7], and the results showed that too many AgNPs brought the photodamage to the cell walls of the bacteria and the leakage of the cellular cytoplasm contents provided the fluorescence background.

People developed several positive-charged metal NPs that can adsorb and enrich the negatively charged biomolecules, for example, deoxyribonucleic acids (DNAs) and ribose nucleic acids (RNAs). The SERS signals of DNAs obtained on the agarose-coated AuNPs were stronger than those obtained on the citrate-coated AuNPs [8]. Sun et al. measured the DNA by using a cetyltrimethylammonium bromide (CTAB)-coated AgNPs and find the structural information about the bases, phosphate backbone, and the conformation of DNA [9]. Poly(L-lysine) and spermine were also used to convert the surface charges of metal NPs for facilitating DNA binding [10,11]. Spermine not only promotes surface adsorption of the negatively charged oligonucleotides but also aggregates the metal NPs to provide the higher surface enhancement [12]. Polyethylenimine (PEI) with rich amino groups that can be protonated to afford positive charges were decorated on metal NPs' surfaces for the SERS measurements of bacteria, in which strong electrostatic interactions between the metal NPs and negatively charged bacterial walls enable bacteria easily captured from the solution in a shorter time [13].

3.1.1.1.2 Random distribution

For large-sized matters, the combination strategy is to distribute the metal NPs on the analyte's surface randomly. One promising practical example is the shell-isolated nanoparticle (SHIN)-enhanced Raman spectroscopy (SHINERS) [14,15]. The plasmonic NP was coated with an isolated SiO_2 or TiO_2 shell, which can avoid agglomerating, separate them from direct contact with the probed material, and allow the NPs to conform to different contours of substrates [16]. Such NPs are spread as "smart dust" over the surface that is to be probed. SHINERS has been used to probe complex biological systems. For instance, the membrane structures of living cells were obtained using SHINERS with Raman signals detecting mannoprotein and other bioactive substances related to protein secretion and movement in living cells [14]. It is also convenient for SERS measurements of tissues by randomly spreading the SHINs on the frozen sections of breast tissues [17].

3.1.1.1.3 In situ reduction of metal nanoparticles

The *in situ* reduction of the SERS-active metal NPs on the protein/cell surface is also a practical way to make them close contact. These techniques refer to a routine transmission electron microscopic (TEM) technique, temperature gradient gel electrophoresis, and in polyacrylamide gels, named "silver staining," which was used to amplify imaging contrast. The reduced silver tends to nucleate on proteins. Thus the newly produced silver NPs also contribute to SERS enhancement. In Han's work, the silver staining combined Western-blot has been used for the separation of several proteins and the SERS detections on the nitrocellulose membrane plates, which is called "Western-SERS" [18]. With the assistance of silver staining, the SERS of proteins becomes detectable. The detection limit of myoglobin arrives at 4 ng band^{-1}. Not limited to proteins, this method is also applicable to the cell and bacteria [19–21]. The cells were soaked in a solution containing sodium borohydride and silver nitrate or chloroauric acid. The produced colloidal NPs are concentrated on the cell wall because the wall served as an efficient nucleation site. The formed Ag is in uncontrollable aggregation and show extremely large roughness. To prepare Ag colloid inside the bacterial cells, the cells were first washed in silver nitrate solution, and then sodium borohydride solution was added. The uniform colloid formed predominantly inside the bacteria [19].

3.1.1.1.4 Active/passive targeting of metal nanoparticles

To detect large-sized biosystems, for example, cells, one of the combination ways is to internalize the metal NPs into the cells. The internalization of plasmonic NPs in the cell can be divided into two strategies, (1) active identifications, and (2) passive targeting.

1. Passive targeting refers to the selective accumulation of NPs or nanocarriers in specific cells or organelles due to physicochemical or pharmacological factors. Such a targeting mechanism mainly depends on the characteristics of membranes (membrane permeability, membrane proteins, potential, and internal environment) and the properties of NPs, such as size, shape, and surface charge. Endocytosis of matters in cells can be different mechanisms, for example, pinocytosis (nonspecific pinocytosis and receptor-mediated endocytosis), and phagocytosis (large extracellular particles, such as cell fragments, bacteria, viruses, etc.). In these cases, no special functional modification is needed for the SERS-active NPs.
2. The elaborate surface decorations of metal NPs allow an active mechanism. These modified surfaces endow them with the abilities of active identifications by cells. Usually, specific ligands pointing at target locations or components in cells, such as targeting peptides, are employed to coat the NPs to assistant the SERS-active NPs cross the cell membrane or pores to access the intracellular components (Fig. 3.1) [22–24]. Cell membrane penetration peptide, RGD (Arg-Gly-AspArg-glycine-Asp acid), is a cancer cell-specific targeted peptide, which can identify the $\alpha v\beta 6$ or αv integrins on the cell surface [25,26]. Nuclear targeting

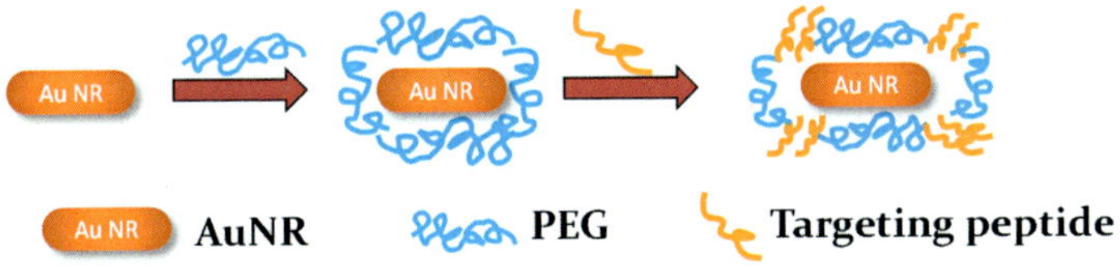

Figure 3.1 The preparation of organelle-targeting nanoprobes.

peptides (NLS) are the most widely used for nuclear active targeting from the simian virus (SV) large T antigen, having a sequence of Lys-Lys-Lys-Arg-Lys (KKKRK), which can bind to the importing protein families in the cytoplasm of the cell, such as importin α and β, faciliting the the nanomaterial trasport to the nucleus [27]. The most widely used nuclear targeting sequence is the TAT (47-57) sequence (Tyr-Gly-Arg-Lys-Lys-Arg-Arg-Gln-Arg-Arg-Arg,YGRKKRRQRRR) from human immunodeficiency virus (HIV-1), which can help plasmonic NPs enter the nucleus through the nuclear pores. In order to arrive in mitochondria, the mitochondrial targeting peptides or targeting sequences (MTS, MLALLGWWWFFSRKKC) that usually consist of 20–40 amino acid sequences can bind to transporters on the inner or outer membrane of the mitochondria to introduce the NPs into the mitochondria [28].

How many NPs are internalized in one cell is always difficult to answer because of the dynamics of import/export and the cell-to-cell heterogeneity. The ICP-Ms method assessed the amount of Au NPs in a single MCF-7 cell is 130–584, and the number of Au NPs increased with the increase of incubation time [29]. Shi et al. [30] employed inductively coupled plasma atomic emission spectroscopy (ICP-AES) to address this issue, in which they evaluated the NLS decorated gold nanorods (10.5 × 40.5 nm) as about 35 μg per 10^6 cells, while the gold nanorods without NLS give a particle density as 10 μg per 10^6 cells.

For *in vivo* SERS detection, delivering these plasmonic nanoprobes from peripheral blood circulation to the solid tumor regions or targeting organs is an essential issue that should be solved first. The penetration of the engineered nanomaterials from tumor blood vessels is mainly by means of the gaps (endothelial spaces) between endothelial cells, which is called "enhanced permeability and retention effect (EPR)" effect [31]. These gaps were found to have a size range up to 2000 nm [32]. So, the designed particles smaller than this size are expected to passively enter the solid tumor for the tumor region accumulation. In a recent study, Chan et al. [33] explored the phenomenon of NP tumor penetration and questioned the mechanism of NP entry into solid tumors. They found that the endothelial space was not the cause of NPs entering solid tumors. Instead, up to 97% of the NPs were found to enter the tumor through an active process in endothelial cells.

3.1.1.2 Biocompatibility

For the intracellular, *in vivo* SERS measurements, the SERS-active nanoprobes are the exogenous substances and should affect cell function as little as possible. So, they are required to be low toxicity and good biocompatibility. Many physicochemical parameters of nanomaterials, for example, size, rigidity, shape, surface modification, will regulate the cellular uptake, subcellular distribution, metabolism, degradation, and finally decide the cytotoxicity of the nanotargeting SERS probes. The biocompatibility of the nanoprobe can be revealed to a certain extent by cell viability after the cells were cocultured with nanoprobes. MTT (3-(4,5-dimethyl-2-thiazolyl)-2,5-diphenyl-2-H-tetrazolium bromide, thiazolyl blue tetrazolium bromide), WST-1 (2-(4-iodophenyl)-3-(4-nitrophenyl)-5-(2, 4-disulfophenyl)-2H, tetrazolium monosodium salt) assays, and flow cytometry, etc. can assess cell viability. A high dose of the targeting nanoprobes for cell culture would do harm to cells, leading to a decrease in cell viability. However, for the consideration of SERS measurements, more SERS nanoprobes will be helpful.

Connor et al. studied a series of defined NPs containing various surface modifiers and stabilizers (e.g., CTAB) with an established human cancer cell line. Their data suggested that spherical gold NPs with a variety of surface modifiers are not inherently toxic to human cells (K562 leukemia cell line), despite being taken up into cells [34].

The incorporation of polyethylene glycol (PEG) offers protection from the chemical and physical environments, minimization of the nonspecific binding [35,36], leading to prolonged blood circulation lifetimes *in vivo* [37]. PEG is readily amenable for bioconjugation to a range of targeting ligands [38]. It can be anchored to noble metal surfaces through a distal thiol functional group. Thiolated PEG ligands with other chemically active distal terminal groups (e.g., carboxyl and amine) can facilitate the functionalization of targeting agents, such as peptides or antibodies. If no specific modification on NPs, the addition amount is usually in the mM level to ensure enough cellular uptake amount [39]. For the organelle-targeting gold nanorod-based nanoprobes, owing to the assistance of PEG and targeting peptides, the acceptable addition concentration is less than 1 nM for HepG2 and MCF-7 cells (Fig. 3.2).

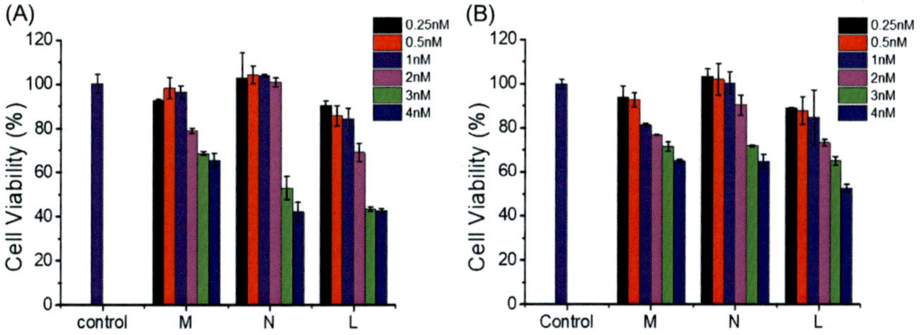

Figure 3.2 The cell viabilities of HepG2 cells (left) and MCF-7 cells (right) incubated with different concentrations (0.25–4 nM) of nanoprobes (Au nanorods decorated with targeting peptides) for 24 h, assessed by WST-1 assay. The control is the cells without nanoprobes. M, N, and L indicate the mitochondrion (MTS), nucleus (NLS), and lysosome-targeting nanoprobes reported in Ref. [24].

3.1.1.3 Tendency of aggregation or monodisperse of metal colloids

3.1.1.3.1 Salt induced aggregation and activation

The aggregation state of colloidal metal NPs is conducive to high SERS activity. Some small analytes that can destroy the static dual-layer of colloids can drive an aggregation by themselves. If not, some salt solutions will be helpful, for example, the salts containing Cl^- [40] and SO_4^{2-}[41,42].

When the salts were added into the colloids, they can weaken the stability of colloids and lead to a certain extent of colloidal aggregation, producing a new plasmonic band at a longer wavelength, which is assigned to the electromagnetic enhancement reason. For Cl^-, there is a different understanding of their role in SERS theory from the chemical enhancement [43]. They believed that a "chloride activation" exists, forming stable surface complexes of atomic-scale roughness, chloride, and the molecular adsorbate. An electronic CT resonant Raman scattering takes place among them [44]. The KCl can provide a further improvement on SERS as high as 100 times [45]. However, Nabiev observed that the enhancement caused the NaCl (0.05 M) activated Ag sols has chemical specificity to adenine when measuring DNA [46]. Du et al. found that the SERS signal intensity increased and then decreased with the dose of KCl [45]. The excess Cl^- caused competitive adsorption with the analytes [43]. So, to avoid the serious precipitation of the metal NPs and competitive adsorption that will go against SERS measurements, the salt volume is preferred to be less than 20 nM [45].

The SO_4^{2-} is also a preaggregation agent for the citrate-reduced metal colloids, which can be conducive to obtain SERS spectra of DNA/RNA mononucleotides [41] (by $MgSO_4$) and proteins [42] (acidified sulfate). With the aid of $MgSO_4$, the SERS spectra of adenine, guanine, thymine, cytosine, and uracil were recorded and analyzed along with their corresponding nucleosides and 5′-deoxynucleotides. Han et al. [42] detected the concentration-dependent SERS spectra of several label-free proteins (lysozyme, ribonuclease B, avidin, catalase, and hemoglobin) in aqueous solutions with acidified sulfate used as an aggregation agent, which can induce high electromagnetic enhancement in SERS. Owing to the pretreatment with the aggregation agent, strong SERS spectra of simple and conjugated protein samples could easily be accessed. The detection limits of the proposed method for lysozyme and catalase were as low as 5 mg mL^{-1} and 50 ng mL^{-1}, respectively.

3.1.1.3.2 Aggregation driven by external factors

Besides the spontaneous aggregation, the aggregation of colloidal metal NPs can be driven by external forces, for example, electric, magnetic fields, filtering, etc., for the purpose of improving SERS signals.

The control of the electric field can confine the locations of particles for SERS detections. In optoelectrofluidics (Fig. 3.3), metal NPs can be

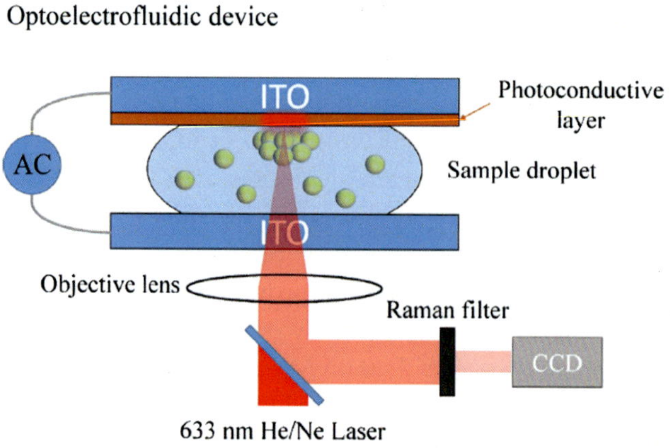

Figure 3.3 Schematic diagram of an optoelectrofluidic device for SERS spectroscopy. *SERS*, Surface-enhanced Raman scattering. *From H. Hwang, H. Chon, J. Choo, J.K. Park, Optoelectrofluidic sandwich immunoassays for detection of human tumor marker using surface-enhanced Raman scattering, Anal. Chem. 82 (2010) 7603–7610. doi:10.1021/ac101325t.*

concentrated on the defined reaction areas for SERS detection based on electrokinetic motions of particles or fluids under a light-induced unsymmetrical electric field [47,48]. When an alternating current (AC) voltage was applied on the indium tin oxid (ITO) electrodes and photoconductive layer, the nonuniform electric field generated in the liquid chamber induced an AC electroosmosis (ACEO)/electrothermal flows and dielectrophoresis force to simultaneously trap Au NPs and analytes into a SERS-active area at the laser spot, leading to a significant increase in the SERS intensity.

The magnetic-SERS substrate has multiple functions in one by the combination of sample separation and SERS capability. The geometry of the magnetic-SERS substrate is the metal NPs decorated magnetic beads. A magnetic bar [49] or the instinct magnetic field [50,51] can accumulate these beads. No matter the external force driving or spontaneous collection are both favorable for high-sensitive SERS detections due to the magnetic field-driven aggregation of metal NPs.

Filtering by a membrane filter can easily collect the large-sized analytes and SERS-active NPs as well, which has been used for the SERS analysis of bacteria [52,53]. White et al. prepared a portable automatic optical flow control microsystem for detecting contaminants in food and water (Fig. 3.4) [54]. The optofluidic system uses silica microspheres to form a porous microfluidic matrix for the adsorption of analyte molecules. The double fiber was placed at 90 degrees and focused at the same point for SERS signal excitation and collections. The analyte and colloidal NPs were mixed off-chip and were injected via the same port to a multiport injector with meander channels. This platform is an integrated competitive displacement unit for DNA sequence detection [55].

3.1.1.3.3 Monodisperse

For intracellular SERS experiments, the aggregation of metal NPs is undesirable because the relatively large particle size of aggregates would be unfavorable for their internalization by cells. Thus, several polymer ligands have been applied to prevent aggregation by electrostatic repulsion or steric forces, for example, sodium citrate, dodecanethiol, PEG, CTAB, tannic acid, hydroxylamine hydrochloride, and polyvinylpyrrolidone (PVP), etc. Cui et al. [56] introduced PVP as a stabilizer as well as a biocompatible shell to keep the SERS nanoprobe resistance to salt effect (Fig. 3.5). PVP can chemically adsorb on the Ag surface and interact with the surface of Ag by

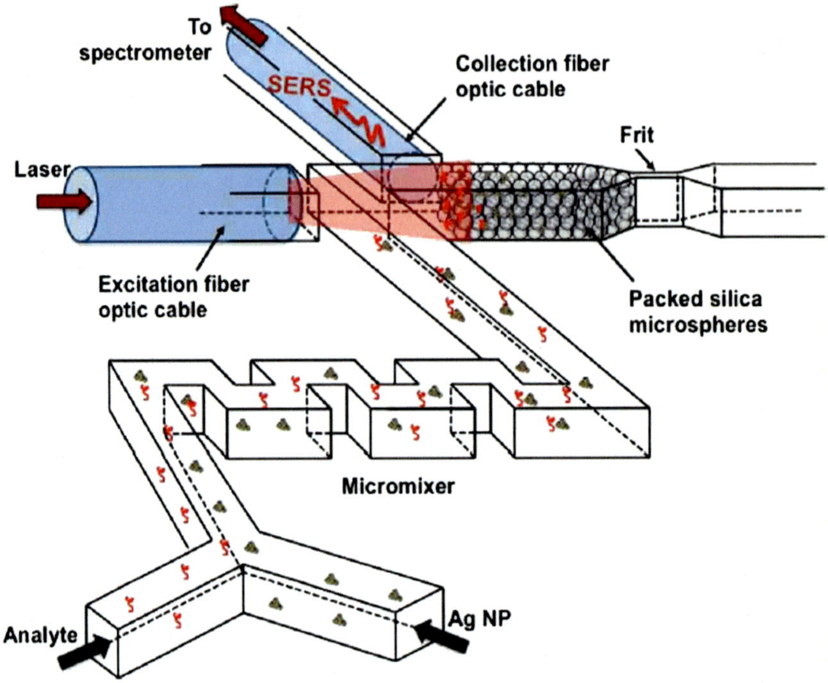

Figure 3.4 Optofluidic SERS microsystem with packed microspheres for passive concentration, an integrated micromixer to promote the adsorption of the target analyte, and integrated fiber optic cables for optical excitation and collection. *SERS*, Surface-enhanced Raman scattering. *From S.H. Yazdi, I.M. White, Optofluidic surface enhanced Raman spectroscopy microsystem for sensitive and repeatable on-site detection of chemical contaminants, Anal. Chem. 84 (2012) 7992–7998. doi:10.1021/ac301747b.*

forming coordinative bonds [57], thus making these silver nanoaggregates well dispersed in solution.

Though the monodisperse metal NPs were used for cell incubation, owing to the accumulation effect, particle aggregations were still found in their targeted specific organelles, as shown in Fig. 3.6.

A silica shell can also be a protective layer to maintain colloidal stability and to prevent interaction with external environments [58]. The main strategy for silica shell decoration was achieved by the mercaptotrimethoxysilane or aminopropyltrimethoxysilane (APTMS) and then reacted with sodium silicate or tetraethylorthosilicate. The silica shell allows the further surface modifications of antibody/aptamer by the surface coating of APTMS that can provide amine groups for bioconjugation of biomolecules [59,60]. In Choo's study [60], the silica encapsulation of

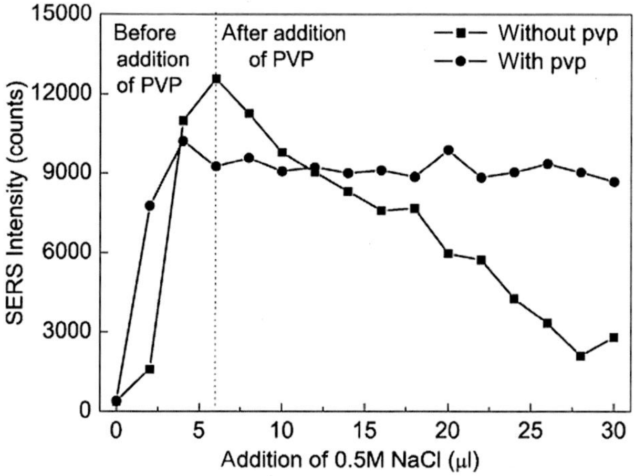

Figure 3.5 Variation of SERS intensities of a PVP-coated SERS probe (circle) and uncoated probe (square) upon addition of NaCl solution (0.5 M) from 0 to 30 μL. The SERS signal at 1510 cm^{-1} was used as a reference. *PVP*, Polyvinylpyrrolidone; *SERS*, surface-enhanced Raman scattering. *From T. Xuebin, W. Zhuyuan, Y. Jing, S. Chunyuan, Z. Ruohu, C. Yiping, Polyvinylpyrrolidone- (PVP-) coated silver aggregates for high performance surface-enhanced Raman scattering in living cells, Nanotechnology (2009) 445102. doi:10.1088/0957−4484/20/44/445102.*

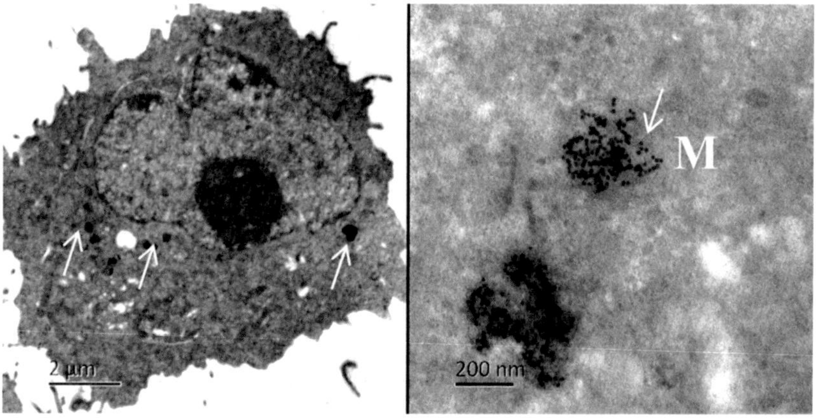

Figure 3.6 The bio-TEM images of a MCF-7 cell treated with the MTS-coated Au@MBN@Ag NPs for 12 h. *TEM*, Transmission electron microscopic. *MTS*, Mitochondrial targeting peptides. *MBN*,4-Mercaptobenzonitrile.

the AuNPs improved the thermal stability of the SERS-based lateral flow assay strips at high temperatures (Fig. 3.7), which made this SERS-based assay becoming an efficient diagnostic platform in tropical areas.

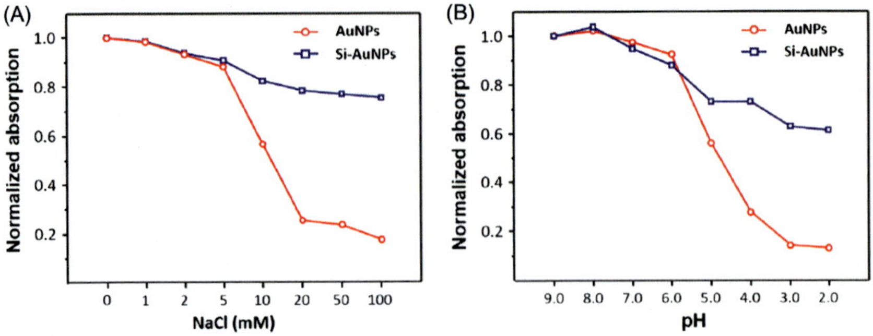

Figure 3.7 Chemical stability evaluation of AuNPs and silica-encapsulated (Si)-AuNPs. Changes in the normalized ultraviolet–visible absorption intensity of AuNPs and Si-AuNPs at various (A) NaCl concentrations and (B) pH values. *From J. Jeon, S.H. Lee, Y. Joung, K. Kim, N. Choi, J. Choo, Improvement of reproducibility and thermal stability of surface-enhanced Raman scattering-based lateral flow assay strips using silica-encapsulated gold nanoparticles, Sens. Actuators B: Chem. 321 (2020). doi:10.1016/j.snb.2020.128521.*

3.1.2 Solid-supported metal nanostructures

The solid-supported metal nanostructures, metal deposited film, SERS-active electrode, colloid-assembled film, nanostructures prepared by micro-nanomanipulation/nanolithography/nanotemplate, etc., are widely available, as well as many commercial SERS substrates. Biosamples involving cells and bacteria are placed or dripped on the surfaces of the SERS-active substrates or grown above the solid–liquid interfaces. Covering the liquids with a glass coverslip can help the sample avoid dry. If the sample is enough, the SERS substrate can be immersed in the liquid to achieve a self-assembled film of the probed molecules on the solid-supported metallic surface. The adsorption of the analytes will arrive in an equilibration for a long enough contacting time.

3.1.2.1 Surface modification for metal nanostructures

Compared with colloidal NPs that are more suitable for measuring liquid samples due to easier mixing according to large collision possibility, the solid-supported SERS substrate is more dependent on the diffusion of analytes to the surface. In order to help the analytes fix on the surface of metal nanostructures and solve the trouble for many molecules with low affinity to metallic surfaces, researchers decorated captures/hunters on the surfaces of metal nanostructures. The hunting of analytes is driven by various interactions, for example, molecular identification, static attraction, supramolecular interaction, and hydrophobic action, etc. Van Duyne et al. [61] decorated the metal film over nanosphere substrates with a straight-chain alkanethiols, for example, (1-mercaptoundeca-11-yl)tri(ethylene glycol), to collect the glucose molecules by this self-assembled monolayer that can partition the analyte of interest in a manner analogous to chromatographic stationary phases. This method is available for quantitative SERS detections of weak or nonexistent binding molecules. The decoration of capture molecules with functional groups for molecular identification is a feasible way to fix analytes above the metal surface. An Ag dendrites substrates modified with 4-mercaptophenylboronic acid was used to capture the bacterial cells according to the boronic acid and glycans via molecular identification [62]. Capture efficiency for *Salmonella enterica* was $84.92 \pm 3.25\%$ at 10^6 CFU mL^{-1} and as high as $99.65 \pm 3.58\%$ at 10^3 CFU mL^{-1}.

Moreover, molecular imprinting that can capture the imprinted molecule analogs by a geometry matching mechanism has been applied to integrate solid-supported SERS nanostructures [63]. Liu group [64] developed a boronate-affinity sandwich assay for the specific and sensitive determination of trace glycoproteins in complex samples. This method relies on the formation of sandwiches between boronate-affinity molecularly imprinted polymers (MIPs), target glycoproteins, and boronate-affinity SERS probes. MIP ensures specificity, and SERS gives high sensitivity.

3.1.2.2 Needle-like surface-enhanced Raman scattering microprobes

Not limited to the flat glass or quartz slide surfaces, some optical fiber tip, needle tip, and microelectrode with a size of less than 500 nm can be solid surfaces for metal NPs loading. The materials of the tip involve single-core quartz fiber, hollow borosilicate glass/quartz capillary, etc., which are conducive to the tip preparation and have enough toughness and hardness for cell insertion.

3.1.2.2.1 Fabrication

The sharp fiber tips were made by the heat and pull procedure of optical fiber or glass capillary *via* a special fiber-pulling device. Some optical fiber tips can be obtained by the HF etching rate difference between the center and the outer edge caused by the cladding layer [65]. The reported diameter of the probe tip is on the order of 100–500 nm. A conic tip with a large apex angle tends to incur cell damage when inserting it in a cell for a prolonged time. A longer and thinner fiber tip by means of laser pulling is preferred for cellular studies [66].

Different surface decoration ways based on physical and chemical approaches were applied on the tips, for example, vacuum metal deposition [67,68], colloidal NP assembly [69], laser-induced in situ deposition [70,71], magnetron sputtering growth [72], and hydrothermal growth [73], etc. Vo-Dinh et al. discussed the coating metals on the fiber tip and pointed out the best performance is from Ag tips [74]. Masson et al. [75] studied the tip surface coverage, aggregation state, and plasmonic properties of different types of NPs to realize the optimal SERS detections. Liu et al. [76] optimized the optimal enhancement of Au nanoneedle and Ag SERS tags and found that the signal enhancement of the combinations of two was 40 times higher than other combination ways, which has been used for tracing low-copy proteins in single cells.

A block copolymer brush layer as a template on submicron diameter optical fibers could make the NPs dispersed evenly and tightly, thus improving the optical performance of the SERS probes by at least an order of magnitude. The simulated electric field of hollow nanotube and solid nanofiber (with a radius of 325 nm) indicated that the electric field of nanofiber is about twice as strong as that of a nanotube, which is explained by the supra-lens effect formed on the optical fiber [77,78].

Gogotsi et al. [79] developed a carbon nanotube (CNT)-based SERS nano-tip probe with the Au NPs attached to the surface of CNTs *via* the electrostatic deposition. Compared with rigid glass nanotubes, CNT has the advantages of high mechanical strength and rheological property, allowing for long-term intracellular measurement.

3.1.2.2.2 Excitation/collection ways

The excitation/collection ways are diverse for the needle-like SERS microprobes. Fiber-based nanotips were developed initially as scanning probes for near-field scanning optical microscopy with subwavelength sizes for high spatial resolution imaging and detections. Incident light diffused from the tip in the form of evanescent waves and attenuated exponentially. Thus only molecules near the tip can be excited. The signal collection can be realized by the same fiber [69], or *via* the microscopic setup [68,80], classified as a near-field excitation/near-field collection and a near-field excitation/far-field collection, respectively. Different from the optical fiber-based nanotips, some nanoneedles are used as nanopipettes that take the analytes away from the cells, and measurements are completed *via* an external setup, belonging to the far-field excitation/far-field collection.

3.1.2.2.3 Platforms for single-cell analysis

The SERS nanotips integrate with the microscopic Raman systems and micro-operation, becoming an in vivo single-cell physiological detection tool [81–83]. The surfaces of the nanofiber tips need to be filled or modified with the corresponding biological capturers to achieve the purpose of specific detection, mainly include antibodies, enzymes, nucleic acids, cell structures, and biomimetic materials. Or, they were decorated with responsive reporters toward the intracellular analytes [84], for example, pH [66,85,86], free iron ions/hemoglobin [87], cancer biomarker [88], and oxygen content [89], etc.

Label-free detection of intracellular components by a nanopipette tip was achieved by Gogotsi et al. [82]. *Via* positioning the nanopipette tip either in the cell nucleus or cytoplasm, they can measure their SERS spectra *in vivo*. Assisted with the principal component analysis (PCA), the fingerprint information of intracellular components can be clearly distinguished (Fig. 3.8). The main characteristic of the nuclei are the high

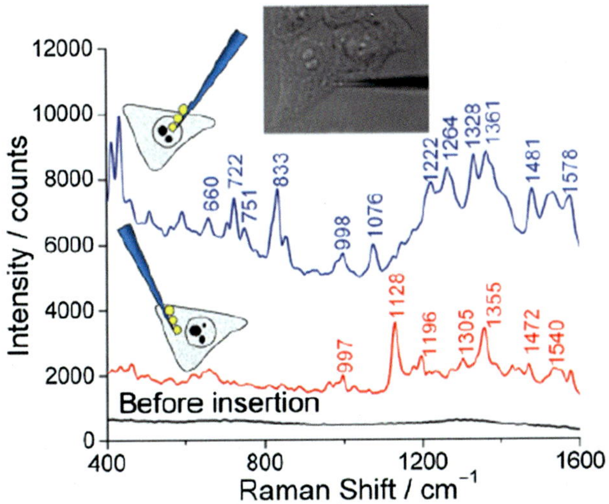

Figure 3.8 SERS spectra from the cell nucleus (upper) and cytoplasm (middle). *SERS*, Surface-enhanced Raman scattering. *From E.A. Vitol, Z. Orynbayeva, M.J. Bouchard, J. Azizkhan-Clifford, G. Friedman, Y. Gogotsi, In situ intracellular spectroscopy with surface enhanced Raman spectroscopy (SERS)-Enabled nanopipettes, ACS Nano 3 (2009) 3529–3536. doi:10.1021/nn9010768.*

contents of proteins and amino acids, and DNAs, while the phenylalanine signal in the cytoplasm is stronger than the nucleus. They also monitored the function of living cells when they were treated with a potassium chloride solution.

Masson et al. [90] implemented the label-free dynamic monitoring of cell secretion events by using SERS nanoprobes. Moreover, the quantitative detections of chemical gradients of metabolites near cells were achieved. SERS probes were positioned at 30 μm near the cells to continuously monitor the secretion of pyruvate, lactic acid, adenosine triphosphate (ATP), glucose, and other substances. Based on these instinct SERS spectra, researchers build a spectral library and then dealt them with the PCA to realize dimensional reduction. Finally, they selected 10 Raman band barcodes to point to the secrete events of living cells. The detection was based on a data-processing method capable of sorting and counting metabolites in reference to a SERS spectra database. In order to achieve a variety of metabolites and increase the probability of positive recognition, they further optimized the standard spectral database using a machine learning method [91]. The gradients of at least eight important metabolites were detected simultaneously near the cell different cell lines. By using this method, they monitored the cells secrete distance curve of events and learned more in-depth about the aerobic glycolysis metabolic pathways.

3.1.2.2.4 Merits and uniqueness

The nanoneedle-based platform is applicable for single-cell analysis. Its merits can be revealed in many aspects: (1) high versatility. It is easy to combine with other optical sensing strategies, such as localized surface plasmon resonance (LSPR) [92,93], field-effect transistor [94], ion current rectification [85], for the electrophysiological analysis of a single cell. (2) High accuracy of the controllable positioning system. These methods broke through the traditional SERS substrates in detections of microregions by the initiative or passive cell internalization of metal NPs. Differently, they can accurately reach the interesting positions of cells. (3) Relatively high stability of the sensing layer. Compared with the intracellular sensing based on the internalization of nanoprobes through long-time cell coculture, the mechanical insertion was carried out in these strategies, which is a minimally invasive way. In addition, the functionalized substances loaded on the tips (for example, linkers, probes, SERS substrates, etc.) are in a relatively fixed and stable state. (4) Living cell measurements and less effect on cell functions. Since the mechanical insertion and removal operations can be completed quickly (usually less than 30 s), these cells do not cause significant damage and can undergo normal division and differentiation.

3.1.3 Other unique surface-enhanced Raman scattering substrates

3.1.3.1 2D surface-enhanced Raman scattering hotspot substrates for high-resolution imaging

SERS microimaging has been used for visualizing the distributions of analytes on a single cell with the help of SERS tags [95]. In contrast to the indirect SERS imaging, Dana et al. [96] reported a label-free SERS high-resolution imaging technique. By using a high-density, uniform hot spot substrate (hot spot density 10^{11} cm^{-2}, 20−35 nm gap), they developed a SERS-stochastic optical reconstruction (STORM) technique that exhibits a spatially uniform distribution of spot locations with no identifiable sharp features, as expected from a uniformly distributed, featureless molecular layer. They obtained the 20 nm resolution super-resolved SERS images on these self-ordered metasurfaces (Fig. 3.9). This technique was applied for several complex biological architectures, for example, the supramolecular self-assembled peptide networks, microalgae membranes, and eukaryotic cells.

Lindquist's group [97] combined SERS direct detection with STORM by a 700 nm hexagonal periodic metal nanopore SERS hotspot array with a unit diameter of 150 nm to match a 660 nm laser, which causes the hotspot scintillation locating at 10 nm regions of the substrate to achieve the SERS-STORM imaging. In his later study, they employed an ultrathin silver island film to identify the molecular structures of the cell walls of two types of bacteria (gram-negative bacteria and gram-positive bacteria) (Fig. 3.10) [98].

3.1.3.2 3D plasmonic colloidosomes for single-cell analysis

Plasmonic colloidosomes are constructed by the self-assembly of hydrophobic ligand-functionalized metal NPs on an immiscible liquid−liquid interface. The colloidosome-based platform exhibits negligible SERS background interference, ultrasensitive SERS response, and excellent signal reproducibility in response to analytes inside/outside the colloidosomes

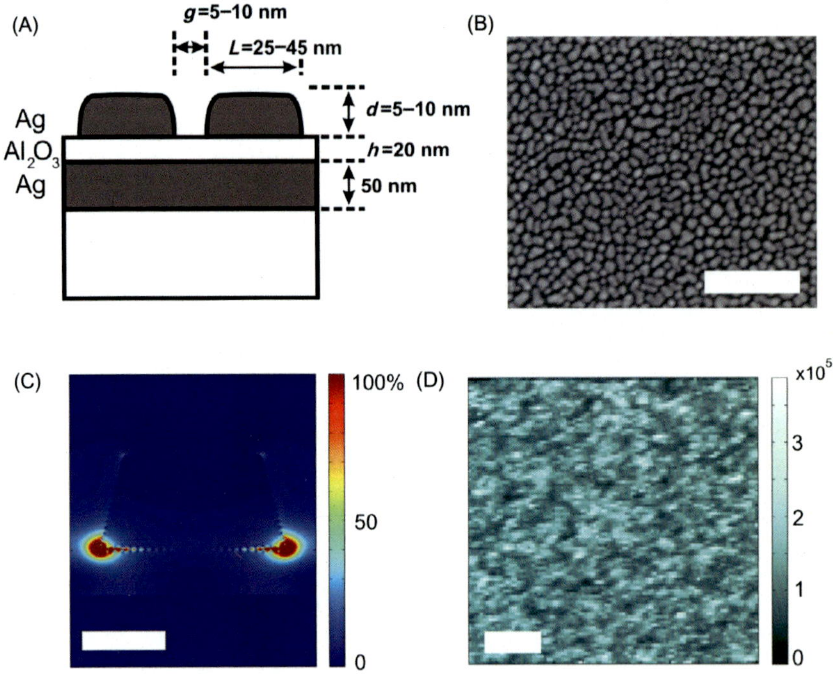

Figure 3.9 (A) Schematic description of self-organized metasurface (SOM), which is developed by Dana et al. and used as a SERS substrate for the SERS-STROM. (B) Scanning electron micrograph of the SOM is shown (scale bar 200 nm). The Ag nanoislands are formed due to dewetting and spontaneous segregation, with approximately 30 nm diameter and 10 nm thicknesses, separated from a continuous Ag film by a 10–20 nm thick dielectric layer (Al_2O_3). (C) Field profiles show confinement and relative enhancement of field along the surface cross-section (scale bar 10 nm). (D) Confocal Raman map on SOM treated with 1 mM methylene blue shows the high uniformity of enhancement (scale bar 2 mm, Raman intensity integrated between 500 and 2000 cm^{-1}). *SERS*, Surface-enhanced Raman scattering. *From H. Zareie, T. Tekinay, A.B. Tekinay, M.O. Guler, A. Dana, D. Kocaay, et al., Label-free nanometer-resolution imaging of biological architectures through surface enhanced Raman scattering, Sci. Rep. 3 (2013). doi:10.1038/srep02624.*

[99–101]. Colloidosomes can encapsulate single cells and aid cell secretion detections. Liu et al. employed the MBA modified AgNPs to construct plasmonic colloidosomes, which have been used for exploring the anomalous acidification of the extracellular microenvironment (Fig. 3.11) [102]. SERS detections can be executed on the entire plasmonic colloidosome shell (with an area of about 0.9×10^{-9} to 1.6×10^{-9} m^2), and this 3D SERS substrate shows high uniformity for the SERS response.

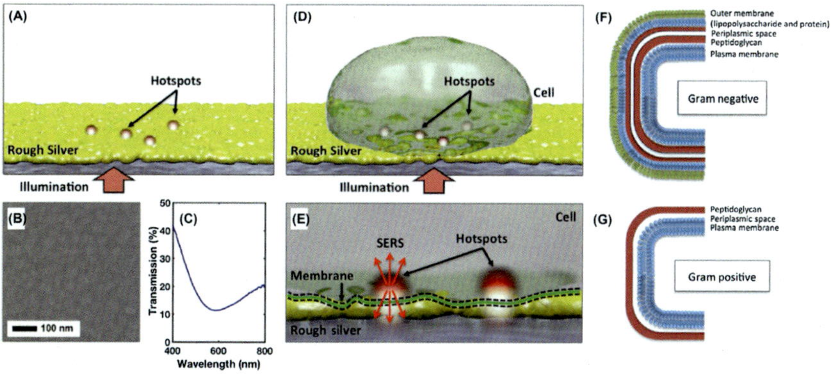

Figure 3.10 Illustration of imaging process and interaction between the plasmonic surface and the cell wall. (A) Depiction showing the creation of enhanced electromagnetic field "hotspots" due to illumination of the silver plasmonic surface from below. (B) SEM image of the plasmonic surface created by depositing 10 nm silver and a 2 nm chromium adhesion layer upon a glass microscope coverslip. (C) Transmission spectrum of an ultrathin silver island film showing a clear plasmonic resonance dip. (D) Depiction of a cell adsorbed to the rough plasmonic surface. (E) Schematic depicting the interaction between the hotspots and the cell wall. SERS is emitted from the sample and collected by the same objective used to illuminate from below. (F) Schematic depicting the molecular structure of a cell wall on a gram-negative and (G) a gram-positive bacterial species. *SERS*, Surface-enhanced Raman scattering. *From A.P. Olson, K.B. Spies, A.C. Browning, P.A.G. Soneral, N.C. Lindquist, Chemically imaging bacteria with super-resolution SERS on ultra-thin silver substrates, Sci. Rep. 7 (2017). doi:10.1038/s41598–017–08915-w.*

3.2 Laser-related issues

3.2.1 Laser wavelength selection according to surface plasmon resonance

The selection of laser wavelength was of great importance in the SERS studies. Based on the SERS electromagnetic enhancement mechanism, the laser wavelength prefers to resonant with the plasmon bands of nanomaterials (NPs) or in their vicinity could obtain a strong enhancement effect. Previous studies have suggested that green light is favorable for the enhancement of silver NPs, while a red laser is more suitable for gold NPs. However, the interference of spontaneous fluorescence from organisms and the penetration depth of organisms are often considered when they are applied in biomedical systems. Owing to the above two reasons, most studies consider the excitation wavelengths of 633 and 785 nm.

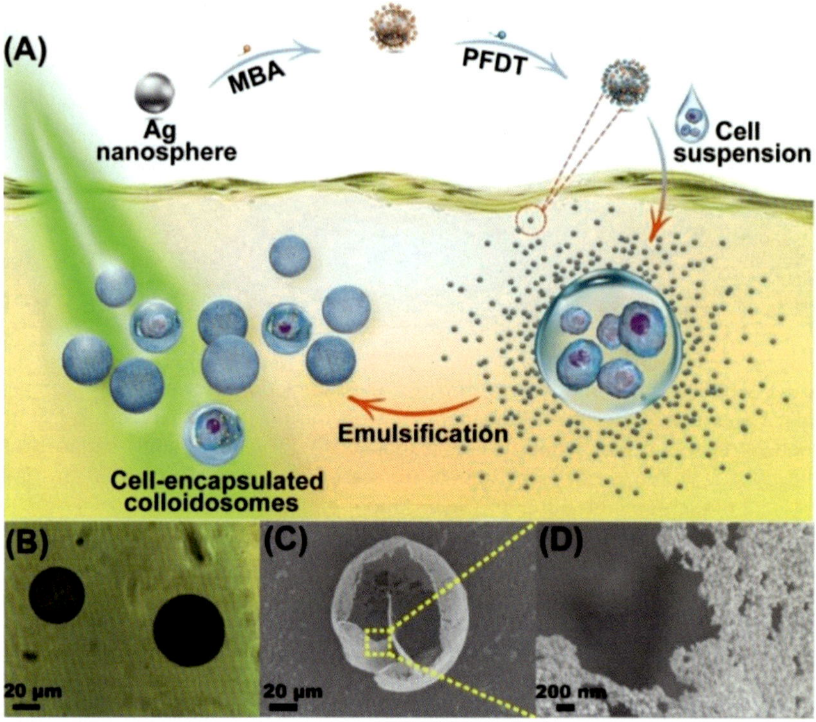

Figure 3.11 Fabrication and characterization of the plasmonic colloidosome-based single cell detectors (PCSDs). (A) Schematic illustrating the formation of PCSDs. (B) Microscopic image and (C) field-emission scanning electron microscope image of the established PCSDs. (D) The magnified segment of the surface of a PCSD. *From X. Wang, J. Ji, T. Liu, Y. Liu, L. Qiao, B. Liu, Plasmonic colloidosome-based single cell detector: a strategy for individual cell secretion sensing, Anal. Chem. 91 (2019) 2260–2265. doi:10.1021/acs.analchem.8b04850.*

SPR effect brings an obvious photothermal effect. This harms the biological system and SERS acquisition as well. Ye's group found that gap-enhanced Raman tags (GERTs) show strong near-infrared (NIR) Raman enhancement due to the combination of the near-field electromagnetic and chemical enhancement in a subnanometer core–shell junction geometry. GERTs only present one LSPR in the visible range [103]. The off-resonance NIR excitation strategy can minimize the excitation laser-induced photothermal effect. Their results indicated that ultrahigh SERS photostability of GERTs during 30 min continuous cell and tumor SERS imaging could be realized without photobleaching, which is available for long-period SERS bioimaging [104,105].

3.2.2 Laser wavelength and surface-enhanced resonance Raman scattering

In many SERS measurements, matching the molecular energy level of reporters or analytes by the laser wavelength can bring the resonance Raman effect, which can cause 10^4-fold of Raman scattering intensity and called surface-enhanced resonance Raman scattering (SERRS). Many proteins with chromophores having visible absorption bands can provide stronger SERRS under the excitation of a wavelength-selective laser [106]. Table 3.1 lists many reported chromoproteins that have been measured by the label-free SERRS spectroscopy.

In many SERS detection strategies, chromophores molecules with the Raman resonance effects were chosen for the homogeneous phase DNA analyses (e.g., polymerase chain reaction to amplify specific DNA sequences) and biomedical assays. Resonating the exciting laser wavelength gives a significant increase in sensitivity. Graham and Faulds [10] summarized the SERRS methods for DNA detections. Many commercially available fluorescent dyes can be attached to oligonucleotide probes and these dye-labeled DNAs were attached to the surfaces of metal NPs based on the electrostatic interactions. The chromophores used for this strategy can be either conventional fluorescence labels whose fluorescence can be quenched on the metal surface, or targeted designer dyes specifically used in SERRS studies. The types of labels used for DNA detections can be either fluorophores or nonfluorophores, and they summarized these dyes in a review paper [10]. These chromophores molecules with the resonance effect can also be used as reporters for constructing SERS tags [116], as well as the responsive sensors according to the generation of the resonance-type chromophores [117].

Table 3.1 Chromoproteins measured by SERRS.

Biomolecules	Chromophore	λ_{ex} (nm^{-1})	Ref.
Hemoglobin	Heme group	514	[107,108]
Cyt c	Heme iron	514	[109,110]
Myoglobin	Heme group	633	[111]
β-Carotene	Π-Conjugated polyene backbone	532	[112]
Anthocyanidins	Benzopyrylium moiety	532	[113]
Rhodopsin II	Retinal	532	[114,115]

SERRS, Surface-enhanced resonance Raman scattering.

3.2.3 Laser power setting and defocusing for avoiding photodamage

Metal NPs can kill cancer cells according to their selective targeting and large absorption cross-sectional area of near-infrared light. Thus photodamage in living cells and bodies often happens and should be paid attention to the laser-based methods. In Lim's study [22], 4.0, 2.0, and 0.2 mW of laser power for Raman imaging (785 nm wavelength Ti: Sapphire laser, 3900S, Spectra-Physics) were compared for human oral cancer cells (HSC-3), the significant change of cell morphology and the cell death (red staining) were observed for the 4.0 and 2.0 mW, while for 0.2 mW of laser power, no significant change of cell morphology was observed.

On a 2D spontaneously self-organized metasurfaces (SOMs), wide-field illumination and resulting video sequence were processed using stochastic optical reconstruction (STORM) imaging. High excitation laser densities (e.g., 10 kW cm^{-2}) were observed to result in chemical and structural damages due to the organic decomposition of biological samples, even in the absence of plasmonic enhancement. Thus a low excitation density (1 kW cm^{-2}) was set for recording Raman bands and this laser power makes the band stable during blink events over durations long enough to record sufficient data to produce stochastic reconstructions [96].

The photothermal effect of metal NPs has been widely used for cancer cell therapy. For example, Xia et al. [118] quantified the photothermal treatment effect of 65 nm immune gold nanocages on SK-BR-3 breast cancer cells with a pulsed near-infrared laser (805 nm with a bandwidth of 54 nm, 4.77 W cm^{-2}). If no Au cages, the cells could stand the laser power of 6.0 W cm^{-2} for 5 min, but they started to die at a laser power of 1.5 W cm^{-2} if the Au cages existed (Fig. 3.12).

In order to reduce photodamage when measuring biological samples, Ren et al. proposed a defocusing method for a Raman microscope system, which can reduce the laser power density and minimize the effects of light damage when the vertical spatial resolution can be ignored. They adjusted the sampling phase in the defocus state with slits and/or magnify the pinhole of the Raman microscope (Fig. 3.13A) [119]. The focusing way supports the stronger signal strength in comparison to reducing laser power (Fig. 3.13B). Although the defocusing method sacrifices the spatial resolution due to the increased laser spot size, it prefers the single-spot detections of photosensitive biological samples.

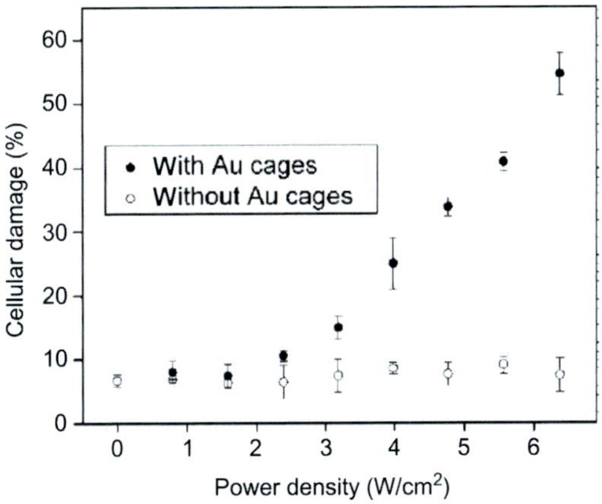

Figure 3.12 Plots of cellular damage versus laser power density for SK-BR-3 cells incubated with immuno Au nanocages (●) and for the control in which cells were not incubated with immuno Au nanocages (○). Cells were irradiated for 5 min and harvested for analysis 3 h after irradiation. *From L. Au, D. Zheng, F. Zhou, Z.Y. Li, X. Li, Y. Xia, A quantitative study on the photothermal effect of immuno gold nanocages targeted to breast cancer cells, ACS Nano 2 (2008) 1645–1652. doi:10.1021/nn800370j.*

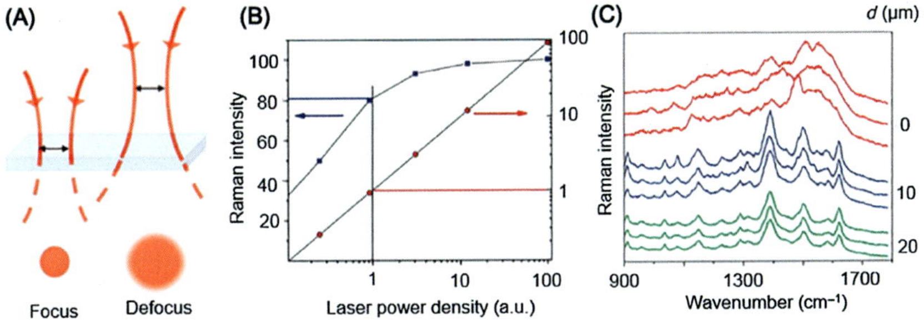

Figure 3.13 (A) Scheme of defocusing method. (B) Dependence of normalized Raman intensity on the laser power density obtained by the defocusing method (left) and by lowering the laser power (right). *From X.M. Lin, Y. Cui, Y.H. Xu, B. Ren, Z. Q. Tian, Surface-enhanced Raman spectroscopy: substrate-related issues. Anal. Bioanal. Chem. 394 (2009) 1729–1745. doi:10.1007/s00216–009–2761–5.*

3.2.4 Light penetration depth for *in vivo* detection

When the light hits the sample, the absorption and scattering of the material cause light attenuation. Under normal operating conditions, the typical tissue penetration depth of SERS signals is 1 – 5 mm [120,121]. This is a problem that must be considered in many optical detections of internal tissues and in vivo imaging. To arrive in deeper locations, people find three biological transparency windows in the NIR ranges, 650 − 950 nm (NIR-I) and 1−1.35 μm (NIR-II); 1.5 to ~1.8 μm (NIR-III) [122,123], which allow maximum light penetration depth and minimized autofluorescence. With the assistance of the NIR-responsive SERS substrates, confocal SERS imaging of biotissues can be achieved.

The traditional SERS can only analyze the area close to the surface of the sample or the subsurface components of the transparent packaging. Spatially offset Raman spectroscopy (SORS) is a new analytical technique that can be used for the analysis of several millimeters thick samples or also for the chemical analyses of materials in opaque packaging [124,125]. SORS can be used in relatively low-energy lasers to identify the Raman spectra of the individual layer in discrete scattering systems, while a longer integration time might be needed to gain signals deeper within the tissue. The Raman spectra were collected from the spatial displacement regions away from the excitation point. The SORS can effectively eliminate fluorescence interference from the surface layer, which is valuable relative to the traditional backscattering Raman spectrum. Under normal operating conditions, the typical tissue penetration depth of SERS signals is 1 − 5 mm. It has been proposed with instrument optimization that depths of 2 − 5 cm are theoretically possible using surface-enhanced spatially offset Raman spectroscopic (SESORS) imaging [126,127].

A perspective toward SERS for *in vivo* and clinic studies may enable image-guided surgery of tumor margins, and the development of the endoscopy-SERS coupled systems to identify and localize internal tumors [128−131]. A unique endoscopic opto-electromechanical Raman device was the first instrument to allow SERS imaging of tissue during gastrointestinal endoscopy [129,132]. Mohs et al. [121] reported the development of a hand-held spectroscopic pen device that operates in the NIR to detect fluorescent and Raman signals based on an integrated filer optic system. With the aid of SERS tags or the NIR fluorescent dye [indocyanine green (ICG)], the central tumor region can be clearly identified. The detection limit of the device for SERS tags was as low as 5×10^{-14} M in comparison to ICG at 2×10^{-11} M. The SERS-guided surgery can also benefit to searching microscopic metastases and residual tumor cells. In the review paper written by Nie et al. [130], they summarized the recent development of intraoperative SERS and image-guided surgery (Section 6.4).

3.3 Reproducibility and reliability

Repeatability and reliability are always two unavoidable issues in each new SERS strategy, which are inseparable from the enhancement issue of SERS substrates. Reasons for poor SERS detection repeatability are as follows: (1) random distribution of SERS hot spots. (2) The complexity of the analyzed materials, the heterogeneity of spatial distribution, and the disorder of adsorption. (3) Interference caused by nonspecific adsorption and fluorescence background when the above two are combined. To solve these problems, people have made a lot of effort.

3.3.1 Mean spectra

For large analytes, for example, cells, they show spot-to-spot difference since they are larger than the last spot (about 1 μm in diameter) (Fig. 3.14A) [133]. Moreover, cells in a culture dish are not identical, because they are dynamic living systems and in different states. So, acquiring spectra from single random points of a cell might result in a false interpretation of the cellular conditions due to the complexity in spectral variations. In order to avoid this, averaging the recorded SERS spectra was applied, which can effectively eliminate the outliers. Following Culha's suggestion [133], the minimum number of cells that need to be scanned for reliable spectral interpretation is at least 20 cells, whereas scanning 30 cells can be preferred. In Fig. 3.14B – F, the effect of sampling size in spectral variation is demonstrated by plotting three randomly chosen average cell spectra (shown as red, black, and blue spectra) of 2, 4, 6, 8, and 10 raster-scanned cells and comparing them to the overall average spectrum obtained from 20 cells (yellow spectrum). Although the spectral pattern is quite similar even when only two cells were averaged, the differences almost disappear when 10 cells were randomly averaged, indicating smaller experimentally introduced variations.

3.3.2 Homogenization of sample

It is easy to uniformly disperse the small biomolecules on the surface of a solid-SERS substrate. However, the large-sized biosamples, for example, cells and tissues, are inhomogeneous. To better obtain the consistent sample features under SERS measurements, in da Silve et al. study [134], the cells were frozen, ground, and stirred to obtain a liquid and homogeneous solution. Then, 10 μL of the cells were deposited onto

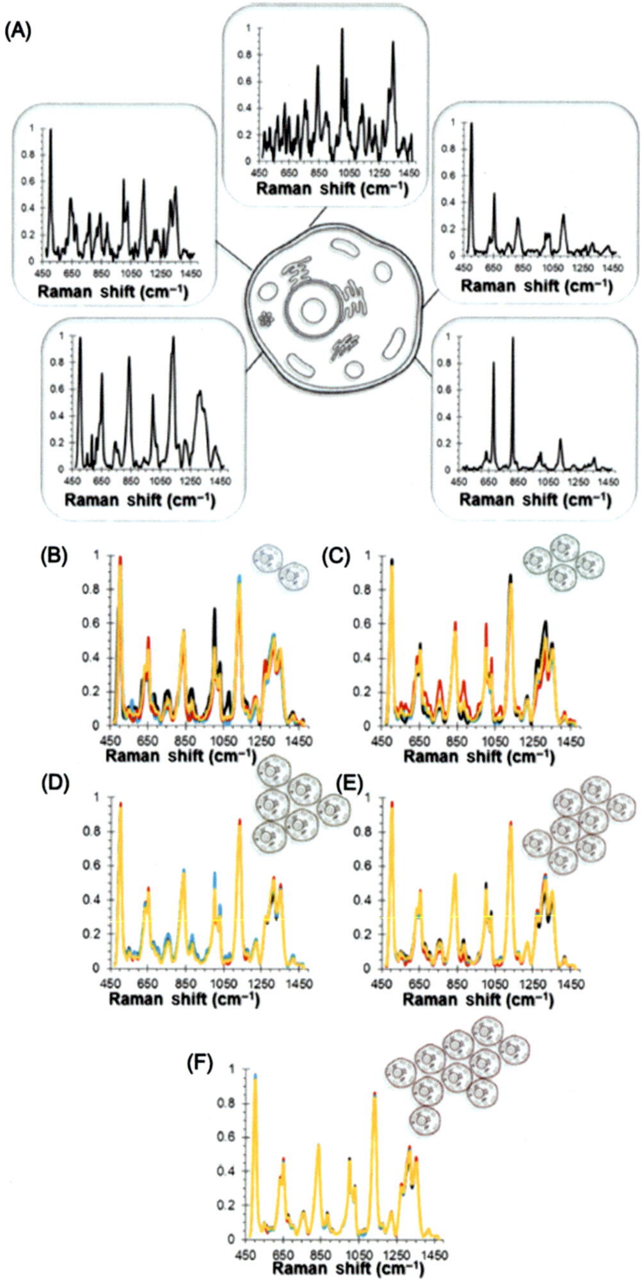

Figure 3.14 Spectral variations (A) within a cell and (B – F) the comparison of average cell spectrum (yellow spectrum in each graph) obtained from 20 cells to the average of (B) 2, (C) 4, (D) 6, (E) 8, and (F) 10 cells chosen randomly among the raster-scanned cells.

the SERS substrate for Raman measurements shortly after the probing samples were prepared (up to 4 h after preparation) to avoid film oxidation and cell degradation.

Barhoumi and Halas investigated the SERS of thiolated single-stranded and double-stranded DNA oligomers bound to Au nanoshells [135]. They found that thermal pretreatment of the DNA prior to adsorption onto the Au nanoshell changed the conformation of the DNA molecules, resulting in improved reproducibility of the SERS spectra.

3.3.3 Controlled immobilization and orientation

According to the surface selection rules, the molecular orientation of the adsorbed molecules can result in the Raman signal intensity changes [136]. The poor spectral reproducibility, originating from random protein immobilization on SERS substrates, still makes it challenging for SERS to probe protein functions without any extrinsic Raman labels. Han, Lu, and Zhao et al. [137] prepared iminodiacetic acid-functionalized silver substrates, which are used for capturing His-tagged proteins *via* nickel – imidazole coordination. The controlled immobilization enables excellent SERS spectral reproducibility, as evidenced by six polypeptides.

During the preparation of SERS tag methods, the binding ways of Raman reporters to metal are based on functional groups, such as thoil, amino, and carboxyl groups. The dynamic desorption of reporter molecules about the metal surfaces under laser irradiation might result in signal blinking. Li and Tang et al. [138] developed the selenol-based Raman reporters, and found that the Au-Se interface is much stronger than the Au-S bond, displaying good resistance to abundant thiol under the biological condition and show better stability.

◀ SERS spectra were acquired from living HeLa cells upon incubation for 24 h with 60 nm diameter sized spherical citrate-reduced AuNPs. An 830 nm NIR laser with 150 mW power and 2 s exposure duration was applied. Measurements were carried out in a cell area of about 20 × 20 μm^2 with 2 μm intervals in raster-scan mode by using a 20× long distance objective (NA, 0.40). *NIR*, near-infrared; *SERS*, surface-enhanced Raman scattering. *From A.B. Veloso, J.P.F. Longo, L.A. Muehlmann, B.F. Tollstadius, P.E.N. Souza, R.B. Azevedo, et al. SERS investigation of cancer cells treated with PDT: quantification of cell survival and follow-up. Sci. Rep. 7 (2017). doi:10.1038/s41598–017–07469–1.*

3.3.4 Purification of the surface of surface-enhanced Raman scattering substrates

Selective extraction of the information from the molecules of interest and minimization of interference from surface impurities and other biomolecules are the main challenges in direct SERS analysis [139]. One approach to overcome these issues is to coat the surface of the NPs with iodide to eliminate surface impurities and improve the reproducibility of SERS signals [140].

3.3.5 Contributions of media and reagents

Biosystems, for example, cells and bacteria species, are grown in various nutrient mediums containing glucose, various salts, a source of amino acids and nitrogen (e.g., beef broth and yeast extract), pH controller (CO_2), etc. They might contribute to SERS signals overlapping with the peaks of cells and bacteria [141]. Some extraction reagents for cell organelles can also remain and provide SERS signals [142]. Ziegler et al. [143] suggested at least three standard buffer-washing/centrifugation cycles to minimize the effect of the cell growth culture medium. Buffers, for example, phosphate-buffered saline and borate buffers, are suggested in the washing steps to avoid the problems of osmolarity and lyse [144].

3.3.6 Integration of surface-enhanced Raman scattering with microfluidics

One promising method to improve reproducibility is to combine SERS with microfluidics by controlling colloid aggregation and uniform analyte adsorption on solid SERS substrates [145,146]. The microfluidic approaches can help precisely control the aggregation time of colloid and the mixing efficiency of colloid and analyte solution *via* the flow rate and functional structure (zigzag shape structure). A fluid dispersion is suppressed due to droplet transport in a plug flow regime, and a fluctuation of less than 5% can be realized [147,148].

The SERS-microfluidic droplet platform allows for high-throughput single-cell analysis. A single cellwas isolated into one water-in-oil droplet along with a lot of SERS-active NPs. With enough nutrition, a long-term off-chip cell incubation can be carried out. Relatively large NPs are preferred based on the consideration of cell cellular endocytosis [50,51]. One droplet provides a nanoliter to picoliter volume, which can accelerate the process of diffusion and adsorption equilibriums. The isolated microenvironment can decrease the likely contamination from external interferences, which is available for the analysis of single-cell metabolites. More information will be introduced in Chapter 7, *Surface-enhanced Raman scattering-based microfluidic devices for in vitro diagnostics.*

3.3.7 Internal standard method

The internal standard method is a commonly used strategy in analysis science. By introducing a molecule not affected by the system as an internal standard signal, SERS ratio-type biosensors can be performed [149].

To trace dipicolinic acid (DPA; pyridine-2,6-dicarboxylic acid) that is unique for quantifying Bacillus spores and widely used as an indicator for evaluating potential anthrax attacks, Bell et al. assessed the temporal SERS signal variation of the symmetric ring stretch from pyridine of DPA with the aggregating agent, while the signal of potassium thiocyanate was used as an internal standard [150]. Similarly, Cowcher et al. quantified DPA from a nitric acid extract from Bacillus spores with glutaric acid as the internal standard [151].

One of the problems using these internal standards is that there is still unequal competition between the target analyte and the internal standard on the metal surface [115]. A core—molecule—shell NPs were developed with an internal standard molecule embedded within the nanogap between an Au core and an Ag shell [152], which overcomes the issue of competitive adsorption. The shell can completely isolate the internal standard from the environment, providing a stable and reliable reference signal. This strategy can correct fluctuations in SERS signals caused by the variation of the aggregation states and the measurement conditions and can be applied to diverse target molecules for the quantitative analysis purpose.

Developments involve the use of isotopologues noted as isotope dilution SERS, in which isotope substitution is used in an internal standard, with the same molecular formula and structure as the determinant [153]. The substitution of hydrogen with deuterium leads to a reduced mass of the C-H to C-D substitution, which results in a shift in CH vibrations from c.2800 to c.2100 cm^{-1}, and band ratios can then be made of the isotopologue to the natural isotope.

Moreover, many substances that appear in the SERS detection systems and exhibit independent bands can also be internal standards, for example, the magnetic beads (664 cm^{-1}) [51], PDMS (709 cm^{-1}) [117], and silicon (520 cm^{-1}) [154], etc.

3.3.8 Reporters having bands in silent range

Traditional Raman reporters possess multiple peaks in the fingerprint region (1000–1700 cm^{-1}), which typically overlap with each other and cause crosstalk. For biosystems, there is a silent range referring to the spectral range of 1800–2800 cm^{-1} [155], where the signals of biospecies are negligible, making these chemical groups highly suitable for multiplex detection [156]. Molecules with special groups give bands in this range, for example, alkynes, nitriles, azides, and deuterium. Metal NPs with the molecules with these groups to form SERS tags have been applied for multiplex targeting and multicolor identification of cancer biomarkers with nonoverlapping Raman signals [157].

3.4 Raman data-related issues

3.4.1 Data processing

Data processing is a critical step in Raman data interpretation. Many small changes can be overshadowed easily by corrupting contributions during measurements, such as intensity variations from changes in laser focusing on uneven surfaces and fluctuations of a fluorescent background, the SERS substrates, and the sample. Therefore, the instrumental issues should be confirmed before Raman and SERS measurements. Fig. 3.15 displays the basic preprocessing steps before and after statistical processing was carried out. Each spectrum experiences the "baseline" correction, "off-set," "smoothing," and "normalization." Cosmic rays are easy to identify and thus removed from the data set *via* the median filter. Then, many treated spectra were dealt with the "averaging" to achieve the typical spectrum of each experimental group. The differential spectrum can be obtained by two spectra subtraction to easily compare the differences between different groups, [133]. In a review paper written by Popp et al. [158], an overview of data processing methods for biomedical Raman spectroscopy has been given.

It is commonly observed that the raw SERS spectra for one same sample have different spectral profiles, in which Raman peaks at different frequencies experience different radiation enhancement. This relative intensity of the peaks in a SERS spectrum is the contribution of plasmonic photoluminescence. The spectral background contribution can be reduced via polynomial fitting or adaptive iteratively reweighted penalized least squares (airPLS). Ren developed a robust method to correct this effect [159]. The native SERS spectra (SERScorr) can be obtained using SERScorr = SERSsub ×

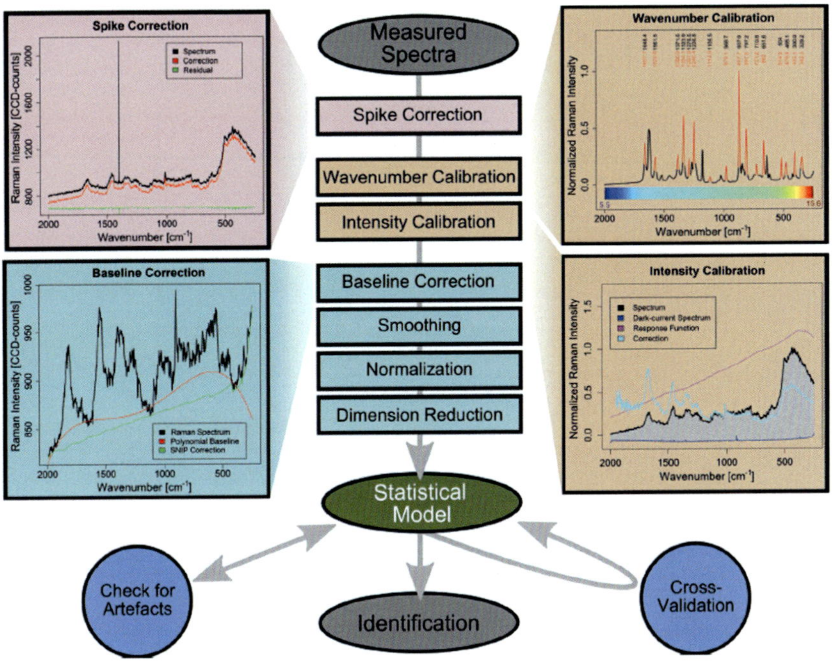

Figure 3.15 Data pipeline for the analysis of Raman spectra. First, a spike correction, a wavenumber and spectrometer calibration, and preprocessing (baseline correction, smoothing, and normalization) are carried out. Then, a dimension reduction is carried out by a factor method (e.g., PCA) or by means of feature extraction. Thereafter, clustering is applied for visualization, or a supervised technique is trained to predict labels or continuous parameters. *PCA*, Principal component analysis. *From A.N. Kuzmin, A. Pliss, P.N. Prasad, Ramanomics: new omics disciplines using micro Raman spectrometry with biomolecular component analysis for molecular profiling of biological structures, Biosensors 7 (2017) 15.*

PLbulk/Bg, where Bg is the fitted background of the raw SERS spectrum (that is, the PL of the metal NPs), SERSsub is the background-subtracted SERS spectrum (that is, the true SERS spectrum of reporters) and PLbulk is the PL of bulk Au. As shown in Fig. 3.16, the correction of the plasmon spectral shaping effect shows a high consistency for the spectra recorded from different nanoaggregates.

3.4.2 Chemometric sorting algorithm

The material and composition of cells and biological tissue samples are too complex to be identified directly. So people propose a concept, "Ramanomics," to explain its complexity, which is a new omics discipline

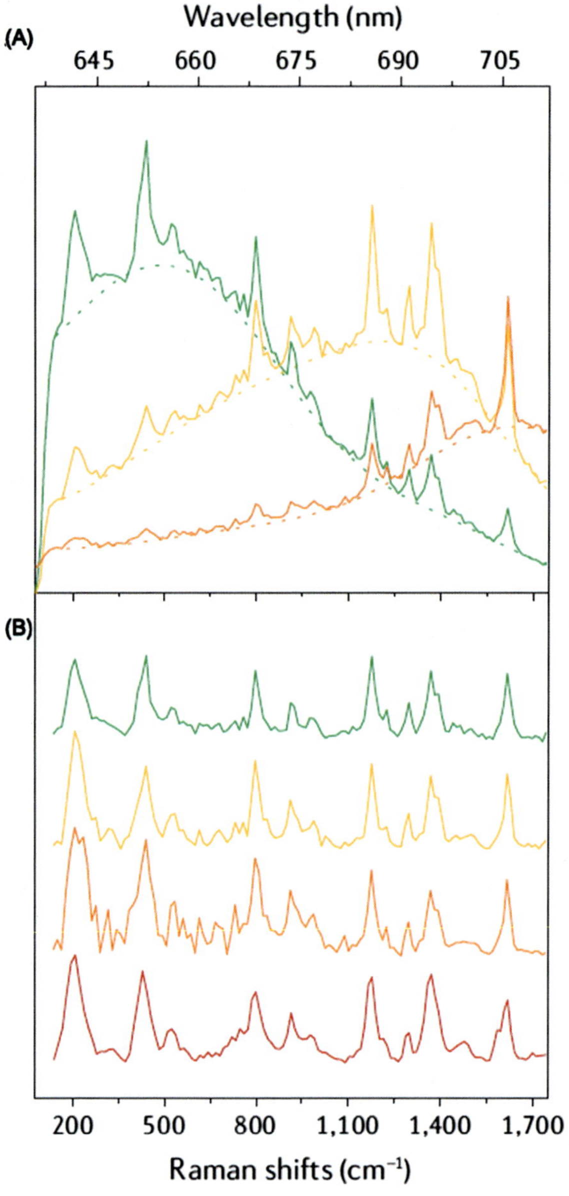

Figure 3.16 Correction of the plasmon spectral shaping effect to recover native chemical information. (A) The raw surface-enhanced Raman spectroscopy (SERS) spectra of malachite green isothiocyanate on Au nanorods; the variation in the relative intensity of the peaks is caused by the plasmon effect. (B) The corrected spectra, which enables the recovery of native chemical information. *From M.D. Li, Y. Cui, M.X. Gao, J. Luo, B. Ren, Z.Q. Tian, Clean substrates prepared by chemical adsorption of iodide followed by electrochemical oxidation for surface-enhanced Raman spectroscopic study of cell membrane, Anal. Chem. 80 (2008) 5118–5125. doi:10.1021/ac8003083.*

by using Raman microspectrometry with biomolecular component analysis for molecular profiling of biological structures and clinic samples [160].

Computational methods are necessary to generate images or model differences between different states of biological samples. PCA, linear discriminate analysis, support vector machine cluster analysis, and multivariate curve resolution are widely applied to identify the maximal chemical information from the spectra without a priori knowledge of the chemical properties of the sample [82,90,135,161]. After the dimension reduction, the computational methods are applied, which can realize the clustering methods, demixing procedures, and classification approaches. The introduction of artificial intelligence algorithms (machine learning) provides solutions to the problems that cannot be effectively classified. In Masson's studies [91] a convolutional neural network was applied for SERS spectra. Preprocessed training data set were fed into the untrained neural network. Each data point was labeled to its corresponding analytes. Weights were then optimized to minimize the error of the output of the neural network. Following the optimization of the model for those specific components, unknown data, acquired and preprocessed in the same way, were fed into the trained model. SERS spectra were all sorted and counted based on this classification process. More details can be referred to Chapter 11, *Multivariate approaches for surface-enhanced Raman scattering data analysis in clinical applications*.

References

[1] T.M. Cotton, J.-H. Kim, G.D. Chumanov, Application of surface-enhanced Raman spectroscopy to biological systems, J. Raman Spectrosc. 22 (1991) 729–742. Available from: https://doi.org/10.1002/jrs.1250221203.

[2] I. Nabiev, I. Chourpa, M. Manfait, Applications of Raman and surface-enhanced Raman scattering spectroscopy in medicine, J. Raman Spectrosc. 25 (1994) 13–23. Available from: https://doi.org/10.1002/jrs.1250250104.

[3] K. Kneipp, H. Kneipp, I. Itzkan, R.R. Dasari, M.S. Feld, Surface-enhanced Raman scattering and biophysics, J. Phys. Condens. Matter 14 (2002) R597–R624. Available from: https://doi.org/10.1088/0953-8984/14/18/202.

[4] P.C. Lee, D. Meisel, Adsorption and surface-enhanced Raman of dyes on silver and gold sols, J. Phys. Chem. 86 (1982) 3391–3395. Available from: https://doi.org/10.1021/j100214a025.

[5] G. Frens, Controlled nucleation for the regulation of the particle size in monodisperse gold suspensions, Nat. Phys. Sci. 241 (1973) 20–22. Available from: https://doi.org/10.1038/physci241020a0.

[6] O. Siiman, L.A. Bumm, R. Callaghan, C.G. Blatchford, M. Kerker, Surface-enhanced Raman scattering by citrate on colloidal silver, J. Phys. Chem. 87 (1983) 1014–1023. Available from: https://doi.org/10.1021/j100229a020.

[7] P.A. Mosier-Boss, K.C. Sorensen, R.D. George, P.C. Sims, A. O'braztsova, SERS substrates fabricated using ceramic filters for the detection of bacteria: eliminating the citrate interference, Spectrochim. Acta A: Mol. Biomol. Spectrosc. 180 (2017) 161–167. Available from: https://doi.org/10.1016/j.saa.2017.03.021.
[8] V. Kattumuri, M. Chandrasekhar, S. Guha, K. Raghuraman, K.V. Katti, K. Ghosh, et al., Agarose-stabilized gold nanoparticles for surface-enhanced Raman spectroscopic detection of DNA nucleosides, Appl. Phys. Lett. 88 (2016) 153114. Available from: https://doi.org/10.1063/1.2192573.
[9] L. Sun, Y. Sun, F. Xu, Y. Zhang, T. Yang, C. Guo, et al., Atomic force microscopy and surface-enhanced Raman scattering detection of DNA based on DNA-nanoparticle complexes, Nanotechnology 20 (2009). Available from: https://doi.org/10.1088/0957-4484/20/12/125502.
[10] D. Graham, K. Faulds, Quantitative SERRS for DNA sequence analysis, Chem. Soc. Rev. 37 (2008) 1042–1051. Available from: https://doi.org/10.1039/b707941a.
[11] Y. Liu, N. Lyu, V.K. Rajendran, J. Piper, A. Rodger, Y. Wang, Sensitive and direct DNA mutation detection by surface-enhanced Raman spectroscopy using rational designed and tunable plasmonic nanostructures, Anal. Chem. 92 (2020) 5708–5716. Available from: https://doi.org/10.1021/acs.analchem.9b04183.
[12] K. Faulds, R.E. Littleford, D. Graham, G. Dent, W.E. Smith, Comparison of surface-enhanced resonance Raman scattering from unaggregated and aggregated nanoparticles, Anal. Chem. 76 (2004) 592–598. Available from: https://doi.org/10.1021/ac035053o.
[13] Y.F. Huang, Y.F. Wang, X.P. Yan, Amine-functionalized magnetic nanoparticles for rapid capture and removal of bacterial pathogens, Environ. Sci. Technol. 44 (2010) 7908–7913. Available from: https://doi.org/10.1021/es102285n.
[14] J.F. Li, Y.F. Huang, Y. Ding, Z.L. Yang, S.B. Li, X.S. Zhou, et al., Shell-isolated nanoparticle-enhanced Raman spectroscopy, Nature 464 (2010) 392–395. Available from: https://doi.org/10.1038/nature08907.
[15] J.F. Li, Y.J. Zhang, S.Y. Ding, R. Panneerselvam, Z.Q. Tian, Core-shell nanoparticle-enhanced Raman spectroscopy, Chem. Rev. 117 (2017) 5002–5069. Available from: https://doi.org/10.1021/acs.chemrev.6b00596.
[16] Z.L. Wang, Z.Q. Tian, S.B. Li, J.R. Anema, Z.L. Yang, Y. Ding, et al., Surface analysis using shell-isolated nanoparticle-enhanced Raman spectroscopy, Nat. Protoc. 8 (2013) 52–65. Available from: https://doi.org/10.1038/nprot.2012.141.
[17] C. Zheng, L. Liang, S. Xu, H. Zhang, C. Hu, L. Bi, et al., The use of Au@SiO_2 shell-isolated nanoparticle-enhanced Raman spectroscopy for human breast cancer detection, Anal. Bioanal. Chem. 406 (2014) 5425–5432. Available from: https://doi.org/10.1007/s00216-014-7967-5.
[18] X.X. Han, H.Y. Jia, Y.F. Wang, Z.C. Lu, C.X. Wang, W.Q. Xu, et al., Analytical technique for label-free multi-protein detection based on Western blot and surface-enhanced Raman scattering, Anal. Chem. 80 (2008) 2799–2804. Available from: https://doi.org/10.1021/ac702390u.
[19] S. Efrima, B.V. Bronk, Silver colloids impregnating or coating bacteria, J. Phys. Chem. B 102 (1998) 5947–5950. Available from: https://doi.org/10.1021/jp9813903.
[20] S. Efrima, L. Zeiri, Understanding SERS of bacteria, J. Raman Spectrosc. 40 (2009) 277–288. Available from: https://doi.org/10.1002/jrs.2121.
[21] P.A. Mosier-Boss, Review on SERS of bacteria, Biosensors 7 (2017). Available from: https://doi.org/10.3390/bios7040051.
[22] J.W. Kang, P.T.C. So, R.R. Dasari, D.K. Lim, High resolution live cell Raman imaging using subcellular organelle-targeting SERS-sensitive gold nanoparticles with highly narrow intra-nanogap, Nano Lett. 15 (2015) 1766–1772. Available from: https://doi.org/10.1021/nl504444w.

[23] Y. Shen, L. Liang, S. Zhang, D. Huang, R. Deng, J. Zhang, et al., Organelle-targeting gold nanorods for macromolecular profiling of subcellular organelles and enhanced cancer cell killing, ACS Appl. Mater. Interfaces 10 (2018) 7910–7918. Available from: https://doi.org/10.1021/acsami.8b01320.

[24] A.G. Tkachenko, H. Xie, D. Coleman, W. Glomm, J. Ryan, M.F. Anderson, et al., Multifunctional gold nanoparticle-peptide complexes for nuclear targeting, J. Am. Chem. Soc. 125 (2003) 4700–4701. Available from: https://doi.org/10.1021/ja0296935.

[25] S. Zitzmann, V. Ehemann, M. Schwab, Arginine-glycine-aspartic acid (RGD)-peptide binds to both tumor and tumor-endothelial cells in vivo, Cancer Res. 62 (2002) 5139–5143. Available from: https://pubmed.ncbi.nlm.nih.gov/12234975/.

[26] A.S.D.S. Indrasekara, B.J. Paladini, D.J. Naczynski, V. Starovoytov, P.V. Moghe, L. Fabris, Dimeric gold nanoparticle assemblies as tags for SERS- based cancer detection, Adv. Healthcare Mater. 2 (2013) 1370–1376. Available from: https://doi.org/10.1002/adhm.201200370.

[27] C. Dingwall, R.A. Laskey, Nuclear targeting sequences - a consensus? Trends Biochem. Sci. 16 (1991) 478–481. Available from: https://doi.org/10.1016/0968-0004(91)90184-W.

[28] E Kawamura, Y Yamada, H Harashima, Mitochondrial targeting functional peptides as potential devices for the mitochondrial delivery of a DF-MITO-Porter, Mitochondrion. 13 (2013) 610–614. Available from: https://doi.org/10.1016/j.mito.2013/08/010.

[29] X. Wei, D.H. Zheng, Y. Cai, R. Jiang, M.L. Chen, T. Yang, et al., High-throughput/high-precision sampling of single cells into ICP-MS for elucidating cellular nanoparticles, Anal. Chem. 90 (2018) 14543–14550. Available from: https://doi.org/10.1021/acs.analchem.8b04471.

[30] L. Pan, J. Liu, J. Shi, Nuclear-targeting gold nanorods for extremely low NIR activated photothermal therapy, ACS Appl. Mater. Interfaces 9 (2017) 15952–15961. Available from: https://doi.org/10.1021/acsami.7b03017.

[31] Y. Matsumura, H. Maeda, A new concept for macromolecular therapeutics in cancer chemotherapy: mechanism of tumoritropic accumulation of proteins and the antitumor agent smancs, Cancer Res. 46 (1986) 6387–6392. Available from: https://pubmed.ncbi.nlm.nih.gov/2946403/.

[32] S.K. Hobbs, W.L. Monsky, F. Yuan, W.G. Roberts, L. Griffith, V.P. Torchilin, et al., Regulation of transport pathways in tumor vessels: role of tumor type and microenvironment, Proc. Natl Acad. Sci. U. S. A. 95 (1998) 4607–4612. Available from: https://doi.org/10.1073/pnas.95.8.4607.

[33] S. Sindhwani, A.M. Syed, J. Nagi, B.R. Kingston, L. Maiorino, J. Rothschild, et al., The entry of nanoparticles into solid tumours, Nat. Mater. 19 (2020) 566–570. Available from: https://doi.org/10.1038/s41563-019-0566-2.

[34] E.E. Connor, J. Mwamuka, A. Gole, C.J. Murphy, M.D. Wyatt, Gold nanoparticles are taken up by human cells but do not cause acute cytotoxicity, Small 1 (2005) 325–327. Available from: https://doi.org/10.1002/smll.200400093.

[35] S. Yu, S.B. Lee, M. Kang, C.R. Martin, Size-based protein separations in poly(ethylene glycol)-derivatized gold nanotubule membranes, Nano Lett. 1 (2001) 495–498. Available from: https://doi.org/10.1021/nl010044l.

[36] X. Qian, X.H. Peng, D.O. Ansari, Q. Yin-Goen, G.Z. Chen, D.M. Shin, et al., In vivo tumor targeting and spectroscopic detection with surface-enhanced Raman nanoparticle tags, Nat. Biotechnol. 26 (2007) 83–90. Available from: https://doi.org/10.1038/nbt1377.

[37] K. Knop, R. Hoogenboom, D. Fischer, U.S. Schubert, Poly(ethylene glycol) in drug delivery: Pros and cons as well as potential alternatives, Angew. Chem. Int. (Ed.) 49 (2010) 6288–6308. Available from: https://doi.org/10.1002/anie.200902672.

[38] C. Jehn, B. Küstner, P. Adam, A. Marx, P. Ströbel, C. Schmuck, S. Schlücker, Water soluble SERS labels comprising a SAM with dual spacers for controlled bioconjugation, Phys. Chem. Chem. Phys. 11 (2009) 7499–7504. Available from: https://doi.org/10.1039/b905092b.

[39] Y. Liu, Y.L. Balachandran, D. Li, Y. Shao, X. Jiang, Polyvinylpyrrolidone-poly(ethylene glycol) modified silver nanorods can be a safe, noncarrier adjuvant for HIV vaccine, ACS Nano 10 (2016) 3589–3596. Available from: https://doi.org/10.1021/acsnano.5b08025.

[40] H. Wetzel, H. Gerischer, Surface enhanced Raman scattering from pyridine and halide ions adsorbed on silver and gold sol particles, Chem. Phys. Lett. 76 (1980) 460–464. Available from: https://doi.org/10.1016/0009-2614(80)80647-6.

[41] S.E.J. Bell, N.M.S. Sirimuthu, Surface-enhanced Raman spectroscopy (SERS) for sub-micromolar detection of DNA/RNA mononucleotides, J. Am. Chem. Soc. 128 (2006) 15580–15581. Available from: https://doi.org/10.1021/ja066263w.

[42] X.X. Han, G.G. Huang, B. Zhao, Y. Ozaki, Label-free highly sensitive detection of proteins in aqueous solutions using surface-enhanced Raman scattering, Anal. Chem. 81 (2009) 3329–3333. Available from: https://doi.org/10.1021/ac900395x.

[43] S.Y. Fu, P.X. Zhang, X.Y. Li, Influerence of chloride ions on SERS in Ag sol, Acta Phys. Sin. 40 (1915).

[44] A. Otto, A. Bruckbauer, Y.X. Chen, On the chloride activation in SERS and single molecule SERS, J. Mol. Struct. 661–662 (2003) 501–514. Available from: https://doi.org/10.1016/j.molstruc.2003.07.026.

[45] S.H. Liu, P. Zhang, Y. Huan, X.L. Li, Q. Hao, The influence of KCl on SERS of fuchsin basic in silver colloids, Spectrosc. Spectr. Anal. 17 (1997) 27–31.

[46] I.R. Nabiev, K.V. SokolovK, O.N. Voloshin, Surface-enhanced Raman spectroscopy of biomolecules. Part III-determination of the local destabilization regions in the double helix, J. Raman Spectrosc. 21 (1990) 333–336. Available from: https://doi.org/10.1002/jrs.1250210603.

[47] H. Hwang, H. Chon, J. Choo, J.K. Park, Optoelectrofluidic sandwich immunoassays for detection of human tumor marker using surface-enhanced Raman scattering, Anal. Chem. 82 (2010) 7603–7610. Available from: https://doi.org/10.1021/ac101325t.

[48] H. Hwang, D. Han, Y.J. Oh, Y.K. Cho, K.H. Jeong, J.K. Park, In situ dynamic measurements of the enhanced SERS signal using an optoelectrofluidic SERS platform, Lab. Chip 11 (2011) 2518–2525.

[49] Z. Liu, Y. Wang, R. Deng, L. Yang, S. Yu, S. Xu, et al., Fe_3O_4@graphene oxide@Ag particles for surface magnet solid-phase extraction surface-enhanced Raman scattering (SMSPE-SERS): from sample pretreatment to detection all-in-one, ACS Appl. Mater. Interfaces 8 (2016) 14160–14168. Available from: https://doi.org/10.1021/acsami.6b02944.

[50] D. Sun, F. Cao, Y. Tian, A. Li, W. Xu, Q. Chen, et al., Label-free detection of multiplexed metabolites at single-cell level via a SERS-microfluidic droplet platform, Anal. Chem. 91 (2019) 15484–15490. Available from: https://doi.org/10.1021/acs.analchem.9b03294.

[51] D. Sun, F. Cao, W. Xu, Q. Chen, W. Shi, S. Xu, Ultrasensitive and simultaneous detection of two cytokines secreted by single cell in microfluidic droplets via magnetic-field amplified SERS, Anal. Chem. 91 (2019) 2551–2558. Available from: https://doi.org/10.1021/acs.analchem.8b05892.

[52] P.A. Mosier-Boss, K.C. Sorensen, R.D. George, A. Obraztsova, SERS substrates fabricated using ceramic filters for the detection of bacteria, Spectrochim. Acta A: Mol. Biomol. Spectrosc. 153 (2016) 591–598. Available from: https://doi.org/10.1016/j.saa.2015.09.012.

[53] E. Witkowska, T. Szymborski, A. Kamińska, J. Waluk, Polymer mat prepared via Forcespinning™ as a SERS platform for immobilization and detection of bacteria from blood plasma, Mater. Sci. Eng. C 71 (2017) 345–350. Available from: https://doi.org/10.1016/j.msec.2016.10.027.
[54] S.H. Yazdi, I.M. White, Optofluidic surface enhanced Raman spectroscopy microsystem for sensitive and repeatable on-site detection of chemical contaminants, Anal. Chem. 84 (2012) 7992–7998. Available from: https://doi.org/10.1021/ac301747b.
[55] S.H. Yazdi, K.L. Giles, I.M. White, Multiplexed detection of DNA sequences using a competitive displacement assay in a microfluidic SERRS-based device, Anal. Chem. 85 (2013) 10605–10611. Available from: https://doi.org/10.1021/ac402744z.
[56] X.B. Tan, Z.Y. Wang, J. Yang, C.Y. Song, R.H. Zhang, Y.P. Cui, Polyvinylpyrrolidone- (PVP-) coated silver aggregates for high performance surface-enhanced Raman scattering in living cells, Nanotechnology 20 (2009) 445102. Available from: https://doi.org/10.1088/0957-4484/20/44/445102.
[57] Z. Zhang, B. Zhao, L. Hu, PVP protective mechanism of ultrafine silver powder synthesized by chemical reduction processes, J. Solid. State Chem. 121 (1996) 105–110. Available from: https://doi.org/10.1006/jssc.1996.0015.
[58] W.E. Doering, S. Nie, Spectroscopic tags using dye-embedded nanoparticles and surface-enhanced Raman scattering, Anal. Chem. 75 (2003) 6171–6176. Available from: https://doi.org/10.1021/ac034672u.
[59] X. Liu, M. Knauer, N.P. Ivleva, R. Niessner, C. Haisch, Synthesis of core-shell surface-enhanced Raman tags for bioimaging, Anal. Chem. 82 (2010) 441–446. Available from: https://doi.org/10.1021/ac902573p.
[60] J. Jeon, S.H. Lee, Y. Joung, K. Kim, N. Choi, J. Choo, Improvement of reproducibility and thermal stability of surface-enhanced Raman scattering-based lateral flow assay strips using silica-encapsulated gold nanoparticles, Sens. Actuators B: Chem. 321 (2020) 128521. Available from: https://doi.org/10.1016/j.snb.2020.128521.
[61] C.R. Yonzon, D.A. Stuart, X. Zhang, A.D. McFarland, C.L. Haynes, R.P. Van Duyne, Towards advanced chemical and biological nanosensors - an overview, Talanta 67 (2005) 438–448. Available from: https://doi.org/10.1016/j.talanta.2005.06.039.
[62] P. Wang, S. Pang, B. Pearson, Y. Chujo, L. McLandsborough, M. Fan, et al., Rapid concentration detection and differentiation of bacteria in skimmed milk using surface enhanced Raman scattering mapping on 4-mercaptophenylboronic acid functionalized silver dendrites, Anal. Bioanal. Chem. 409 (2017) 2229–2238. Available from: https://doi.org/10.1007/s00216-016-0167-8.
[63] X. Guo, J. Li, M. Arabi, X. Wang, Y. Wang, L. Chen, Molecular-imprinting-based surface-enhanced Raman scattering sensors, ACS Sens. 5 (2020) 601–619. Available from: https://doi.org/10.1021/acssensors.9b02039.
[64] J. Ye, Y. Chen, Z. Liu, A boronate affinity sandwich assay: an appealing alternative to immunoassays for the determination of glycoproteins, Angew. Chem. Int. (Ed.) 53 (2014) 10386–10389. Available from: https://doi.org/10.1002/anie.201405525.
[65] P. Lambelet, A. Sayah, M. Pfeffer, C. Philipona, F. Marquis-Weible, Chemically etched fiber tips for near-field optical microscopy: a process for smoother tips, Appl. Opt. 37 (1998) 7289–7292. Available from: https://doi.org/10.1364/AO.37.007289.
[66] J.P. Scaffidi, M.K. Gregas, V. Seewaldt, T. Vo-Dinh, SERS-based plasmonic nanobiosensing in single living cells, Anal. Bioanal. Chem. 393 (2009) 1135–1141. Available from: https://doi.org/10.1007/s00216-008-2521-y.
[67] C. Viets, W. Hill, Comparison of fibre-optic SERS sensors with differently prepared tips, Sens. Actuators, B: Chem. 51 (1998) 92–99. Available from: https://doi.org/10.1016/S0925-4005(98)00170-1.

[68] D. Zeisel, V. Deckert, R. Zenobi, T. Vo-Dinh, Near-field surface-enhanced Raman spectroscopy of dye molecules adsorbed on silver island films, Chem. Phys. Lett. 283 (1998) 381–385. Available from: https://doi.org/10.1016/S0009-2614(97)01391-2.
[69] E. Polwart, R.L. Keir, C.M. Davidson, W.E. Smith, D.A. Sadler, Novel SERS-active optical fibers prepared by the immobilization of silver colloidal particles, Appl. Spectrosc. 54 (2000) 522–527. Available from: https://doi.org/10.1366/0003702001949690.
[70] S.J. Jia, S.P. Xu, X.L. Zheng, B. Zhao, W.Q. Xu, Preparation of SERS optical fiber sensor via laser-induced deposition of Ag film on the surface of fiber tip, Chem. J. Chin. Universities 27 (2006) 523–526.
[71] S. Wang, C. Liu, H. Wang, G. Chen, M. Cong, W. Song, et al., A surface-enhanced Raman scattering optrode prepared by in situ photoinduced reactions and its application for highly sensitive on-chip detection, ACS Appl. Mater. Interfaces 6 (2014) 11706–11713. Available from: https://doi.org/10.1021/am503881h.
[72] Y.B. Tan, J.M. Zou, N. Gu, Preparation of stabilizer-free silver nanoparticle-coated micropipettes as surface-enhanced Raman scattering substrate for single cell detection, Nanoscale Res. Lett. 10 (2015). Available from: https://doi.org/10.1186/s11671-015-1122-x.
[73] J. Cao, D. Zhao, Q. Mao, A highly reproducible and sensitive fiber SERS probe fabricated by direct synthesis of closely packed AgNPs on the silanized fiber taper, Analyst 142 (2017) 596–602. Available from: https://doi.org/10.1039/c6an02414a.
[74] T. Vo-Dinh, J. Scaffidi, M. Gregas, Y. Zhang, V. Seewaldt, Applications of fiber-optics-based nanosensors to drug discovery, Expert. Opin. Drug. Discovery 4 (2009) 889–900. Available from: https://doi.org/10.1517/17460440903085112.
[75] J.F. Masson, J. Breault-Turcot, R. Faid, H.P. Poirier-Richard, H. Yockell-Lelièvre, F. Lussier, et al., Plasmonic nanopipette biosensor, Anal. Chem. 86 (2014) 8998–9005. Available from: https://doi.org/10.1021/ac501473c.
[76] J. Liu, D. Yin, S. Wang, H.Y. Chen, Z. Liu, Probing low-copy-number proteins in a single living cell, Angew. Chem. Int. (Ed.) 55 (2016) 13215–13218. Available from: https://doi.org/10.1002/anie.201608237.
[77] J.F. Masson, F. Lussier, C. Ducrot, M.J. Bourque, J.P. Spatz, W. Cui, et al., Block copolymer brush layer-templated gold nanoparticles on nanofibers for surface-enhanced Raman scattering optophysiology, ACS Appl. Mater. Interfaces 11 (2019) 4373–4384. Available from: https://doi.org/10.1021/acsami.8b19161.
[78] Z. Hu, M. Jean-François, B.C. Geraldine, Templating gold nanoparticles on nanofibers coated with a block copolymer brush for nanosensor applications, ACS Appl. Nano Mater. 3 (2019) 516–529. Available from: https://doi.org/10.1021/acsanm.9b02081.
[79] J.J. Niu, M.G. Schrlau, G. Friedman, Y. Gogotsi, Carbon nanotube-tipped endoscope for in situ intracellular surface-enhanced Raman spectroscopy, Small 7 (2011) 540–545. Available from: https://doi.org/10.1002/smll.201001757.
[80] V. Deckert, D. Zeisel, R. Zenobi, T. Vo-Dinh, Near-field surface-enhanced Raman imaging of dye-labeled DNA with 100-nm resolution, Anal. Chem. 70 (1998) 2646–2650. Available from: https://doi.org/10.1021/ac971304f.
[81] T. Vo-Dinh, Nanosensing at the single cell level, Spectrochim. Acta B At. Spectrosc. 63 (2008) 95–103. Available from: https://doi.org/10.1016/j.sab.2007.11.027.
[82] E.A. Vitol, Z. Orynbayeva, M.J. Bouchard, J. Azizkhan-Clifford, G. Friedman, Y. Gogotsi, In situ intracellular spectroscopy with surface enhanced Raman spectroscopy (SERS)-enabled nanopipettes, ACS Nano 3 (2009) 3529–3536. Available from: https://doi.org/10.1021/nn9010768.
[83] T. Vo-Dinh, Y. Zhang, Single-cell monitoring using fiberoptic nanosensors, Nanomed. Nanobiotechnol. 3 (2011) 79–85. Available from: https://doi.org/10.1002/wnan.112.

[84] R. Gessner, P. Rosch, R. Petry, M. Schmitt, M.A. Strehle, W. Kiefer, et al., The application of a SERS fiber probe for the investigation of sensitive biological samples, Analyst 129 (2004) 1193–1199.

[85] H.L. Liu, Q.C. Jiang, J. Pang, Z.Y. Jiang, J. Cao, L.N. Ji, et al., A multiparameter pH-sensitive nanodevice based on plasmonic nanopores, Adv. Funct. Mater. 28 (2018).

[86] J.Q. Wang, Y.J. Geng, Y.T. Shen, W. Shi, W.Q. Xu, S.P. Xu, SERS-active fiber tip for intracellular and extracellular pH sensing in living single cells, Sens. Actuators B: Chem. 290 (2019) 527–534. Available from: https://doi.org/10.1016/j.snb.2019.03.149.

[87] K. Wang, H. Liu, M. Chen, P. Muhammad, Y. Zhou, J. Cao, et al., Organic cyanide decorated SERS active nanopipettes for quantitative detection of hemeproteins and Fe3 + in single cells, Anal. Chem. 89 (2017) 2522–2530. Available from: https://doi.org/10.1021/acs.analchem.6b04689.

[88] S. Hanif, H.L. Liu, S.A. Ahmed, J.M. Yang, Y. Zhou, J. Pang, et al., Nanopipette-based SERS aptasensor for subcellular localization of cancer biomarker in single cells, Anal. Chem. 89 (2017) 9911–9917. Available from: https://doi.org/10.1021/acs.analchem.7b02147.

[89] T.D. Nguyen, M.S. Song, N.H. Ly, S.Y. Lee, S.W. Joo, Nanostars on nanopipette tips: a Raman probe for quantifying oxygen levels in hypoxic single cells and tumours, Angew. Chem. Int. (Ed.) 58 (2019) 2710–2714. Available from: https://doi.org/10.1002/anie.201812677.

[90] F. Lussier, T. Brulé, M. Vishwakarma, T. Das, J.P. Spatz, J.F. Masson, Dynamic-SERS optophysiology: a nanosensor for monitoring cell secretion events, Nano Lett. 16 (2016) 3866–3871. Available from: https://doi.org/10.1021/acs.nanolett.6b01371.

[91] F. Lussier, D. Missirlis, J.P. Spatz, J.F. Masson, Machine-learning-driven surface-enhanced Raman scattering optophysiology reveals multiplexed metabolite gradients near cells, ACS Nano 13 (2019) 1403–1411. Available from: https://doi.org/10.1021/acsnano.8b07024.

[92] W. Hong, F. Liang, D. Schaak, M. Loncar, Q. Quan, Nanoscale label-free bioprobes to detect intracellular proteins in single living cells, Sci. Rep. 4 (2014). Available from: https://doi.org/10.1038/srep06179.

[93] F. Liang, Y. Zhang, W. Hong, Y. Dong, Z. Xie, Q. Quan, Direct tracking of amyloid and Tu dynamics in neuroblastoma cells using nanoplasmonic fiber tip probes, Nano Lett. 16 (2016) 3989–3994. Available from: https://doi.org/10.1021/acs.nanolett.6b00320.

[94] R.M. Wightman, Probing cellular chemistry in biological systems with microelectrodes, Science 311 (2006) 1570–1574. Available from: https://doi.org/10.1126/science.1120027.

[95] A.I. Henry, B. Sharma, M.F. Cardinal, D. Kurouski, R.P. Van Duyne, Surface-enhanced Raman spectroscopy biosensing: in vivo diagnostics and multimodal imaging, Anal. Chem. 88 (2016) 6638–6647. Available from: https://doi.org/10.1021/acs.analchem.6b01597.

[96] H. Zareie, T. Tekinay, A.B. Tekinay, M.O. Guler, A. Dana, D. Kocaay, et al., Label-free nanometer-resolution imaging of biological architectures through surface enhanced Raman scattering, Sci. Rep. 3 (2013). Available from: https://doi.org/10.1038/srep02624.

[97] C.T. Ertsgaard, R.M. McKoskey, I.S. Rich, N.C. Lindquist, Dynamic placement of plasmonic hotspots for super-resolution surface-enhanced Raman scattering, ACS Nano 8 (2014) 10941–10946. Available from: https://doi.org/10.1021/nn504776b.

[98] A.P. Olson, K.B. Spies, A.C. Browning, P.A.G. Soneral, N.C. Lindquist, Chemically imaging bacteria with super-resolution SERS on ultra-thin silver substrates, Sci. Rep. 7 (2017). Available from: https://doi.org/10.1038/s41598-017-08915-w.

[99] H.K. Lee, Y.H. Lee, I.Y. Phang, J. Wei, Y.E. Miao, T. Liu, et al., Plasmonic liquid marbles: a miniature substrate-less SERS platform for quantitative and multiplex ultratrace molecular detection, Angew. Chem. Int. (Ed.) 53 (2014) 5054–5058. Available from: https://doi.org/10.1002/anie.201401026.

[100] P.-Q.G. Chuong, L.H. Kwee, P.I. Yee, L.X. Yi, Plasmonic colloidosomes as three-dimensional SERS platforms with enhanced surface area for multiphase sub-microliter toxin sensing, Angew. Chem. 127 (2015) 9827–9831. Available from: https://doi.org/10.1002/ange.201504027.

[101] G.C. Phan-Quang, H.K. Lee, X.Y. Ling, Isolating reactions at the picoliter scale: parallel control of reaction kinetics at the liquid–liquid interface, Angew. Chem. Int. (Ed.) 55 (2016) 8304–8308. Available from: https://doi.org/10.1002/anie.201602565.

[102] X. Wang, J. Ji, T. Liu, Y. Liu, L. Qiao, B. Liu, Plasmonic colloidosome-based single cell detector: a strategy for individual cell secretion sensing, Anal. Chem. 91 (2019) 2260–2265. Available from: https://doi.org/10.1021/acs.analchem.8b04850.

[103] J. Ye, M. Zapata, M. Xiong, Z. Liu, S. Wang, H. Xu, et al., Nanooptics of plasmonic nanomatryoshkas: shrinking the size of a core-shell junction to subnanometer, Nano Lett. 15 (2015) 6419–6428. Available from: https://doi.org/10.1021/acs.nanolett.5b02931.

[104] Y. Gu, Y. Zhang, Y. Li, X. Jin, C. Huang, S.A. Maier, et al., Raman photostability of off-resonant gap-enhanced Raman tags, RSC Adv. 8 (2018) 14434–14444. Available from: https://doi.org/10.1039/c8ra02260g.

[105] Y. Zhang, Y. Qiu, L. Lin, H. Gu, Z. Xiao, J. Ye, Ultraphotostable mesoporous silica-coated gap-enhanced Raman tags (GERTs) for high-speed bioimaging, ACS Appl. Mater. Interfaces 9 (2017) 3995–4005. Available from: https://doi.org/10.1021/acsami.6b15170.

[106] W.E. Smith, Practical understanding and use of surface enhanced Raman scattering/surface enhanced resonance Raman scattering in chemical and biological analysis, Chem. Soc. Rev. 37 (2008) 955–964. Available from: https://doi.org/10.1039/b708841h.

[107] A. Urumese, R.N. Jenjeti, S. Sampath, B.R. Jagirdar, Colloidal europium nanoparticles via a solvated metal atom dispersion approach and their surface enhanced Raman scattering studies, J. Colloid Interface Sci. 476 (2016) 177–183. Available from: https://doi.org/10.1016/j.jcis.2016.05.015.

[108] M. Casella, A. Lucotti, M. Tommasini, M. Bedoni, E. Forvi, F. Gramatica, et al., Raman and SERS recognition of β-carotene and haemoglobin fingerprints in human whole blood, Spectrochim. Acta A: Mol. Biomol. Spectrosc. 79 (2011) 915–919. Available from: https://doi.org/10.1016/j.saa.2011.03.048.

[109] P. Hildebrandt, M. Stockburger, Cytochrome c at charged interfaces. 1. Conformational and redox equilibria at the electrode/electrolyte interface probed by surface-enhanced resonance Raman spectroscopy, Biochemistry 28 (1989) 6710–6721. Available from: https://doi.org/10.1021/bi00442a026.

[110] A. Królikowska, J. Bukowska, Surface-enhanced resonance Raman spectroscopic characterization ofcytochrome c immobilized on 2-mercaptoethanesulfonate monolayers on silver, J. Raman Spectrosc. 41 (2010) 1621–1631. Available from: https://doi.org/10.1002/jrs.2618.

[111] A.R. Bizzarri, S. Cannistraro, Surface-enhanced resonance Raman spectroscopy signals from single myoglobin molecules, Appl. Spectrosc. 56 (2002) 1531–1537. Available from: https://doi.org/10.1366/000370202321115977.

[112] M. Gühlke, Z. Heiner, J. Kneipp, Surface-enhanced Raman and surface-enhanced hyper-Raman scattering of thiol-functionalized carotene, J. Phys. Chem. C 120 (2016) 20702–20709. Available from: https://doi.org/10.1021/acs.jpcc.6b01895.

[113] C. Zaffino, B. Russo, S. Bruni, Surface-enhanced Raman scattering (SERS) study of anthocyanidins, Spectrochim. Acta A: Mol. Biomol. Spectrosc. 149 (2015) 41–47. Available from: https://doi.org/10.1016/j.saa.2015.04.039.
[114] K. Kajimoto, T. Kikukawa, H. Nakashima, H. Yamaryo, Y. Saito, T. Fujisawa, et al., Transient resonance Raman spectroscopy of a light-driven sodium-ion-pump rhodopsin from indibacter alkaliphilus, J. Phys. Chem. B 121 (2017) 4431–4437. Available from: https://doi.org/10.1021/acs.jpcb.7b02421.
[115] A.V. Vlasov, N.L. Maliar, S.V. Bazhenov, R. Scattering, From structural biology to medical applications, Crystals 10 (2020) 38.
[116] T.E. Rohr, T. Cotton, N. Fan, P.J. Tarcha, Immunoassay employing surface-enhanced Raman spectroscopy, Anal. Biochem. 182 (1989) 388–398. Available from: https://doi.org/10.1016/0003-2697(89)90613-1.
[117] D. Sun, F. Cao, L. Cong, W. Xu, Q. Chen, W. Shi, et al., Cellular heterogeneity identified by single-cell alkaline phosphatase (ALP) via a SERRS-microfluidic droplet platform, Lab. Chip 19 (2019) 335–342. Available from: https://doi.org/10.1039/C8LC01006D.
[118] L. Au, D. Zheng, F. Zhou, Z.Y. Li, X. Li, Y. Xia, A quantitative study on the photothermal effect of immuno gold nanocages targeted to breast cancer cells, ACS Nano 2 (2008) 1645–1652. Available from: https://doi.org/10.1021/nn800370j.
[119] X.M. Lin, Y. Cui, Y.H. Xu, B. Ren, Z.Q. Tian, Surface-enhanced Raman spectroscopy: substrate-related issues, Anal. Bioanal. Chem. 394 (2009) 1729–1745. Available from: https://doi.org/10.1007/s00216-009-2761-5.
[120] S. Keren, C. Zavaleta, Z. Cheng, A. De La Zerda, O. Gheysens, S.S. Gambhir, Noninvasive molecular imaging of small living subjects using Raman spectroscopy, Proc. Natl Acad. Sci. U. S. A. 105 (2008) 5844–5849. Available from: https://doi.org/10.1073/pnas.0710575105.
[121] A.M. Mohs, M.C. Mancini, S. Singhal, J.M. Provenzale, B. Leyland-Jones, M.D. Wang, et al., Hand-held spectroscopic device for in vivo and intraoperative tumor detection: contrast enhancement, detection sensitivity, and tissue penetration, Anal. Chem. 82 (2010) 9058–9065. Available from: https://doi.org/10.1021/ac102058k.
[122] E. Hemmer, N. Venkatachalam, H. Hyodo, A. Hattori, Y. Ebina, H. Kishimoto, et al., Upconverting and NIR emitting rare earth based nanostructures for NIR-bioimaging, Nanoscale 5 (2013) 11339–11361. Available from: https://doi.org/10.1039/c3nr02286b.
[123] E. Hemmer, A. Benayas, F. Légaré, F. Vetrone, Exploiting the biological windows: current perspectives on fluorescent bioprobes emitting above 1000nm, Nanoscale Horiz. 1 (2016) 168–184. Available from: https://doi.org/10.1039/c5nh00073d.
[124] R.A. Odion, P. Strobbia, B.M. Crawford, T. Vo-Dinh, Inverse surface-enhanced spatially offset Raman spectroscopy (SESORS) through a monkey skull, J. Raman Spectrosc. 49 (2018) 1452–1460. Available from: https://doi.org/10.1002/jrs.5402.
[125] F. Nicolson, L.E. Jamieson, S. Mabbott, N.C. Shand, D. Graham, K. Faulds, Through barrier detection of ethanol using handheld Raman spectroscopy—conventional Raman versus spatially offset Raman spectroscopy (SORS), J. Raman Spectrosc. 48 (2017) 1828–1838. Available from: https://doi.org/10.1002/jrs.5258.
[126] S.M. Asiala, N.C. Shand, K. Faulds, D. Graham, Surface-enhanced, spatially offset Raman spectroscopy (SESORS) in tissue analogues, ACS Appl. Mater. Interfaces 9 (2017) 25488–25494. Available from: https://doi.org/10.1021/acsami.7b09197.
[127] F. Nicolson, L.E. Jamieson, S. Mabbott, K. Plakas, N.C. Shand, M.R. Detty, et al., Through tissue imaging of a live breast cancer tumour model using handheld surface enhanced spatially offset resonance Raman spectroscopy (SESORRS), Chem. Sci. 9 (2018) 3788–3792. Available from: https://doi.org/10.1039/c8sc00994e.

[128] Y.W. Wang, S. Kang, A. Khan, P.Q. Bao, J.T.C. Liu, In vivo multiplexed molecular imaging of esophageal cancer via spectral endoscopy of topically applied SERS nanoparticles, Biomed. Opt. Express 6 (2015) 3714–3723. Available from: https://doi.org/10.1364/BOE.6.003714.

[129] E. Garai, S. Sensarn, C.L. Zavaleta, N.O. Loewke, S. Rogalla, M.J. Mandella, A real-time clinical endoscopic system for intraluminal, multiplexed imaging of surface enhanced Raman scattering nanoparticles, PLOS One 10 (2015).

[130] L.A. Lane, X. Qian, S. Nie, SERS nanoparticles in medicine: from label-free detection to spectroscopic tagging, Chem. Rev. 115 (2015) 10489–10529. Available from: https://doi.org/10.1021/acs.chemrev.5b00265.

[131] D. Cialla-May, X.-S. Zhang, K. Weber, J. Popp, Recent progress in surface-enhanced Raman spectroscopy for biological and biomedical applications: from cells to clinics, Chem. Soc. Rev. 46 (2017) 3945–3961. Available from: https://doi.org/10.1039/C7CS00172J.

[132] S. Laing, L.E. Jamieson, K. Faulds, D. Graham, Surface-enhanced Raman spectroscopy for in vivo biosensing, Nat. Rev. Chem. 1 (2017). Available from: https://doi.org/10.1038/s41570-017-0060.

[133] G. Kuku, M. Altunbek, M. Culha, Surface-enhanced Raman scattering for label-free living single cell analysis, Anal. Chem. 89 (2017) 11160–11166. Available from: https://doi.org/10.1021/acs.analchem.7b03211.

[134] A.B. Veloso, J.P.F. Longo, L.A. Muehlmann, B.F. Tollstadius, P.E.N. Souza, R.B. Azevedo, et al., SERS Investigation of Cancer Cells Treated with PDT: Quantification of Cell Survival and Follow-up, Sci. Rep. 7 (2017). Available from: https://doi.org/10.1038/s41598-017-07469-1.

[135] A. Barhoumi, D. Zhang, F. Tam, N.J. Halas, Surface-enhanced Raman spectroscopy of DNA, J. Am. Chem. Soc. 130 (2008) 5523–5529. Available from: https://doi.org/10.1021/ja800023j.

[136] M. Moskovits, Surface selection rules, J. Chem. Phys. 77 (1982) 4408–4416. Available from: https://doi.org/10.1063/1.444442.

[137] H. Ma, X. Tang, Y. Liu, X.X. Han, C. He, H. Lu, et al., Surface-enhanced Raman scattering for direct protein function investigation: controlled immobilization and orientation, Anal. Chem. 91 (2019) 8767–8771. Available from: https://doi.org/10.1021/acs.analchem.9b01956.

[138] X Li, X Duan, L Li, S Ye, B Tang, An accurate and ultrasensitive SERS sensor with Au-Se interface for bioimaging and in situ quantitation, Chem. Comm. 56 (2020) 9320–9323. Available from: https://doi.org/10.1039/D0CC02068K.

[139] X. Wang, S.C. Huang, S. Hu, S. Yan, B. Ren, Fundamental understanding and applications of plasmon-enhanced Raman spectroscopy, Nat. Rev. Phys. 2 (2020) 253–271. Available from: https://doi.org/10.1038/s42254-020-0171-y.

[140] M.D. Li, Y. Cui, M.X. Gao, J. Luo, B. Ren, Z.Q. Tian, Clean substrates prepared by chemical adsorption of iodide followed by electrochemical oxidation for surface-enhanced Raman spectroscopic study of cell membrane, Anal. Chem. 80 (2008) 5118–5125. Available from: https://doi.org/10.1021/ac8003083.

[141] N.E. Marotta, L.A. Bottomley, Surface-enhanced Raman scattering of bacterial cell culture growth media, Appl. Spectrosc. 64 (2010) 601–606. Available from: https://doi.org/10.1366/000370210791414326.

[142] Y. Shen, L. Yang, L. Liang, Z. Li, J. Zhang, W. Shi, et al., Ex situ and in situ surface-enhanced Raman spectroscopy for macromolecular profiles of cell nucleus, Anal. Bioanal. Chem. 411 (2019) 6021–6029. Available from: https://doi.org/10.1007/s00216-019-01981-1.

[143] W.R. Premasiri, Y. Gebregziabher, L.D. Ziegler, On the difference between surface-enhanced Raman scattering (SERS) spectra of cell growth media and whole

bacterial cells, Appl. Spectrosc. 65 (2011) 493–499. Available from: https://doi.org/10.1366/10-06173.
[144] P.A. Mosier-Boss, K.C. Sorensen, R.D. George, A. Obraztsova, SERS substrates fabricated using ceramic filters for the detection of bacteria, Spectrochim. Acta A Mol. Biomol. Spectrosc. 15 (2016) 591–598. Available from: https://doi.org/10.1016/j.saa.2015.09.012.
[145] R. Keir, E. Igata, M. Arundell, W.E. Smith, D. Graham, C. McHugh, et al., In situ substrate formation and improved detection using microfluidicsSERRSAnal. Chem. 74 (2002) 1503–1508. Available from: https://doi.org/10.1021/ac015625 + .
[146] I.J. Jahn, O. Žukovskaja, X.S. Zheng, K. Weber, T.W. Bocklitz, D. Cialla-May, et al., Surface-enhanced Raman spectroscopy and microfluidic platforms: challenges, solutions and potential applications, Analyst 142 (2017) 1022–1047. Available from: https://doi.org/10.1039/c7an00118e.
[147] K.R. Strehle, D. Cialla, P. Rösch, T. Henkel, M. Köhler, J. Popp, A reproducible surface-enhanced Raman spectroscopy approach. Online SERS measurements in a segmented microfluidic system, Anal. Chem. 79 (2007) 1542–1547. Available from: https://doi.org/10.1021/ac0615246.
[148] A. März, T. Henkel, D. Cialla, M. Schmitt, J. Popp, Droplet formation via flow-through microdevices in Raman and surface enhanced Raman spectroscopy - concepts and applications, Lab. Chip 11 (2011) 3584–3592. Available from: https://doi.org/10.1039/c1lc20638a.
[149] A. Lorén, J. Engelbrektsson, C. Eliasson, M. Josefson, J. Abrahamsson, M. Johansson, et al., Internal standard in surface-enhanced Raman spectroscopy, Anal. Chem. 76 (2004) 7391–7395. Available from: https://doi.org/10.1021/ac0491298.
[150] S.E.J. Bell, J.N. Mackle, N.M.S. Sirimuthu, Quantitative surface-enhanced Raman spectroscopy of dipicolinic acid - towards rapid anthrax endospore detection, Analyst 130 (2005) 545–549. Available from: https://doi.org/10.1039/b415290e.
[151] D.P. Cowcher, Y. Xu, R. Goodacre, Portable, quantitative detection of bacillus bacterial spores using surface-enhanced Raman scattering, Anal. Chem. 85 (2013) 3297–3302. Available from: https://doi.org/10.1021/ac303657k.
[152] B. Ren, X. Lin, C. Jiang, C. Li, H. Lin, J. Huang, et al., Reliable quantitative SERS analysis facilitated by core-shell nanoparticles with embedded internal standards, Angew. Chem. Int. (Ed.) 54 (2015) 7308–7312. Available from: https://doi.org/10.1002/anie.201502171.
[153] D. Zhang, Y. Xie, S.K. Deb, V.J. Davison, D. Ben-Amotz, Isotope edited internal standard method for quantitative surface-enhanced Raman spectroscopy, Anal. Chem. 77 (2005) 3563–3569. Available from: https://doi.org/10.1021/ac050338h.
[154] S.P. Xu, X.H. Ji, W.Q. Xu, X.L. Li, L.Y. Wang, B. Zhao, et al., Immunoassay using probe-labeling immunogold nanoparticles with silver staining enhancement via surface-enhanced Raman scattering, Analyst 129 (2004).
[155] H. Yamakoshi, K. Dodo, M. Okada, J. Okada, A. Palonpon, K. Fujita, et al., Imaging of EdU, an alkyne-tagged cell proliferation probe, by Raman microscopy, Angew. Chem. Int. (Ed.) 133 (2011) 6102–6105. Available from: https://doi.org/10.1021/ja108404p.
[156] J. Wu, D. Liang, Q. Jin, J. Liu, M. Zheng, X. Duan, et al., Bioorthogonal SERS nanoprobes for mulitplex spectroscopic detection, tumor cell targeting, and tissue imaging, Chem. Eur. J. 21 (2015) 12914–12918. Available from: https://doi.org/10.1002/chem.201501942.
[157] J. Ando, M. Asanuma, K. Dodo, H. Yamakoshi, S. Kawata, K. Fujita, et al., Alkyne-tag SERS screening and identification of small-molecule-binding sites in protein, J. Am. Chem. Soc. 138 (2016) 13901–13910. Available from: https://doi.org/10.1021/jacs.6b06003.

[158] C. Krafft, M. Schmitt, I.W. Schie, D. Cialla-May, C. Matthäus, T. Bocklitz, et al., Label-free molecular imaging of biological cells and tissues by linear and nonlinear Raman spectroscopic approaches, Angew. Chem. Int. (Ed.) 56 (2017) 4392–4430. Available from: https://doi.org/10.1002/anie.201607604.
[159] R. Esteban, B. Ren, J.H. Zhong, S. Hu, B.J. Liu, J.Y. Liu, et al., Plasmonic photoluminescence for recovering native chemical information from surface-enhanced Raman scattering, Nat. Commun. 8 (2017). Available from: https://doi.org/10.1038/ncomms14891.
[160] A.N. Kuzmin, A. Pliss, P.N. Prasad, Ramanomics: new omics disciplines using micro Raman spectrometry with biomolecular component analysis for molecular profiling of biological structures, Biosensors 7 (2017) 15.
[161] Y. Zhang, X. Mi, X. Tan, R. Xiang, Recent progress on liquid biopsy analysis using surface-enhanced Raman spectroscopy, Theranostics 9 (2019) 491–525. Available from: https://doi.org/10.7150/thno.29875.

CHAPTER 4

Label-free surface-enhanced Raman scattering for clinical applications

Alois Bonifacio
Department of Engineering and Architecture (DIA), University of Trieste, Trieste, Italy

4.1 General aspects

4.1.1 Defining *label-free* surface-enhanced Raman scattering

Should one try to systematize the vast literature on analytical surface-enhanced Raman scattering (SERS) applications, the simplest and more general criterion would not be the metal used as a substrate nor the wavelength used for excitation. A broad and heterogeneous assortment of substrates is used by different research groups [1−3], and quite a number of different excitation wavelengths are commonly employed, spanning the whole visible and near-infrared (NIR) region of the spectrum. The simplest criterion to organize SERS applications would perhaps address the *detection strategy* adopted: *direct* versus *indirect* detection [4] as illustrated in Fig. 4.1. The detection and quantification of one or more analytes with

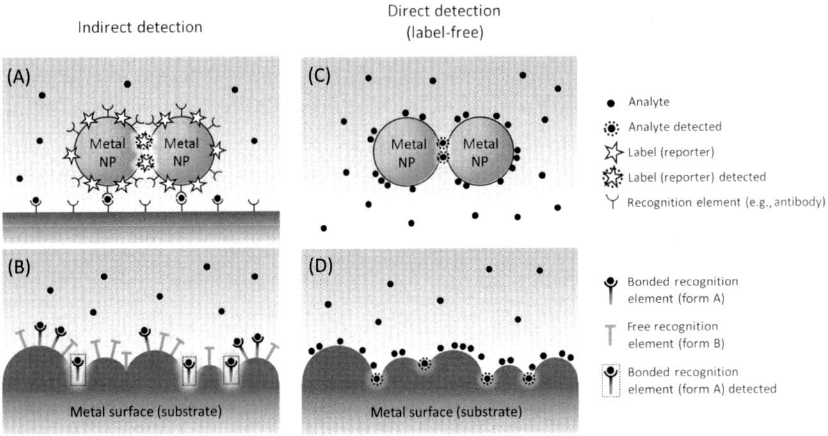

Figure 4.1 Schematic examples of direct (i.e., LF, A,B) versus indirect (C,D) SERS detection on colloidal (A,C) and solid (B,D) substrates. *LF*, Label-free; *SERS*, surface-enhanced Raman scattering.

Principles and Clinical Diagnostic Applications of Surface-Enhanced Raman Spectroscopy
DOI: https://doi.org/10.1016/B978-0-12-821121-2.00001-9

SERS can be done either by directly looking for the bands due to the analyte itself (i.e., direct detection) or by observing the bands of an entirely different molecular species somehow connected with the analyte (i.e., indirect detection). In the latter case, the molecule observed, usually named as "reporter" or "label," is most often a well-characterized species having well-known, intense SERS bands. The bands of these *labels* or *reporters* can simply appear or disappear in presence of the analyte, or shift in case the structure of the reporting molecule changes because of a direct interaction with the analyte. Readers who are interested in the indirect method (label or reporter) are encouraged to follow Chapter 5, *Surface-enhanced Raman scattering nanotags design and synthesis*. The direct approach is often referred to as "label-free" (LF), since it does not require the use of "labels" as in the indirect approach. This broad criterion has the advantage of being simple and straightforward: are the bands observed in the spectrum due to the analyte? Then it is LF-SERS. Otherwise, it is not.

Both detection strategies have their own advantages and disadvantages. SERS is intrinsically complex: the observation of a signal is the result of many factors such as the specific interaction between the analyte and the substrate surface and the interaction between the laser light and the *surface plasmons* on the substrate (see the detailed discussion in Chapter 1, *Principles of surface-enhanced Raman spectroscopy)*. The indirect approach is a way to cope with the complexity intrinsic to SERS: the use of known reporters having predictable SERS spectra gives the researchers a tighter control on the system. This better control, however, comes at a cost: the reporter needs to be linked to the targeted analyte by chemical means (e.g., by a direct covalent bond or by the mediation of a recognition element such as an antibody).

LF-SERS is conceptually simpler, as analytes are directly observed. However, this approach heavily relies on a suitable interaction between the analytes and the SERS substrates, which is mostly ensured by the occurrence of some characteristic structural features of the analytes. There are no exact rules to reliably predict if a molecule, upon given conditions, will yield an intense SERS spectrum. However, experienced SERS researchers know that some classes of molecules have better chances than others to be detected with LF-SERS. Thiols and N-containing heterocyclic aromatic compounds (a very broad class of structures) usually yield intense SERS spectra. SERS spectra might be also observed from substances having amine or carboxylate groups, especially associated with an aromatic structure, as in many drugs. All these molecules have functional groups that can chemisorb on Ag and Au surfaces, allowing a stable interaction with the metal surface and thus a longer *residence time* within *hot spots*, those regions of a substrate in which the enhancement effect is larger and that are responsible for most of the SERS signal [5,6]. As far as biological molecules are concerned (or, in general, molecules which might be targeted for clinical applications), carbohydrates and lipids are more difficult to detect in LF-SERS, while polypeptides, proteins, and nucleic acids have better chances to be observed. Drugs with a thiol group such as 6-mercaptopurine, metabolites such as glutathione, or purine derivatives such as hypoxanthine are readily detected by LF-SERS. In summary, LF-SERS is a simple, straightforward approach that will work well only on a limited set of targets when bare Ag and Au surfaces are used as SERS substrates. Differently put, LF-SERS requires quite a bit of luck, besides a carefully planned experimental strategy. For targets with a low affinity for bare Au and Ag surfaces, one must try to optimize the analyte—surface interaction or move to an indirect detection strategy. Despite this limits, LF-SERS has been successfully applied to detect clinically relevant biomolecules (e.g., metabolites, nucleic acids, and proteins) and drugs in biofluids, exosomes, bacterial and eukaryotic cells, and tissues [4,7,8]. The aim of this chapter, however, is not to systematically present all these achievements, but rather to present key challenges and critical issues of LF-SERS when used to develop clinical applications. Many of these challenges arise from the biochemical complexity of most common clinical samples.

4.2 Label-free SERS and the complexity of biological samples

The possibilities and limitations of a LF-SERS method depend on the chemical complexity of the sample, as well as on what is to be detected. From the discovery of this technique to its most recent applications, the chemical complexity of samples analyzed by SERS constantly increased. In its infancy, SERS has been mostly used by physical chemists to study solutions of pure substances in pure solvents. Today, in biomedical SERS applications, researchers with different backgrounds analyze samples with varying degrees of complexity, from relatively simple aqueous extracts to incredibly complex mixtures. When considering clinical applications, samples such as common biofluids (e.g., urine, blood, and saliva) are extremely complex, from a chemical perspective. Blood serum contains many different proteins and several thousands of metabolites [9], dissolved in a buffered aqueous solution containing a variety of ions. This is still a simplified picture, not mentioning exosomes, circulating-free nucleic acids, and lipid vesicles, for instance. Still, serum and other biofluids can be the object of a LF-SERS analysis [10]. Since LF-SERS spectra are only showing bands due to strongly adsorbed species, SERS spectra of biochemically complex samples usually look very different from the corresponding normal Raman spectra (Fig. 4.2). The two spectra carry different information, and the bands observed can belong to a different set of analytes. From the analytical perspective, this biochemical complexity can be both a nuisance and a treasure trove, depending on the scope of the LF-SERS method and the analytical strategy adopted. Moreover, even when considering subjects having similar demographic and clinical characteristics, clinical samples will always present differences in their biochemical composition due to diet, metabolism, habits, and many other factors. These *interindividual* differences can further complicate the LF-SERS analysis (see later in the chapter), adding another level of complexity.

4.3 Clinical needs and analytical strategies

The development of a LF-SERS method should always stem from a clear *clinical need*, keeping well in mind which specific features of SERS can constitute a real advantage over established methods. Diagnosis, prognosis, therapeutic efficacy assessment, and *therapeutic drug monitoring* (TDM) are

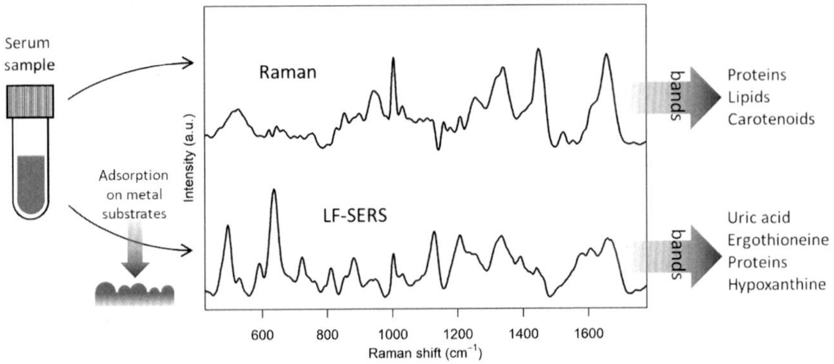

Figure 4.2 The two spectra, relying on different processes, show bands from a different set of analytes. Analogous differences between Raman and LF-SERS are observed for most biofluids. *LF*, Label-free; *SERS*, surface-enhanced Raman scattering.

the most common clinical needs that can be tackled with a LF-SERS method. Once the clinical need is clearly established, an *analytical strategy* must be chosen, and the first decision is about what is actually to be detected by LF-SERS. Depending on the answer to this question, one has then to consider which kind of clinical samples are or will be available and how to deal with the problems posed by their biochemical complexity.

In many cases, the aim of the LF-SERS analysis is to detect or quantify a specific *target*, such as a specific metabolite, protein, or drug. This is called a "targeted" analysis (Fig. 4.3). A *targeted* analysis can be used for *qualitative* or *quantitative* purposes. For instance, a targeted diagnostic LF-SERS method could detect the presence of a specific biomarker, classifying the sample as "disease" or "nondisease" (*classification* task), or it can quantify the marker concentration (*regression* task), relating it to a diagnosis. A targeted LF-SERS method could be also used in TDM to quantify (regression task) a specific drug in a biofluid [11]. In both classification or regression approaches used for targeted LF-SERS, in the original biological specimen (e.g., biofluids, tissues or cells) the target is present together with hundreds or thousands of other chemical species (i.e., the "matrix"), so that we are in the classical "the needle in a haystack" situation. Often, the situation is further complicated by the fact that the target concentration is usually much lower than many other components. A viable solution to tackle this issue is to dramatically decrease the chemical

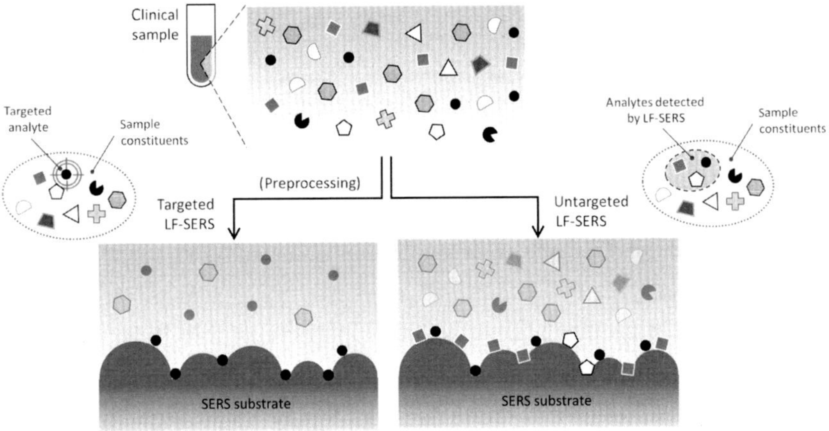

Figure 4.3 Targeted and untargeted LF-SERS analysis. *LF*, Label-free; *SERS*, surface-enhanced Raman scattering.

complexity by processing the sample before the LF-SERS analysis. When considering clinical applications of LF-SERS, samples can undergo a broad variety of preprocessing steps: from simple one-step operations (e.g., dilution or centrifugation) to several steps leading to a "pure" extract (e.g., "pure" extracts containing miRNA molecules in buffer). Sometimes, the preprocessing can even involve amplification (e.g., PCR) or enrichment stages, or the use of highly specialized separation techniques (e.g., HPLC).

Heavier preprocessing thus leads to samples in which the target is found in a simple *matrix* (e.g., a buffer solution or a pure solvent) so that it is much easier to analyze by LF-SERS. It should be stressed, however, that even the purest target solution obtained by a heavy preprocessing does not guarantee the observation of a SERS spectrum of the analyte if the target molecule is not capable of yielding a SERS signal. Analytical approaches based on a heavy preprocessing shift the complexity from the actual LF-SERS measurement to a preanalytical step. The development of such preprocessing steps, which for some analytes is sometimes the only viable solution, can be very challenging in itself, being expensive and time consuming. Simpler one- or two-step preprocessing methods, such as dilution, centrifugation, filtration, or solvent extraction are often a reasonable compromise to decrease the complexity of the matrix to a level allowing for a LF-SERS detection (see later in the chapter). Moreover, by relying on the adsorption on nanostructured metal surfaces, SERS intrinsically leads to a *simplified representation* of the sample. SERS substrates always act as "spectroscopic filters" (Fig. 4.4), conveying the spectroscopic information only of those molecules which are adsorbed on the metal

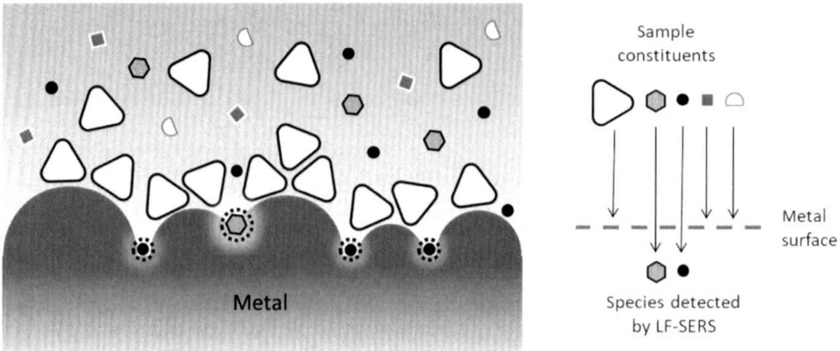

Figure 4.4 Metal substrates act as filters, selecting those with a stronger affinity for the metal as candidates for LF-SERS detection. *LF*, Label-free; *SERS*, surface-enhanced Raman scattering.

surface and yield a SERS signal (see later in the chapter). These requirements often lead to a dramatic simplification of the sample (i.e., target + matrix) as seen by a SERS perspective, where only a few species (hopefully including the target) out of hundreds or thousands are actually observed. Even considering this *spectroscopic filter* effect, a clinical sample is likely to yield a SERS spectrum with several bands due to the matrix constituents, besides those due to the targeted analyte. The bands due to the matrix often overlap with those of the target, interfering with its detection or quantification based upon the intensity or area of few characteristic bands. Moreover, interindividual differences in the matrix composition further complicate the analysis, being a source of spectral variability overlapping that due to the target analyte. *Multivariate analysis* techniques (see later in the chapter) are a viable solution to extract useful spectroscopic information (i.e., that on the target) out of a varying matrix background, constituting an indispensable tool of a targeted LF-SERS method [12,13].

From a rather different perspective, however, the biochemical complexity of clinical samples is not seen as trouble, but as an valuable source of clinically useful information. Looking for a specific target is only one way to approach some clinical problems, especially in diagnosis. Often, a *multimarker* approach, in which more markers are considered at the same time, is found to be more effective than a single-marker strategy. For many diseases, there are no specific or reliable diagnostic markers available, so that a targeted analysis is impossible, and alternative approaches must be used to discover some markers. But even in those cases for which markers are available, broadening the research by considering other biomolecules can be a good strategy. In an *untargeted* LF-SERS approach, all the biochemical information available in a spectrum is used for diagnostic or prognostic purposes. Thus an untargeted LF-SERS method is mostly used for classification problems, using multivariate analysis techniques capable of dealing with many variables at the same time (see later in the chapter). Once some SERS bands are found by multivariate analysis to be diagnostic of a condition, a correct spectral interpretation might tell us if those bands are due to one or more molecular species, thus leading to the identification of new diagnostic markers. This untargeted approach to diagnostics is the strategy used in many LF-SERS studies [8,10], where different metabolites are observed in the spectra of biofluids. Ultimately, the untargeted LF-SERS strategy relies on the assumption that the biochemical (or more specifically, metabolic) differences between the samples belonging to subjects with different medical conditions can be revealed by the spectra. In other words, it is a "high risk, high stakes" game of chance. The role of chance in scientific discoveries is well established, but, as Pasteur said, "chance only favors the prepared mind." For LF-SERS, preparing the mind translates to using all the possible available a priori information about the system, using the correct multivariate analysis methods and, most importantly, to achieve a better understanding of what information is actually present in LF-SERS spectra (i.e., band interpretation, see later in the chapter).

4.4 Experimental aspects

4.4.1 Preanalytical sample processing

As previously stated, the chemical complexity of biological samples can be decreased by some processing steps, thus facilitating SERS analysis. Depending on the specific aim of the study, a wide variety of preprocessing methods can be employed, from relatively simple operations (e.g., dilution) to lengthy and complex protocols (e.g., HPLC separation). One should always select the best compromise between the complexity of the sample and that due to the preprocessing, depending on the SERS strategy adopted. For biofluids, the most common preprocessing adopted before SERS analysis include dilution, acidification/basification, centrifugation, or filtration (usually to separate solid or cellular fractions from liquid fractions), deproteinization (by ultrafiltration, dialysis, or precipitation), and solvent extraction.

Blood is an excellent example to illustrate the effect of simple preprocessing steps such as centrifugation and deproteinization (Fig. 4.5), being constituted by a cellular fraction and by a protein-rich liquid fraction (i.e., plasma). When the target of the analysis is found in the plasma fraction, a simple centrifugation step can separate the cellular fraction. Depending on the type of SERS substrate used, the proteins present in plasma can also be a nuisance for SERS analysis, for example, by hindering the functional aggregation of Ag and Au nanoparticles, impeding the formation of hot spots [14] (Fig. 4.5A). Moreover, even in the case of *noncolloidal* substrates proteins such as the human serum albumin readily adsorb on metal surfaces ("protein fouling"), potentially competing with the analyte for adsorption, or even blocking it from adsorbing, to the detriment of a SERS detection (Fig. 4.5B). Filtration with a proper cutoff (or other deproteinization methods, from dialysis to protein precipitation) will make a LF-SERS analysis easier, allowing the formation of hot spots (in case of *colloidal* substrates) and by avoiding *protein fouling* of the

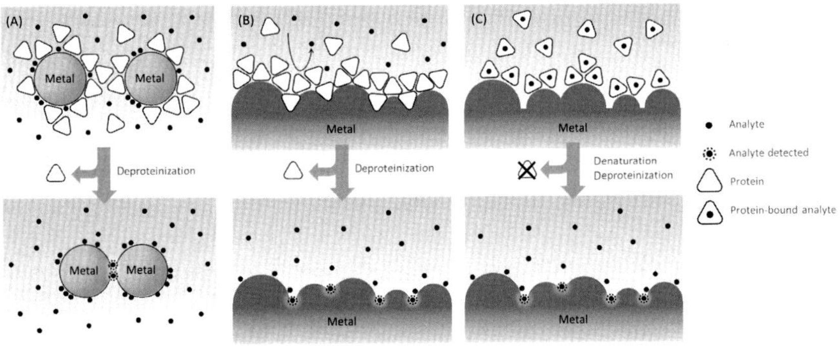

Figure 4.5 Deproteinization can allow colloidal aggregation and hot-spots formation (A), avoid protein fouling (B), and (if preceded by protein denaturation) free protein-bound analytes (C).

nanostructured metal surface. A preprocessing including a deproteinization step is also useful for a *untargeted* LF-SERS approach, in which low-molecular-weight metabolites are to be observed. As many targets (e.g., drugs) might be tightly bound to plasma proteins up to considerable fractions (in some cases more than 90%), a deproteinization step which causes protein denaturation (e.g., the addition of cold methanol) is essential [15], as it "frees" the analytes from the proteins, making them available for the interaction with the SERS substrates (Fig. 4.5C). Since many drugs are more soluble in organic solvents than in water, this solvent-based deproteinization works also as a solvent extraction step, and it is widely used as sample preprocessing in HPLC-based methods. Some other biofluids (e.g., urine) do not contain proteins (or, in the case of some diseases, low amounts of proteins) and thus do not need any deproteinization steps. Urine, however, presents other challenges for a LF-SERS analysis, such as a high *inter*- and *intraindividual* variation in density and composition.

Since the interaction between the SERS substrate and the target depends on the chemical structure of the latter, slight variations in the pH might induce important changes in the analyte, favoring its adsorption on the nanostructured metal surface and thus the observation of an intense SERS spectrum [16,17]. Therefore acidification or basification should be also considered as sample preprocessing, keeping in mind that biofluids are buffered. A significant change in pH can also lead to the precipitation of proteins, so that pH change and deproteinization can be also achieved in a single step. If the target is not found in the liquid fraction of a biofluid, but within cells or tissues, then the sample might benefit from a *lysis* step, where cell walls are ruptured and the cytosolic content, together with subcellular organelles, is released and made available for SERS analysis upon adsorption on nanostructured metal substrates [18–20].

In general, there is no limit to the complexity of the preprocessing before a LF-SERS analysis, and more structured approaches include separation steps requiring the use (and relative expertise) of different techniques (e.g., gel- or capillary electrophoresis, TLC, HPLC, etc.), or the use of commercial extraction kits (e.g., for DNA/RNA). As with other analytical techniques, one should be aware that it is always a trade-off: the more complex the preprocessing, the easier the analysis, and vice versa. For LF-SERS, one should always consider if and how the preprocessing will affect the spectra, in terms of both the target and matrix bands. To better evaluate the impact of preprocessing, one should learn as much as possible about how the target and the matrix are interacting with the nanostructured metal surfaces of the substrate.

4.4.2 SERS substrates and the nano–bio interface

There is a huge number of different SERS substrates described in the literature (see the detailed discussion in Chapter 2, *Nanoplasmonic materials for surface-enhanced Raman scattering*), and they can be classified using different criteria [1,2,21,22]. A simple criterion is the nature of the metal used: gold, silver, copper, or other metals. Another criterion, which is especially important with respect to the analysis of biological samples, can be applied to most SERS substrates classifying them as *colloidal* or *noncolloidal.*

Colloidal substrates are well defined as being constituted by metallic nanoparticles of various shapes and sized dispersed in a liquid matrix (a pure solvent or a solution). These substrates are very popular, as they are relatively easy to prepare via simple chemical reactions without requiring expensive or large instrumentation, or the need for specific expertise. Moreover, the volume produced by a single synthesis is usually sufficient for many SERS measurements, making these substrates inexpensive and easily affordable. Although many different recipes are available to produce colloidal substrates with various shapes and sizes, and no one has been yet established as a standard, few recipes are particularly popular among SERS researchers, especially those relying on a reduction of silver and gold salts with citrate or hydroxylamine. All the substrates in this category, with the exception perhaps of nanostars, must be somehow induced to form small nanoparticle aggregates with nanogaps between particles acting as hot spots. The literature clearly showed that the enhancement of the signal for isolated nanoparticles is too small to be useful for detection purposes so that an aggregation step is required [23–25].

Noncolloidal substrates, also referred to as *solid* substrates, constitute a much broader and heterogeneous class, including nanostructured metal surfaces obtained from various bottom-up or top-down methods, from simple chemical or electrochemical etching to extremely precise nanolithographic methods. Arrays of gold nanopores, nanoparticles deposited on paper and potential-controlled roughened silver electrodes are all part of this broad category, where nanostructures acting as hot spots are already formed, so that no aggregation or other kind of nanostructure-forming processes are required. Some exceptions are those flexible nanopillars-based substrates that can benefit from aggregation as well, generating hot spots in aggregated structures formed by bending nanopillars.

For both colloidal and solid substrates, the capability to enhance the Raman signal is usually considered as the most important characteristic. This ability is usually summarized by a single parameter known as the "enhancement factor" [26] (EF). This factor, however, presents some problems and it is not an intrinsic characteristic of the substrate itself, as it depends on the analyte and the laser used. Moreover, especially in the perspective of clinical applications, other characteristics such as *batch-to-batch* reproducibility, stability with respect to different chemical environments (e.g., ionic strength and pH), shelf-life, and related costs are also very important. Clinical applications usually require the use of hundreds of substrates, since such number of spectra is needed to set-up and validate a method (see later in the chapter), so that substrate-related costs are an important issue. Also, clinical samples are likely to be analyzed in several sessions spanning over a time range of weeks or months, so that *batch-to-batch* reproducibility and shelf-life should be carefully considered. For many clinical applications, especially for point-of-care devices, factors such as the sample—substrate incubation time or practical aspects related to substrate handling can also be important factors. Fragile or tiny substrates are more difficult to handle, and the analysis of biofluids or aqueous samples might be easier and faster on porous, hydrophilic solid substrates (e.g., based on plasmonic paper) than on hydrophobic ones. When selecting a SERS substrate for a clinical application, one should carefully consider all these aspects as well, possibly trading off plasmonic properties for other characteristics.

Most importantly, the choice of a SERS substrate should be made by considering the specific aspects of the *nano–biointerface*, that is, the interaction between the biological sample and the nanostructured metal surface. Even a general knowledge of the chemical characteristics and composition of the sample and of the analyte is useful in selecting a substrate with adequate properties. Since proteins hinder nanoparticle aggregation, a LF-SERS analysis of samples with a high protein concentration (e.g., whole blood, serum, or plasma) can be directly performed on solid substrates, but it requires sample deproteinization if colloidal substrates are used. Alternatively, a preaggregation or concentration of the colloidal substrates can be used to form hot spots before incubation with the sample. Electrostatic interactions are also important at the nano–bio interface, being a problem or an asset depending on the circumstances. Many colloidal substrates (e.g., citrate-reduced colloids) are electrostatically stabilized by a layer of negatively charged ions, so that the interaction with negatively charged analytes (and thus the observation of their LF-SERS signal) is difficult [27]. This problem can be addressed in several ways: by changing the surface charge of the substrates (i.e., substituting the coating), by shielding electrostatic interactions upon increasing the ionic strength or by changing the electrostatic charges of the molecular species involved (i.e., coating or analytes) upon varying the pH. Nucleic acids, for instance, can be detected by LF-SERS when positively charged substrates are used or by the addition of positively charged Mg (II) ions [28].

These are just a few examples, and many others can be found in the literature. It is crucial to realize that there is no such thing as a "universal" SERS substrate to be generally used for the LF-SERS analysis of any clinical sample. Each LF-SERS clinical application, depending on the particular clinical needs, on the sample characteristics and on the preprocessing steps involved, asks for substrates with specific characteristics. This is a further reason why clinical LF-SERS needs an in-depth, systematic, and detailed investigation of the nano–bio interface between plasmonic structures most commonly used as substrates and different types of clinical samples. While most nano–bio interface studies address proteins and their interaction with nanoparticles [29–31], the understanding of how smaller, low-molecular-weight metabolites interact with different kinds of Ag and Au substrates is even more important for LF-SERS. The importance of an adequate understanding of the nano–bio interface for LF-SERS cannot be understated.

4.4.3 Excitation wavelengths

Different excitation wavelengths will yield different results, in terms of both intensity and bands observed, depending on the type of SERS substrate used and the characteristics of the sample investigated [32]. The success of a SERS measurement depends on the choice of an adequate excitation as much as it depends on the characteristics of the nanostructured metal substrate. The most common commercially available Raman instruments are usually equipped with either a 785 or a 532 nm diode lasers. Diode lasers are the only viable solution for compact, affordable Raman setups, whose characteristics are often an advantage for clinical applications. Older instruments or custom setups might include less common excitation wavelength, and in principle, lasers emitting from the UV to the NIR are available. For SERS, excitation in the green (i.e., 514 or 532 nm), red (i.e., 633 nm), or NIR (i.e., 785 or 830 nm) are the norm.

These wavelengths adequately match the *surface plasmon* frequencies of most common Ag and Au SERS substrates, an essential requirement for the electromagnetic mechanism to take place and concur to the SERS effect. Ideally, the surface plasmon frequencies of a SERS substrate should be determined as precisely as possible, thus enabling the selection of the optimum excitation wavelength. In practice, however, this is not always feasible. Determining the surface plasmons using far-field approaches has been demonstrated to give an incomplete picture. Some plasmonic modes ("dark modes") are not observed in the far-field (e.g., by using UV−vis absorption to observe extinction spectra), while they can play an important role in determining the SERS effect [33,34]. The intense SERS spectra obtained from citrate or hydroxylamine Ag colloids upon 785 nm excitation, an observation that puzzled many SERS researchers, are an excellent example of how misleading the UV−vis extinction spectrum can be to guide the choice of the excitation wavelength. These colloids show an intense extinction peak around 400 nm, which shifts and broadens upon aggregation toward 500 and 600 nm, but no maxima are present around 780 nm in the extinction spectra of the isolated or aggregated Ag nanoparticles, in spite of the intense SERS spectra observed upon using excitation in that region. A rational selection of the exciting wavelength can thus be done either by modeling the optical properties with theoretical approaches or by using characterization techniques such as EELS [35]. If those are not available, however, an empirical rule of thumb that is suitable for most common SERS metals and excitation wavelengths is that

Ag substrates work well with excitation in the 514, 532, or 785 nm, while Au substrates can be investigated with a 633 or 785 nm excitation.

Another issue to take into account when selecting the excitation wavelength for a SERS experiment is the occurrence of a *resonance* between the wavelength used and an electronic transition of a species present in the sample (i.e., target or matrix component). When using an excitation wavelength falling within an absorption band of a sample component, a resonance (or preresonance) Raman (RR) effect will take place, combining with the SERS effect to yield a *surface-enhanced resonance Raman scattering* (SERRS) spectrum. In RR, the band intensity of the species in resonance with the exciting laser is enhanced or several orders of magnitude. The combination of the RR ad SERS is a distinctive advantage if the RR effect is achieved for the target: SERRS spectra are extremely intense, effectively increasing the sensitivity of the analysis (examples of SERRS spectra of single molecules are not uncommon in literature). If the target (e.g., a drug) has an absorption band in the visible or NIR region of the spectrum, then the achievement of a RR enhancement might guide the choice of the excitation wavelength, determining which metal should be used as SERS substrate. Achieving SERRS conditions should be always considered, if possible, for the detection of a target: they might help to achieve better sensitivity (the signal is more intense compared with nonresonant SERS) and selectivity (the bands of the target will be selectively enhanced). Note, however, that the RR effect is beneficial only for those cases in which the target is directly adsorbed on the metal surface, allowing for the quenching of potential fluorescence, which can be a problem for electronically excited systems. For systems requiring a coating or spacer between the metal and the analyte (e.g., a *recognition element* such an *antibody*), fluorescence could be enhanced as well, interfering with the observation of the Raman and SERS bands. On the other hand, the RR effect can also be a nuisance if accidentally occurring for some matrix constituents. Plasma and serum, for instance, might contain beta-carotene molecules or hemoglobin (released upon hemolysis), so that a SERS spectrum of these biofluids obtained with a blue or green excitation (on Ag substrates, e.g.) is likely to be dominated by intense carotenoids or hemoglobin RR bands, interfering with the observation of those of the target. Since these carotenoids are usually bound to proteins, a deproteinization by filtration will help to partially solve the problem, although some heme groups can be released by hemoglobin and can be found, together with some residual carotenoid molecules, even in deproteinized samples (Fig. 4.6). The most sensible solution would be using an excitation at 785 nm, far from the absorption bands of beta-carotene and heme.

Even more problematic is the interference due to the normal fluorescence of matrix components, which can be minimized by using an excitation wavelength in the NIR region (i.e., at 785 or 830 nm), even though a residual fluorescence due to two-photon excitation processes can still be present. The minimization of fluorescence is the main reason behind the huge success of the 785 nm diode lasers, which should be always considered as the first choice for clinical SERS applications unless SERRS conditions for the target are feasible. Moreover, a 785 nm excitation proved

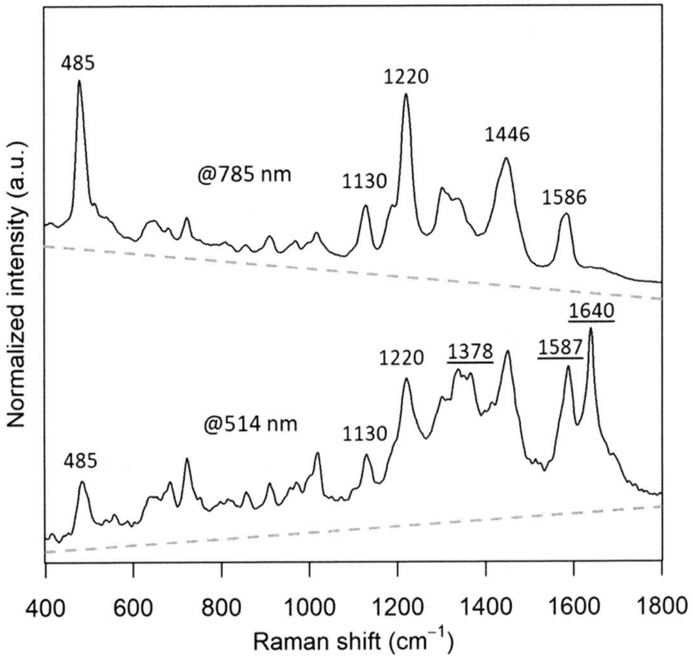

Figure 4.6 Spectra of the same sample of deproteinized serum on the same Ag colloids excited with different wavelengths. SERRS bands due to heme (underlined bands, due to hemoglobin released upon hemolysis) are observed in the spectra excited at 514 nm, but not in the one at 785 nm, where ergothioneine bands are mostly observed. The relative intensity of the bands present in both spectra are different because of the plasmonic spectral-shaping effect. Sloping backgrounds (indicated as dashed lines) are different because of the diverse position of spectral regions with respect to fluorescence (in the case of 785 nm laser, due to a two-photon excitation). *SERRS*, Surface-enhanced resonance Raman scattering.

to be viable for both Ag and Au nanostructured surfaces, allowing some flexibility in the choice of the SERS substrates. Diode lasers, however, suffer from possible wavelength drifts during a measurement session, so that spectra of reference standards (e.g., cyclohexane, paracetamol, and silicon) should be often collected (possibly before or after every measurement) to monitor such drifts.

4.4.4 Common artifacts and anomalous bands

All those researchers approaching the use of SERS for tackling bioanalytical and clinical problems should become well acquainted with the spectral patterns of some common artifacts and interferents, whose bands are often mistaken for the analyte signal, undermining any meaningful conclusion inferred from the dataset [36]. Carotenoids bands due to an unwanted RR effect of matrix components are not the only undesired interfering bands in SERS spectra of biological samples, nor the most common. Intense and broad bands around 1350 cm^{-1} (*D-band*) and between 1500 and 1600 cm^{-1} (*G-band*) due to the *sp^2-carbons* of amorphous carbon material are often observed in SERS spectra (Fig. 4.7), also as a broad background onto which other bands are present. Such bands are associated with the occurrence, at least to some degree, of thermal *photodegradation* of the species adsorbed on the metal substrate [32]. The intense electromagnetic field generated at the hot spots can be absorbed by some of the

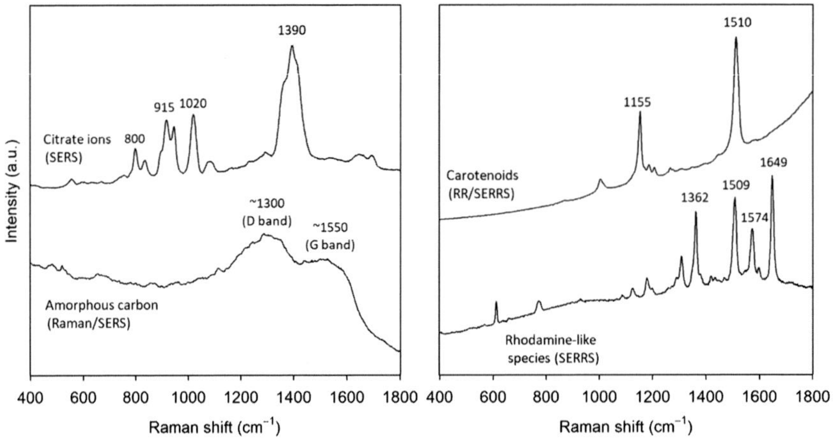

Figure 4.7 Anomalous bands often observed in label-free SERS spectra: (left) citrate ions and amorphous carbon bands and (right) carotenoids and rhodamine-like bands (observed upon 514–532 nm excitation). For a complete discussion of these bands see [36].

species present on the metal surface (i.e., target or matrix components), and thus converted into thermal energy. The heat generated might induce the transformation of organic molecules to amorphous carbon. This carbon will efficiently adsorb radiation, converting it into thermal energy, promoting the further formation of other amorphous carbon, in an autocatalytic process. The occurrence of these bands can be minimized by making heat dissipation more efficient (e.g., by working in presence of water) or by decreasing the amount of energy delivered by the laser (e.g., by decreasing the laser power or by moving the sample under the laser).

Bands due to citrate ions are also commonly observed in SERS spectra (Fig. 4.7), since Ag and Au colloids stabilized by a capping of citrate ions are very common and popular substrates. Since citrate ions are already present on the metal nanoparticles, the observation of SERS bands of an analyte requires the displacement of a fraction of the citrate ions by the analyte molecules. Sometimes this displacement is not efficient, but nanoparticle aggregation is induced nonetheless by the presence of salts in the analytes solution. Biofluids with a low protein content usually contain enough ions to induce nanoparticle aggregation. If aggregation occurs without the displacement of the citrate ions from the metal surface, then intense SERS bands due to citrate will be observed, and they might be mistaken for the analyte bands.

A set of intense "rhodamine-like" bands [36] are another common artifact occurring in SERS spectra obtained from Ag substrates upon excitation with a wavelength in the green region (Fig. 4.7). The origin of these bands is still debated, but the most accredited hypothesis is that they are due to some impurities resonant with the green excitation which might be omnipresent in the water or in the Ag salts used for the colloids preparation. Considering the intensity of the SERRS effect, just trace amounts of these impurities would be sufficient to observe these bands.

4.5 Study design and data analysis

4.5.1 Sources of variability

Besides the usual sources of variability which are expected in chemical analysis (e.g., sampling, weight, and volume measurements, *inter- and intraindividual* differences, instrumental responses, operators, etc.), SERS has additional sources of variability related to the fact that it relies on nanostructured metal substrates interacting with the sample in specific ways. The SERS signal is actually originating from just a small fraction

($<1\%$) of the population of molecules adsorbed on the substrate, those residing in a *hot spot* during the measurement time interval [3]. Moreover, depending on the strength of the molecule–substrate interaction, this "hot spot population" is not fixed but can vary over a time interval, as molecules diffuse in or out of a hot spot. For low analyte concentrations, this might translate in a signal intensity fluctuation over time and it could be an important source of variability. Moreover, if the hot spots are not homogeneously or regularly distributed on the substrate, different locations might have a different hot spot density, leading to differences in overall intensity (i.e., intrasubstrate variability).

The characteristics of a SERS spectrum also depend on how a molecule is interacting with the metal substrate: which molecular moieties are interacting with the metal, and which is the overall spatial orientation of the molecule with respect to the surface. Slight variation in pH or ionic strength might translate into appreciable spectral differences, introducing a source of variability. Furthermore, since crowding at the surface might affect the molecular orientation of the analyte, the overall shape of SERS spectra often changes with the analyte concentration. Orientation at the surface can also be affected by the presence of other molecules, so that the SERS spectrum of the same analyte might look different depending on the other components of the system.

To summarize, a SERS spectrum is the result of a complex network of interactions and factors which are difficult or impossible to disentangle. This complexity translates in many sources of variability, so that the same sample can yield different SERS spectra if these are collected on different substrates, on even different spots or at different times on the same substrate. *Reproducibility* is thus widely considered an issue for SERS, mainly in terms of substrates. However, reproducibility is a term that requires a bit of explanation, in the context of SERS. Reproducibility is usually referred to a method and concerns *the closeness of results obtained by different laboratories using different experimental setups*. In this sense, reproducibility is a misused term when applied to SERS substrates: substrates are just a part of a method, and they are usually evaluated within one lab. *Repeatability* expresses *the closeness of results obtained within one lab using the same setup*, and would be a better choice than reproducibility, but it is used for methods, while substrates are not the only source of variability in a SERS method (intended as the whole experimental protocol to obtain a SERS spectrum for a given sample). A somewhat neutral but perhaps more correct term would be *variability*, specifying if it has been evaluated within a single lab (i.e., *within-lab variability*) or among different labs (i.e., *between-lab variability*).

Moreover, SERS substrates can have different types of *variability*, depending on their characteristics and on the specific aspects observed in the spectra. For noncolloidal substrates, spectral characteristics might vary from one spot to another within the same substrate (i.e., *intrasubstrate* variability), or, as already mentioned, from one substrate to another (i.e., *intersubstrate* variability). The intrasubstrate variability for noncolloidal substrates might also depend on the specific protocol used for putting the sample in contact with the nanostructure. Depositing a sample drop on the substrate and letting it dry, for example, might lead to a dried spot with a so-called "coffee-ring" pattern, increasing the intrasubstrate variability. For colloidal substrates, spectral characteristics might vary from one batch of colloid to another. This kind of variability could be referred to as *batch-to-batch variability*, or intersubstrate variability if each batch of colloid is considered as a distinct substrate. However, the type of variability to consider will depend on the protocol used also for colloidal substrates. If the sample is mixed with the colloid, drop-casted on a solid surface (e.g., metal, glass, or fluorite slide) and let dry, this dried aggregated colloid might present a high spot-to-spot (intrasubstrate) variability. Intra- and intersubstrate variabilities are nested, since the latter is evaluated from measurements collected from single substrates, and thus is affected by the former. The intrasubstrate variability should be considered first, as it will inevitably propagate to the intersubstrate variability. A high intrasubstrate variability will inevitably lead to high intersubstrate variability (while a low intrasubstrate variability does not necessarily lead to a low intersubstrate variability).

For clinical LF-SERS studies where tens or hundreds of samples are measured, it is crucial to consider intersubstrate or batch-to-batch variability when designing a study and planning the measurements. Using a single batch of colloidal substrate or merging in a single flask several batches would certainly eliminate this source of variability from the study. This approach, although advisable for pilot studies, is not giving a realistic assessment of the overall method variability. If more batches of substrates are used, the measurements should be always planned so that each *class* (for classification studies) or *analyte concentration level* (for regression studies) is equally represented the measurements from each batch (Fig. 4.8). Otherwise, the differences observed might be due to the differing characteristics of the batch rather than those of the samples.

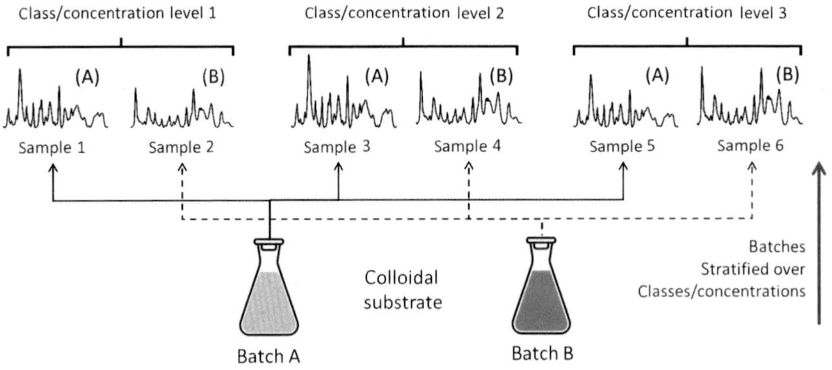

Figure 4.8 Correct stratification of different substrate batches over different classes or concentration levels.

Depending on the specific application, different aspects of a SERS spectrum are considered as varying. For SERS applications where the absolute intensity is relevant (e.g., the quantification of a target analyte for diagnosis or TDM), the focus is usually on the latter for one or more bands (i.e., *absolute intensity variability*). This type of variability can be easily expressed as the *variance* or *(relative) standard deviation* of the absolute intensity of one well-defined and intense band (before or after baseline subtraction, depending on the approach adopted), and it is the type of variability most commonly reported in SERS studies. Many research groups and companies struggle to achieve SERS substrates with low absolute intensity variability. Most studies in the literature on SERS substrates report absolute intensity variabilities (both inter- and intrasubstrate) from 10% to 20% of relative standard deviation, with values below 10% considered as very good and values below 5% as excellent. However, this is not the only kind of variability to be considered. In many situations, such as in diagnostic LF-SERS applications relying on classification models where intensity is normalized, one is usually interested in minimizing the substrate-related variability of the relative intensities among different bands (*relative intensity variability*), rather than the absolute intensity. Since in those applications SERS spectra are meant to be classified on the basis of their relative intensity pattern, the substrate-related absolute intensity variability is not an issue, provided that a spectrum is observed.

Most important, however, is the fact that all these substrate-related variabilities are not intrinsic characteristics of the substrate itself, but rather of a specific substrate—sample pair when investigated with a specific excitation wavelength. The same substrate can in fact show different variabilities when investigated with different samples or different excitation wavelengths. This is always due to the fact that in SERS the signal originates from the interaction of these three elements: laser light, sample, and substrate. The contribution of each element cannot be easily disentangled from the others, so that the variability of the signal is only partially due to the substrate. For this very reason, the variability of a substrate should be always considered as specifically limited to the analyte and excitation wavelength used for the evaluation.

In absence of a standard, well-defined terminology for SERS methods, one should still try to be as correct as possible. Stating that "this substrate has 8% of relative standard deviation" in relation to its variability is an incomplete and misleading statement. A more correct statement would be: "this substrate, when applied to the analysis of the X analyte using the Y excitation, has a *within-lab intra- and intersubstrate variability* of XX% and YY% of relative standard deviation (absolute intensity), respectively." The statement, of course, should be accompanied by a detailed protocol with the description of how the variabilities were evaluated. All these aspects related to a SERS method should be adequately considered when planning a clinical SERS application, and possibly addressed by setting up a preliminary study to evaluate the different types of variability with the sample of interest (e.g., a specific biofluid or a target). Another source of variability to be considered is the interindividual variability, that is, the difference in the sample biochemical composition as seen by LF-SERS from one subject to another (Fig. 4.9). The interindividual variability is a nuisance in targeted LF-SERS studies, as a varying background (i.e., matrix signal) can pose a serious challenge for data analysis, but is the main source of biochemical information for untargeted LF-SERS studies. Variability due to instrumental factors (e.g., laser power fluctuations, wavenumber calibration changes) or operators should be taken into account, so that data collection should be carefully randomized, avoiding data collection from samples sharing similar metadata (e.g., class or analyte concentration) on the same day. The evident risk is that the differences observed in the LF-SERS dataset are due to day-to-day *instrumental variability* or different operators, rather than to real biochemical differences.

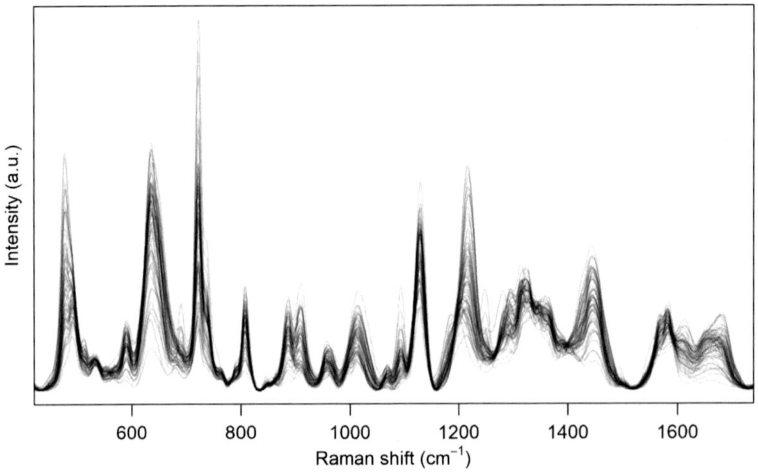

Figure 4.9 Interindividual variability in a dataset of LF-SERS spectra of deproteinized serum samples from 60 subjects (1 spectrum/subject). Spectra were obtained from Ag colloids using a 785 nm excitation. *LF*, Label-free; *SERS*, surface-enhanced Raman scattering.

4.5.2 Data structure and sample size

Because of the many sources of variability in LF-SERS, datasets in clinical studies often have a "nested" (or "hierarchical") structure (Fig. 4.10). Typically, more replicate spectra are collected for each substrate (e.g., collected at different locations, to tackle intrasubstrate variability), more replicate spectra on separate substrates or batches for each clinical sample (to tackle intersubstrate variability) and sometimes even spectra from different samples from the same subject, patient, or analyte concentration. Thus such nested datasets have many spectra for each subject/patient or concentration level. However, using the datasets as they are to set-up classification or regression models would be a mistake, since the spectra are not *statistically independent.* Only one spectrum per subject/patient or, in most cases, per concentration level should be used to set-up and validate models, so that all spectra from lower levels of the nested data structure should be used to calculate a mean or median spectrum for each subject/concentration. The information contained in the lower levels of the nested structure might still be used to get information about the intersubstrate and intrasubstrate variability, but they should absolutely not be used as independent elements in a model. Ideally, intra- and intersubstrate variability should be previously explored in a separate study, possibly

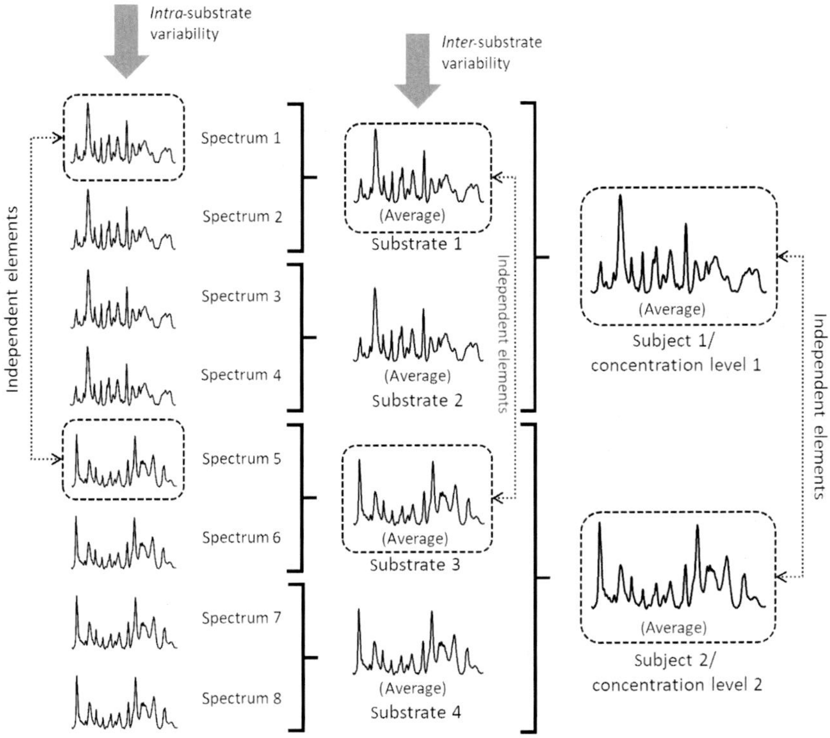

Figure 4.10 An example of a nested data structure for a LF-SERS dataset. Example of independent elements at each level of the structure is shown. *LF*, Label-free; *SERS*, surface-enhanced Raman scattering.

using a *design of experiments* approach [37,38]. Once these variabilities are known, the necessary number of replicates per substrate and the number of substrates can be estimated before the actual data collection from clinical samples. If these substrate-related variabilities are low, a single measurement for each sample could be sufficient. Thus the "sample size" (i.e., the number of samples to be included in a study) should be counted only using the independent samples (e.g., the number of patients), without taking into account the different replicate spectra.

For LF-SERS regression studies, the number of samples necessary to set-up a model might be difficult to estimate in general, as it depends on the regression model used, and on the substrate-related variabilities. For univariate models, at least ten regularly spaced concentration levels are recommended, with a number of replicate spectra for each concentration depending on the substrate characteristics. For multivariate regression, the number of independent samples to be used will depend on the algorithm used and on a number of case-specific features (e.g., the number of data points per spectrum, the signal-to-noise, the number of latent variables used). Usually, multivariate models require a larger sample size than for the univariate case. In general, larger sample sizes allow a better validation, narrowing the performance-related uncertainties. For example, in a study by our group, a multivariate regression model using 3 latent variables from a dataset having spectra with about 450 data points, a sample size of 84 independent samples as used to set-up and validate the model [15]. Moreover, to have a realistic picture of how the regression model will behave on new samples it is very important to eventually validate it against a *reference method*, by directly comparing the predictions of the two methods.

For LF-SERS classification studies, the sample size cannot be exactly determined a priori, as it is done for statistical tests with a predefinite "statistical power." The more independent samples are collected for each class, the more stable the classification model and the narrower the confidence intervals of the *figures of merit* (Fig. 4.11). Some studies suggested

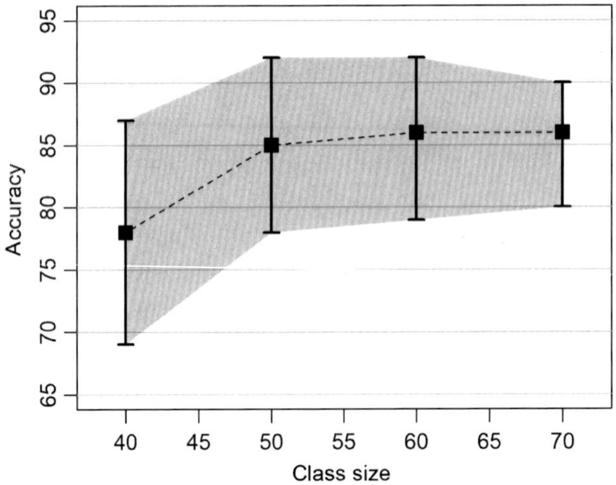

Figure 4.11 Classification accuracy (including 95% confidence intervals) as a function of number of samples per class (two classes: disease vs control). Data are derived from a RDCV of a PCA-LDA algorithm applied to a dataset of LF-SERS spectra of serum on solid Ag substrates (i.e., Ag colloids deposited on paper) collected with a 785 nm excitation. The data clearly show how the accuracy confidence interval is getting narrower by increasing the number of samples. *LF*, Label-free; *PCA*, principal component analysis; *RDCV*, repeated double cross-validation; *SERS*, surface-enhanced Raman scattering.

that at least 70–100 samples per class are required to get reasonable confidence intervals [39]. Such sample size will also allow for a more reliable *resampling strategy*, such as a *repeated k-fold cross-validation* [CV, or even a *repeated double cross-validation* (RDCV), see later in the chapter]. Pilot studies with less than 70 patients per class are possible, but at the price of a high uncertainty about results, which should always be confirmed by successive studies on more samples. In LF-SERS classification studies, a particular attention should be paid to the cohort-related study design. Subjects enrolled in each class should be matched as much as possible by age, body mass index, and other relevant characteristics (e.g., smoking habits, therapies, surgical interventions, etc.). Untargeted LF-SERS studies aiming at diagnosing a disease should take care that the main spectral differences observed in the datasets are not just due to differences in population characteristics.

4.5.3 Data analysis: preprocessing, representation, and modeling

The way LF-SERS data are preprocessed, represented, and analyzed is a crucial part of a clinical LF-SERS application. A correct preprocessing and analysis can make the difference between success and failure, whereas a correct data representation might be able to truthfully convey much information in a compact figure or graph. Few books, chapters, and extensive reviews are dedicated to data analysis, even specifically applied to spectroscopy [12,13,40,41], and we encourage the reader to read them for a detailed overview of different methods and follow Chapter 11, *Multivariate approaches for surface-enhanced Raman Spectroscopy data analysis in clinics*. This section will instead focus on general aspects that are important for data analysis and representation.

Preprocessing methods are used to correct SERS data before representation or analysis, compensating for those features which are not directly correlated with those bands carrying the biochemical information of interest. Baseline subtraction, noise removal, and intensity normalization are all examples of widely used preprocessing procedures, but there are others as well (e.g., *outlier* detection and elimination, data *centering* and *scaling*, etc.). There are several algorithms and methods to perform these preprocessing steps, but there is no such thing as a universally valid procedure to be used with any data. Each dataset has its own specific characteristics and purpose, and requires a specific preprocessing. In particular, preprocessing is intrinsically linked to the purpose of a study, and it should ideally enhance those spectral features which are relevant. By modifying or transforming the original data, all preprocessing methods carry the inevitable risk of altering or deleting relevant information from the dataset. Or even worse, to introduce misleading artifacts in the dataset. For this reason, a good rule of thumb is to keep preprocessing as simple as possible.

For most LF-SERS data, this will include *baseline subtraction* and possibly *intensity normalization*. Linear or *polynomial-fitted* baselines work well for most cases, although more complex algorithms such as the *asymmetric least squares* [42,43] is also very effective for SERS data. Intensity normalization very much depends on the specific purpose. LF-SERS classification studies aiming at diagnosis usually benefit from an intensity normalization such as a *vector normalization* (e.g., total intensity normalization), while for regression studies an intensity normalization might be detrimental unless an internal standard is used. As an alternative, a model-based preprocessing such as the *extended multiplicative scatter correction* [44,45] (EMSC) can be used. The EMSC has the advantage of performing both baseline-correction and normalization in a single step, and it is increasingly being used for SERS datasets. Irrespectively of the methods used, a detailed description of the procedures used, possibly together with a figure illustrating one or a few examples of original spectra fitted with baselines, should be included in reports and publications.

Another important aspect when dealing with LF-SERS datasets, especially in the context of clinical applications requiring many spectra, is the detection of *outliers* (i.e., anomalous spectra which are very different from the other spectra in the dataset) and their subsequent elimination from the dataset. Outliers containing anomalous bands and artifacts, due to impurities or unwanted substances adsorbing on the metal substrates, are very common in LF-SERS datasets. Furthermore, clinical LF-SERS datasets might also contain outliers due to the specific characteristics of a sample or patient. Unfortunately, many data analysis algorithms and approaches are particularly sensitive to the presence of outliers, so that few anomalous spectra can significantly impact the overall performance of the classification or regression model. As for the baseline subtraction and intensity normalization, however, also for the outlier detection there are many different and equally efficient methods, and there is such thing as a universally valid approach. A direct visual inspection of each spectrum is highly advisable for small datasets (<100 spectra), but for larger datasets, other approaches must be adopted. A simple method to spot outliers is to inspect the score plots of a *principal component analysis* [46] (PCA). Another promising method is to use a so-called functional boxplot representation of the dataset, relying on *functional data analysis* [47], where outliers are shown and can be easily compared with the "median" spectrum. The number of outliers detected and removed from the dataset and the method used for their detection should be always specified in reports and publications.

Often, especially prior to multivariate analysis such as a PCA, data are *centered* (i.e., the mean of the dataset is subtracted from each spectrum), and sometimes *scaled* [48]. While there is a wide consensus on the usefulness of centering the spectra data before analyses such as PCA, scaling a Raman and SERS dataset is more controversial. Centered data can still be related to the original data, with all bands still visible (although some of them as "negative" bands). Scaled data, however, can be more difficult to relate to the original spectra, and the importance of spectral regions with just noise can be inadvertently amplified (Fig. 4.12). For this reason, the advice is to center but not scale LF-SERS data prior PCA or other multivariate analyses.

Once the LF-SERS dataset has been preprocessed, it should be correctly represented in a figure (or more) capable of conveying most of the relevant information. How do the spectra look like? What are the most intense and most characteristic bands? Are the spectra similar or different? Reporting few spectra can be easily done by stacking or overlaying them, possibly using colors to differentiate them. Each spectrum can be directly observed, and no information is lost to the reader. This is clearly impossible for datasets with more than a few spectra. A very common practice is to rely on *descriptive statistics* concepts by representing the *average* spectrum of the dataset as the *central tendency*, sometimes together with the *standard deviation* of the intensity for each wavenumber to convey the information about the data dispersion. Since in most cases the intensity for each wavenumber is not normally distributed in the dataset, a more robust representation would be plotting the *median* spectrum, together with the first and third *quartiles* (thus representing the *interquartile* or IQR) for the intensity at each wavenumber (Fig. 4.13). Median and IQR are also less sensitive to outliers, in the case these are still present in the dataset, and thus are to be preferred with respect to mean and standard deviation. An interesting and much simpler alternative is to plot all the spectra as overlaid but as semitransparent lines (as in Fig. 4.9), so that single spectra are still visible,

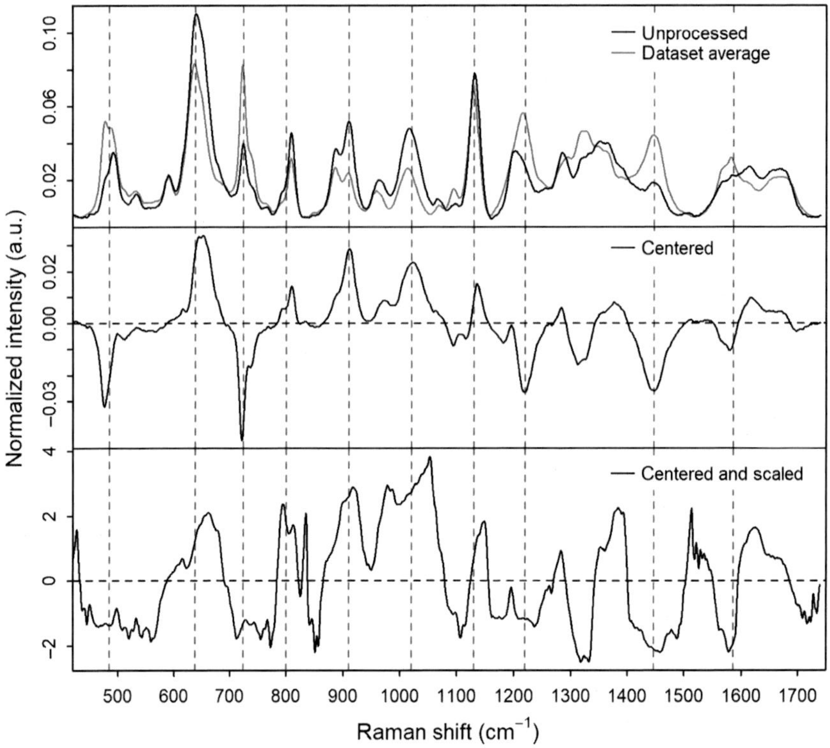

Figure 4.12 An example of the effect of centering only and centering followed by scaling on a LF-SERS spectrum of serum (obtained on Ag colloids upon 785 nm excitation). The position of the main bands is marked with dashed lines. *LF*, Label-free; *SERS*, surface-enhanced Raman scattering.

while the overlap over the spectra is conveying the information about the most frequent spectral profile.

A figure depicting the mean or average spectrum alone is lacking the information about the data variability, so that the intensity (normalized or not) standard deviation or IQR should be always included as well. This is especially useful when comparing two or more groups of spectra, for example, disease versus control, or different analyte concentration. Inferring a meaningful difference among two or more sets of spectra by comparing the mean o median can be misleading, if spectral variabilities are not represented (Fig. 4.14). When comparing two groups or classes, sometimes it can be useful to calculate the difference between all the spectra of one class and those of the other, and then represent the median and IQR of these difference spectra. By doing so, the spread of the IQR

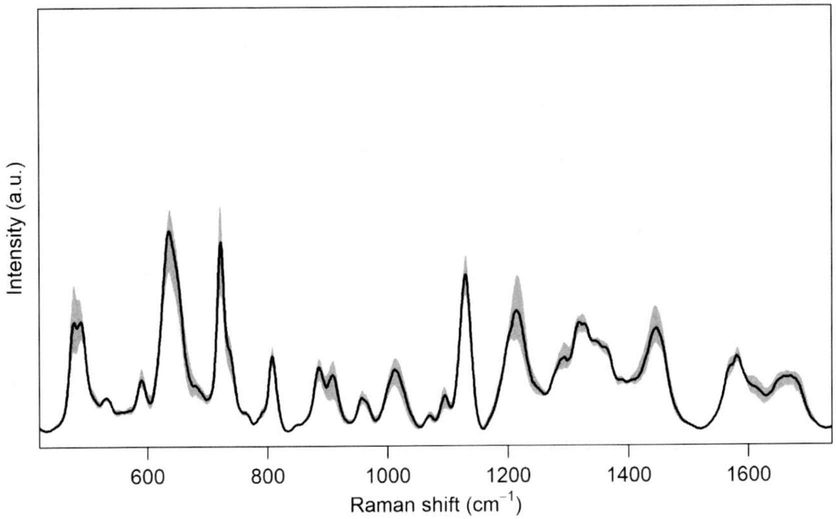

Figure 4.13 A representation of the dataset of Fig. 4.9 as median (as a measure of central tendency) and interquartile range (IQR, as a measure of dispersion). The dataset is composed of LF-SERS of serum obtained on Ag colloids using a 785 nm excitation. *LF*, Label-free; *SERS*, surface-enhanced Raman scattering.

will tell us how significant the difference between the two groups is for each band considered (Fig. 4.14.)

Once the spectral characteristics of the LF-SERS dataset have been correctly reported, the final part of the data analysis can be performed, where consistent spectral differences can be explored and eventually used to set-up classification (e.g., diagnosis) or regression (e.g., quantification of a metabolite or a drug) models. It should be emphasized that the study design (see above) is entirely part of the *data analysis workflow*. Each model works best on datasets having specific characteristics, so that setting up a regression or classification model on a dataset collected without a related study design will very likely lead to weak or meaningless results. Another important aspect is that the whole data analysis is not a linear process seeing preprocessing and model setup (and validation) as independent, subsequent steps. In many cases, setting up a model suggests a better preprocessing, and one must go back to the previous step in a sort of *self-consistent* process.

There is a great variety of possible approaches to data analysis, so that addressing them one by one is out of the scope of this chapter. Despite the great variety of models and algorithms available to analyze data, one

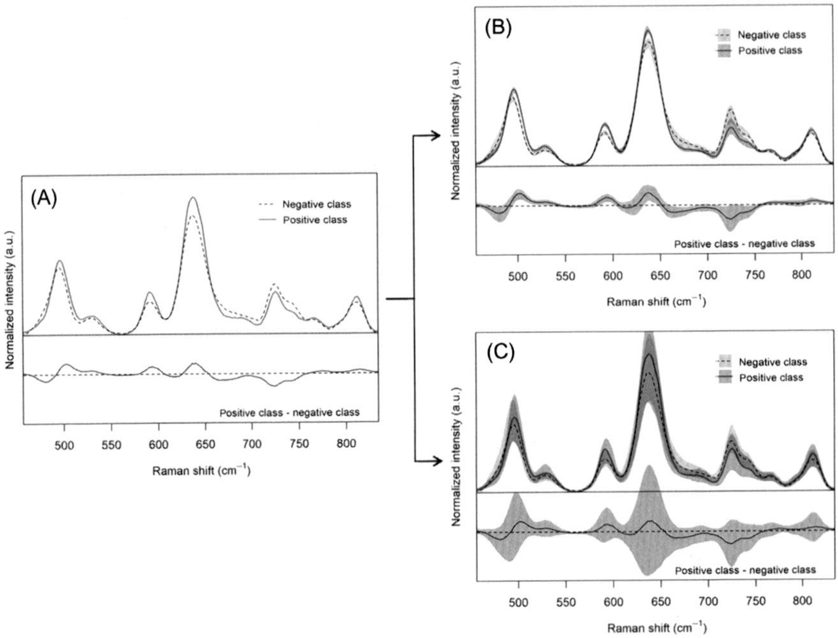

Figure 4.14 An example of how different data visualization can convey more or less information. In (A) only the median spectra of two classes ("positive" and "negative") and their difference are shown. In (B and C) median spectra and IQR intervals are reported for each class, together with the median and IQR of all the difference spectra (between the two classes). (B) and (C) are very different situations: in (B) the difference between classes is significant for some bands, whereas in (C) it is not. The difference between (B) and (C) cannot be appreciated by looking at (A), however. *IQR*, Interquartile range.

should always adopt the *parsimony principle*, aiming at the simplest approach that works. Wherever possible, univariate approaches, where a single variable (e.g., the intensity or area of a single band, or the intensity ratio between two bands) is derived from each spectrum as a representative for a sample, should be first considered. Univariate methods are well established and easier to use. For instance, univariate linear regression models are straightforward to set up and characterize by well-known figures of merit (e.g., limit of detection and quantification, sensitivity, etc.). For classification purposes, differences between two groups of spectra can be detected by plotting intensities, areas, or intensity ratios as boxplots, applying adequate statistical tests.

When univariate methods fail, multivariate methods should be considered, where many variables are taken into account. After all, spectra are intrinsically multivariate data, and multivariate methods better deal with their complexity by using all the information available. These methods are powerful tools that can cope with the different sources of variability found in clinical LF-SERS datasets, focusing on those spectral features which are important and disregarding the others. However, powerful tools can be also dangerous, and should only be handled by competent, trained users. For nonexperts, multivariate methods are in fact as powerful as treacherous. Poorly conducted experiments usually lead to poor results, but multivariate data analysis, if incorrectly applied, can easily produce excellent results that might be wrong, misleading, or meaningless. Researchers, as human beings, are prone to the so-called "confirmation bias," that is, the tendency to favor results which confirm their expectations [49]. Unfortunately, this inclination is easily seconded by multivariate methods, unless extreme careful steps are taken to oppose it. The devious mechanism by which these methods can easily deliver overoptimistic results is called *overfitting*.

Overfitting occurs when the model is more complex than necessary, so that it will fit any situation: an overfitted model will always give the expected result [50,51] (Fig. 4.15). With the intrinsically multivariate spectroscopic data, the risk of overfitting is very high. For *partial least squares* [51,52] (PLS) regression models or for classification models built on a *discriminant analysis* by reducing the number of variables (e.g., PCA-LDA and PLS-LDA), for instance, retaining a high number of *principal components* (or *latent variables*, in PLS) increases the risk of overfitting. Since all multivariate models require the specification of at least one parameter (also called *hyperparameter*), an inconsiderate choice of the latter can easily lead to overfitting. The main problem with multivariate models is how to make a sensible choice of this *hyperparameter* (e.g., the number of principal components or latent variables). Whenever possible, all a priori, external knowledge about the system (e.g., biochemical, biological, or clinical knowledge about the samples or medical conditions studied as well as specific physicochemical and spectroscopic knowledge about LF-SERS data) should be used to choose the value of the hyperparameter. For instance, if a metabolite is known (or suspected) to be a likely marker for a disease, its bands should be present as features in the loadings of the principal components used for a PCA-LDA classification model, whereas those principal components whose loadings are very noisy should be excluded. Although markers are usually not known in untargeted approaches, checking the *signal-to-noise ratio* of the loadings of the last principal component used to set-up the model is advisable to prevent overfitting. Similarly, in a PLS

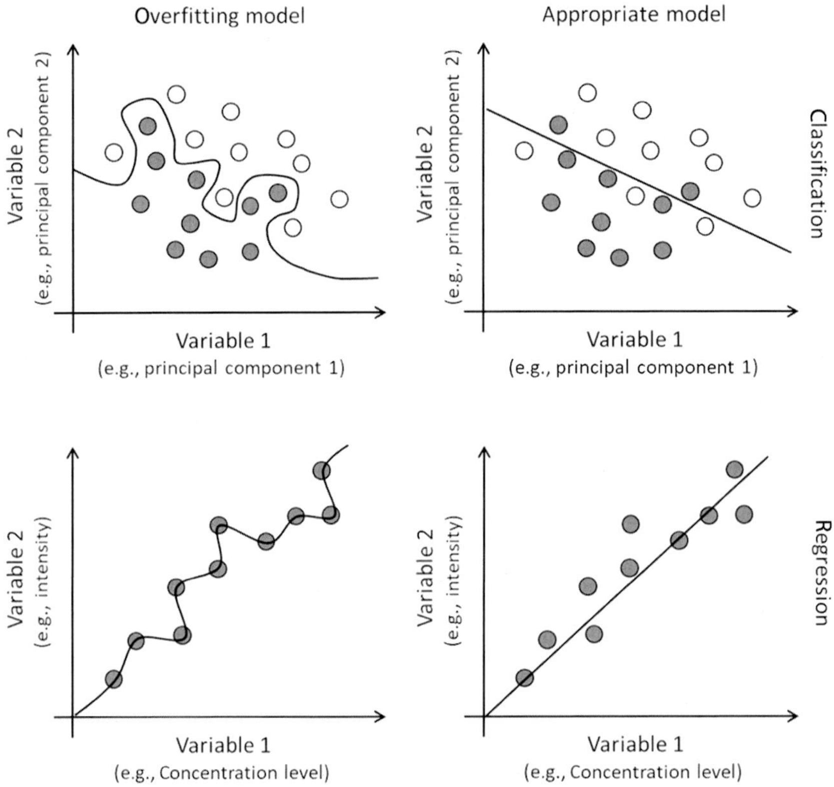

Figure 4.15 Idealized examples of overfitting (left graphs) and correct (right graphs) models for classification (top graphs) and regression (bottom graphs).

regression model, a latent variable whose regression coefficient has features resembling the target analyte (e.g., a drug) should be included, while that resembling noise should be excluded.

The use of this a priori knowledge is not always possible but, fortunately, overfitting can be detected. An overfitted model will work very well on the data used to train it (*autoprediction*), but it will likely fail when other data are used. This is why *validation* is a crucial step in multivariate data analysis (Fig. 4.16). Ideally, models should always be validated using an independent, external "test" dataset (i.e., data which were *not* used to train the model). CV, where the original dataset is split into *train* and *test* sets [51], is a common validation procedure for LF-SERS studies having a limited number of samples. *Repeated k-fold* CV is advisable for datasets with an adequate number of samples, while the *leave-one-out* CV should

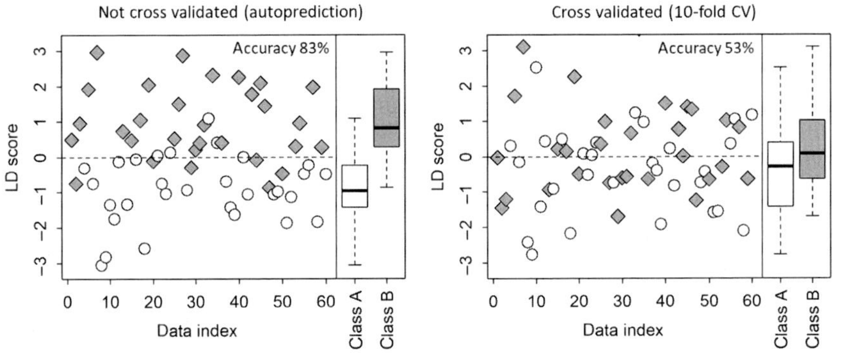

Figure 4.16 Predictions (i.e., LD scores) of a PCA-LDA model from a dataset where classes were randomly assigned. Results are shown with and without cross-validation, where autoprediction leads to overoptimistic results. Dataset composed of LF-SERS spectra of deproteinized serum ($n = 60$, dataset of Fig. 4.9) on Ag colloids upon 785 nm excitation. *LF*, Label-free; *PCA*, principal component analysis; *SERS*, surface-enhanced Raman scattering.

be avoided, and only used for very small datasets (i.e., pilot studies) for which a repeated k-fold CV is not feasible. A crucial aspect of a correct CV is that no information about the test should be used to set-up the model. In the CV of PCA-LDA classification models, for instance, it is not correct to perform the PCA over the whole dataset to select the principal components to be included, but only over the train set. Ideally, an integrated approach such as the RDCV or *double CV* should be adopted [53,54], where the model hyperparameter is optimized using a "data-driven" approach (i.e., without any a priori knowledge) and at the same time the all the optimized models are *cross-validated* with a repeated k-fold CV. One advantage of these methods is that a single dataset is used to generate and cross-validate many slightly different models, so that confidence intervals can be calculated for the figures of merit (e.g., accuracy, sensitivity, specificity, etc. for classification models, or RMSEP for regression models) over all the models. Since overfitted models, for which complexity is too high, will be "unstable" and show a broader variety of performance, these confidence intervals will give an indication about how much results can be trusted. To summarize, multivariate models are powerful tools that must be used responsibly. They should always be carefully set-up, optimized, and validated, possibly using a repeated (double) CV approach yielding an indication about the stability of their performance.

4.6 Spectral interpretation

Once a LF-SERS spectrum is obtained, then it should be correctly *interpreted*: the direct experimental information present in the spectrum must be *translated into biochemical information* to be used for clinical purposes. In other words, the LF-SERS bands observed in the spectrum need to be interpreted in terms of specific molecular species (e.g., a metabolite, a protein, or a nucleic acid). Note that this interpretation is something different than the "band assignment" usually performed in vibrational spectroscopy of pure analytes, where each band in the SERS spectrum can be assigned to specific *vibrational modes* (e.g., C = O stretching). Band assignment usually requires a deeper knowledge and understanding of the system under study: it can be done for LF-SERS only once a correct spectral interpretation is achieved and only if sufficient experimental (e.g., isotopic substitution) or theoretical (e.g., vibrational frequencies calculation from first principles) data are available from literature for the analytes observed in the spectrum.

Since one is usually interested in identifying only the bands due to the target analytes and not in those due to the matrix, spectral interpretation for targeted LF-SERS studies is straightforward. This identification can be readily achieved by a *direct comparison* with the SERS spectra of the pure analytes in a simple matrix (e.g., buffer or solvent) with characteristics as close as possible to those of real samples after preprocessing. In TDM studies based on LF-SERS, for instance, bands due to the drug can be identified by a direct comparison with the SERS spectrum of the pure drug (Fig. 4.17). This approach necessarily assumes that the matrix is not significantly affecting the way the analyte is interacting with the metal substrate. This assumption, however, is not always correct, so that sometimes the SERS spectrum of the pure analyte can show some differences with respect to that of the analyte in a biological matrix, making the band identification more difficult.

For untargeted LF-SERS studies, the situation is different, since the relevant bands (as emerging from data analysis) can in principle be due to any biological molecule present in the sample. Luckily, however, the situation for LF-SERS is not as complicated as in other techniques use for untargeted analyses. Even biochemically complex matrices such as plasma or serum, containing thousands of different chemical species, often lead to LF-SERS spectra displaying relatively few well-defined intense bands. Since SERS is a surface technique, only those species adsorbed on the

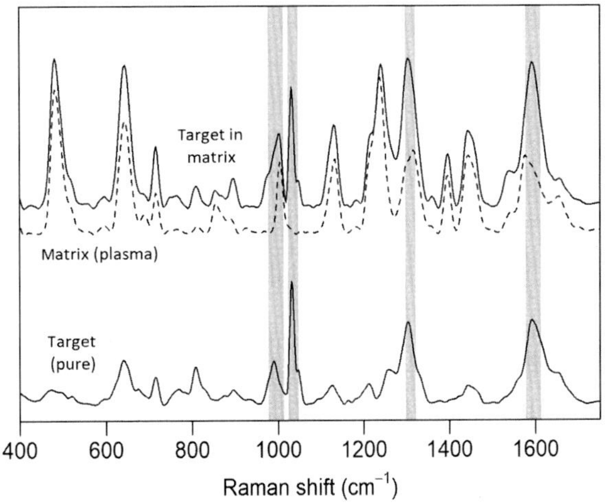

Figure 4.17 Spectral interpretation as identification of the LF-SERS bands due to the target. Such identification of the target in a complex matrix can be achieved by a direct comparison with the spectrum of the pure target. Spectra in the figure are of the drug imatinib in deproteinized plasma methanolic extracts and in pure solvent, obtained on solid Ag substrates upon 785 nm excitation. *LF*, Label-free; *SERS*, surface-enhanced Raman scattering.

surface to a sufficient amount will yield an observable signal. Thus as previously stated the sample—surface interaction is modulated by the chemical and physical characteristics of the substrate. In other words, a SERS spectrum obtained from a biochemically complex sample is always an extremely simplified representation of that sample, and it is a function of the substrate and wavelength used for excitation. For instance, all most intense bands in the LF-SERS spectrum of serum or plasma obtained with a NIR laser on most commonly used substrates can be attributed to just about few metabolites, out of more than 4000 metabolites identified in those biofluids [9], and a similar situation is found for other biofluids as well. Tear fluid [55] and seminal plasma [56,57], for instance, present bands similar to those of serum and plasma.

This "oversimplification" of the actual biochemical information contained in LF-SERS spectra can be an advantage or a disadvantage depending on the context. For targeted analysis, where the matrix signal is an interference to be avoided, having a simplified picture of a complex matrix is likely to be an advantage: the fewer bands from the matrix, the more "space" is available on the wavenumber axis for the observation of the target bands. On the other hand, for an untargeted approach aiming at getting the most out of the wealth of information in biofluids, the simplified picture given by SERS might not contain enough of that information. The limited set of metabolites whose bands are observed in a LF-SERS spectrum of serum might not have a correlation to a specific pathology. In this sense, the simplified representation of a biological sample given by LF-SERS sets the limits of this approach.

Performing a correct band interpretation is thus essential to plan a clinical LF-SERS study and to critically evaluate the results. A correct, careful biochemical interpretation of spectra is often overlooked in untargeted LF-SERS studies, focusing on the fact that the diagnostic algorithms work even without knowing the origin of each band, and relying on the belief that somehow the biochemical information present in the spectra is being correctly exploited by the algorithm used. The intensity or frequency of some bands in LF-SERS spectra might work well as markers for a disease, in spite of the fact that we cannot label these makers with a name. This attitude, however, can lead be misleading. If not carefully used by skilled data analysts, the algorithms employed in multivariate analysis can easily lead to incorrect conclusions. A proper interpretation of LF-SERS bands, however, can help in assessing the validity of the results by considering them from a biochemical perspective. Moreover, a correct interpretation of LF-SERS data might even help to advance the biochemical insight on a particular disease, suggesting further studies or research directions.

Then how can be LF-SERS spectra interpreted? Let me state clearly that *there are no defined, rigorous algorithms* to assign a band to a specific analyte in a LF-SERS spectrum. Spectral interpretation relies on the intuition of a researcher, as well as on her biochemical and spectroscopic knowledge of the sample. A common practice is to compare a band wavenumber with those already available in the literature for SERS or Raman spectra. This approach requires an adequate knowledge of literature and the availability of systematic studies reporting a collection of spectra from various biological molecules. Such systematic studies, however, are available for Raman, but not for yet for SERS. Assigning a SERS band on the basis of the closeness of its frequency to that of a Raman band is very risky. SERS and Raman spectra of a substance can have remarkable differences, both in terms of band frequency and intensity. This is due to the mechanisms intrinsic to SERS, so that the bands related to chemical bonds affected by the interaction with the metal will be shifted up several tens of wavenumbers. Their intensity will be enhanced with respect to the corresponding Raman, leading to a different relative intensity pattern for the

SERS spectrum. As a result, a SERS band might not have its Raman equivalent visible in the same wavenumber region (Fig. 4.18).

Moreover, many biological molecules can display intense Raman spectra but are notoriously difficult to observe with SERS (see the beginning of the chapter). Even SERS spectra of the same substance can present significant differences when obtained upon different experimental conditions (e.g., substrate, excitation wavelength, pH, etc.), further complicating a comparison with normal Raman spectra. Browsing a collection of Raman spectra in literature or in a database could be useful to select some candidates, but the tentative interpretation of a band in a LF-SERS spectrum should be always confirmed by a direct comparison with SERS spectra of pure species obtained in the same or very similar experimental conditions (e.g., the same metal as substrate, the same excitation wavelength, etc.).

Such a direct comparison allowed us to interpret SERS spectra of plasma and serum collected on a variety of Ag and Au substrates as mainly

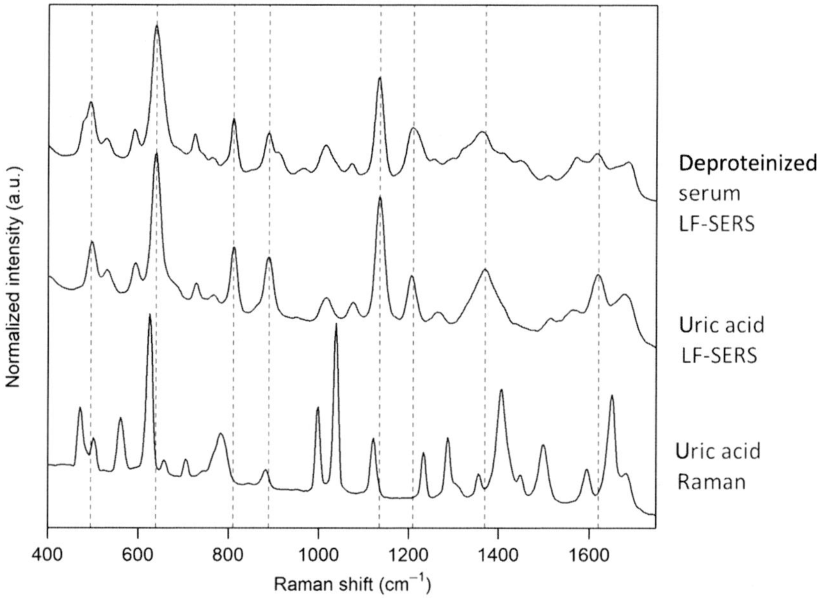

Figure 4.18 LF-SERS spectra of deproteinized serum (top), uric acid (middle), and normal Raman (bottom) spectra of uric acid. Raman and LF-SERS spectra of uric acid present marked differences. Raman shifts of main LF-SERS serum bands are marked with dashed lines. All spectra were obtained using a 785 nm excitation. SERS spectra were obtained on Ag colloids. *LF*, Label-free; *SERS*, surface-enhanced Raman scattering.

constituted of bands due to three metabolites: uric acid, hypoxanthine, and ergothioneine [10,58] (Fig. 4.19), besides proteins (for samples of unfiltered, whole biofluids). The same metabolites were reported for tear fluid [55], and some of them can be clearly observed as well in the spectra of seminal plasma [56,57] and tissue extracts [59–61]. Similarly, LF-SERS spectra of cell lysates could be deconvoluted in terms of few metabolites such as glutathione, adenine, and adenosine [18]. Note that in all these cases most of the bands observed in LF-SERS spectra of biological samples could be interpreted in terms of few metabolites, exemplifying how LF-SERS can dramatically reduce biochemical complexity (see above). Systematic studies aiming at the biochemical interpretation of saliva, urine, synovial, ascetic, and cerebrospinal fluids as well as other biofluids will be hopefully carried out in the future.

On the other hand, relying on Raman shift values obtained from Raman spectra to interpret LF-SERS spectra often lead to misinterpretations. In many LF-SERS studies on serum or plasma, for instance, the band at 638 cm^{-1} due to uric acid has been often misinterpreted as due to L-tyrosine and lactose, while the one at 1134 cm^{-1} to D-mannose [62–69]. The intense band around 490 cm^{-1}, due to overlapping bands of uric acid and ergothioneine, has been often reported as due to L-arginine or glycogen [62,64–67,69,70]. In some cases, on the sole basis of closeness between Raman shift values, some LF-SERS serum bands have been interpreted as due to collagen [62,63,66,68,69,71], although native

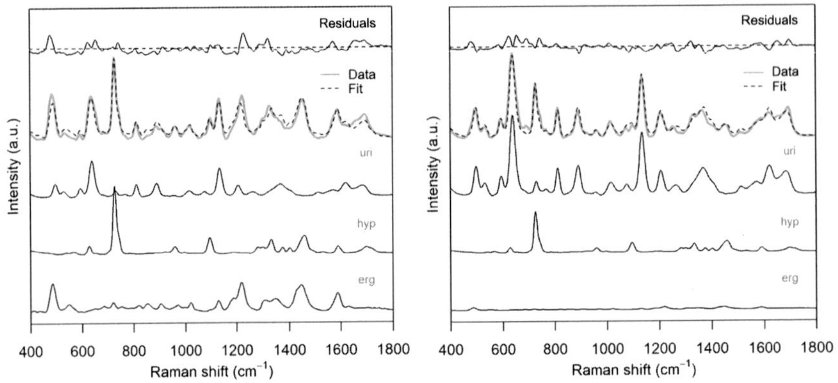

Figure 4.19 LF-SERS spectra of deproteinized serum on Ag colloids at 785 nm from two different subjects (left and right). Spectra are interpreted by a direct comparison with SERS spectra of uric acid (uri), hypoxanthine (hyp), and ergothioneine (erg). *LF*, Label-free; *SERS*, surface-enhanced Raman scattering.

collagen proteins are not supposed to be present in these biofluids. Bands due to ergothioneine, being observed in SERS spectra of erythrocytes (and their lysates) where this metabolite is present in high concentrations [58], have been often misinterpreted as due to hemoglobin [72–75]. Such examples show that a more careful methodological approach for spectral interpretation relying on a direct comparison with LF-SERS spectra of pure analytes is thus required to avoid misinterpretations.

4.7 Perspectives and challenges

LF-SERS holds great promise for the development of clinical applications. Advances in LF-SERS methods for quantification of drugs in biofluids (thus aiming at TDM), for diagnostics and prognostics based on untargeted analysis of biofluids, and for bacteria identification in biological samples are particularly promising. Especially in view of their light sample preprocessing, and thus their fast response, low costs, and ease of use. Other diagnostic or prognostic approaches targeting specific systems and relying on more complex and articulated sample preprocessing (e.g., extraction kits, microfluidics, etc.), such as LF-SERS analysis of miRNA, exosomes, or cells, are also extremely interesting. However, in spite of these advances, LF-SERS still needs to find its way into real clinics. Several challenges need to be addressed to effectively bridge the gap between ongoing LF-SERS research and its use in real clinical practices.

The first challenge is the need to increase the statistical confidence of LF-SERS studies. That is, to carefully design studies planning an adequate number of samples and a robust validation. Pilot studies are welcome to explore new applications, but to convince clinicians about the reliability of a LF-SERS method we need to provide solid numbers complemented by confidence intervals or other measures of uncertainty. Furthermore, interlaboratory studies are needed to validate methods, showing that they can be used by different operators in different labs without affecting their performance.

These efforts can only be sustained by having at disposal an adequate number of SERS substrates. Thus, the availability of reliable, robust, durable, and inexpensive commercial substrates with low inter- and intra-substrate variability is a major requirement to set up large LF-SERS studies aiming at clinical applications. At the same time, the SERS community should agree on setting some standard substrate characteristics, developing shared protocols on how to characterize substrates. Even if not directly related to specific clinical problems, methodological research efforts in this direction would certainly accelerate the development of LF-SERS in clinical applications. Rather than focusing on EFs, more attention should be given to the characterization of batch-to-batch variability, especially over longer periods of time. Low batch-to-batch variability is especially important for quantitative, targeted SERS analyses, where absolute intensity is used. SERS "internal standards" could be added to the sample [76], or perhaps already integrated into the substrates, to compensate for the batch-to-batch or other substrate-related variability. The development of general-purpose SERS internal standards, however, is unlikely, so that each specific application will require some research to establish its own standard.

Another challenge concerns the selective extraction of information about the analyte(s) of interest while minimizing the interference from the matrix. Better, faster, simpler, and more efficient sample preprocessing methods or devices are needed to take out of the analysis unrelevant matrix components, allowing a safer detection of the analytes. Alongside, novel physical and chemical modifications of the substrates are needed to address the key challenge of capturing and detecting analytes whose affinity for the bare metal surface is low. Mixed self-assembled monolayers [20,77], molecularly imprinted polymers films [78], and molecular cages relying on the host—guest interaction (e.g., cucurbiturils, cyclodextrins, etc.) [79—81] showed some promise in addressing this challenge.

Last but not least, fundamental research is needed to better understand and characterize the nano—bio interface between SERS substrates and biological samples used in clinics. Such a better understanding will translate in (1) the rational design of better substrates, (2) better and new ways to cope with the problem of matrix interference in targeted studies, and (3) a more robust biochemical interpretation of LF-SERS spectra. The latter is especially a priority for untargeted diagnostic or prognostic studies, as it sets the boundaries of applicability of a LF-SERS method, allowing to focus on those unmet clinical needs amenable to be solved with this approach. For instance, since LF-SERS spectra of serum and plasma contain bands of uric acid and hypoxanthine, pathologies correlated with those metabolites such as gout [82], Parkinson [83], cardiovascular [70,84—86], and liver [68,87,88] diseases are more likely to be successfully addressed by a LF-SERS approach. On the contrary, developing LF-SERS applications without a proper understanding of what is actually observed in spectra, and why, is a risky stance which can easily lead down a wrong path. With a proper understanding of the nano–bio interface as observed by LF-SERS, however, a wide range of reliable and ambitious clinical applications can be achieved.

References

[1] J. Langer, D. Jimenez de Aberasturi, J. Aizpurua, R.A. Alvarez-Puebla, B. Auguié, J. J. Baumberg, et al., Present and future of surface-enhanced Raman scattering, ACS Nano 14 (2020) 28–117. Available from: https://doi.org/10.1021/acsnano.9b04224.

[2] J. Wang, K.M. Koo, Y. Wang, M. Trau, Engineering state-of-the-art plasmonic nanomaterials for SERS-based clinical liquid biopsy applications, Adv. Sci. 6 (2019) 1900730. Available from: https://doi.org/10.1002/advs.201900730.

[3] X. Wang, S.-C. Huang, S. Hu, S. Yan, B. Ren, Fundamental understanding and applications of plasmon-enhanced Raman spectroscopy, Nat. Rev. Phys. 2 (2020) 253–271. Available from: https://doi.org/10.1038/s42254-020-0171-y.

[4] C. Zong, M. Xu, L.-J. Xu, T. Wei, X. Ma, X.-S. Zheng, et al., Surface-enhanced Raman spectroscopy for bioanalysis: reliability and challenges, Chem. Rev. 118 (2018) 4946–4980. Available from: https://doi.org/10.1021/acs.chemrev.7b00668.

[5] S.L. Kleinman, R.R. Frontiera, A.-I. Henry, J.A. Dieringer, R.P.V. Duyne, Creating, characterizing, and controlling chemistry with SERS hot spots, Phys. Chem. Chem. Phys. 15 (2012) 21–36. Available from: https://doi.org/10.1039/C2CP42598J.

[6] H. Wei, H. Xu, Hot spots in different metal nanostructures for plasmon-enhanced Raman spectroscopy, Nanoscale 5 (2013) 10794–10805. Available from: https://doi.org/10.1039/C3NR02924G.

[7] M. Procházka, Surface-Enhanced Raman Spectroscopy: Bioanalytical, Biomolecular and Medical Applications, Springer International Publishing, Switzerland, 2016.

[8] X.-S. Zheng, I.J. Jahn, K. Weber, D. Cialla-May, J. Popp, Label-free SERS in biological and biomedical applications: Recent progress, current challenges and opportunities, Spectrochim. Acta A: Mol. Biomol. Spectrosc. 197 (2018) 56–77. Available from: https://doi.org/10.1016/j.saa.2018.01.063.

[9] N. Psychogios, D.D. Hau, J. Peng, A.C. Guo, R. Mandal, S. Bouatra, et al., The human serum metabolome, PLOS ONE 6 (2011) e16957. Available from: https://doi.org/10.1371/journal.pone.0016957.

[10] A. Bonifacio, S. Cervo, V. Sergo, Label-free surface-enhanced Raman spectroscopy of biofluids: fundamental aspects and diagnostic applications, Anal. Bioanal. Chem. 407 (2015) 8265–8277. Available from: https://doi.org/10.1007/s00216-015-8697-z.

[11] A. Jaworska, S. Fornasaro, V. Sergo, A. Bonifacio, Potential of surface enhanced Raman spectroscopy (SERS) in therapeutic drug monitoring (TDM). A critical review, Biosensors 6 (2016) 47. Available from: https://doi.org/10.3390/bios6030047.

[12] F. Lussier, V. Thibault, B. Charron, G.Q. Wallace, J.-F. Masson, Deep learning and artificial intelligence methods for Raman and surface-enhanced Raman scattering, TrAC. Trends Anal. Chem. 124 (2020) 115796. Available from: https://doi.org/10.1016/j.trac.2019.115796.

[13] J. Yang, J. Xu, X. Zhang, C. Wu, T. Lin, Y. Ying, Deep learning for vibrational spectral analysis: recent progress and a practical guide, Anal. Chim. Acta 1081 (2019) 6–17. Available from: https://doi.org/10.1016/j.aca.2019.06.012.

[14] A. Bonifacio, S. Dalla Marta, R. Spizzo, S. Cervo, A. Steffan, A. Colombatti, et al., Surface-enhanced Raman spectroscopy of blood plasma and serum using Ag and Au nanoparticles: a systematic study, Anal. Bioanal. Chem. 406 (2014) 2355–2365. Available from: https://doi.org/10.1007/s00216-014-7622-1.
[15] S. Fornasaro, A. Bonifacio, E. Marangon, M. Buzzo, G. Toffoli, T. Rindzevicius, et al., Label-free quantification of anticancer drug imatinib in human plasma with surface enhanced raman spectroscopy, Anal. Chem. 90 (2018) 12670–12677. Available from: https://doi.org/10.1021/acs.analchem.8b02901.
[16] C. Garrido, T. Aguayo, E. Clavijo, J.S. Gómez-Jeria, M.M. Campos-Vallette, The effect of the pH on the interaction of L-arginine with colloidal silver nanoparticles. A Raman and SERS study, J. Raman Spectrosc. 44 (2013) 1105–1110. Available from: https://doi.org/10.1002/jrs.4331.
[17] I.J. Hidi, A. Mühlig, M. Jahn, F. Liebold, D. Cialla, K. Weber, et al., LOC-SERS: towards point-of-care diagnostic of methotrexate, Anal. Methods 6 (2014) 3943–3947. Available from: https://doi.org/10.1039/C3AY42240B.
[18] E. Genova, M. Pelin, G. Decorti, G. Stocco, V. Sergo, A. Ventura, et al., SERS of cells: what can we learn from cell lysates? Anal. Chim. Acta 1005 (2018) 93–100. Available from: https://doi.org/10.1016/j.aca.2017.12.002.
[19] M. Hassoun, W.I. Schie, T. Tolstik, S.E. Stanca, C. Krafft, J. Popp, Surface-enhanced Raman spectroscopy of cell lysates mixed with silver nanoparticles for tumor classification, Beilstein J. Nanotechnol. 8 (2017) 1183–1190. Available from: https://doi.org/10.3762/bjnano.8.120.
[20] N. Kim, M.R. Thomas, M.S. Bergholt, I.J. Pence, H. Seong, P. Charchar, et al., Surface enhanced Raman scattering artificial nose for high dimensionality fingerprinting, Nat. Commun. 11 (2020) 207. Available from: https://doi.org/10.1038/s41467-019-13615-2.
[21] W. Li, X. Zhao, Z. Yi, A.M. Glushenkov, L. Kong, Plasmonic substrates for surface enhanced Raman scattering, Anal. Chim. Acta 984 (2017) 19–41. Available from: https://doi.org/10.1016/j.aca.2017.06.002.
[22] P.A. Mosier-Boss, Review of SERS substrates for chemical sensing, Nanomaterials 7 (2017) 142. Available from: https://doi.org/10.3390/nano7060142.
[23] G.B. Braun, S.J. Lee, T. Laurence, N. Fera, L. Fabris, G.C. Bazan, et al., Generalized approach to SERS-active nanomaterials via controlled nanoparticle linking, polymer encapsulation, and small-molecule infusion, J. Phys. Chem. C 113 (2009) 13622–13629. Available from: https://doi.org/10.1021/jp903399p.
[24] M. Moskovits, Persistent misconceptions regarding SERS, Phys. Chem. Chem. Phys. 15 (2013) 5301–5311. Available from: https://doi.org/10.1039/C2CP44030J.
[25] Y. Zhang, B. Walkenfort, J.H. Yoon, S. Schlücker, W. Xie, Gold and silver nanoparticle monomers are non-SERS-active: a negative experimental study with silica-encapsulated Raman-reporter-coated metal colloids, Phys. Chem. Chem. Phys. 17 (2015) 21120–21126. Available from: https://doi.org/10.1039/C4CP05073H.
[26] R.E. Le, E. Blackie, M. Meyer, P.G. Etchegoin, Surface enhanced Raman scattering enhancement factors: a comprehensive study, J. Phys. Chem. C 111 (2007) 13794–13803. Available from: https://doi.org/10.1021/jp0687908.
[27] R.A. Alvarez-Puebla, E. Arceo, P.J.G. Goulet, J.J. Garrido, R.F. Aroca, Role of nanoparticle surface charge in surface-enhanced Raman scattering, J. Phys. Chem. B 109 (2005) 3787–3792. Available from: https://doi.org/10.1021/jp045015o.
[28] E. Garcia-Rico, R.A. Alvarez-Puebla, L. Guerrini, Direct surface-enhanced Raman scattering (SERS) spectroscopy of nucleic acids: from fundamental studies to real-life applications, Chem. Soc. Rev. 47 (2018) 4909–4923. Available from: https://doi.org/10.1039/C7CS00809K.

[29] C. Carrillo-Carrion, M. Carril, W.J. Parak, Techniques for the experimental investigation of the protein corona, Curr. Opin. Biotechnol. 46 (2017) 106–113. Available from: https://doi.org/10.1016/j.copbio.2017.02.009.
[30] K. Nienhaus, H. Wang, G.U. Nienhaus, Nanoparticles for biomedical applications: exploring and exploiting molecular interactions at the nano-bio interface, Mater. Today Adv. 5 (2020) 100036. Available from: https://doi.org/10.1016/j.mtadv.2019.100036.
[31] G.P. Szekeres, M. Montes-Bayón, J. Bettmer, J. Kneipp, Fragmentation of proteins in the corona of gold nanoparticles as observed in live cell surface-enhanced Raman scattering, Anal. Chem. 92 (2020) 8553–8560. Available from: https://doi.org/10.1021/acs.analchem.0c01404.
[32] R.A. Álvarez-Puebla, Effects of the excitation wavelength on the SERS spectrum, J. Phys. Chem. Lett. 3 (2012) 857–866. Available from: https://doi.org/10.1021/jz201625j.
[33] D.E. Gómez, Z.Q. Teo, M. Altissimo, T.J. Davis, S. Earl, A. Roberts, The dark side of plasmonics, Nano Lett. 13 (2013) 3722–3728. Available from: https://doi.org/10.1021/nl401656e.
[34] S.L. Kleinman, B. Sharma, M.G. Blaber, A.-I. Henry, N. Valley, R.G. Freeman, et al., Structure enhancement factor relationships in single gold nanoantennas by surface-enhanced Raman excitation spectroscopy, J. Am. Chem. Soc. 135 (2013) 301–308. Available from: https://doi.org/10.1021/ja309300d.
[35] K. Kneipp, H. Kneipp, J. Kneipp, Probing plasmonic nanostructures by photons and electrons, Chem. Sci. 6 (2015) 2721–2726. Available from: https://doi.org/10.1039/C4SC03508A.
[36] S. Sánchez-Cortés, J.V. García-Ramos, Anomalous Raman bands appearing in surface-enhanced Raman spectra, J. Raman Spectrosc. 29 (1998) 365–371. doi:10.1002/(SICI)1097-4555(199805)29:5 < 365::AID-JRS247 > 3.0.CO;2-Y.
[37] H. Fisk, C. Westley, N.J. Turner, R. Goodacre, Achieving optimal SERS through enhanced experimental design, J. Raman Spectrosc. 47 (2016) 59–66. Available from: https://doi.org/10.1002/jrs.4855.
[38] R. Leardi, Experimental design in chemistry: a tutorial, Anal. Chim. Acta 652 (2009) 161–172. Available from: https://doi.org/10.1016/j.aca.2009.06.015.
[39] C. Beleites, U. Neugebauer, T. Bocklitz, C. Krafft, J. Popp, Sample size planning for classification models, Anal. Chim. Acta 760 (2013) 25–33. Available from: https://doi.org/10.1016/j.aca.2012.11.007.
[40] S. Fornasaro, C. Beleites, V. Sergo, A. Bonifacio, Data analysis in SERS diagnostics, SERS for Point-of-Care and Clinical Applications, Elsevier, 2021.
[41] H. Mark, J. Workman Jr, Chemometrics in Spectroscopy, Academic Press, 2018.
[42] P.H.C. Eilers, Parametric time warping, Anal. Chem. 76 (2004) 404–411. Available from: https://doi.org/10.1021/ac034800e.
[43] S. He, W. Zhang, L. Liu, Y. Huang, J. He, W. Xie, et al., Baseline correction for Raman spectra using an improved asymmetric least squares method, Anal. Methods 6 (2014) 4402–4407. Available from: https://doi.org/10.1039/C4AY00068D.
[44] N.K. Afseth, V.H. Segtnan, J.P. Wold, Raman spectra of biological samples: a study of preprocessing methods, Appl. Spectrosc. (2016). Available from: https://doi.org/10.1366/000370206779321454.
[45] H. Martens, E. Stark, Extended multiplicative signal correction and spectral interference subtraction: new preprocessing methods for near infrared spectroscopy, J. Pharm. Biomed. Anal. 9 (1991) 625–635. Available from: https://doi.org/10.1016/0731-7085(91)80188-F.
[46] R. Bro, A.K. Smilde, Principal component analysis, Anal. Methods 6 (2014) 2812–2831. Available from: https://doi.org/10.1039/C3AY41907J.

[47] T. Fearn, Functional boxplots, NIR N. (2011) 19–20. Available from: https://doi.org/10.1255/nirn.1260.
[48] R. Bro, A.K. Smilde, Centering and scaling in component analysis, J. Chemom. 17 (2003) 16–33. Available from: https://doi.org/10.1002/cem.773.
[49] R. Nuzzo, How scientists fool themselves – and how they can stop, Nature 526 (2015) 182. Available from: https://doi.org/10.1038/526182a.
[50] J. Lever, M. Krzywinski, N. Altman, Model selection and overfitting, Nat. Methods 13 (2016) 703–704. Available from: https://doi.org/10.1038/nmeth.3968.
[51] K. Varmuza, P. Filzmoser, Introduction to Multivariate Statistical Analysis in Chemometrics, CRC Press, 2016.
[52] M. Barker, W. Rayens, Partial least squares for discrimination, J. Chemom. 17 (2003) 166–173. Available from: https://doi.org/10.1002/cem.785.
[53] P. Filzmoser, B. Liebmann, K. Varmuza, Repeated double cross validation, J. Chemom. 23 (2009) 160–171. Available from: https://doi.org/10.1002/cem.1225.
[54] D. Krstajic, L.J. Buturovic, D.E. Leahy, S. Thomas, Cross-validation pitfalls when selecting and assessing regression and classification models, J. Cheminformatics 6 (2014) 10. Available from: https://doi.org/10.1186/1758-2946-6-10.
[55] P. Hu, X.-S. Zheng, C. Zong, M.-H. Li, L.-Y. Zhang, W. Li, et al., Drop-coating deposition and surface-enhanced Raman spectroscopies (DCDRS and SERS) provide complementary information of whole human tears, J. Raman Spectrosc. 45 (2014) 565–573. Available from: https://doi.org/10.1002/jrs.4499.
[56] X. Chen, Z. Huang, S. Feng, J. Chen, L. Wang, P. Lu, et al., Analysis and differentiation of seminal plasma via polarized SERS spectroscopy, Int. J. Nanomed. 7 (2012). Available from: https://www.dovepress.com/analysis-and-differentiation-of-seminal-plasma-via-polarized-sers-spec-peer-reviewed-article-IJN. accessed 27.07.2020.
[57] Z. Huang, G. Cao, Y. Sun, S. Du, Y. Li, S. Feng, et al., Evaluation and optimization of paper-based SERS substrate for potential label-free Raman analysis of seminal plasma, J. Nanomater. (2017). Available from: https://www.hindawi.com/journals/jnm/2017/4807064/. accessed 27.07.2020.
[58] S. Fornasaro, E. Gurian, S. Pagarin, E. Genova, G. Stocco, G. Decorti, et al., Ergothioneine, a dietary amino acid with a high relevance for the interpretation of label-free surface enhanced Raman scattering (SERS) spectra of many biological samples, Spectrochim. Acta A 246 (2021) 119024. Available from: https://doi.org/10.1016/j.saa.2020.119024.
[59] S. Feng, J. Lin, Z. Huang, G. Chen, W. Chen, Y. Wang, et al., Esophageal cancer detection based on tissue surface-enhanced Raman spectroscopy and multivariate analysis, Appl. Phys. Lett. 102 (2013) 043702. Available from: https://doi.org/10.1063/1.4789996.
[60] J. Li, C. Wang, Y. Yao, Y. Zhu, C. Yan, Q. Zhuge, et al., Label-free discrimination of glioma brain tumors in different stages by surface enhanced Raman scattering, Talanta 216 (2020) 120983. Available from: https://doi.org/10.1016/j.talanta.2020.120983.
[61] S.C. Pinzaru, C.A. Dehelean, A. Falamas, N. Leopold, C. Lehene, Cancer tissue screening using surface enhanced Raman scattering, Laser Appl. Life Sci. 7376 (2010) 73760T. Available from: https://doi.org/10.1117/12.871378.
[62] S. Feng, R. Chen, J. Lin, J. Pan, G. Chen, Y. Li, et al., Nasopharyngeal cancer detection based on blood plasma surface-enhanced Raman spectroscopy and multivariate analysis, Biosens. Bioelectron. 25 (2010) 2414–2419. Available from: https://doi.org/10.1016/j.bios.2010.03.033.
[63] S. Feng, W. Wang, I.T. Tai, G. Chen, R. Chen, H. Zeng, Label-free surface-enhanced Raman spectroscopy for detection of colorectal cancer and precursor

lesions using blood plasma, Biomed. Opt. Express 6 (2015) 3494–3502. Available from: https://doi.org/10.1364/BOE.6.003494.
[64] J. Li, J. Ding, X. Liu, B. Tang, X. Bai, Y. Wang, et al., Label-free serum detection of *Trichinella spiralis* using surface-enhanced Raman spectroscopy combined with multivariate analysis, Acta Tropica 203 (2020) 105314. Available from: https://doi.org/10.1016/j.actatropica.2019.105314.
[65] D. Lin, Y. Wang, T. Wang, Y. Zhu, X. Lin, Y. Lin, et al., Metabolite profiling of human blood by surface-enhanced Raman spectroscopy for surgery assessment and tumor screening in breast cancer, Anal. Bioanal. Chem. 412 (2020) 1611–1618. Available from: https://doi.org/10.1007/s00216-020-02391-4.
[66] R. Liu, Y. Xiong, Y. Guo, M. Si, W. Tang, Label-free and non-invasive BS-SERS detection of liver cancer based on the solid device of silver nanofilm, J. Raman Spectrosc. 49 (2018) 1426–1434. Available from: https://doi.org/10.1002/jrs.5408.
[67] Y. Lu, Y. Lin, Z. Zheng, X. Tang, J. Lin, X. Liu, et al., Label free hepatitis B detection based on serum derivative surface enhanced Raman spectroscopy combined with multivariate analysis, Biomed. Opt. Express 9 (2018) 4755–4766. Available from: https://doi.org/10.1364/BOE.9.004755.
[68] L. Shao, A. Zhang, Z. Rong, C. Wang, X. Jia, K. Zhang, et al., Fast and non-invasive serum detection technology based on surface-enhanced Raman spectroscopy and multivariate statistical analysis for liver disease, Nanomed. Nanotechnol. Biol. Med. 14 (2018) 451–459. Available from: https://doi.org/10.1016/j.nano.2017.11.022.
[69] Q. Wu, S. Qiu, Y. Yu, W. Chen, H. Lin, D. Lin, et al., Assessment of the radiotherapy effect for nasopharyngeal cancer using plasma surface-enhanced Raman spectroscopy technology, Biomed. Opt. Express 9 (2018) 3413–3423. Available from: https://doi.org/10.1364/BOE.9.003413.
[70] Y.X. Chen, M.W. Chen, J.Y. Lin, W.Q. Lai, W. Huang, H.Y. Chen, et al., Label-free optical detection of acute myocardial infarction based on blood plasma surface-enhanced Raman spectroscopy, J. Appl. Spectrosc. 83 (2016) 798–804. Available from: https://doi.org/10.1007/s10812-016-0366-2.
[71] K. Liu, S. Jin, Z. Song, L. Jiang, L. Ma, Z. Zhang, Label-free surface-enhanced Raman spectroscopy of serum based on multivariate statistical analysis for the diagnosis and staging of lung adenocarcinoma, Vib. Spectrosc. 100 (2019) 177–184. Available from: https://doi.org/10.1016/j.vibspec.2018.12.007.
[72] Y. Kang, M. Si, R. Liu, S. Qiao, Surface-enhanced Raman scattering (SERS) spectra of hemoglobin on nano silver film prepared by electrolysis method, J. Raman Spectrosc. 41 (2010) 614–617. Available from: https://doi.org/10.1002/jrs.2489.
[73] Y. Kang, M. Si, Y. Zhu, L. Miao, G. Xu, Surface-enhanced Raman scattering (SERS) spectra of hemoglobin of mouse and rabbit with self-assembled nano-silver film, Spectrochim. Acta A: Mol. Biomol. Spectrosc. 108 (2013) 177–180. Available from: https://doi.org/10.1016/j.saa.2013.01.098.
[74] R. Liu, M. Si, Y. Kang, X. Zi, Z. Liu, D. Zhang, A simple method for preparation of Ag nanofilm used as active, stable, and biocompatible SERS substrate by using electrostatic self-assembly, J. Colloid Interface Sci. 343 (2010) 52–57. Available from: https://doi.org/10.1016/j.jcis.2009.11.042.
[75] R. Liu, Y. Xiong, W. Tang, Y. Guo, X. Yan, M. Si, Near-infrared surface-enhanced Raman spectroscopy (NIR-SERS) studies on oxyheamoglobin (OxyHb) of liver cancer based on PVA-Ag nanofilm, J. Raman Spectrosc. 44 (2013) 362–369. Available from: https://doi.org/10.1002/jrs.4216.
[76] A. Lorén, J. Engelbrektsson, C. Eliasson, M. Josefson, J. Abrahamsson, M. Johansson, et al., Internal standard in surface-enhanced Raman spectroscopy, Anal. Chem. 76 (2004) 7391–7395. Available from: https://doi.org/10.1021/ac0491298.

[77] F. Sun, H.-C. Hung, A. Sinclair, P. Zhang, T. Bai, D.D. Galvan, et al., Hierarchical zwitterionic modification of a SERS substrate enables real-time drug monitoring in blood plasma, Nat. Commun. 7 (2016) 13437. Available from: https://doi.org/10.1038/ncomms13437.
[78] X. Guo, J. Li, M. Arabi, X. Wang, Y. Wang, L. Chen, Molecular-imprinting-based surface-enhanced raman scattering sensors, ACS Sensors 5 (2020) 601–619. Available from: https://doi.org/10.1021/acssensors.9b02039.
[79] W.-I.K. Chio, S. Moorthy, J. Perumal, S. Du, I.P. Parkin, M. Olivo, et al., Dual-triggered nanoaggregates of cucurbit[7]uril and gold nanoparticles for multi-spectroscopic quantification of creatinine in urinalysis, J. Mater. Chem. C 8 (2020) 7051–7058. Available from: https://doi.org/10.1039/D0TC00931H.
[80] J.C. Fraire, V.N. Sueldo Ocello, L.G. Allende, A.V. Veglia, E.A. Coronado, Toward the design of highly stable small colloidal SERS substrates with supramolecular host–guest interactions for ultrasensitive detection, J. Phys. Chem. C 119 (2015) 8876–8888. Available from: https://doi.org/10.1021/acs.jpcc.5b01647.
[81] S. Kasera, F. Biedermann, J.J. Baumberg, O.A. Scherman, S. Mahajan, Quantitative SERS using the sequestration of small molecules inside precise plasmonic nanoconstructs, Nano Lett. 12 (2012) 5924–5928. Available from: https://doi.org/10.1021/nl303345z.
[82] M. Park, H. Jung, Y. Jeong, K.-H. Jeong, Plasmonic Schirmer strip for human tear-based Gouty arthritis diagnosis using surface-enhanced Raman scattering, ACS Nano 11 (2017) 438–443. Available from: https://doi.org/10.1021/acsnano.6b06196.
[83] M. Wen, B. Zhou, Y.-H. Chen, Z.-L. Ma, Y. Gou, C.-L. Zhang, et al., Serum uric acid levels in patients with Parkinson's disease: a meta-analysis, PLOS ONE 12 (2017) e0173731. Available from: https://doi.org/10.1371/journal.pone.0173731.
[84] C.-C. Chang, C.-H. Wu, L.-K. Liu, R.-H. Chou, C.-S. Kuo, P.-H. Huang, et al., Association between serum uric acid and cardiovascular risk in nonhypertensive and nondiabetic individuals: the Taiwan I-Lan longitudinal aging study, Sci. Rep. 8 (2018) 5234. Available from: https://doi.org/10.1038/s41598-018-22997-0.
[85] M.L. Muiesan, C. Agabiti-Rosei, A. Paini, M. Salvetti, Uric acid and cardiovascular disease: an update, Eur. Cardiol. Rev. 11 (2016) 54–59. Available from: https://doi.org/10.15420/ecr.2016:4:2.
[86] H. Yang, C. Zhao, R. Li, C. Shen, X. Cai, L. Sun, et al., Noninvasive and prospective diagnosis of coronary heart disease with urine using surface-enhanced Raman spectroscopy, Analyst 143 (2018) 2235–2242. Available from: https://doi.org/10.1039/C7AN02022H.
[87] E. Gurian, P. Giraudi, N. Rosso, C. Tiribelli, D. Bonazza, F. Zanconati, et al., Differentiation between stages of non-alcoholic fatty liver diseases using surface-enhanced Raman spectroscopy, Anal. Chim. Acta 1110 (2020) 190–198. Available from: https://doi.org/10.1016/j.aca.2020.02.040.
[88] X. Li, T. Yang, S. Li, L. Jin, D. Wang, D. Guan, et al., Noninvasive liver diseases detection based on serum surface enhanced Raman spectroscopy and statistical analysis, Opt. Express 23 (2015) 18361–18372. Available from: https://doi.org/10.1364/OE.23.018361.

CHAPTER 5

Surface-enhanced Raman scattering nanotags design and synthesis

Xiao-Dong Zhou[1], Xue Li[2] and Ai-Guo Shen[3]
[1]The Centre of Analysis and Measurement of Wuhan University, Wuhan University, Wuhan, P.R. China
[2]College of Chemistry and Molecular Sciences, Wuhan University, Wuhan, P.R. China
[3]School of Printing and Packaging, Wuhan University, Wuhan, P.R. China

5.1 SERS nanotags and its optical properties

SERS technique has the advantages of high sensitivity, high spectral resolution and suitable for the application of aqueous system, especially different from many optical analysis methods as discussed in Chapter 1, *Principles of surface-enhanced Raman spectroscopy*. SERS can not only detect the multicomponent directly without separation through the spectral fingerprint of the target [1–6], but also detect the SERS signals of the nanotags through the principle of derivation or sensing, so as to achieve indirect detection of the target [7]. For the latter, the primary problem of indirect measurement is to "rent-seeking" the substitute signal of the target, which not only can be uniquely identified, but also has the quantization condition.

In general, SERS nanotags could be prepared by binding or bonding the molecules with strong Raman scattering effect (i.e., Raman reporter molecules) to the surface of plasmonic gold or silver nanoparticles (NPs), and the expected SERS spectrum could be generated. Once this basic structure was formed and a protective shell layer and biometric elements (such as antibodies) were added, the SERS nanocabber was endowed with features such as spectral/physical stability and biocompatibility, as well as the function of specific identification and binding of the target object [8,9].

The blowout development of SERS nanotags was an important step forward for the spectroscopic analysis of biological samples. Compared with other optical probes (e.g., organic fluorescent dyes, fluorescent quantum dots, etc.) [10,11], it has four fascinating advantages. First, the

Principles and Clinical Diagnostic Applications of Surface-Enhanced Raman Spectroscopy
DOI: https://doi.org/10.1016/B978-0-12-821121-2.00011-1

sensitivity of SERS nanotags was sufficient for trace analysis. Several studies showed that if the substrate had high enhancement factor, the intensity of Raman scattering signals generated by SERS nanotags would be equal to or even stronger than fluorescence [12–14]. Second, the peak width of Raman scattering light is very narrow (~1 nm) [15], while the peak width of fluorescence is as wide as 50 nm [16]. Therefore, Raman scattering light as the output signal of optical label is particularly suitable for multiplex and nonpreprocessing detection. Third, the extremely short lifetime of Raman scattering light effectively prevented photobleaching, energy transfer and quenching of signal molecules in the excited state [17], which made SERS nanotags have high optical stability and suitable for long-term dynamic monitoring of biological processes [18]. Fourth, the use of infrared or near-infrared excitation light can weaken the interference of spontaneous fluorescence of cells and tissues [19], enabling SERS nanotags to achieve nondestructive imaging with high signal-to-noise ratio in vivo. As shown in Table 5.1, SERS nanotags, quantum dots and common dye molecules were comprehensively compared in terms of composition, size, luminescence principle, optical properties and biocompatibility, etc.

Table 5.1 Comprehensive comparison among SERS nanotags, quantum dots, and common dye molecules [12–19].

Properties	SERS nanotags	Quantum dots	Dye molecules
Physical principle	Raman scattering	Fluorescence-emission	Electron absorption/Fluorescenc-e-emission
Core composition	Au and Ag based NPs	Carbon, silicon and selenides-based NPs	Organic compounds
Size	~20–50 nm	~10 nm	~1 nm
Monochromaticity of emission (bandwidth)	<2 nm	~30–50 nm	Usually more than 50 nm
Multiplexing capacity of emission	~10–100	~3–10	~1–3
Photostability of emission	Antiphotobleaching	Decay under strong laser	Decay under weak excitation
Biocompatibility	Good	Poor	Poor

5.2 Clinical application of SERS nanotags: strategies and essence

With the emergence of a wide variety of noble metal NPs such as gold and silver, SERS fully reflected its advantages of high sensitivity and high selectivity to the research fields, for example, label or label-free detection [13,20], biochemical sensing [21,22], cellular imaging [23–25] and in vivo analysis [19,26]. Generally, the implementation strategies of biochemical or clinical analysis using SERS nanotags as the signal output were flexible and could be generally divided into the following three categories. In the first category, it is similar to the traditional fluorescent tags, with the help of SERS nanotags and relying on the high-resolution property of Raman signal to realize simultaneous detection of multiple species [27]. Such strategy is commonly seen in surface/interface label detection, or in label imaging studies of cells, biological tissues [28], etc. (see Fig. 5.1A). Secondly, the concentration of NPs is usually very low in the solution phase, therefore, the sensitivity is insufficient and it is difficult to achieve quantitative detection. When a special analyte in the solution can trigger the aggregation/dispersion of noble metal NPs, the Raman signal of reporters will show the response of "turn on/off" modes with the change of the surface coupling effect of particles and particles [29–32]. The quantitative relationship between the concentration of the analyte and the strength of a certain Raman band of the surface reporters can be established to achieve quantitative detection (see Fig. 5.1B). This strategy was

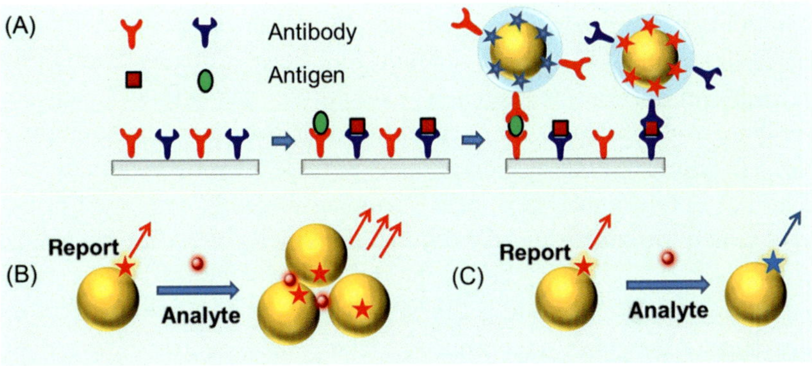

Figure 5.1 The strategies and schematics of SERS nanotags used for detection and imaging analysis. (A) immunosandwich assay using SERS tags for protein analyte detection; (B) Analyte-induced SERS tag aggregation; (C) analytes changing the Raman signature of the reporter.

derived from the visual colorimetric detection method based on the change of surface plasmon resonance (SPR) absorption of sol system caused by the change of dispersion state of noble metal NPs. The difference between SPR and SERS lies in that SERS has higher sensitivity and the quantitative basis of single narrow band is more accurate. The third category is based on the chemical/physical interaction between surface reporters and the analyte. When the intensity or displacement of individual Raman band changes, a quantitative relationship can be established between the variable and the concentration of the analyte [33] (see Fig. 5.1C). The difference between this detection strategy and the second one is that the former is to seek the quantitative evidence of the analyte from the dimension of intensity ratio or Raman shift change of Raman bands. The latter establishes the quantitative relationship only from the dimension of intensity ratio of Raman bands. It is very similar to the application system of the first kind of strategy and is suitable for sensing detection (or in-direct analysis) on the solid surface/interface. Whether it is detection strategies such as "turn on/off," Raman shift change and intensity ratio, or traditional label detection and imaging, the essence of SERS nanotags-based application is to make qualitative analysis based on the whole fingerprint information of reporters, and to make quantitative analysis based on the intensity/shift changes of certain Raman bands [34].

5.3 SERS nanotags design and synthesis

What exactly can a Raman spectrometer do? Researchers in different fields think differently. Nonanalysts consider Raman spectrometer as a characterization tool to identify different molecular structures through full spectrum information. Researchers working in the field of traditional optical analysis consider it more as a kind of signal output technology. A kind of typical SERS nanotags can provide sensitive, specific and stable (*3S*) optical signals [35]. SERS nanotag is mainly composed of two parts, namely the noble metal NPs and Raman reporter (usually a small organic molecules). After the excitation laser interacts with the noble metal nanostructure to produce the SPR effect, the near field space of the metal surface generates an enhanced electric field, which enhances the Raman signal of the molecules in this space [34]. Therefore the chemical composition, morphology, and size of metal nanostructures are closely related to the brightness of SERS nanotags [36]. In addition, the Raman reporters should be adsorbed or bonded to the surface of metal nanostructure in order to generate SERS fingerprint signals. However, the structure of this simple metal NP-Raman reporter nanotag is generally unstable and vulnerable to ambient interference [37,38], requiring (but not necessarily requiring) the additional protective shells and the "addition" of the outermost targeted molecules [39–41]. The former can improve the biocompatibility of the nanotags and reduce the nonspecific adsorption and irreversible aggregation, while the latter can give the nanotags biological targeting function. In this chapter, we will first introduce the research progress on how to improve the brightness, stability (small blinking) and signal-to-noise ratio of SERS nanotags from three aspects of the substrates, reporters and shell materials. In addition, the design and application of SERS nanotags in various aspects such as super multivariate analysis, multimode imaging and adjuvant therapy were introduced from the perspectives of signal coding and functional expansion.

5.3.1 Highly bright SERS nanotags: substrate construction

Noble metal nanomaterials are not only the signal amplifier of SERS nanotags, but also the structural support. Generally, the size, geometry and chemical composition of the material will affect the Raman enhancement. With the rise of nanoscience, many nanoparticles with different morphologies and clusters of nanoparticles have been used as SERS substrate (see Chapter 2, Nanoplasmonic materials for surface-enhanced Raman scattering), which greatly expands the type of SERS substrates and provides convenience for the analysis of various biological samples.

5.3.1.1 Single gold/silver NPs

According to the electromagnetic enhancement theory [42,43], the strength of SERS signal is closely related to the vibration frequency of SPR of the metal substrate and the frequency of incident laser. When the two are close, the phenomenon of SPR occurres, and the field enhancement effect reaches the maximum at this time, which is conducive to the generation of SERS signals [34]. In practical research, in order to avoid the interference of spontaneous fluorescence of biological endogenous substances, near-infrared laser with wavelength of 632.8 or 785 nm is used as excitation laser. Therefore, the preparation of noble metal nanomaterials with suitable SPR absorption wavelength has become the focus of SERS research. In this section, several typical noble metal nanoparticles will be listed and the effects of size, morphology, and composition on their SPR will be briefly discussed. For the detailed discussion on the nanoplasmoinic materials, please refer to Chapter 2.

Gold/silver nanospheres. Spherical Au/Ag NPs are the most widely used SERS substrates because their preparation methods are relatively simple and have some obvious advantages, such as easy control of particle size, uniformity, good stability and so on. Therefore, this kind of SERS substrate is usually used for simple and primary research on biological samples [44–46]. After years of exploration and practice, the preparation methods of spherical gold/silver NPs have been gradually unified: gold NPs can be prepared by reducing $HAuCl_4$ with trisodium citrate at high temperature (Frens method) [47]; silver NPs can be obtained by reducing silver nitrate with trisodium citrate at high temperature (Lee-Meisel method) [48], and also can be prepared by reducing silver nitrate with hydroxylamine hydrochloride at room temperature (Leopold method) [49]. As shown in Fig. 5.2, the surface plasmon vibration frequency of gold and silver NPs prepared by these methods is generally in the range of 400–600 nm, and red shift occurs with the increase of particle size [50]. Objectively, silver is a more efficient SERS substrate material, and the SERS signal generated by silver NPs is 10–100 times stronger than that of gold NPs [51]. However, in most cases, people prefer to use gold NPs as SERS substrate, the main reason is that the chemical properties of silver are active, and the stability of SERS enhancement on silver is poor. In addition, the biocompatibility of silver is poor, and there are potential safety risks in vivo detection [52].

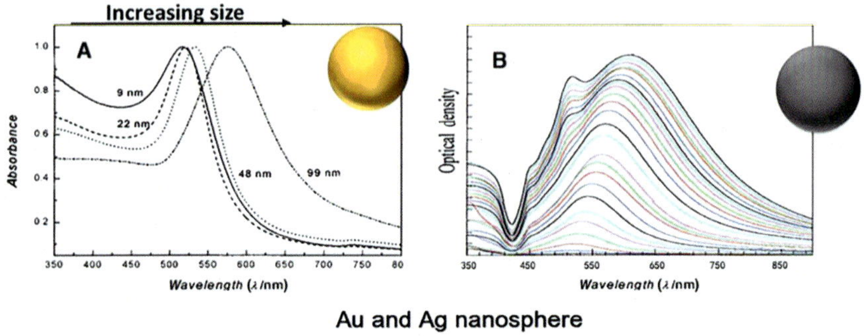

Figure 5.2 Surface plasmon resonance absorption wavelengths of Au and Ag nanospheres (350–800 nm). (A) UV-vis absorption spectra of different sized gold nanoparticles (NPs) in water. The particle size is 9, 22, 48, and 99 nm, respectively. (B) Extinction spectra of different sized silver NPs. The particle size is 29, 34, 37, 44, 48, 52, 58, 61, 75, 78, 92, 97, 105, 113, 120, and 136 nm, respectively.

The size of spherical Au/Ag NPs also has an impact on the enhancement effect because the strength of the electromagnetic field depends largely on the number of excited electrons. In this regard, the increase of particle size is conducive to signal enhancement [51]. On the other hand, the increase of particle size will lead to greater radiation damping effect, thus, reducing the enhancement factor. Therefore, the selection of particle size needs to be verified by experiments. It has been found that only spherical Au/Ag NPs with a particle size of 30–100 nm have good enhancement ability [53].

In view of the above, the appearance of gold/ silver composite NPs well balances the problems of strengthening effect and stability. The SERS enhancement effect of gold/silver composite NPs is better than that of silver NPs and has better stability. Gold/silver composite NPs generally include three types: gold core and silver shell NPs [54], silver core and gold shell NPs [55] and gold/silver alloy NPs [56]. There are many studies on the synthesis of gold /silver alloy NPs, mainly because the lattice parameters of gold and silver are very similar. It is easy to obtain the alloy structure when the two metal salts are reduced at the same time with strong reducing agent.

Gold nanoshells. Gold nanoshells are also SERS substrates that have been studied more in recent years, such as spherical gold coated silicon NPs [57], hollow gold NPs [58] and hollow cage NPs [59], etc. Strong local electromagnetic field can be generated inside such hollow NPs, which is similar to the SERS "hot spot" formed between dimer gold NPs and has a great promotion effect on the enhancement of Raman signal [60]. Moreover, the SPR can be tuned by changing the ratio of the inner and outer radius of the hollow gold NPs, and the tuning range can cover most from ultraviolet to near-infrared wavelength region [61]. As shown in Fig. 5.3A, when the wall thickness of hollow gold NPs with a diameter of 40 nm is reduced from 6.8 to 3 nm, its SPR gradually shifts from 610 to 820 nm. For in vivo biological experiments, the selection of near-infrared excitation laser can not only obtain the high enhancement activity of the substrate, but also effectively avoid the interference of spontaneous fluorescence of biological molecules [62].

Gold nanorods. The spherical or hollow gold NPs mentioned above have only one SPR peak, while the gold nanorods have two SPR peaks: one in the visible region, which is close to the absorption peak of the gold NPs. This band corresponds to the oscillation of electrons in the transverse direction. The other one can be adjusted in the visible and

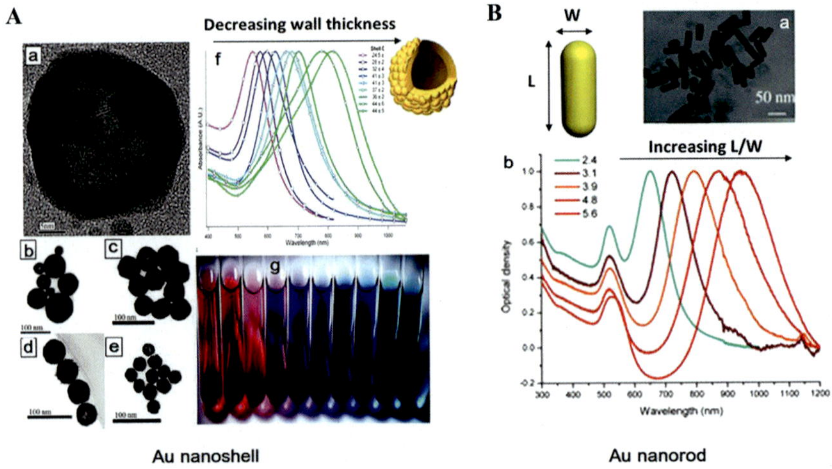

Figure 5.3 Synthesis and characterization of gold nanoshell and gold nanorod. (A) (a) High-resolution transmission electron micrographs (TEM) of a single 30 nm hollow Au nanoshell (HGN) with a wall thickness of approximately 4 nm. (b–e) Low-resolution TEM images of particles of 71 ± 17 nm (b), 50 ± 5 nm (c), 40 ± 3.5 nm (d), and 28 ± 2.3 nm (e), respectively. (f) UV-vis absorption spectra of nine HGN samples with varying diameters and wall thicknesses, and (g) the color range of HGN solutions. (B) (a) TEM image of gold nanorods (AuNRs) of aspect ratio 3.9. Surface plasmon absorption spectra of AuNRs of different aspect ratios (b), showing the sensitivity of the strong longitudinal band to the aspect ratios of the nanorods.

near-infrared regions, and this peak corresponds to the oscillation of electrons in the longitudinal direction [63]. As shown in Fig. 5.3B, the aspect ratio of gold nanorods can be adjusted from 2.4 to 5.6 by changing the amount of silver nitrate in the synthesis, and the SPR peak corresponding to the aspect ratio direction will move from the visible region to the near-infrared region [64]. In theory, the absorption coefficient of gold nanorods is one order of magnitude higher than that of gold nanospheres [65]. Therefore, in addition to being used as SERS substrates, gold nanorods can also be used for photothermal therapy (PTT) of tumor cells [64,66,67].

5.3.1.2 Gold nanoparticles rich in tips or gaps

The surface electromagnetic field intensity of noble metal nanoparticles is often unevenly distributed, which is highly related to the morphology of metal nanoparticles. In the narrow space region, such as the tip of noble metal nanoparticles or the gap between nanoparticles, the electromagnetic field strength is very high. In other words, these regions will have much better SERS enhancement effect than other regions, which are commonly known as SERS "hot spots" [12,68–70]. In order to improve the brightness of SERS nanotags, it is necessary to build high-density SERS "hot spots" on a single nanostructure. There are two ways to realize SERS substrate with this characteristic: one is to control the growth of metal single crystal, polycrystalline and twin to obtain anisotropic noble metal nanoparticles, which often have tips or intragranular voids; Second, we control/prepare clusters rich in intergap hotspots in two-dimensional/three-dimensional space by physical and chemical methods as reinforcement substrate. The first method is introduced here, and the substrate of the nanoparticle cluster will be discussed separately.

Gold nanostars, namely gold NPs rich in tips. As for anisotropic particles, localized surface plasmon resonance (LSPR) effect is not uniformly distributed on the whole particle surface, but causes the concentration of electromagnetic field density in several specific regions of the NPs [71]. Such electromagnetic field concentration was observed in the corner of the nano triangle, the end of the nanorod, the middle depression of the nano peanut, and the corner of the nanorod and nano cube. In the SERS substrate of monodisperse NPs, the NPs with multiple branched structures have been paid close attention, such as nanostar (Fig. 5.4A) [72,73], nanoflower [74,75] and so on. However, the synthesis with high efficiency, good biocompatibility and high yield is still a challenge. Shen et al. [76] added and mixed chitosan, chloroauric acid, silver nitrate, and ascorbic acid in turn at room temperature to prepare chitosan coated gold/silver

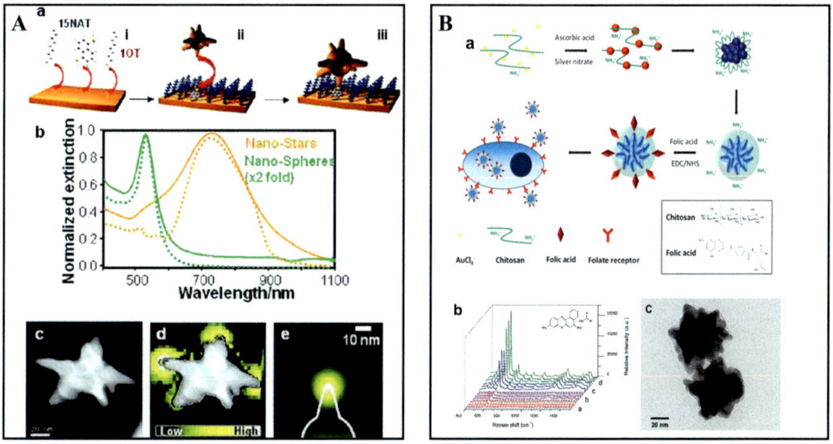

Figure 5.4 Synthesis and characterization of gold nanostar (A) and gold nanoflower (B).

nanoflowers (Fig. 5.4B). Chitosan was used to adjust the size of NPs in this process, while $AgNO_3$ was used to improve the dispersion of NPs. The results show that the anisotropic flower like nanomaterials have stronger and more stable SERS signals than spherical NPs. One reason is that the radius of curvature at the ends of their branches shrinks sharply, and the resulting super local electromagnetic field has a very large enhancement factor, which is usually called "tip effect" [77–79]. The other reason is that the ends of their branches have a very large specific surface area, which can connect more molecules and cause local signal enhancement. After the systematic synthesis by Liz Marzán et al. and the theoretical explanation of its plasmonic effect by Nordlander et al., the morphology of gold nanostar was established [80]. Nevertheless, these particles still appear in the literature with various names. For example, they are defined as "branched" gold NPs, gold nanoflowers, gold "nanoseaurchin," "hedgehog" gold NPs, "dendritic" gold NPs, gold "nanopopcorn," and even gold "nanospines." The different synthetic methods are the main reasons for the different morphology, which will not be listed here.

Single gold NPs rich in "inner gap." According to the above discussion of SERS "hot spots," the narrow areas in space include not only the tips of "lightning rod effect," but also the porous, concave structure and other nanogaps on the surface of single NPs. In addition, the nano "inner gap" is an ultra-small gap formed in a single nanoparticle. In the process of synthesizing NPs with ultra-small gaps, Raman reporters can be placed in the gap, which can enhance the Raman signal efficiently. Lim et al. [81] synthesized a structure with ultra-small (less than 1 nm) nanogaps: this structure takes gold NPs as the core, where DNA were modified to act as a bridge, and grows into gold shell on the outside through reduction of $HAuCl_4$ by hydroxylamine hydrochloride. In order to enhance the Raman signal effectively, three different lengths of DNA were modified to regulate the size of the nanogaps. The smaller the gap, the stronger the Raman signal (Fig. 5.5A). Li et al. [82] used gold NPs as the core, modified mercapto polyethylene glycol and Raman reporters on the surface, and then coated with the Pluronic F127 gel layer to play a protective role. After the growth of the silver shell, $HAuCl_4$ was used to corrode the silver shell. At the same time, ascorbic acid (AA) was used to reduce the gold atoms, forming an ultra small nanogap inside the particles, which has super strong Raman signal (Fig. 5.5B).

Based on the consideration of fast imaging, Ye and his team from Shanghai Jiaotong University has designed and synthesized a series of

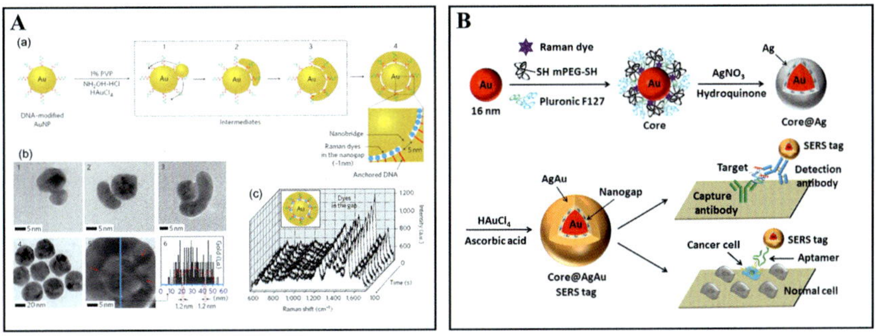

Figure 5.5 (A) Surface DNA-mediated synthesis and characterization of DNA-anchored nanobridged nanogap particles; (B) Schematic illustration of the synthesis and application of SERS-active plasmonic nanoparticles with interior nanogap.

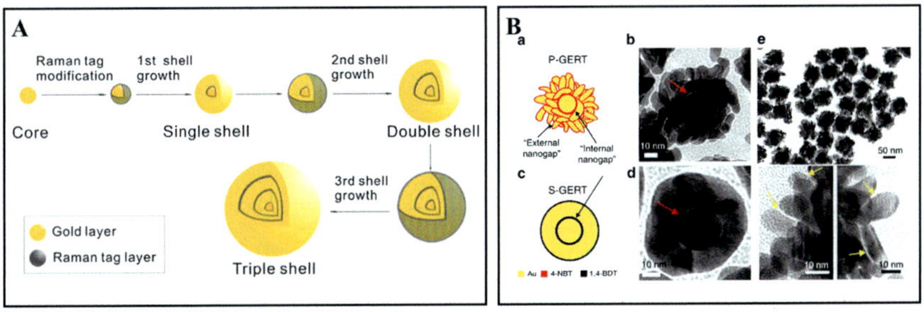

Figure 5.6 (A) Schematic illustration of the synthesis of nanomatryoshkas (NMs) with single, double, and triple Au shell layers; (B) The comparison of P-GERTs and S-GERTs. Schematic diagrams and representative TEM images of GERTs with a petal-like shell (P-GERTs, a, b), and GERTs with a smooth-surface shell (S-GERTs, c, d).

nanogap enhanced Raman tags (GERTs), one of which is multishell nanostructure [83]. First, the uniform gold cores were coated with a layer of Raman reporters as seeds to grow the first layer of gold shell. Next, the obtained particles were coated with a Raman reporter layer again, and then underwent a second gold shell growth to form a double shell nanostructure. By repeating this process, these particles become nanostructures with three or more shells. In this ingenious design, 1,4-phenyldithiol (BDT), as a Raman reporter, can contribute more Raman signals than general SERS nanotags in terms of the content. At the same time, BDT, as an insulating layer, provides conditions for generating more shells (Fig. 5.6A).

Another kind of ultra bright nanogap enhanced Raman tags (p-GERTs) structure is also from Ye's group. It has strong electromagnetic hot spots from the inner subnano gap and the outer petal like shell structure, large surface area, and Raman scattering cross section of reporter molecules [84]. The design and preparation process are as follows: the uniform 22 nm gold core was modified by 4-BNT through Au-S bond, and the excess molecules are removed by centrifugation, and then the obtained 4-NBT modified gold core was used as seed to further grow the gold shell. TEM images confirmed that by there was a uniform internal gap between gold core and rough shell structure (shown by the red arrow in Fig. 5.6B(b)) in a single p-GERTs, which was determined by the thickness of the monolayer of embedded 4-NBT molecules. Different from s-GERTs (Fig. 5.6B(d)) 2 with a complete and smooth outer gold shell, a large number of 1–3 nm nanogaps (shown by the yellow arrow in Fig. 5.6B(e)) were formed in the rough petal like structure within the gold shell of p-GERTs, which were defined as external nanogaps. Therefore 4-NBT molecules were further modified on the external nanogap through the self-assembly process. The total diameter of p-GERTs was 66 ± 4 nm, with monodisperse particle size and morphology (Fig. 5.6B(e)). The Raman enhancement factor of these GERTs is more than 5×10^9, and the detection sensitivity is as low as that of single NPs. Only 370 μW laser was used to achieve high-resolution cell imaging within 6 s, and sentinel lymph node imaging with high contrast (SNR 80) and wide area (3.2×2.8 cm^2) within 52 s.

5.3.1.3 Gold/silver nanoaggregates

Nanoaggregates rich in "inter gap." When the NPs aggregate with each other, there will present a large number of gaps called "hot spots". In these areas, the physical enhancement factor of SERS can reach above 10^{10}, even reach the level of single molecule detection. Therefore, people have done a lot of research on the aggregation of gold/silver NPs. The aggregation of gold/silver NPs will have obvious absorption in the near-infrared region, the absorption peak of Ag clusters is about 500–600 nm, and that of Au clusters is about 700–900 nm [85]. This is very similar to the wavelength of near-infrared laser, which is often used in the detection of biological samples, and can promote the generation of high-intensity Raman signal. Aggregation does bring ultra-high enhancement factor, but at the same time, the order of aggregation, the homogeneity of materials and the stability of signal have been the focus and difficulty of this type's research.

When the gold or silver NPs were prepared by sodium citrate reduction, the surface of NPs was covered with citrate protective layer and had

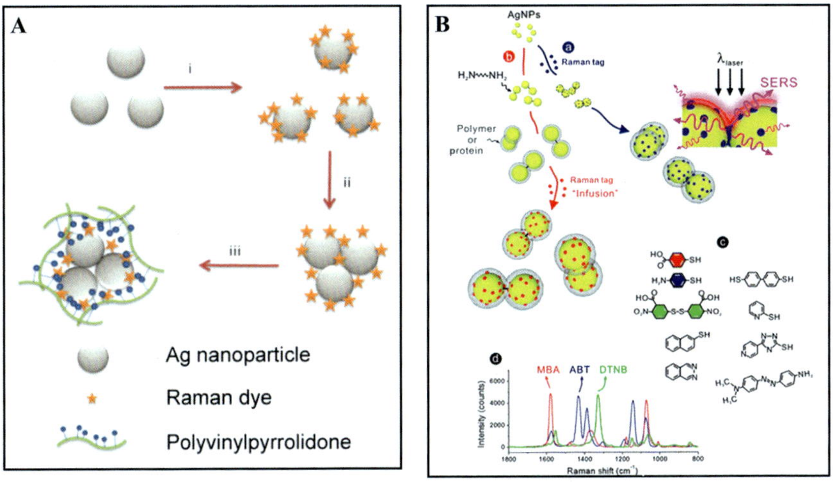

Figure 5.7 (A) Preparation of PVP-coated, dye-tagged Ag nanoparticles; (B) SERS nanocapsule synthesis. Ag nanoparticles are cross-linked with the bifunctional linker 4-aminobenzenethiol (ABT, blue) or (b) 1,6-hexamethylenediamine (HMD, black), each then coated with a layer of PVPA.

negative charge. When strong electrolytes such as salt were added, the electrostatic repulsion between NPs will be destroyed, and the aggregation phenomenon will occur [86,87]. By precisely controlling the amount of salt to control the aggregation rate, and then adding polymer to wrap the NPs, the aggregation process can be terminated. In 2009, Tan et al. [87] successfully controlled the aggregation of gold NPs by using polyvinylpyrrolidone (PVP), and finally synthesized SERS probe with super Raman signal (Fig. 5.7A); Braun [88] also achieved the same effect by using polyvinylpyrrolidone polyacrylic acid (PVPA) (Fig. 5.7B). Not only the strong electrolyte, but also the substitution of citrate will affect the stability of gold NPs when Raman reporter molecules were added [89,90]. If the added Raman reporter molecules have strong positive charge, the above aggregation phenomenon will be more obvious. In addition, this process can also be realized by adjusting the pH: when x-rhodamine-5-(-6-) isothiocyanate (xritc) was added to the gold colloid, a slight aggregation occurred when the pH of the system was between 7 and 10, and a large amount of aggregation occurred rapidly when the pH was below 5 [91].

As shown in Fig. 5.8A, it is a more controllable method to load the smaller noble metal NPs on the large-size materials to form the aggregation coating layer. In this method, silica, polystyrene and polymer spheres were

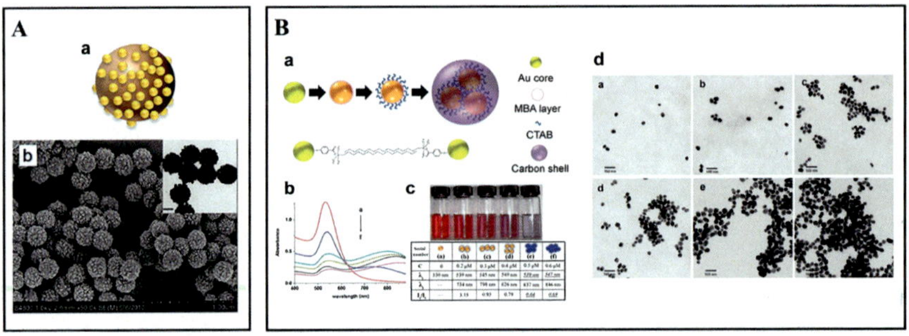

Figure 5.8 (A) TEM images of typical PSA@Ag NPs composite microspheres; (B) Schematic illustration of the synthesis of CEGNAs and the reaction mode between CTAB and 4-MBA (a). The UV-vis spectra of 4-MBA coated gold NPs (b). Gold nanoaggregates mediated by CTAB at different concentrations (c). Corresponding TEM images of gold NPs (d).

often used as carriers for gold NPs aggregation [92–94]. When the surface of polymer support was treated with 3-mercaptopropyltrimethoxysilicon (MPTMS) or other methods to produce a large number of −SH groups, a layer of noble metal NPs could directly grow on the surface by in situ synthesis, and the number and size of the aggregates could also be adjusted by the amount of $HAuCl_4$ or $AgNO_3$ [95]. The SERS nanotags prepared by this method usually have strong SERS response and good signal stability. However, the size of tags is usually hundreds of nanometers or even larger, which has been limited to in vivo experiments. Similarly, Shen et al. [96] utilized a simple hydrothermal one pot method to prepare carbon materials and used them as the coating layer of the aggregates. Under the mediation of cetyltrimethylammonium bromide, the size of gold nanoaggregates was controlled by electrostatic force and hydrophobic interaction. With the increase of the number of particles in the aggregate, the SERS intensity of a single SERS nanotags increased greatly. Although biocompatibility is excellent, it was only used in cell imaging. The in vivo application of this tag has not been reported (Fig. 5.8B).

5.3.2 Weak-background SERS nanotags: signal output

The second step of SERS nanotags preparation is the selection and modification of Raman reporter molecules. When the molecules are located in the SPR region of SERS substrate, ultra-high intensity SERS fingerprint signals will be generated. In order to prepare highly sensitive and stable SERS nanotags, we need to screen the reporter molecules according to the following principles: (1) the selected reporter molecules should directly bond to the surface of SERS substrates, because the chemical enhancement effect requires direct contact between the reporter molecules and the substrates, and the electromagnetic field will increase sharply with the distance between the reporter molecules and the SERS substrates [43]. In addition, the bonding between the molecules and the substrates can also avoid the shedding of the reporter molecules in the subsequent modification process, which is very important for the signal strength and stability of SERS nanotags [97,98]. Therefore, the more commonly used Raman reporter molecules contain sulfhydryl or amino groups, which have strong binding affinity with noble metal elements [99,100]. (2) The selected reporter molecule should have a large Raman scattering cross section to improve the intensity of SERS signal [101]. (3) The absorption wavelength of the selected reporter molecule should be close to the wavelength of the excitation light, resulting in the appearance of surface-enhanced resonance Raman scattering (SERRS), which will increase the signal intensity by 2–3 orders of magnitude [102]. (4) The selected reporter molecule should not have too many characteristic peaks, which can effectively reduce the spectral overlap between different SERS characteristic peaks in multichannel detection and improve the accuracy of detection. After years of research and practice, a large number of Raman reporter molecules have been selected, which are usually divided into three categories: nitrogen-containing cationic molecules, sulfur-containing molecules, and sulfhydryl small molecules. Their advantages and disadvantages are summarized in Table 5.2.

5.3.2.1 New generation of Raman reporters in "bio-silent" region

In the past decade, SERS has been widely used in various analytical fields, especially in biochemistry and life sciences [103,104]. As an indispensable part of the research, SERS nanotags have always been a research hot topic. In order to get better experimental results, a lot of research have been done on the selection and preparation of SERS substrates [105,106]. However, as another major factor determining the quality of the nanotags, the selection and design of Raman reporter molecules have not attracted enough attention. Novel reporter molecules usually need reasonable design, careful screening and systematic characterization. Compared with the "ready-made" Raman reporter molecules, it is undoubtedly too cumbersome. As a result, there are few articles on the design of Raman reporter molecules nowadays [107,108]. However, in the complex

Table 5.2 Classification and characteristics of typical Raman reporter molecules for SERS.

Category	Connection mode	Advantages	Disadvantages	Examples
Nitrogen containing cationic molecule	Electrostatic adsorption and N–Au (Ag) bonding	Low cost, large Raman cross section and easy to implement SERRS	Weak affinity, poor signal stability and difficulty in subsequent modification	Crystal violet (CV) [132], rhodamine 6G (R6G) [154], Nile Blue (NB) [155]
Sulfur molecule	S–Au (Ag) bonding	strong affinity, easy to implement SERRS and simple subsequent modification	High price, few species, difficult to realize self-assembly monolayer	Malachite green (mg) [157] and 3.3'diethylthioaldehyde tricarbocyanine iodide [65]
Sulfhydryl molecule	S–Au (Ag) bonding	Low price, strong affinity and simple Raman characteristic peak	Small scattering cross section, No SERRS potential	*p*-mercaptobenzoic acid [129], *p*-mercaptoaniline [156]

research system, the random selection of Raman reporter molecules in SERS nanotags may become an obstacle to the further development ofSERS applications, because interference molecules or target molecules and traditional Raman reporter molecules are prone to spectral overlap, which will affect the accuracy of the results. With the emergence of the concept of multitarget simultaneous detection, this problem will even extend to all kinds of SERS nanotags. At this time, the difficulty of traditional Raman reporter molecules selection will be further magnified [109]. In the face of such difficulties, a large number of novel Raman reporter molecules without overlapping are urgently needed for anti-interference and multitarget analysis and detection of complex systems.

Encouragingly, researchers have soon discovered that all mammalian cells exhibit a Raman silent region and it ranged from 1800 to 2800 cm^{-1}, which was further regarded as biological Raman transparent window that was almost all biological endogenous molecules do not produce Raman scattering signals. And then, this discovery prompted that the alkynes were gradually served as attractive Raman bioorthogonal tags with specific Raman signals in Raman silent region, of which "bioorthogonal" emerged as a hybrid strategy between the Raman spectroscopy and the unique Raman vibration of some tiny chemical bonds, thereby endowing the development of Raman labeling in biological Raman transparent window.

5.3.2.1.1 Small triple-bond containing molecules

On basis of EM theory, when molecules containing triple bond located onto the metallic surface, the Raman vibration of triple bond could be amplified. Therefore, the vibration of triple bond with the amplified capability of SERS has also got extremely development in biological labeling, performing higher sensitivity than spontaneous Raman scattering. Metal-carbonyl complexes have been recommended as drug candidates, which could be against cancer and infectious diseases, and could be served as "solid storage forms" for carbon monoxide, thus metal-carbonyl complexes are significant by direct monitoring. Varga et al. [110] first reported that metal carbonyl fragments could be served as a new class of markers in molecular biology, but they used the infrared spectra of $v_{(CO)}$ peaks in the 2150–1800 cm^{-1} region to monitor these active molecules. However, the infrared spectroscopy is not suitable in biological application because of the wide spectral absorption from the background of H_2O. Significantly, Meister et al. [111] first investigated the uptake and cellular distribution of metal-carbonyl compound through non-invasive Raman micro-spectroscopy. They proved that strong metal-C≡O stretching vibration with Raman signals at 1963 cm^{-1} of $[Mn(tpm)(CO)_3]Cl$ (tpm = tris (1-pyrazolyl) methane) was an ideal marker for Raman imaging. In this study, they referred that the Raman vibrational signal of metal-C≡O between 1800 and 2200 cm^{-1}, where the vibrational signals from the constituents of the cell were negligible. The as-designed metal carbonyl complex has suggested the futuristic potential of metal carbonyls for Raman labeling in cellular silent region. However, most metal carbonyl compounds faced the problems of the low Raman cross section along with a high concentration requirement, and low solubility in water, which limited its utility for biomedical applications. In order to overcome these shortcomings, Olivo et al. [112] first explored metal carbonyl-gold NPs conjugates for enhancing the metal carbonyl Raman intensity with the help of SERS. They reported that an organometallic osmium carbonyl cluster, $Os_3(CO)_{10}(m\text{-}H)_2$, onto the surface of gold NPs (henceforth designated as "OM-NP"), could significantly enhance the inherently weak Raman signal of the metal carbonyl CO stretching vibrations, and the CO intensity was increased by over four orders of magnitude (a factor of ≈15,000). Meanwhile, OM-NP conjugates were conjugated with Polyethylene glycol (PEG) and anti-EGFR antibody to ensure the water solubility and targeting capability of the conjugates. Good imaging potential for living cancer cells using the Raman signal at 2030 cm^{-1} was verified in this system. After that, Olivo et al. [113] further reported metal carbonyl probe for biological application with SERS-based assay. They synthesized a triosmium carbonyl cluster-boronic acid conjugate for glucose quantification with highly specific and sensitive mode. According to their design, they utilized the C≡O stretching vibrations of the metal carbonyl, which lied in a silent region of the SERS spectrum (1800–2200 cm^{-1}), for signal output to reflect the level of glucose, even in a human urine sample doped condition (Fig. 5.9).

According to the previous reports, cyano bond (C≡N) was served as one of the decent reporters for biological transparent labeling by using infrared absorption spectra. In 2006, Suydam et al. [114] incorporated nitriles into inhibitor to deliver a unique probe vibration to the active site of human aldose reductase, and the observed frequency shifts also demonstrated the utility of nitrile vibrations as probes of electrostatic fields in proteins. In their paper, they referred that the nitrile stretching mode has nearly ideal properties for a vibrational probe, because its frequency is distinct from those of any protein absorption. Besides, cyano bond is the

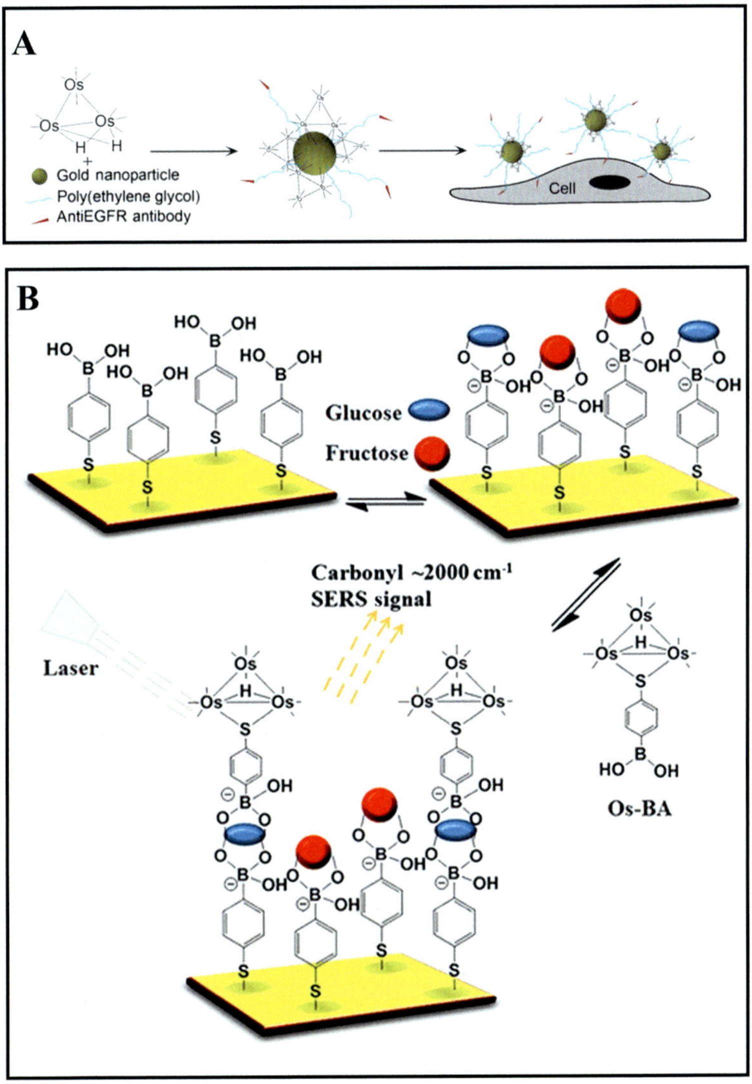

Figure 5.9 (A) Conjugates of organometallic osmium carbonyl clusters and gold nanoparticles. (B) Schematic for the Fabrication of BA-Functionalized BMFON. A glucose molecule brings Os-BA to the substrate via formation of a bidentate complex. Inset image: photograph of BMFON.

chemical group of nitriles, which has received a great deal of attention as another important class of Raman reporters and is widely used for Raman labeling in biological Raman transparent window. Noestheden et al. [115] used the N-hydroxy succinimide (NHS) bioconjugation reaction to incorporate nitrile (CN) modes onto proteins, that was 4-CN-benzoyl-NHS (4-CN-NHS) modified HVHP428 (HVHP428-CN). Among them, nitrile was served as stable reagents for obtaining in vivo track by Raman and coherent anti-Stokes Raman spectroscopy (CARS) microscopies. Notably, they also referred that nitrile groups were chosen because they possessed orthogonal vibrational modes when compared to endogenous cellular materials. In addition, apart from this direct incorporation of CN into the targeted molecular structure, an exogenous Raman probe-based CN group has also attracted more attentions, especially in the SERS and stimulated Raman scattering (SRS) tags. As a CN-based SERS reporter, commercialized mercaptobenzonitrile (MBN) is popular because of the nitrile and thiol group, which gives Raman signal at the interference free region and endows the strong chemisorption capability with plasmonic metal core such as gold NPs. For example, Liu et al. [116] developed MBN-embedded SERS nanotags as the background free Raman reporter for cancer biomarkers imaging at the single-cell level. These tags exhibited a sharp and strong peak at 2232 cm^{-1} (nitrile vibration) in the cellular Raman-silent region (1800–2800 cm^{-1}), and possessed excellent Raman imaging capability in MCF-7 cell lines. Of course, other synthesized Raman reporters containing nitrile and the chemisorbed groups can be similarly designed, of which nitrile group as zero background Raman signal using for bio-orthogonal labeling and chemisorbed groups can endow the strong affinity for the metal surface to avoid the desorption of Raman reporters in physiological environment (Fig. 5.10).

In the early studies about alkynes-based Raman scattering, Raman spectroscopy has been demonstrated for directly detecting small molecule containing alkynes in several samples. Yamakhoshi and coworkers [117] made their great contributions for promoting alkynes as Raman tags in living biological labeling. They first designed the alkynes moiety to be conjugated with the thymidine analog, 5-Ethenyl-2-deoxy uridine (EDU), and then achieved direct Raman imaging of an alkyne-modified molecule in living cell (Fig. 5.11C). Notably, EDU was reported as an alternative for 5-bromo-2′-deoxyuridine (BrdU), which is used for the measurement of DNA synthesis in cells. As expected, this alkyne-tagged EDU could also be demonstrated by real-time monitoring of active DNA synthesis in living HeLa cells, and its localization in the nucleus was clearly visualized by Raman scattering technology. Alkyne labeling has unique non-interference Raman emission, but its further application in the field of biochemical labeling is hindered due to its very weak spontaneous Raman scattering. Shen et al. [118] used DFT

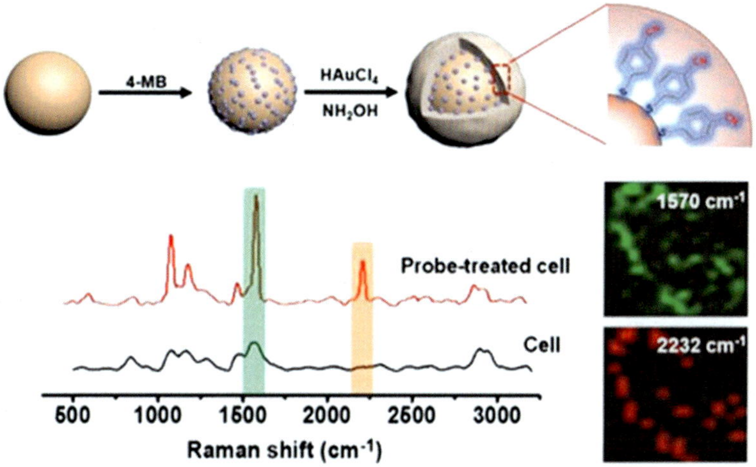

Figure 5.10 Schematic illustration of procedure for the preparation of Au@4-MB@Au NPs. Raman spectra of single MCF-7 cells before (*black*) and after (*red*) treating with the as-prepared IF-SERS nanotags (0.15×10^{-9} m). Bright-field (BF), mapping images, and merged images of a single MCF-7 cell treated with the IF-SERS nanotags, which were acquired in 1570 cm^{-1} channel (strong background noise) and 2232 cm^{-1} channel (negligible background noise).

calculation method to construct the SERS palette modulated by alkyne based on the reasonably designed 4-ethynylphenylmercaptan derivatives (Fig. 5.11A). Among them, Au@Ag NPs were used as the core of optical reinforcement, polyacrylamine was used as the shell for protecting and bonding. Even for pigment rich plant cells (such as pollen), alkyne encoded SERS nanotags can be highly recognized in two-dimensional distribution without strong organic interference caused by resonance-enhanced Raman scattering or autofluorescence. In addition, when the alkynyl group was located in the para position of mercaptobenzene ring, the alkynyl containing Raman reporter has narrow emission, tunable (2100–2300 cm^{-1}) and greatly enhanced Raman signal. The proposed alkyne modulated SERS palette may provide a more effective solution for multiplex cellular imaging with bright colors when hyperspectral and strong optical noise in low wavenumber region (< 1800 cm^{-1}) are inevitably generated under complex environmental conditions. Therefore, more and more researchers have paid their attention to the design and synthesis of ideal Raman reporter molecules, which will bring new opportunities for the development and application of SERS nanotags and other Raman scattering tags. Similar work have also been done to detect heavy metal ions in organic polluted water without pretreatment

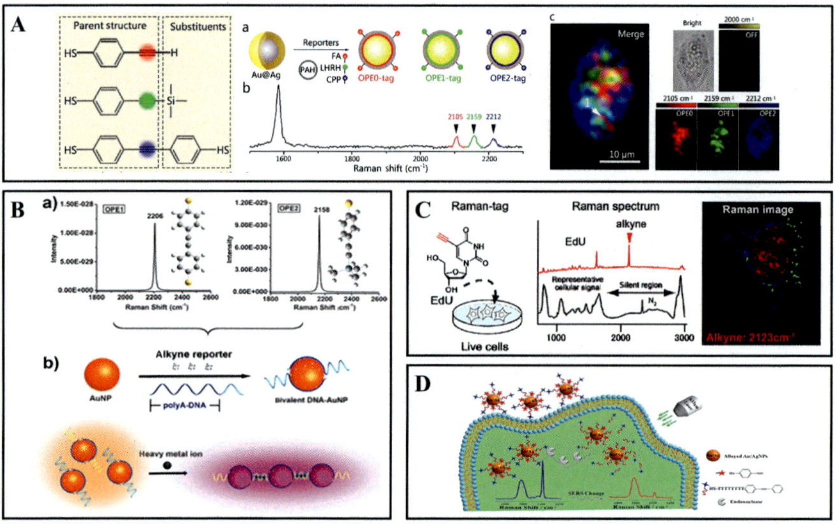

Figure 5.11 (A) Three-color SERS imaging using SERS probes of alkyne SERS palette. (B) Schematic illustration for the design of alkyne-coded SERS NPs for metal-ion detection. (C) Raman spectra of thymidine analogs and a representative cellular spectrum. The band in the silent region of the cellular signal (2334 cm^{-1}) is derived from N2. And Raman image obtained from HeLa cells treated with EdU. (D) Schematic illustration of DNA-alkyne-functionalized alloyed Au/Ag nanoparticle-based SERS nanosensor for ratiometric detection of endonuclease.

with alkyne encoded SERS kit (Fig. 5.11B) [119], and to use alkyne DNA functionalized alloy Au/Ag nanospheres to measure the activity of endonuclease in living cells by proportional SERS imaging (Fig. 5.11D) [120].

5.3.2.1.2 Graphitic nano capsules

Carbon materials are still an active field which has attracted many interests in Raman scattering. Since single-walled carbon nanotubes were studied as resonance Raman tags, which give potential in molecular imaging of living subjects. Remarkably, graphene emerged as novel Raman tags with the feature of 2D bond that its Raman shift was located at ~2700 cm^{-1} (also the cellular Raman-silent region with zero background), expanding the biological development of carbon materials in Raman scattering. Tan et al. [121] first reported the graphene-isolate-Au-nanocrystal (GIAN) for SERS analysis (Fig. 5.12A), they demonstrated that GIAN possessed decent chemical stability with no interference in strong acid, oxidation, and even mimic bio-environment. Furthermore, the accuracy of SERS analysis was greatly improved with 2D band internal standard integrated

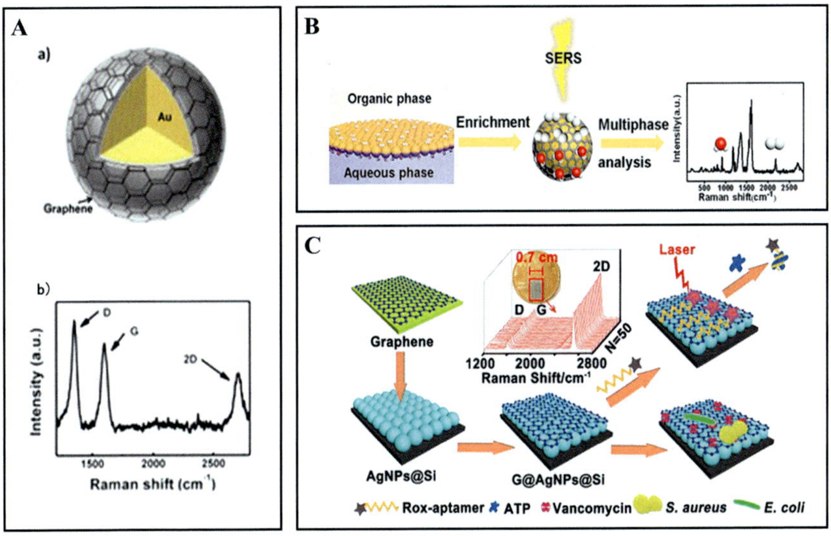

Figure 5.12 (A) Schematic diagram of GIAN and Raman spectrum of GIAN under 532 nm laser. (B) A graphene-isolated-Au-nanocrystal based multiphase analysis system. (C) Synthesis of high-performance SERS sensor chip based on graphene (G)-Ag NP-Si sandwich nano hybrid material.

in system. Tan et al. [122] further discovered that the synthetic GIAN owned the simultaneous multiphase detection capability and GIAN could suspend at the liquid–liquid interface and it could adsorb or attach lipid- and water-soluble analytes into SERS detecting system. Thismainly attributed to a large surface area and unique π–π electrostatic stacking properties of graphene. In particular, 2D bond Raman shift from the graphene as the internal standard, which performed a unique vibration band located in the Raman biological silence region, was indeed improving analytical accuracy and reproducibility (Fig. 5.12B). By using 2D bond of graphene as internal standard, reliable quantitative SERS analysis has been successfully conducted in common fluids, and high-performance SERS sensor chip based on graphene (G)-Ag NP-silicon (Si) sandwich nanohybrids (G@AgNPs@Si) has achieved quantitative detection ranging from the molecular to cellular (e.g., bacteria) level (Fig. 5.12C) [123].

5.3.2.1.3 Prussian-blue and its analogs shells

Whether it is small organic molecule or graphene shell, the adsorption, assembly or bonding process on the surface of noble metal NPs is cumbersome, complex and of poor reproducibility, which will lead to the

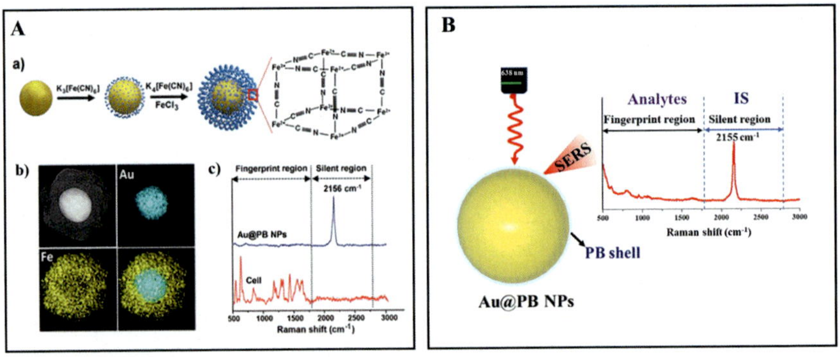

Figure 5.13 Schematic illustration and characterization of the procedure for preparation of Au@PB NPs. Biological species exhibit complex multiple bands in the fingerprint region whereas the Au@PB NPs possess an intense and sharp single band (2156 cm^{-1}) in the cellular Raman-silent region throughout the whole spectrum.

deviation of signal uniformity of SERS nanotags, and thus affect the quantitative detection. Interestingly, Liu et al. [124] discovered that Prussian blue (PB) can be used as a highly sensitive and background free resonance Raman signal. The traditional Raman reporter spectrum shows multispectral bands in the fingerprint region, which usually overlaps with the spectral bands of endogenous dominant biomolecules and is difficult to distinguish. They found that PB only had a strong and sharp single band in the Raman silent region, and also can be easily distinguished from the Raman spectra of endogenous biomolecules (Fig. 5.13A). In addition, PB has strong absorption in the visible near-infrared region of 500–900 nm, which resonates with the incident laser and produces extremely high resonance Raman scattering signals. A high signal-to-noise ratio (SBR) SERRS nanotags were fabricated by assembling PB on the plasmonic nano core. The performance of PB based SERRS nanotags were demonstrated for high sensitivity immunoassay and cancer cell imaging.

SERS is an ultra sensitive label free analytical technique, which can provide unique chemical and structural fingerprint information. However, due to the instability and poor reproducibility of nanostructured SERS active surfaces, it is still a challenge to use SERS for reliable quantitative analysis. In order to detect the concentration of analytes in complex systems, the PB-based SERS nanotags provided a solution to this problem. Li et al. [125] developed an effective route to synthesize gold core and ultra-thin and uniform PB shell (Au@PB) NPs for the quantitative detection of dopamine (DA) in serum and crystal violet (CV) in lake water.

The only strong Raman band of PB at 2155 cm^{-1} was regarded as an ideal interference-free internal standard to correct the Raman intensity fluctuation of analytes (Fig. 5.13B). At the same time, this paper studied the stability of Au@PB NPs, which showed good functionality and thermal stability even at 100°C in strong acid solution.

5.3.2.2 Spectral coding on SERS nanotags

5.3.2.2.1 Click, mixing and combined SERS emission

With assistance of triple bond vibration as novel Raman reporters, these SERS emissions have become one of novel readout techniques. Remarkably, their representative and selective Raman shifts (in biological Raman transparent window, 1800–2800 cm^{-1}) have been used for multiplex labeling that allowed for simultaneous detection of multiplex biotargets without spectral overlay. Besides, these novel readouts could provide precise biomolecular labeling by multiple judgments with logical judgments, and it would be great potential for clinical diagnostics. Recently, our groups have proposed new SERS readout techniques strategies, including "Click," "Mixing," and "Combined" SERS emissions for obtaining more scanning messages, aiming at reducing false positives/negatives results. With the help of the concept of click chemistry, "Click" SERS readout was employed by DNA hybridization or other binding force to mediate the generation of hotspots, and triple bonded SERS-active NPs with several single and narrow emissions were assembled together, thus realized the dynamic combinatorial signal outputs with artificial method [126]. In detail, we chosen four alkyne and nitrile molecules as Raman reporters, then synthesized four kinds of SERS tags with distinct Raman emissions. Masterly, the rationally designed polyadenine (polyA)-DNA and the capture DNA were precisely modified on the triple bond-labeled Au NPs (one strand per NP). When input specific DNA was added, the dimer of SERS active hotspots was immediately formed by their corresponding DNA hybridization, thereby generating two target-specific combinatorial Raman peaks. The amplification of Raman signal by each addition of distinguishable emissions could be acted as new judgment for detection that was "Click" spectra output by the typical double code. By this "Click" SERS readout, we have successful obtained simultaneous 10-plex DNA target detection (Fig. 5.14A). Moreover, we further achieved more accurate positioning of targeted protein in cancer cell by analyzing dynamic combinatorial output of bioorthogonal Raman reporters after the double exposure treatment, avoiding the false positive

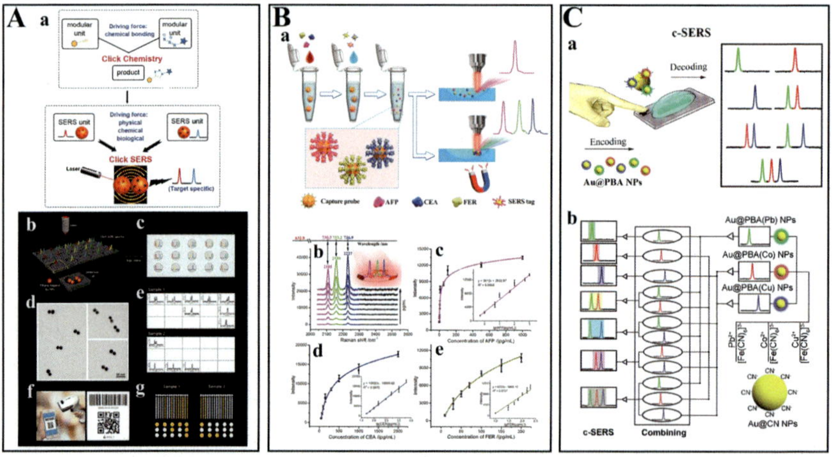

Figure 5.14 (A) "Click" SERS emissions for 10-plex synchronous biomarkers detection. (B) "Mixing" SERS emissions for multiplex detection of liver cancer antigens. (C) "Combined" SERS emissions for high-throughput optical labels on microscale objects, such as bacterium.

signals from NPs endocytosis. Therefore, "Click" SERS readout technology has provided an authentic optical analysis method for multiple biotargets by analyzing the quantity and position of resultant combinatorial SERS emissions.

For the assay in a liquid phase to detect multiplex biomarkers, our group has proposed a "Mixing" SERS emission readout in one-pot, which was achieved by using the magnetic bead to effectively separate the biotargets in the microscale spot and then enrich biotargets to obtain the uniform and specific signals. In our previous study, three specific liver cancer antigens, including α-fetoprotein (AFP), carcinoembryonic antigen (CEA), and ferritin (FER), have been successfully detected by this m-SERS readout [127]. Typically, three triple bonds (C≡N and C≡C) coded SERS tags with well separated Raman shifts were synthesized, and the corresponding antibodies (anti-AFP1, anti-CEA1, and anti-FER1) were further modified into SERS tags. Moreover, antibodies (anti-AFP2, anti-CEA2, and anti-FER2) were bitched conjugated on the surface of magnetic beads (MBs). As shown in Fig. 5.14B, in present of AFP, CEA or FER markers, a microscale core-satellite assembly (double antibody sandwich) structure would be formed by the mature immunoreactions. Before measuring, the extra magnetic force was used to generate an ideal platform based on the uniform 3D packing of SERS tags absorbed on

MBs. Finally, the signal of the reporters represented the specific antigens of liver cancer, and the limit of detection of AFP, CEA, and FER was counted at 0.15, 20, and 4 pg mL^{-1}. Significantly, this simultaneous m-SERS detection has been demonstrated to improve the accuracy of liver cancer diagnosis up to 86.7% in 39 clinical serum samples, and this result was superior to the existing clinical diagnosis method. With such a rapidly growing body of research, continuously expanding methodologies and applications of high-throughput optical labeling technologies-based Raman spectroscopy, playing important in targeted identification, disease diagnosis and so forth. Notably, our group has proposed a "Combined" SERS emission readout strategy for high throughput micrometer-size objects labeling (Fig. 5.14C). In detail, we synthesized a series of CN-bridged coordination polymer encapsulated gold nanoparticles with three distinct Raman frequencies. Interestingly, we demonstrated that eight micrometer-size objects could be well discriminated using the combination of only three distinct Raman frequencies under the focused laser of Raman detection. Furthermore, we concluded that 2n−1 micrometer-size objects could be labeled by using n single emissions into this "Combined" SERS system [128].

By far, the successful studies of SERS innovations of "Click," "Mixing," "Combined" SERS emissions have shown promise for biological application, and these new readout techniques have been demonstrated to perform the multiplexing capability of SERS labeling analysis.

5.3.2.2.2 Joint-encoding by fluorescence-SERS emission

Before the reporter molecules whose characteristic peaks are located in the Raman silent region have not been proposed or widely used, most SERS coding work uses the common small organic molecules. It is well known that they have similar chemical structures and their characteristic peaks are usually located in the spectral fingerprint region, that is, 400−1800 cm^{-1}. Although the Raman band is narrow and sharp, it will inevitably lead to spectral overlap when more fingerprint reporter molecules are used in order to generate more codes, which will cause inconvenience to the decoding process. These factors greatly limit the amount of code that can be implemented. To solve the above problems, Cui et al. proposed a smart strategy, that is, introducing another optical signal called fluorescence into the SERS encoder, forming a joint-encoding by fluorescence-SERS emission [129]. As shown in Fig. 5.15, two Raman reporter molecules (4-mercaptobenzoic acid (4-MBA) and 5, 5′-dithiobis

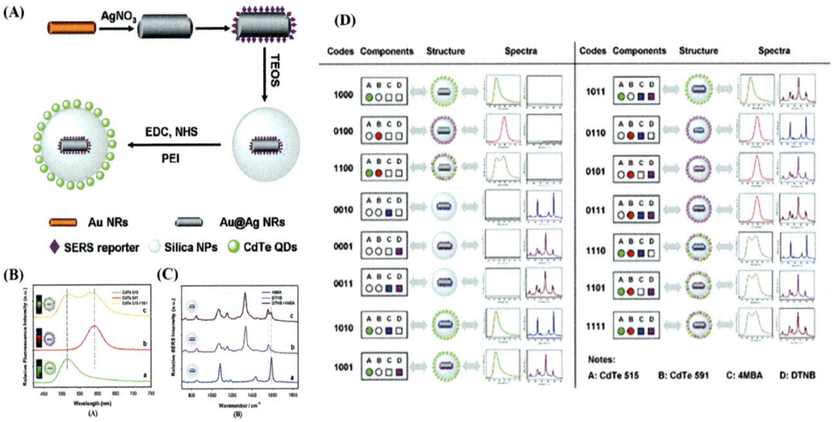

Figure 5.15 (A) Preparation of Organic-Metal-QD Hybrid Nanoparticles (OMQ NPs). PL (B) and (C) SERS spectra of the composite nanoparticles using CdTe 515, CdTe 591, and CdTe 515/591 (volume ratio is 9:1) as the fluorescence indicators and 4MBA, DTNB, 4MBA/DTNB (molar ratio is 5:2) as the SERS reporters. (D) Composition, measured spectra, and codes of the synthesized 15 nanoparticles.

(2-nitrobenzoic acid) (DTNB)), and two fluorescent quantum dots (cyclodextrin Te 515 and cyclodextrin Te 591), were used in this study. The excitation wavelength of fluorescence was 400 nm, and that of SERS was 632.8 nm. In addition, the encoder was a multilayer core—shell structure, and the organic molecules and quantum dots were assembled in different layers to reduce the crosstalk. By combining SERS with fluorescence emission, the number of coding elements was increased, and 15 kinds of coding emissions were obtained. Compared with the encoder using single SERS or fluorescence emission, the encoder with SERS-fluorescence emission can improve the coding ability and expand the number of codes to a certain extent. However, this joint mode of optical signal from two kinds of analytical instrument undoubtedly increases the complexity of the decoding process, resulting in the post-decoding analysis, in which the decoding results can not be displayed directly and clearly.

5.3.3 Low-blinking SERS nanotags: surface coating

In theory, it is only necessary to connect Raman molecules on the surface of gold nanoparticles to generate specific SERS fingerprint signals, but in practice, it is found that such a simple structure is extremely unreliable in biological analysis. There are two main reasons: first, endogenous biomolecules in physiological environment can gather on the surface of SERS substrates due to electrostatic adsorption, and even replace the reporter molecules, which will not only affect the stability of Raman signal of nanotags, but also cause harm to organisms. Second, in biological system, SERS nanotags will collide, adhere or agglomerate, and the near-field coupling of the local surface plasmon will occur, and the SERS signal of nanotags will flicker intermittently. In order to overcome these problems, special materials are usually used to protect the surface of NPs modified with reporter molecules to improve the stability and biocompatibility of the SERS nanotags. Nowadays, a variety of new protective shells have become the research hotspot of SERS nanotags. The types of protective shells include biomolecular shell, polymer shell, silica shell, and carbon shell.

5.3.3.1 Biomolecule coating

The shell of biomacromolecule usually improves the biocompatibility of SERS nanotags and is beneficial to the subsequent modification of biological functional molecules. Bovine serum albumin (BSA) is the most frequently used biomacromolecule for coating, which can bind to gold nanoparticles through electrostatic interaction. If glutaraldehyde was added, the amino groups in BSA would be cross-linked, which could further enhance the stability of SERS nanotags [130]. On the other hand, due to the large consumption of amino groups, the exposed carboxyl groups will make the surface of SERS nanotags negatively charged, and the electrostatic repulsion between them can effectively prevent the tags from aggregating. In addition, the inactivated denatured bovine serum albumin (d-BSA) is also a satisfactory protective shell of SERS nanotag. The reason is that the disulfide bond will be broken and a large number of cysteine residues will be produced. The strong S—Au bond increases the affinity between d-BSA and gold NPs. Xie et al. [106] used d-BSA as the coating shell of gold nanoflower substrate, and also encapsulated a large number of Rhodamine B as Raman reporter molecules in the pores of three-dimensional structure protein to prepare SERS nanotags with strong stability and excellent biocompatibility, as shown in Fig. 5.16A.

Liposome is a kind of synthetic phospholipid bilayer similar to biomass membrane, which has good biocompatibility. It can fuse with the biological plasma membrane and rapidly release the contents into the cytoplasm, which has been regarded as an excellent drug delivery carrier. In recent years, there are many studies trying to use liposomes as the protective shell of SERS nanotags. As shown in Fig. 5.16B, Tam et al. [107] prepared

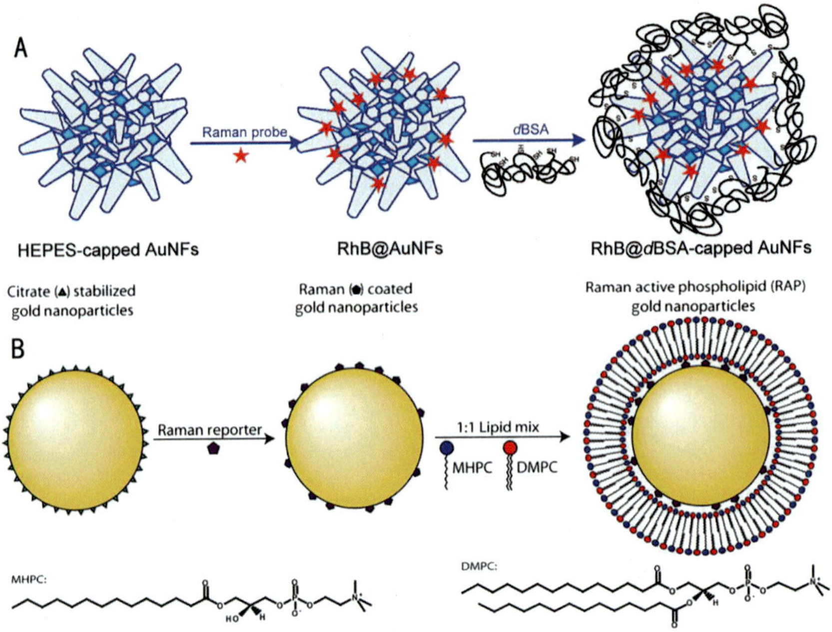

Figure 5.16 (A) Preparation of dBSA-protected AuNFs as SERS tags. (B) Synthesis of Raman-active phospholipid gold nanoparticles.

liposomes using double chain di-myristoylphosphatidylcholine and single chain methyl-hydroxypropyl cellulose phospholipid as precursors, and coated 60 nm gold NPs modified with reporter molecules to prepare SERS nanotags with good biocompatibility.

5.3.3.2 Polymer coating

People like to use HS-PEG polymer to make protective shell, because PEG modified SERS nanotags have good biocompatibility and can avoid the adsorption of interfering molecules, and sulfhydryl group can greatly increase the affinity between shell and gold core. Moveover, it is easy to control the thickness of coating shell. So far, the SERS nanotags protected by HS-PEG have performed well in the analysis and detection of complex biological systems and pharmacokinetic experiments [19]. In addition to HS-PEG, polymer molecules such as PVP and chitosan are often used as the protective shell of SERS nanotags due to their good biocompatibility and biodegradability. In 2009, Yang et al. [131] used a novel amphiphilic block polymer (PS-PAA60, PS154-b-PAA60) as the coating materials of SERS nanotags. The self-assembly process of this polymer was very

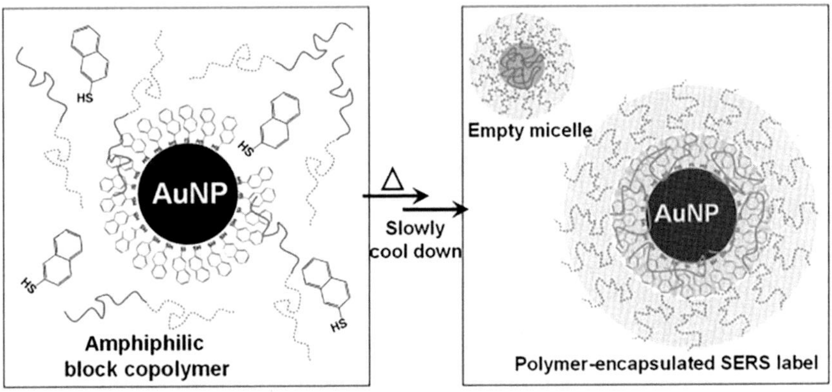

Figure 5.17. Synthesis of PS154-b-PAA60-coated nano label.

simple, and a protective shell with uniform thickness was formed by simple operation of heating and cooling in a reaction vessel (as shown in Fig. 5.17).

Shen et al. [132] utilized label and label-free SERS nanotags for three-dimensional imaging of double organelles (nucleus and cell membrane) in a single HeLa cell. The results showed that SERS could describe the intrinsic chemical properties of the nucleus under high confocal conditions, and accurately locate the positions of folic acid (FA) and luteinizing hormone releasing hormone (LHRH) on the membrane. In this work, the only difference between label and label-free SERS nanotags is that the former has Raman reporter molecules, while the latter does not. In order to improve the stability of SERS nanotags in complex systems and facilitate the connection of targeted functional molecules, polyacrylamine was used as a protective shell. The polymer molecules are easy to form films on the surface of NPs, and the amino groups on the branched chains are highly active, which is conducive to surface functionalization (as shown in Fig. 5.18).

The carbon shell is a kind of new hydrophilic shell material obtained by special molecular crosslinking and further carbonization, which is similar as polymer coating. Shen et al. [133,134] developed a monodisperse, easily grafted and biocompatible SERS nanotag , which was composed of reinforced bimetallic substrates Au@Ag NPs, three kinds of self-assembly organic monolayers and the outermost carbon shell. The Au-Ag-C core–shell NPs were synthesized by layer-by-layer approach. The self-assembled Raman reporter molecules were introduced into the outer layer

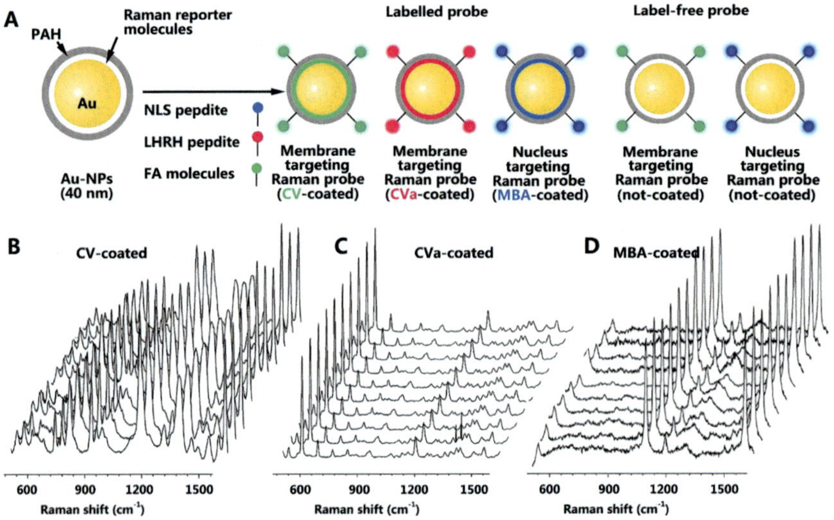

Figure 5.18 (A) Scheme for the PAHylation and peptide/micromolecule modifications on the Raman dye-coated Au-NPs: CV-coated Au-NPs were modified with PAH and FA for targeting the membrane; CVa-coated Au-NPs were modified with PAH and LHRH peptide for also targeting the membrane; MBA-coated Au-NPs were modified with PAH and NLS peptide for targeting the nucleus. (B–D) Representative Raman spectra of CV-coated AuNPs (B), CVa-coated AuNPs (C) and MBA-coated AuNPs (D) obtained from particle solution with an incident laser power of 3.0 mW and 1 s exposure time per spectrum.

of Au@Ag NPs. The obtained NPs were further coated with a layer of controllable thickness carbon shell to form monodisperse colloidal carbon coated Au_{core}/Ag_{shell} spheres with uniform particle size distribution. In addition, these spherical NPs with SERS activity showed interesting properties as a new Raman tag for quantitative immunoassay. The results showed that the SERS nanotags can be used for multiple and ultra sensitive detection of biomolecules, as well as nontoxic and in vivo molecular imaging of animal and plant tissues (Fig. 5.19).

5.3.3.3 Silica coating

Silica is one of the most widely used SERS nanotag coating materials, which has high stability, good water solubility, and can be modified by a large number of $-NH_2$ and $-COOH$ functional groups on its surface, providing great convenience for the subsequent connection of biomolecules. In 2003, McCreery and Mulvaney [17] prepared silica coated SERS nanotags by hydrolysis of sodium silicate (Fig. 5.20). Since then, there has

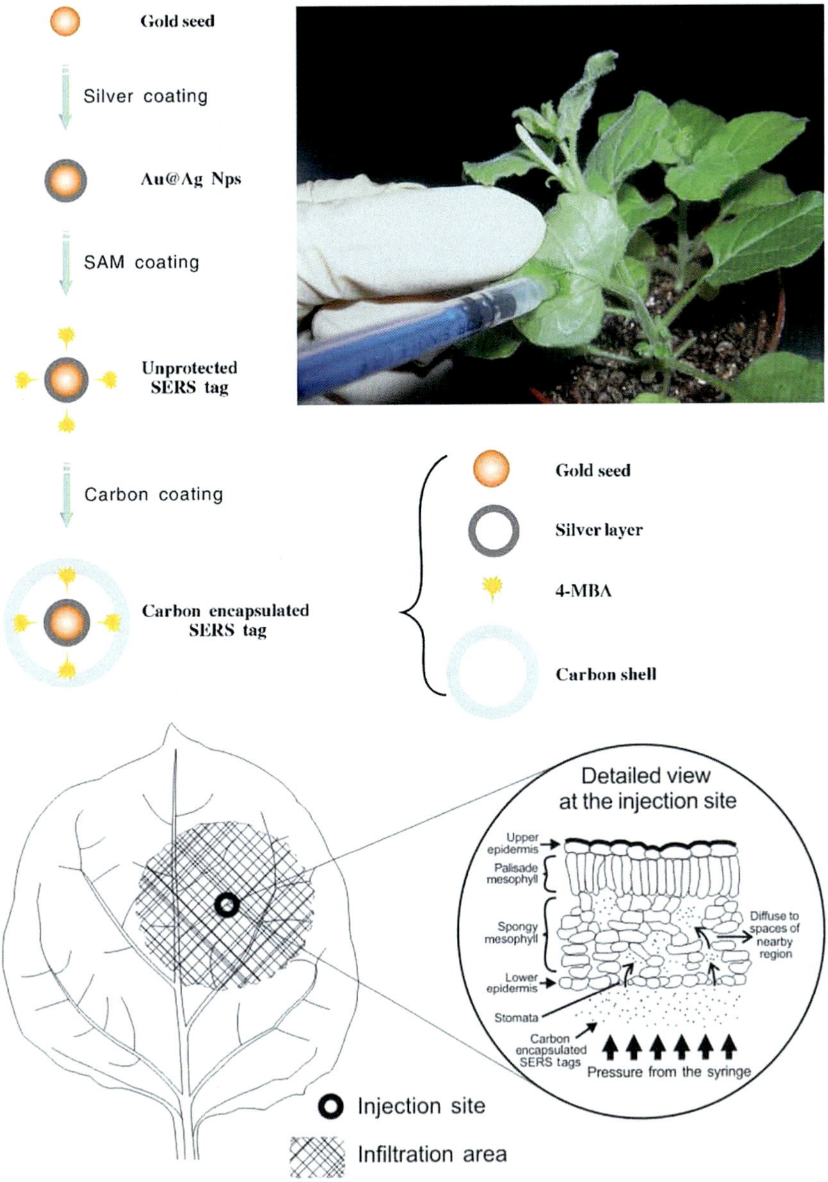

Figure 5.19 Schematic illustration of the preparation of carbon-encapsulated SERS tags; photograph of injection and its schematic diagram.

been an upsurge of works on silica coating. In silica coated experiments, silane coupling agents, such as (3-Aminopropyl)trimethoxysilane (APTMS) or (3-mercaptopropyl) trimethoxysilane (MPTMS), are usually connected with

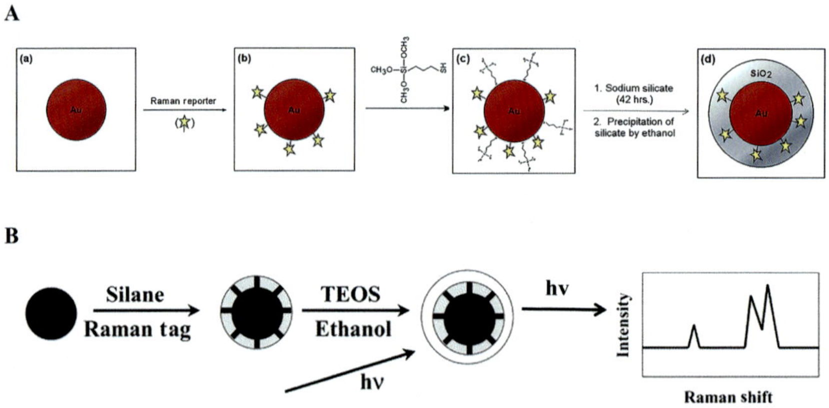

Figure 5.20 Schematic illustration of the core–shell nanoparticle structure (A) and the procedure for preparing silica-coated SERS-active gold colloids (B).

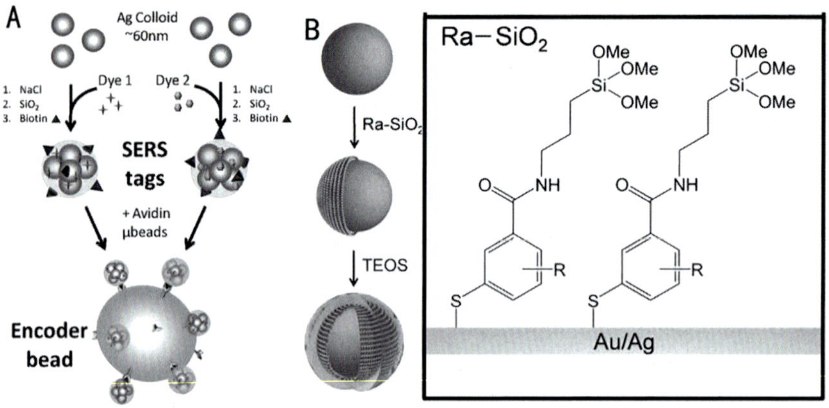

Figure 5.21 (A) Schematic of the process for generating encoder bead assemblies from SERS nanoparticle spectral tags. (B) Silica as a probe for shell preparation.

gold NPs through amino or sulfhydryl groups. The trimethoxysilane group of the above reagent molecules are exposed on the surface of the NPs. Sodium silicate is then added to form a protective shell of silica. In order to obtain stronger SERS signal, Brady [135] also tried to use silica to encapsulate the aggregates of NPs, and achieved good experimental results (as shown in Fig. 5.21A).

If sodium silicate is used as silicon source, it is easy to form a thin silica shell, which requires dialysis, ion exchange and other operation steps. In order to solve these problems, Haisch [136] proposed a rapid and simple method by using tetraethyl orthosilicate (TEOS) in alkaline condition to prepare silica shell, therefore, silane coupling reagent was not required to participate in the surface pretreatment of nanoparticles. However, Raman reporter molecules, such as *p*-mercaptobenzoic acid or isothiocyano malachite green and other molecules with strong affinity are needed for gold NPs, otherwise the reporter molecules will be easy to fall off. In 2009, Schlücker et al. [137] proposed a different method for the preparation of silica shell, which made the SERS nanotags stable and reproducible. First, polyacrylamine hydrochloride (PAH) was coated on the surface of gold NPs with self-assembled monolayer of MBA molecules by electrostatic interaction. Then, PVP molecules were continuously adsorbed on the surface of the monolayer (which was helpful for the growth of SiO_2 on the surface of NPs). Finally, TEOS was hydrolyzed under alkaline conditions to realize the epitaxial growth of SiO_2 shell (as shown in Fig. 5.21B).

5.3.4 Multifunctional SERS nanotags: materials combination

The emerging multifunctional nanotags show great potential in the innovation of traditional biological analysis, imaging, and diagnosis. As the synthesis and application of SERS nanotags continue to warm up, the focus of SERS technology has shifted from the synthesis to the design and application of complex nanostructures with multiple targeting functions and organic combination with separation, imaging, and treatment functions.

5.3.4.1 Magnetic materials for separation

Magnetic NPs combined with SERS nanotags (M-SERS) have been used for optical labeling of target molecules or cells, and can also be separated from complex systems, which has become a new biological analysis tool. The magnetic and SERS activity are integrated into the nanotags, and the magnetic force is used to induce the rapid aggregation of the nanotags, so as to generate SERS hot spots in a narrow space. Even if the concentration of the analyte is low, the high-quality SERS signal can be obtained [138]. For example, Jun et al. [139] prepared multifunctional labeling materials for cancer cell localization and separation. As shown in Fig. 5.22, the nanotags used 18 nm magnetic nanoparticle as the core and 16 nm thick silica as the shell. The outer surface of the silica shell was loaded with silver NPs, and the surface adsorbed aromatic compounds as Raman reporter molecules. The outermost layer was silica shell again. The silica encapsulated M-SERS generated a strong SERS signal and had magnetic

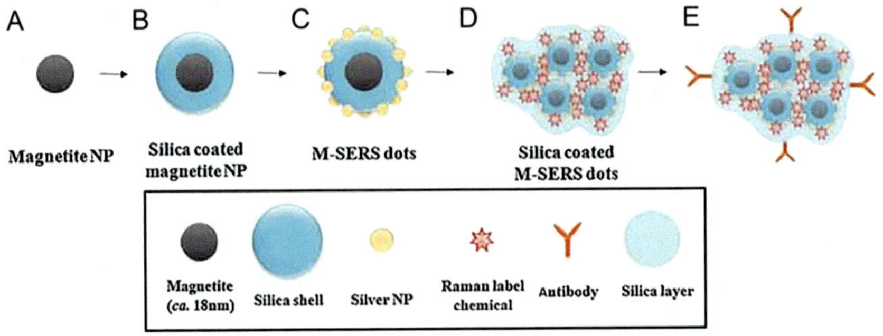

Figure 5.22 Illustration of synthesis of magnetic active SERS tagging material (M-SERS dots).

properties. It was successfully used to locate thymic cancer cells (SKBR3) and floating white blood cells (SP2/O).

Gole et al. [140] designed an one-pot synthesis route of M-SERS, in which the prepared iron oxide NPs and the Raman reporter molecules, *p*-MBA were dispersed in the microemulsion system, and the silver NPs were formed in the shell when the silica NPs were formed. In this composite structure, both types of NPs were embedded in silica particles. In addition, they also prepared M-SERS by depositing silver and magnetic NPs on silica or polystyrene microspheres.

5.3.4.2 Multimodal imaging materials

In recent years, people have been interested in developing multimodal imaging methods. Different imaging modes can be integrated into a nanotag to make up for the shortcomings of a single imaging mode. A considerable amount of research work has been involved in the synthesis and characterization of SERS related multimodal imaging nanotags, including nanomaterials suitable for fluorescence, X-ray computed tomography (CT) and MRI imaging.

Fluorescent-SERS nanotags (F-SERS). To measure SERS and fluorescence signals is of high sensitivity and convenience, and the dual function nanotags have been developed for a variety of applications. SERS signal has good multichannel detection ability because of its narrow bandwidth, high spectral resolution, however, its slow imaging speed is the main obstacle for fast identification of specific tags. In order to overcome this limitation, a combination strategy of fluorescence SERS (F-SERS) dual-mode nanotags were proposed. The fluorescence signal is more intuitive

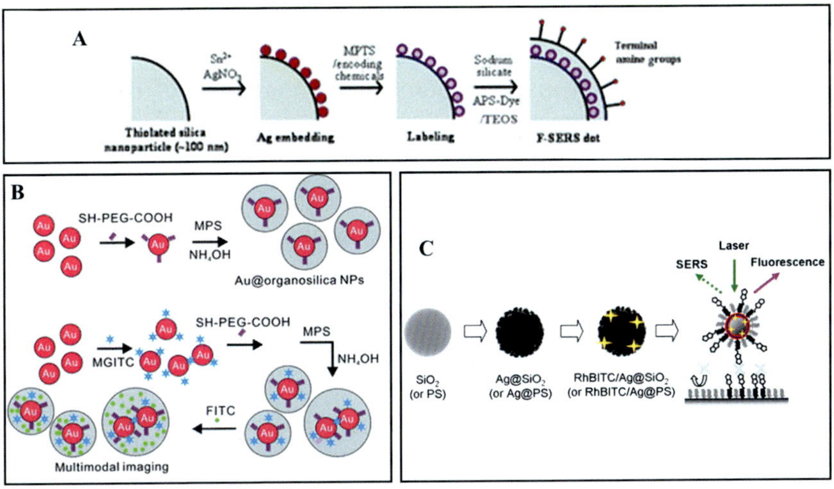

Figure 5.23 (A) Synthetic scheme of F-SERS dots. (B) Schematic illustration of the synthetic procedure for Au@organosilica nanoparticles. (C) Fabrication of Ag-coated silica (or Ag-coated polystyrene) beads usable as a template of biosensors operating via SERS/SERRS and fluorescence.

and faster than SERS signal in imaging, so it was used as a direct indicator of molecular recognition, and then the SERS signal was used as the recognition signal of specific molecules in multiple interactions, which was more specific than fluorescence.

Fluorescent tags can be generated in different ways. Cho et al. [141] deposited fluorescent conjugated molecules of 3-aminopropyltriethoxysilane and fluorescein isothiocyanate (FITC) or Alexa fluoro onto silica coated SERS nanotags by silane chemistry to make the F-SERS nanotags (Fig. 5.23A). Tian et al. [142] combined fluorescent molecules such as FITC with the silicone shell of the F-SERS by simple mixing (Fig. 5.23B). In another method, the electrostatic force made the RBITC labeled polyallylamine hydrochloride [143] or fluorescent quantum dots [144] deposite layer-by-layer on the SERS substrates to realize the modification of the fluorescent label. RBITC has also been reported to be directly adsorbed on silver NPs, which were loaded on the surface of silica and polystyrene spheres, and used as both Raman reporter molecule and fluorescence generator (Fig. 5.23C) [145].

F-SERS nanotags have been used in cancer cell sensing and in vivo imaging, in which the nanotags have shown the advantages for the bispectral function. For example, tags with different fluorescence and SERS signals were used to detect three proteins in cancer cells, which were expressed simultaneously in mouse lung bronchioles stem cells. The results of quantitative comparison of various protein expressions in cells and tissues suggested that the sensitivity and selectivity of immunoassay using F-SERS have been greatly improved [146].

X-ray computed tomography-SERS nanotags (CT-SERS). Compared with traditional iodine compounds, gold has higher electron density and atomic number, so it has higher X-ray attenuation coefficient. Inspired by the ability of Au NPs to enhance SERS, Xiao et al. [147] synthesized a kind of multicolor nanotags combining SERS and CT imaging modes for the first time. Through the synthesis of a series of Au NPs with different sizes, the optimal particle size of contrast agent was selected. Its SERS intensity and X-ray attenuation were higher than those of iodized CT contrast agent commonly used in clinic. The results of in vivo imaging principal verification of the nanotag showed the dual imaging ability of SERS and CT possessed by a single NP.

Magnetic resonance imaging-SERS nanotags (MRI-SERS). MRI is a wide range of medical diagnostic method, which has high spatial and temporal resolution, unlimited tissue penetration ability and tomography ability. The advantages of SERS and MRI will complement each other and has become a new breakthrough in the field of cancer diagnosis and treatment. Shen et al. [148] synthesized the monodisperse PB encapsulated gold (Au@PB) NPs. The plasmonic gold core and cyanide coordination polymer (C≡N) and iron ion coincidentally became excellent contrast agents for MRI and zero background SERS imaging. PB, as the signal source of both MR and SERS, can be easily assembled on a single Au NP. Among them, iron ions usually have high in vivo MRI relaxation efficiency, for example, the longitudinal and transverse relaxation efficiency are $0.86\ mM^{-1}S^{-1}$ (R1) and $5.42\ mM^{-1}s^{-1}$ (R2), respectively. With the help of plasmonic enhancement of gold core, the C≡N group showed the specific, strong and stable (3 S) SERS emission within the Raman silent region ($1800-2800\ cm^{-1}$), which makes it possible for accurate in vivo imaging at the level of single cell and even subcellular level (as shown in Fig. 5.24).

5.3.4.3 Therapeutic materials

In recent years, a lot of attention has been paid to noninvasive PTT, which can selectively treat tumor cells. This type of treatment utilizes the large absorption cross-sectional area of nanomaterials in the near-infrared

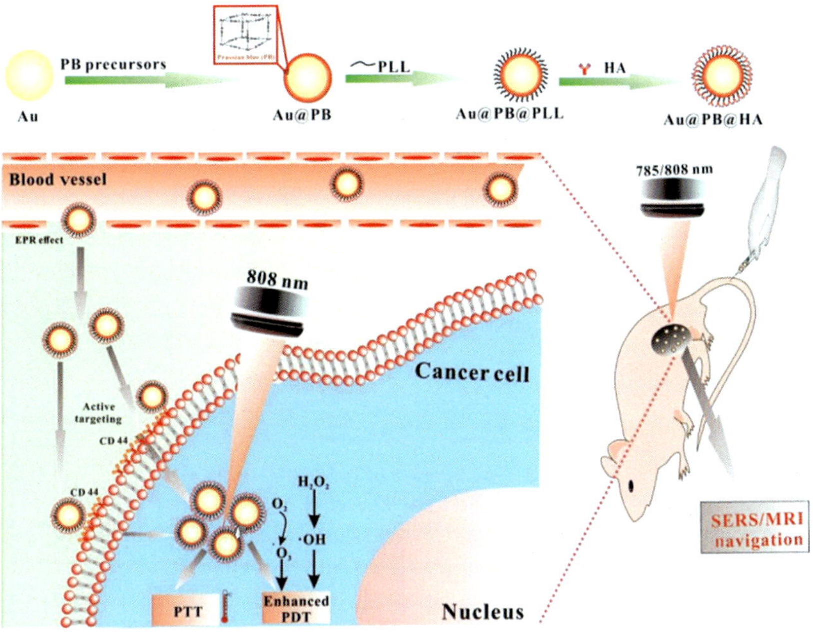

Figure 5.24 Schematic illustration of the stepwise synthesis of plasmonic Au@PB@HA core–shell NPs.

region. Due to the weak absorption of near-infrared light by tissues, near-infrared light can penetrate the skin without causing a lot of energy loss and damage to normal tissues. Therefore, it can be used to treat specific cells labeled by thermogenic nanomaterials. Taking advantage of the near-infrared absorption of some SERS active noble metal NPs, several researchers have developed a multifunctional nanoplatform, which integrates SERS imaging and PTT.

von Maltzahn et al. [65] constructed a dual function SERS nanotags based on Au nanorods for PTT and remote control of cancer therapy based on in vivo imaging. Three kinds of near-infrared absorption molecules: IR-792, DTTC, and DTDC were selected as Raman reporter molecules. Because of the SERRS effect, each molecule showed $10-10^3$ times signal greater than those absorbed in visible light. After intratumoral injection of IR-792 modified nanorods into thymus free mice with bilateral human MDA-MB-435 tumor, obvious SERS signals were observed. Once irradiated by a diode laser (810 nm, 2 W m^{-2}), the temperature of the mouse surface increased to 70°C (Fig. 5.25).

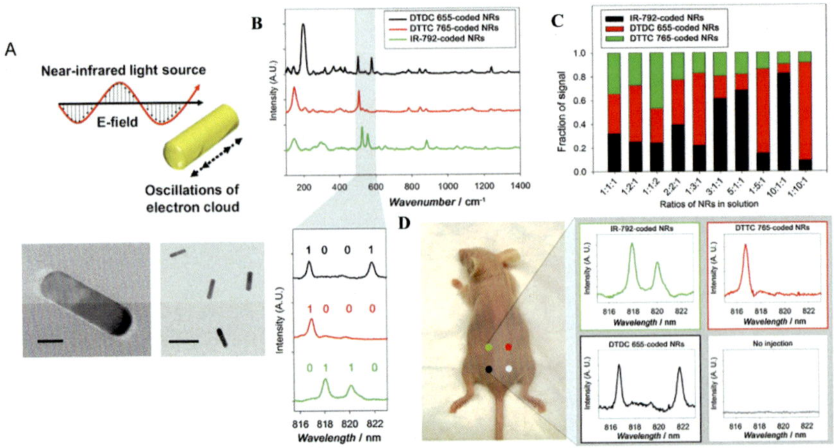

Figure 5.25 (A) Schematic and TEM images of gold NRs. (B) In vitro SERS spectra of DTDC 655- (*black*), DTDC 765-(*red*), and IR-792-coded gold NRs after baseline correction. (C) Quantitative in vitro spectral multiplexing of tertiary SERS-coded NR solutions. (D) In vivo distinction of three SERS-coded NR populations.

Accurate tumor navigation and nanostructures with high PT conversion efficiency are important, but they are still challenging in current biomedical applications. Shen et al. [149] proposed a new method for simple and green synthesis of Au@$Cu_{2\text{-}x}S$ core−shell NPs based on anion exchange in aqueous solution. In addition to the mentioned PT effect of NPs, the SERS signal was also significantly improved due to the local SPR coupling between Au metal core and $Cu_{2\text{-}x}S$ semiconductor shell. Based on the improved epitaxial growth strategy, cuprous oxide was synthesized by in situ reduction on the gold core coated with cresol violet acetate Au@Cu_2O NPs. And then, the Au@$Cu_{2\text{-}x}S$ NPs were finally obtained by anion exchange between S^{2-} and Cu_2O shell (as shown in Fig. 5.26). Cu/S atomic ratio and $Cu_{2\text{-}x}S$ shell thickness could be adjusted. Therefore, the ideal integration of plasma Au core and $Cu_{2\text{-}x}S$ shell into one unit was not only conducive to efficient PT conversion, but also conducive to the construction of SERS based navigator. This novel SERS nanotag has enhanced photoacoustic signal and was an important candidate for precise tumor navigation and in vivo nondestructive PTT. Different with the spatial configuration of Au@$Cu_{2\text{-}x}S$, the outer layer of CuS@Au core−shell NPs was gold shell with controllable morphology. The LSPR coupling between the gold shell and the copper sulfide core can be affected by controlling the morphology of gold shell,

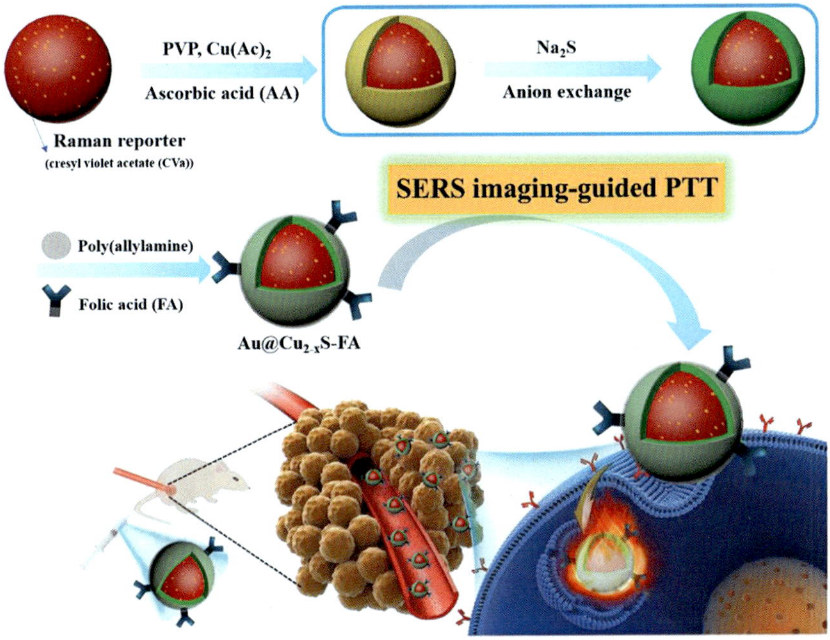

Figure 5.26 Schematic illustration of the synthetic route of Au@$Cu_{2\text{-}x}$S—FA core—-shell NPs and their in vitro and in vivo applications for SERS/PA imaging-guided PTT in cancer.

and the strongest SERS and photothermal effect can be obtained when the h-CuS@s-Au (h: hollow, s: spiky) core—shell nanostructures have been achieved, especially the photothermal conversion efficiency, reach a rare 62.5%. After being introduced into living cells, the nanostructures can easily induce apoptosis through Pt effect [150]. What's more, the outer layer of gold shell can be used as a SERS substrate to directly monitor the chemical changes in the process of apoptosis, such as the expression of heat shock proteins stimulated by PT (as shown in Fig. 5.27).

5.4 Summary and prospect

As a new optical nanotags, SERS nanotags have been synthesized by combining different types of metal nanomaterials, organic Raman reporter molecules and functional groups. A considerable amount of research has been reported about the development of SERS nanotags for molecular multivariate detection or biological imaging at the level of microorganisms, cells, tissues, and living animals. However, it still remains a large

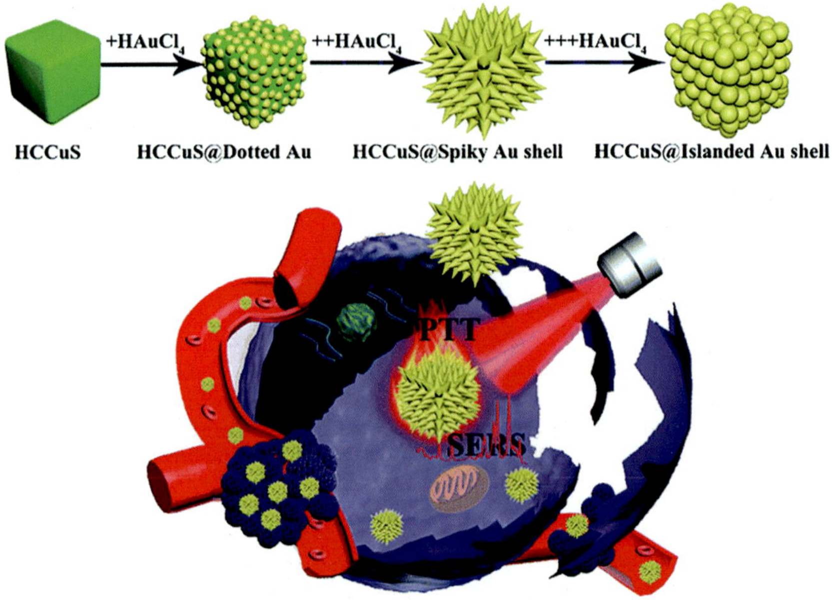

Figure 5.27 Schematic illustration of the synthesis of HCCuS@Au core–shell NPs with different morphologies of Au shell, and label-free SERS intracellular imaging guided photothermal cancer therapy based on HCCuS@Spiky Au core–shell NPs.

number of opportunities and challenges. For example, to synthesize SERS nanotags with high sensitivity and reproducibility, researchers must be able to produce high-performance SERS substrates based on single NP SERS activity; accurately control the location of hot spots between particles; and create tiny tags for single molecule labeling and subcellular imaging.

In recent years, despite the rapid development of SERS technology in the field of biochemical sensing and analysis, it still cannot get rid of the role of "minority" means in the field of optical analysis. The main reason is due to the imperfect SERS theory and experimental technology, as well as the lack of standard SERS substrate and SERS measurement specification. If researchers can fully consider the advantages of SERS and understand the bottleneck/disadvantages of insufficient reliability caused by materials and measurement, SERS biochemical sensing analysis technology will surely develop to the deep field of practicality, popularization and standardization. First, the background noise of SERS measurement sometime is high and the interference is strong. There are two sources of background noise, one is the random noise when the system coexists on the enhanced substrate surface; the other one is the system noise caused by the interference of various signal molecules on the substrate surface. The two types of noise are caused by overlapping spectral signals in fingerprint area, which makes it difficult to identify and affects the quantitative detection, especially in the SERS spectra of biological systems, such as cells, serum and common dyes. Second, the random error of SERS is large and the reproducibility is poor. The detection mode suitable for homogeneous solution detection system is relatively simple, and the signal stability is restricted by many factors. The controllable aggregation and dispersion of nanoparticles in complex systems is a challenge in the field of nanoscience, such as the false-positive interference caused by physical/chemical factors in the detection system. Third, the systematic error of SERS detection is obvious. In homogeneous detection system, the interaction strength and reaction time among SERS probe (signal molecule modified nanoparticles), target and mediator may affect the "capture" rate of target, the specificity of detection results and detection efficiency.

In practical application, how to select the reporter molecules, how to effectively control the dispersion state of SERS nanotags, and how to construct an efficient and highly selective reaction pathway to mediate the coupling of tagging materials need to be discussed by spectral analysis and nanoscience researchers.

The SERS nanotags are endowed with pure and highly recognizable optical signals: spectral "fingerprint area" and biological "silent area" reporter molecules. For SERS sensing analysis, the abundant fingerprint information of SERS is a double-edged sword. On the one hand, it can provide the maximum degree of freedom for the selection of Raman reporter molecules; on the other hand, the fingerprint signal is only suitable for qualitative research. In complex systems, the SERS signal of multiple reporter molecules or coexisting substances will cause interference to the quantification of a certain band used for quantitative detection, resulting in false positive or inaccurate measurement results. Therefore, pure and high identification SERS signal is the key to solve this problem. In our previous work, we have explored the advantages of triple bond coded SERS nanotags in multicolor, nonbackground interference cell imaging [118], and simultaneous detection of various heavy metal ions in organic rich water [119]. In addition, the triple bond SERS nanotags were again effective in the simultaneous quantitative detection of multicomponent in complex systems. Three kinds of antigen detection were distinguished and quantitatively detected by narrow-band, single peak and the only recognizable triple bond Raman reporter molecules. Even if the antibody molecules and reporter molecules were modified on a single NPs at the same time, the NPs aggregated on a micron-sized magnetic bead. In addition, the detection system was a complex serum sample, but there was no background noise interference in SERS detection, and the signal-to-noise ratio was very high [127].

The SERS nanotags are endowed with limited and high stability of aggregation: irreversible aggregation and controllable and ordered coupling. In order to solve the problem of controllable and ordered aggregation or controllable coupling of SERS nanotags in homogeneous system, it is necessary to promote the limited and ordered controllable aggregation of SERS nanotags under the induction of targets. For example, CTAB on the surface of NPs will make NPs close to each other due to the interaction of long hydrophobic chains, forming a limited and stable network structure, which can well control the aggregation of SERS nanotags [96]. In addition, the SERS nanotags are limited to the magnetic bead carrier material through the specific action of antigen-antibody, rather than immune reaction in the solution phase, and its aggregation can be regulated stably [127]. In the early stage, our group constructed a highly stable and reversible aggregation SERS nanotags by using two chitosan and heparin molecules with highly matched positive and negative charge density. The surface of the tags was modified with chitosan coating with positive charge. When heparin with negative charge was added continuously, the charge property of the tag surface was reversed, and the probe experienced a controllable process of first aggregation and then dispersion, which also belongs to the same category to improve the reliability of SERS detection [151].

The SERS nanotags are endowed with specific and high time-dependent mediating force: "weak" interaction and "strong" interaction. In SERS sensing analysis, specific recognition, mediation, and efficient response have a decisive impact on the reliability of detection strategy. Hydrogen bond, van der Waals force, salt bond, hydrophobic force, aromatic ring stacking force, halogen bond and other secondary bonds (or intermolecular force) are basically intermolecular weak interactions, which can not achieve the complete capture of the probe or target, and are also vulnerable to the interference of other substances in the system. On the contrary, strong intermolecular interactions such as coordination bonds and covalent bonds are popular in the field of SERS sensing and analysis. For example, the strong coordination between Hg^{2+} and DPA/papain, and between Ag^+ and MBN greatly improved the selectivity of the detection method itself. It is worth mentioning that Hg^{2+} plays the role of "scavenger" in biological system, which cleans up DPA blend molecules at one stroke, which is impressive [152]. In the early stage of the research group, there was another case in which the SERS nanotags were used to detect heparin molecules with high selectivity through the difference in the cooperative ability of heparin with pentapeptide and undecanopeptide bonds [153]. Therefore, the binding reaction of peptides and other biomolecules, heavy metal ions with mild reaction conditions and high reaction efficiency, is suitable for the sensing system and improves the detection reliability; on the contrary, some biomolecules' immune recognition, DNA hybridization (base complementary pairing) and other reactions need more stringent reaction conditions, and the nonspecific adsorption should also be considered.

References

[1] T.M. Cotton, S.G. Schultz, R.P. Van Duyne, Surface-enhanced resonance Raman scattering from cytochrome C and myoglobin adsorbed on a silver electrode, J. Am. Chem. Soc. 102 (1980) 7960–7962.

[2] P. Negri, Z.D. Schultz, Online SERS detection of the 20 proteinogenic L-amino acids separated by capillary zone electro-phoresis, Analyst 139 (2014) 5989–5998.

[3] A. Barhoumi, D. Zhang, F. Tam, N.J. Halas, Surface-enhanced Raman spectroscopy of DNA, J. Am. Chem. Soc. 130 (2008) 5523–5529.

[4] K. Singhal, A.K. Kalkan, Surface-enhanced Raman scattering captures conformational changes of single photoactive yellow protein molecules under photoexcitation, J. Am. Chem. Soc. 132 (2010) 429–431.

[5] K. Kneipp, H. Kneipp, V.B. Kartha, R. Manoharan, G. Deinum, I. Itzkan, et al., Detection and identification of a single DNA base molecule using surface-enhanced Raman scattering (SERS), Phys. Rev. E 57 (1998) R6281–R6284.

[6] K. Ma, J.M. Yuen, N.C. Shah, J.T. Walsh, M.R. Glucksberg, R.P. Van Duyne, In Vivo, transcutaneous glucose sensing using surface-enhanced spatially offset Raman spectroscopy: multiple rats, improved hypoglycemic accuracy, low incident power, and continuous monitoring for greater than 17 days, Anal. Chem. 83 (2011) 9146–9152.

[7] J. Ando, M. Asanuma, K. Dodo, H. Yamakoshi, S. Kawata, K. Fujita, et al., Alkyne-tag SERS screening and identification of small-molecule-binding sites in protein, J. Am. Chem. Soc. 138 (2016) 13901–13910.

[8] L.A. Lane, X.M. Qian, S.M. Nie, SERS nanoparticles in medicine: from label-free detection to spectroscopic tagging, Chem. Rev. 115 (2015) 10489–10529.

[9] W.E. Doering, M.E. Piotti, M.J. Natan, R.G. Freeman, Sers as a foundation for nanoscale, optically detected biological labels, Adv. Mater. 19 (2007) 3100–3108.

[10] R. Wang, C.W. Yu, F.B.A. Yu, L.X. Chen, Molecular fluorescent probes for monitoring pH changes in living cells, TrAC. Trends Anal. Chem. 29 (2010) 1004–1013.

[11] Y.Q. Wang, L.X. Chen, Quantum dots, lighting up the research and development of nanomedicine, Nanomed. Nanotechnol. Biol. Med. 7 (2011) 385–402.

[12] A.M. Michaels, J. Jiang, L. Brus, Ag nanocrystal junctions as the site for surface-enhanced Raman scattering of single rhodamine 6G molecules, J. Phys. Chem. B 104 (2000) 11965–11971.
[13] K. Kneipp, Y. Wang, H. Kneipp, L.T. Perelman, I. Itzkan, R.R. Dasari, et al., Single molecule detection using surface-enhanced Raman scattering (SERS), Phys. Rev. Lett. 78 (1997) 1667.
[14] Z.Y. Li, Y. Xia, Metal nanoparticles with gain toward single-molecule detection by surface-enhanced Raman scattering, Nano Lett. 10 (2010) 243–249.
[15] R.L. McCreery, Raman Spectroscopy for Chemical Analysis, 1, John Wiley, New York, 2000.
[16] J.R. Lakowicz, Principles of Fluorescence Spectroscopy, Springer, New York, 2006.
[17] W.E. Doering, S. Nie, Spectroscopic tags using dye-embedded nanoparticles and surface-enhanced Raman scattering, Anal. Chem. 75 (2003) 6171–6176.
[18] W. Xie, B. Walkenfort, S. Schlücker, Label-free SERS monitoring of chemical reactions catalyzed by small gold nanoparticles using 3D plasmonic superstructures, J. Am. Chem. Soc. 135 (2013) 1657–1660.
[19] X. Qian, X.H. Peng, D.O. Ansari, Q.Y. Goen, G.Z. Chen, D.M. Shin, et al., In vivo tumor targeting and spectroscopic detection with surface-enhanced Raman nanoparticle tags, Nat. Biotechnol. 26 (2008) 83–90.
[20] S. Nie, S.R. Emory, Probing single molecules and single nanoparticles by surface-enhanced Raman scattering, Science 275 (1997) 1102–1106.
[21] V.M. Zamarion, R.A. Timm, K. Araki, H.E. Toma, Ultrasensitive SERS nanoprobes for hazardous metal ions based on trimercaptotriazine-modified gold nanoparticles, Inorg. Chem. 47 (2008) 2934–2936.
[22] D. Tsoutsi, J.M. Montenegro, F. Dommershausen, U. Koert, L.M. Liz-Marzán, W.J. Parak, et al., Quantitative surface-enhanced Raman scattering ultradetection of atomic inorganic ions: the case of chloride, ACS Nano 5 (2011) 7539–7546.
[23] J. Kneipp, H. Kneipp, A. Rajadurai, R.W. Redmond, K.J. Kneipp, Optical probing and imaging of live cells using SERS labels, J. Raman Spectrosc. 40 (2009) 1–5.
[24] J. Kneipp, H. Kneipp, B. Wittig, K. Kneipp, Novel optical nanosensors for probing and imaging live cells, Nanomed. Nanotechnol. Biol. Med. 6 (2010) 214–226.
[25] J. Kneipp, Nanosensors based on SERS for applications in living cells, Surface-Enhanced Raman Scattering, 103, 2006, pp. 335–349.
[26] D. Naumann, FT-infrared and FT-Raman spectroscopy in biomedical research, Appl. Spectrosc. Rev. 36 (2001) 239–298.
[27] K.L. Rule, P.J. Vikesland, Surface-enhanced resonance Raman spectroscopy for the rapid detection of *Cryptosporidium parvum* and *Giardia lamblia*, Environ. Sci. Technol. 43 (2009) 1147–1152.
[28] K.K. Maiti, U.S. Dinish, A. Samanta, M. Vendrell, K.S. Soh, S.J. Park, et al., Multiplex targeted in vivo cancer detection using sensitive near-infrared SERS nanotags, Nano Today 7 (2012) 85–93.
[29] X.M. Qian, S. Nie, Single-molecule and single-nanoparticle SERS: from fundamental mechanisms to biomedical applications, Chem. Soc. Rev. 37 (2008) 912–920.
[30] P.G. Etchegoin, E. Le Ru, A perspective on single molecule SERS: current status and future challenges, Phys. Chem. Chem. Phys. 10 (2008) 6079–6089.
[31] M. Moskovits, L.L. Tay, J. Yang, T. Haslett, SERS and the single molecule, Optical Properties of Nanostructured Random Media, 82, Springer-Verlag, Berlin, 2002, pp. 215–227.
[32] L. Qin, S. Zou, C. Xue, A. Atkinson, G.C. Schatz, C.A. Mirkin, Designing, fabricating, and imaging Raman hot spots, in: Proceedings of the National Academy of Sciences 103 (2006) 13300–13303.

[33] G.G. Huang, M.K. Hossain, X.X. Han, Y. Ozaki, A. Novel, Reversed reporting agent method for surface-enhanced Raman scattering; highly sensitive detection of glutathione in aqueous solutions, Analyst 134 (2009) 2468–2474.
[34] M.J. Banholzer, J.E. Millstone, L. Qin, C.A. Mirkin, Rationally designed nanostructures for surface-enhanced Raman spectroscopy, Chem. Soc. Rev. 37 (2008) 885–897.
[35] S. Schlücker, SERS microscopy: nanoparticle probes and biomedical applications, Chemphyschem 10 (2009) 1344–1354.
[36] W. Xie, P. Qiu, C. Mao, Bio-imaging, detection and analysis by using nanostructures as SERS substrates, J. Mater. Chem. 21 (2011) 5190–5202.
[37] C.H. Lee, L.M. Tian, S. Singamaneni, Paper-based SERS swab for rapid trace detection on real-world surfaces, ACS Appl. Mater. Interfaces 2 (2010) 3429–3435.
[38] C.L. Zhang, K.P. Lv, H.P. Cong, S.H. Yu, Controlled assemblies of gold nanorods in PVA nanofiber matrix as flexible free-standing SERS substrates by electrospinning, Small 8 (2012) 648–653.
[39] P. Pinkhasova, L. Yang, Y. Zhang, S. Sukhishvili, H. Du, Differential SERS activity of gold and silver nanostructures enabled by adsorbed poly(vinylpyrrolidone), Langmuir 28 (2012) 2529–2535.
[40] T.C. Chiu, C.C. Huang, Aptamer-functionalized nano-biosensors, Sensors 9 (2009) 10356–10388.
[41] G. Wang, Y. Wang, L. Chen, J. Choo, Nanomaterial-assisted aptamers for optical sensing, Biosens. Bioelectron. 25 (2010) 1859–1868.
[42] K. Kneipp, H. Kneipp, J. Kneipp, Surface-enhanced Raman scattering in local optical fields of silver and gold nanoaggregates from single-molecule Raman spectroscopy to ultrasensitive probing in live cells, Acc. Chem. Res. 39 (2006) 443–450.
[43] P.L. Stiles, J.A. Dieringer, N.C. Shah, Surface-enhanced Raman spectroscopy, Annu. Rev. Anal. Chem. 1 (2008) 601–626.
[44] Y. Cheng, A.C. Samia, J.D. Meyers, I. Panagopoulos, F. Baowei, C. Burda, Highly efficient drug delivery with gold nanoparticle vectors for in vivo photodynamic therapy of cancer, J. Am. Chem. Soc. 130 (2008) 10643–10647.
[45] P. Ghosh, G. Han, M. De, C.K. Kim, V.M. Rotello, Gold nanoparticles in delivery applications, Adv. Drug. Deliv. Rev. 60 (2008) 1307–1315.
[46] Q. Zhang, N. Iwakuma, P. Sharma, B.M. Moudgil, C. Wu, J. McNeill, et al., Gold nanoparticles as a contrast agent for in vivo tumor imaging with photoacoustic tomography, Nanotechnology 20 (2009) 395102.
[47] G. Frens, Controlled nucleation for the regulation of the particle size in monodisperse gold suspensions, Nature 241 (1973) 20–22.
[48] P.C. Lee, D. Meisel, Adsorption and surface-enhanced Raman of dyes on silver and gold sols, J. Phys. Chem. 86 (1982) 3391–3395.
[49] N. Leopold, B. Lendl, A. New, Method for fast preparation of highly surface-enhanced Raman scattering (SERS) active silver colloids at room temperature by reduction of silver nitrate with hydroxylamine hydrochloride, J. Phys. Chem. B 107 (2003) 5723–5727.
[50] S. Link, M.A. El-Sayed, Size and temperature dependence of the plasmon absorption of colloidal gold nanoparticles, J. Phys. Chem. B 103 (1999) 4212–4217.
[51] S. Abalde-Cela, P. Aldeanueva-Potel, C. Mateo-Mateo, I. Rodríguez-Lorenzo, R.A. Alvarez-Puebla, L.M.J.R. Liz-Marzán, Surface-enhanced Raman scattering biomedical applications of plasmonic colloidal particles, J. R. Soc. Interface 7 (Suppl 4) (2010) S435–S450.
[52] S. Lee, H. Chon, M. Lee, J. Choo, S.Y. Shin, Y.H. Lee, et al., Surface-enhanced Raman scattering imaging of HER2 cancer markers overexpressed in single Mcf7 cells using antibody conjugated hollow gold nanospheres, Biosens. Bioelectron. 24 (2009) 2260–2263.

[53] D.D. Evanoff, G.J. Chumanov, Size-Controlled Synthesis of Nanoparticles. 1. "Silver-Only" Aqueous Suspensions via Hydrogen Reduction, J. Phys. Chem. B 108 (2004) 13948–13956.
[54] R.G. Freeman, M.B. Hommer, K.C. Grabar, M.A. Jackson, M.J. Natan, Ag-clad Au nanoparticles: novel aggregation, optical, and surface-enhanced Raman scattering properties, J. Phys. Chem. 100 (1996) 718–724.
[55] M. Moskovits, I. Srnová-Šloufová, B. Vlčková, Bimetallic Ag–Au nanoparticles: extracting meaningful optical constants from the surface-plasmon extinction spectrum, J. Chem. Phys. 116 (2002) 10435–10446.
[56] S. Devarajan, B. Vimalan, S. Sampath, Phase transfer of Au–Ag alloy nanoparticles from aqueous medium to an organic solvent: effect of aging of surfactant on the formation of Ag-rich alloy compositions, J. Colloid Interface Sci. 278 (2004) 126–132.
[57] S. Lal, N.K. Grady, J. Kundu, C.S. Levin, S. Carly, J.B. Lassiter, et al., Tailoring plasmonic substrates for surface enhanced spectroscopies, Chem. Soc. Rev. 37 (2008) 898–911.
[58] M.A. Ochsenkuhn, P.R.T. Jess, H. Stoquert, K. Dholakia, C.J. Campbell, Nanoshells for surface-enhanced Raman spectroscopy in eukaryotic cells: cellular response and sensor development, ACS Nano 3 (2009) 3613–3621.
[59] J.X. Fang, S. Lebedkin, S. Yang, H. Hahn, A. New, Route for the synthesis of polyhedral gold mesocages and shape effect in single-particle surface-enhanced Raman spectroscopy, Chem. Commun. 47 (2011) 5157–5159.
[60] J. Huang, K.H. Kim, N. Choi, H. Chon, S. Lee, J. Choo, Preparation of silica-encapsulated hollow gold nanosphere tags using layer-by-layer method for multiplex surface-enhanced Raman scattering detection, Langmuir 27 (2011) 10228–10233.
[61] M. Gellner, B. Küstner, S. Schlücker, Optical properties and SERS efficiency of tunable gold/silver nanoshells, Vib. Spectrosc. 50 (2009) 43–47.
[62] A.M. Schwartzberg, Y.O. Tammy, E.T. Chad, Z.Z. Jin, Synthesis, characterization, and tunable optical properties of hollow gold nanospheres, J. Phys. Chem. B 110 (2006) 19935–19944.
[63] C. Song, Z. Wang, R. Zhang, J. Yang, X. Tan, Y. Cui, Highly sensitive immunoassay based on Raman reporter-labeled immuno-Au aggregates and SERS-active immune substrate, Biosens. Bioelectron. 25 (2009) 826–831.
[64] X. Huang, I.H. El-Sayed, W. Qian, M.A. El-Sayed, Cancer cell imaging and photothermal therapy in the near-infrared region by using gold nanorods, J. Am. Chem. Soc. 128 (2006) 2115–2120.
[65] G. von Maltzahn, A. Centrone, J.H. Park, R. Ramanathan, M.J. Sailor, T.A. Hatton, et al., SERS-coded gold nanorods as a multifunctional platform for densely multiplexed near-infrared imaging and photothermal heating, Adv. Mater. 21 (2009) 3175–3180.
[66] L. Jiang, J. Qian, F. Cai, S. He, Raman reporter-coated gold nanorods and their applications in multimodal optical imaging of cancer cells, Anal. Bioanal. Chem. 400 (2011) 2793–2800.
[67] Z. Wang, S. Zong, J. Yang, C. Song, J. Li, Y. Cui, One-step functionalized gold nanorods as intracellular probe with improved SERS performance and reduced cytotoxicity, Biosens. Bioelectron. 26 (2010) 241–247.
[68] D.O. Ansari, Raman-encoded nanoparticles for biomolecular detection and cancer diagnostics, ProQuest, 2008.
[69] B. Lutz, C. Dentinger, L. Sun, L. Nguyen, J. Zhang, A. Chmura, et al., Raman nanoparticle probes for antibody-based protein detection in tissues, J. Histochem. Cytochem. 56 (2008) 371–379.

[70] B.R. Lutz, C.E. Dentinger, L.N. Nguyen, L. Sun, J. Zhang, A. Allen, et al., Spectral analysis of multiplex Raman probe signatures, ACS Nano 2 (2008) 2306–2314.
[71] 莫, 和平, 陈娟, 等. 表面增强拉曼光谱 (SERS) 及其在定量测量中的研究进展 [J]. 光散射学报 25 (2013) 219–234.
[72] L. Rodríguez-Lorenzo, Ramón A. Álvarez-Puebla, I. Pastoriza-Santos, S. Mazzucco, O. Stéphan, M. Kociak, Luis M. Liz-Marzán, F. Javier García de Abajo, Zeptomol detection through controlled ultrasensitive surface-enhanced Raman scattering, J. Am. Chem. Soc. 131 (2009) 4616–4618.
[73] L. Rodríguez-Lorenzo, Z. Krpetic, S. Barbosa, R. Alvarez-Puebla, L. Liz-Marzán, I. Prior, et al., Intracellular mapping with SERS-encoded gold nanostars, Integr. Biol. 3 (2011) 922–926.
[74] D. Xu, J. Gu, W. Wang, X. Yu, K. Xi, Development of chitosan-coated gold nanoflowers as SERS-active probes, Nanotechnology 21 (2010) 375101.
[75] M. Bechelany, P. Brodard, J. Elias, A. Brioude, J. Michler, L. Philippe, Simple synthetic route for SERS-active gold nanoparticles substrate with controlled shape and organization, Langmuir 26 (2010) 14364–14371.
[76] G.N. Zhang, J.R. Li, A.G. Shen, J.M. Hu, Synthesis of size-tunable chitosan encapsulated gold–silver nanoflowers and their application in SERS imaging of living cells, Phys. Chem. Chem. Phys. 17 (2015) 21261–21267.
[77] P.S. Kumar, I. Pastoriza-Santos, B. Rodríguez-González, G.D.A.F. Javier, L. Liz-Marzán, High-yield synthesis and optical response of gold nanostars, Nanotechnology 19 (2007) 015606.
[78] S. Barbosa, A. Agrawal, L. Rodríguez-Lorenzo, I. Pastoriza-Santos, L. Liz-Marzán, Tuning size and sensing properties in colloidal gold nanostars, Langmuir 26 (2010) 14943–14950.
[79] A. Guerrero-Martínez, S. Barbosa, I. Pastoriza-Santos, L. Liz-Marzán, Nanostars shine bright for you: colloidal synthesis, properties and applications of branched metallic nanoparticles, Curr. Opin. Colloid Interface Sci. 16 (2011) 118–127.
[80] A.M. Goodman, Y. Cao, C. Urban, O. Neumann, C.A. Orozco, M.W. Knight, et al., The Surprising in vivo instability of near-IR-absorbing hollow Au–Ag nanoshells, ACS Nano 8 (2014) 3222–3231.
[81] D.K. Lim, K.S. Jeon, J.H. Hwang, H. Kim, S. Kwon, Y.D. Suh, et al., Highly uniform and reproducible surface-enhanced Raman scattering from DNA-tailorable nanoparticles with 1-nm interior gap, Nat. Nanotechnol. 6 (2011) 452–460.
[82] J.X. Li, Z. Zhu, B.Q. Zhu, Y.L. Ma, B.Q. Lin, R.D. Liu, et al., Surface-enhanced Raman scattering active plasmonic nanoparticles with ultrasmall interior nanogap for multiplex quantitative detection and cancer cell imaging, Anal. Chem. 88 (2016) 7828–7836.
[83] L. Lin, H.C. Gu, J. Ye, Plasmonic multi-shell nanomatryoshka particles as highly tunable SERS tags with built-in reporters, Chem. Commun. 51 (2015) 17740–17743.
[84] Y.Q. Zhang, Y.Q. Gu, J. He, B.D. Thackray, J. Ye, Ultrabright gap-enhanced Raman tags for high-speed bioimaging, Nat. Commun. 10 (2019) 3905–3916.
[85] D. Philip, K.G. Gopchandran, C. Unni, K.M. Nissamudeen, Synthesis, characterization and SERS activity of Au–Ag nanorods, Spectrochim. Acta Part A Mol. Biomol. Spectrosc. 70 (2008) 780–784.
[86] H. Chen, Y. Wang, J. Xu, J.Z. Ji, J. Zhang, Y.Z. Hu, et al., Non-invasive near infrared fluorescence imaging of CdHgTe quantum dots in mouse model, J. Fluoresc. 18 (2008) 801–811.
[87] X. Tan, Z. Wang, J. Yang, Polyvinylpyrrolidone-(PVP-) coated silver aggregates for high performance surface-enhanced Raman scattering in living cells, Nanotechnology 20 (2009) 445102.

[88] G.B. Braun, S.J. Lee, T. Laurence, N. Fera, L. Fabris, G.C. Bazan, et al., Generalized approach to SERS-active nanomaterials via controlled nanoparticle linking, polymer encapsulation, and small-molecule infusion, J. Phys. Chem. C 113 (2009) 13622–13629.
[89] M. Futamata, Y.Y. Yu, T. Yanatori, T. Kokubun, Closely adjacent Ag nanoparticles formed by cationic dyes in solution generating enormous SERS enhancement, J. Phys. Chem. C 114 (2010) 7502–7508.
[90] L.L. Tay, P.J. Huang, J. Tanha, S. Ryan, X.H. Wu, J. Hulse, et al., Silica encapsulated SERS nanoprobe conjugated to the bacteriophage tailspike protein for targeted detection of Salmonella, Chem. Commun. 48 (2012) 1024–1026.
[91] P.J. Huang, L.K. Chau, T.S. Yang, L.L. Tay, T.T. Lin, Nanoaggregate-embedded beads as novel Raman labels for biodetection, Adv. Funct. Mater. 19 (2009) 242–248.
[92] B.H. Jun, J.H. Kim, H. Park, Surface-enhanced Raman spectroscopic-encoded beads for multiplex immunoassay, J. Comb. Chem. 9 (2007) 237–244.
[93] K. Kim, H.B. Lee, H.K. Park, K.S. Shin, Easy deposition of Ag onto polystyrene beads for developing surface-enhanced-Raman-scattering-based molecular sensors, J. Colloid Interface Sci. 318 (2008) 195–201.
[94] J.M. Li, W.F. Ma, C. Wei, J. Guo, C.C. Wang, Poly (styrene-co-acrylic acid) core and silver nanoparticle/silica shell composite microspheres as high performance surface-enhanced Raman spectroscopy (SERS) substrate and molecular barcode label, J. Mater. Chem. 21 (2011) 5992–5998.
[95] J.H. Kim, J.S. Kim, H. Choi, S.M. Lee, B.H. Jun, K.N. Yu, et al., Nanoparticle probes with surface enhanced Raman spectroscopic tags for cellular cancer targeting, Anal. Chem. 78 (2006) 6967–6973.
[96] G.N. Zhang, G. Qu, Y. Chen, A.G. Shen, W. Xie, X.D. Zhou, et al., Controlling carbon encapsulation of gold nanoaggregates as highly sensitive and spectrally stable SERS tags for live cell imaging, J. Mater. Chem. B 1 (2013) 4364–4369.
[97] F. Shao, Z.C. Lu, C. Liu, H. Han, K. Chen, W.T. Li, et al., Hierarchical nanogaps within bioscaffold arrays as a high-performance SERS substrate for animal virus biosensing, ACS Appl. Mater. Interfaces 6 (2014) 6281–6289.
[98] H.L. Liu, J. Cao, S. Hanif, C. Yuan, J. Pang, R. Levicky, et al., Size-controllable gold nanopores with high SERS activity, Anal. Chem. 89 (2017) 10407–10413.
[99] Y. Huang, V.P. Swarup, S.W. Bishnoi, Rapid Raman imaging of stable, functionalized nanoshells in mammalian cell cultures, Nano Lett. 9 (2009) 2914–2920.
[100] L.C. Martin, I.A. Larmour, K. Faulds, Turning up the lights-fabrication of brighter SERRS nanotags, Chem. Commun. 46 (2010) 5247–5249.
[101] K. Kneipp, H. Kneipp, I. Itzkan, R.R. Dasari, M.S. Feld, Ultrasensitive chemical analysis by Raman spectroscopy, Chem. Rev. 99 (1999) 2957–2976.
[102] G. McNay, D. Eustace, W.E. Smith, Surface-enhanced Raman scattering (SERS) and surface-enhanced resonance Raman scattering (SERRS): a review of applications, Appl. Spectrosc. 65 (2011) 825–837.
[103] R.M. Jarvis, R. Goodacre, Characterisation and identification of bacteria using SERS, Chem. Soc. Rev. 37 (2008) 931–936.
[104] H. Xu, Q. Li, L. Wang, Y. He, J.Y. Shi, B. Tang, et al., Nanoscale optical probes for cellular imaging, Chem. Soc. Rev. 43 (2014) 2650–2661.
[105] X.H. Huang, I.H. El-Sayed, W. Qian, M.A. El-Sayed, Cancer cell imaging and photothermal therapy in the near-infrared region by using gold nanorods, J. Am. Chem. Soc. 128 (2006) 2115–2120.
[106] J.P. Xie, Q.B. Zhang, J.Y. Lee, D.I.C. Wang, The synthesis of SERS-active gold nanoflower tags for in vivo applications, ACS Nano 2 (2008) 2473–2480.

[107] N.C.M. Tam, B.M.T. Scott, D. Voicu, B.C. Wilson, G. Zheng, Facile synthesis of Raman active phospholipid gold nanoparticles, Bioconjug. Chem. 21 (2010) 2178–2182.
[108] K.K. Maiti, A. Samanta, M. Vendrell, K.S. Soh, M. Olivoacd, Y.T. Chang, Multiplex cancer cell detection by SERS nanotags with cyanine and triphenylmethine Raman reporters, Chem. Commun. 47 (2011) 3514–3516.
[109] C.L. Zavaleta, B.R. Smith, I. Walton, W. Doering, G. Davis, B. Shojaei, et al., Multiplexed imaging of surface enhanced Raman scattering nanotags in living mice using noninvasive Raman spectroscopy, Proc. Natl. Acad. Sci. U S A 106 (2009) 13511–13516.
[110] G. Jaouen, A. Vessikres, S. Top, Metal carbonyl fragments as a new class of markers in molecular biology, J. Am. Chem. Soc. 107 (1985) 4778–4780.
[111] K. Meister, J. Niesel, U. Schatzschneider, N.N. Metzler, D.A. Schmidt, M. Havenith, Label-free imaging of metal-carbonyl complexes in live cells by Raman microspectroscopy, Angew. Chem. Int. Ed. 49 (2010) 3310–3312.
[112] K.V. Kong, Z. Lam, W.D. Goh, W.K. Leong, M. Olivo, Metal carbonyl-gold nanoparticle conjugates for live-cell SERS imaging, Angew. Chem. Int. Ed. 51 (2012) 9796–9799.
[113] K.V. Kong, Z. Lam, W.K.O. Lau, W.K. Leong, M. Olivo, A transition metal carbonyl probe for use in a highly specific and sensitive SERS-based assay for glucose, J. Am. Chem. Soc. 135 (2013) 18028–18031.
[114] I.T. Suydam, C.D. Snow, V.S. Pande, S.G. Boxer, Electric fields at the active site of an enzyme: direct comparison of experiment with theory, Science 313 (2006) 200–204.
[115] M. Noestheden, Q. Hu, L.L. Tay, A.M. Tonary, A. Stolow, R. MacKenzie, et al., Electric fields at the active site of an enzyme: direct comparison of experiment with theory, Bioorg. Chem. 35 (2007) 284–293.
[116] S. Ma, Q. Li, Y. Yin, J. Yang, D. Liu, Interference-free surface-enhanced Raman scattering tags for single-cell molecular imaging with a high signal-to-background ratio, Small 13 (2017) 1603340.
[117] H. Yamakoshi, K. Dodo, M. Okada, J. Ando, A. Palonpon, K. Fujita, et al., Imaging of EdU, an alkyne-tagged cell proliferation probe, by Raman microscopy, J. Am. Chem. Soc. 133 (2011) 6102–6105.
[118] Y. Chen, J.Q. Ren, X.G. Zhang, D.Y. Wu, A.G. Shen, J.M. Hu, Alkyne-modulated surface-enhanced Raman scattering-palette for optical interference-free and multiplex cellular imaging, Anal. Chem. 88 (2016) 6115–6119.
[119] Y. Zeng, J.Q. Ren, A.G. Shen, J.M. Hu, Field and pretreatment-free detection of heavy-metal ions in organic polluted water through an alkyne-coded SERS test kit, ACS Appl. Mater. Interfaces 8 (2016) 27772–27778.
[120] Y.M. Si, Y.C. Bai, X.J. Qin, J. Li, W.W. Zhong, Z.J. Xiao, et al., Alkyne – DNA-functionalized alloyed Au/Ag nanospheres for ratiometric surface-enhanced Raman scattering imaging assay of endonuclease activity in live cells, Anal. Chem. 90 (2018) 3898–3905.
[121] Y. Zhang, Y. Zou, F. Liu, Y. Xu, X. Wang, Y. Li, et al., Stable graphene-isolated-Au-nanocrystal for accurate and rapid surface enhancement Raman scattering analysis, Anal. Chem. 88 (2016) 10611–10616.
[122] L. Zhang, F. Liu, Y. Zou, X. Hu, S. Huang, Y. Xu, et al., Surfactant-free interface suspended gold graphitic surface-enhanced Raman spectroscopy substrate for simultaneous multiphase analysis, Anal. Chem. 90 (2018) 11183–11187.
[123] X. Meng, H. Wang, N. Chen, P. Ding, H. Shi, X. Zhai, et al., Nanoparticle-silicon sandwich SERS chip for quantitative detection of molecules and capture, discrimination, and inactivation of bacteria, Anal. Chem. 90 (2018) 5646–5653.

[124] Y.M. Yin, Q. Li, S.S. Ma, H.Q. Liu, B. Dong, J. Yang, et al., Prussian blue as a highly sensitive and background-free resonant Raman reporter, Anal. Chem. 89 (2017) 1551−1557.
[125] M. Li, J.Y. Wang, Q.Q. Chen, L.H. Lin, P. Radjenovic, H. Zhang, et al., Background-free quantitative surface enhanced Raman spectroscopy analysis using core−shell nanoparticles with an inherent internal standard, Anal. Chem. 91 (2019) 15025−15031.
[126] Y. Zeng, J.Q. Ren, A.G. Shen, J.M. Hu, Splicing nanoparticles-based "Click" SERS could aid multiplex liquid biopsy and accurate cellular imaging, J. Am. Chem. Soc. 140 (2018) 10649−10652.
[127] X.R. Bai, L.H. Wang, J.Q. Ren, X.W. Bai, L.W. Zeng, A.G. Shen, et al., Accurate clinical diagnosis of liver cancer based on simultaneous detection of ternary specific antigens by magnetic induced mixing surface-enhanced Raman scattering emissions, Anal. Chem. 91 (2019) 2955−2963.
[128] M.Y. Gao, Q. Chen, W. Li, A.G. Shen, J.M. Hu, Combined surface-enhanced Raman scattering emissions for high-throughput optical labels on micrometer-scale objects, Anal. Chem. 91 (2019) 13866−13873.
[129] Z.Y. Wang, S.F. Zong, W. Li, C.L. Wang, S.H. Xu, H. Chen, et al., SERS-fluorescence joint spectral encoding using organic-metal-QD hybrid nanoparticles with a huge encoding capacity for high-throughput biodetection: putting theory into practice, J. Am. Chem. Soc. 134 (2012) 2993−3000.
[130] L. Sun, K.B. Sung, C. Dentinger, B. Lutz, L. Nguyen, J. Zhang, et al., Composite organic−inorganic nanoparticles as Raman labels for tissue analysis, Nano Lett. 7 (2007) 351−356.
[131] M. Yang, T. Chen, W.S. Lau, Y. Wang, Q.H. Tang, Y.H. Yang, et al., Development of polymer-encapsulated metal nanoparticles as surface-enhanced Raman scattering probes, Small 5 (2009) 198−202.
[132] Y. Chen, X.R. Bai, L. Su, Z.W. Du, A.G. Shen, A. Materny, et al., Combined labelled and label-free SERS probes for triplex three-dimensional cellular imaging, Sci. Rep. 6 (2016) 19173.
[133] A.G. Shen, L.F. Chen, W. Xie, J.C. Hu, A. Zeng, R. Richards, et al., Triplex Au-Ag-C core−shell nanoparticles as a novel Raman label, Adv. Funct. Mater. 20 (2010) 969−975.
[134] A.G. Shen, J.Z. Guo, W. Xie, M.X. Sun, R. Richardsc, J.M. Hua, Surface-enhanced Raman spectroscopy in living plant using triplex Au-Ag-C core−shell nanoparticles, J. Raman Spectrosc. 42 (2011) 879−884.
[135] C.I. Brady, N.H. Mack, L.O. Brown, S.K. Doorn, Self-assembly approach to multiplexed surface-enhanced Raman spectral-encoder beads, Anal. Chem. 81 (2009) 7181−7188.
[136] X. Liu, M. Knauer, N.P. Ivleva, R. Niessner, C. Haisch, Synthesis of Core−Shell Surface-Enhanced Raman Tags for Bioimaging, Anal. Chem. 82 (2010) 441−446.
[137] M. Schütz, B. Küstner, M. Bauer, C. Schmuck, S. Schlücker, Synthesis of glass-coated SERS nanoparticle probes via SAMs with terminal SiO_2 precursors, Small 6 (2010) 733−737.
[138] R. Contreras-Caceres, S. Abalde-Cela, P. Guardia-Giros, A. Fernandez-Barbero, J. Pe´rez-Juste, R.A. Alvarez-Puebla, et al., Multifunctional microgel magnetic/optical traps for SERS ultradetection, Langmuir 27 (2011) 4520−4525.
[139] B.H. Jun, M.S. Noh, J. Kim, G. Kim, H. Kang, M.S. Kim, et al., Multifunctional silver-embedded magnetic nanoparticles as SERS nanoprobes and their applications, Small 6 (2010) 119−125.

[140] A. Gole, N. Agarwal, P. Nagaria, M.D. Wyatt, C.J. Murphy, One-pot synthesis of silica-coated magnetic plasmonic tracer nanoparticles, Chem. Commun. 46 (2008) 6140–6142.
[141] K.N. Yu, S.M. Lee, J.Y. Han, H. Park, M.A. Woo, M.S. Noh, et al., Multiplex targeting, tracking, and imaging of apoptosis by fluorescent surface enhanced Raman spectroscopic dots, Bioconjug. Chem. 18 (2007) 1155–1162.
[142] Y. Cui, X.S. Zheng, B. Ren, R. Wang, J. Zhang, N.S. Xia, et al., Au@organosilica multifunctional nanoparticles for the multimodal imaging, Chem. Sci. 2 (2011) 1463–1469.
[143] K. Kim, Y.M. Lee, H.B. Lee, K.S. Shin, Silver salts of aromatic thiols applicable as core materials of molecular sensors operating via SERS and fluorescence, Biosens. Bioelectron. 24 (2009) 3615–3621.
[144] Z.Y. Wang, H. Wu, C.L. Wang, S.H. Xu, Y.P. Cui, Gold aggregates- and quantum dots-embedded nanospheres: switchable dual-mode image probes for living cells, J. Mater. Chem. 21 (2011) 4307–4313.
[145] K. Kim, H.B. Lee, Y.M. Lee, K.S. Shin, Rhodamine B isothiocyanate-modified Ag nanoaggregates on dielectric beads: a novel surface-enhanced Raman scattering and fluorescent imaging material, Biosens. Bioelectron. 24 (2009) 1864–1869.
[146] M.A. Woo, S.M. Lee, G. Kim, J. Baek, M.S. Noh, J.E. Kim, et al., Multiplex immunoassay using fluorescent-surface enhanced Raman spectroscopic dots for the detection of bronchioalveolar stem cells in murine lung, Anal. Chem. 81 (2009) 1008–1015.
[147] M. Xiao, J. Nyagilo, V. Arora, P. Kulkarni, D. Xu, X. Sun, et al., Gold nanotags for combined multi-colored Raman spectroscopy and x-ray computed tomography, Nanotechnology 21 (2010) 035101.
[148] W. Zhu, M.Y. Gao, Q. Zhu, B. Chi, L.W. Zeng, A.G. Shen, et al., Monodispersed plasmonic prussian blue nanoparticles for zerobackground SERS/MRI-guided phototherapy (2012) 1–3.
[149] Q. Lv, H. Min, D.B. Duan, W. Fang, G.M. Pan, A.G. Shen, et al., Total aqueous synthesis of Au@$Cu_{2-x}S$ core–shell nanoparticles for in vitro and in vivo SERS/PA imaging guided photothermal cancer therapy, Adv. Healthc. Mater. (2018) 1801257.
[150] Q. Lv, M.Y. Gao, Z.H. Cheng, Q. Chen, A.G. Shen, J.M. Hu, Rational synthesis of hollow cubic CuS@Spiky Au core–shell nanoparticles for enhanced photothermal and SERS effects, Chem. Commun. 54 (2018) 13399–13402.
[151] Y. Zeng, J.J. Pei, L.H. Wang, A.G. Shen, J.M. Hu, A sensitive sequential 'on/off SERS assay for heparin with wider detection window and higher reliability based on the reversed surface charge changes of functionalized Au@Ag nanoparticles, Biosens. Bioelectron. 66 (2015) 55–61.
[152] N.H. Ly, S.W. Joo, Raman spectroscopy of di-(2-picolyl)amine on gold nanoparticles for Hg (II) detection, Bull. Korean Chem. Soc. 36 (2015) 226–229.
[153] G. Qu, G.N. Zhang, Z.T. Wu, A.G. Shen, J.B. Wang, J.M. Hu, A "turn-off" SERS assay of heparin with high selectivity based on heparin-peptide complex and Raman labelled gold nanoparticles, Biosens. Bioelectron. 60 (2014) 124–129.
[154] D. Graham, B.J. Mallinder, W.E. Smith, Detection and identification of labeled DNA by surface enhanced resonance Raman scattering, Biopolymers. 57 (2000) 85–91.
[155] N.M.S. Sirimuthu, Investigation of the stability of labelled nanoparticles for SE(R)RS measurements in cells, Chem. Commun. 47 (2011) 4099–4101.
[156] Y. Cui, B. Ren, J.L. Yao, R.A. Gu, Z.Q. Tian, Synthesis of $Ag_{core}Au_{shell}$ bimetallic nanoparticles for immunoassay based on surface-enhanced Raman spectroscopy, J. Phys. Chem. B 110 (2006) 4002–4006.
[157] M. Li, J. Zhang, S. Suri, L.J. Sooter, D. Ma, N. Wu, Detection of adenosine triphosphate with an aptamer biosensor based on surface-enhanced Raman scattering, Anal. Chem. 84 (2012) 2837–2842.

CHAPTER 6

Surface-enhanced Raman spectroscopy for circulating biomarkers detection in clinical diagnosis

Yuan Liu, Nana Lyu, Alison Rodger and Yuling Wang
Department of Molecular Sciences, ARC Center of Excellence for Nanoscale BioPhotonics, Faculty of Science and Engineering, Macquarie University, Sydney, NSW, Australia

6.1 Introduction

Cancer encompasses a broad group of diseases characterized by abnormal cells with uncontrollable division, growth, and spreading out from a primary site of origin into different tissues and organs. Cancers cause enormous suffering and some have a high mortality rate. Plenty of evidence has demonstrated that early-stage diagnosis can increase therapeutic efficiency, raise survival rate, lower the cost, and improve patients' life quality [1,2]. A biopsy is the most frequently chosen tool to diagnose the occurrence and the stage of cancer. For solid tumors, it is usually performed by taking a small piece of tissue from a patient (Fig. 6.1, left). This invasive procedure delivers spatially and temporally limited information regarding tumor heterogeneity and metastasis (spreading) and treatment response, even giving false-positive or false-negative results [3]. In the past few decades, as the value of precision medicine has been realized, a large variety of biomaterials circulating in body fluids (blood, urine, and saliva) have been discovered and their presence/level been found to indicate disease status as well as being a useful approach to monitoring therapeutic effect. This situation has led to a transferring of focus from conventional tissue biopsy to liquid biopsy. A liquid biopsy involves easily accessible and minimally invasive sampling of circulating body fluids to detect, for example, cancer biomarkers. The approach has the major advantage that repeated sampling at multiple time points is practicable. Thus, it achieves dynamic tracking of disease evolution or therapy response, and sensitively recognizes the sign of minimal residual disease, all of which are unachievable by limited tissue biopsies

Principles and Clinical Diagnostic Applications of Surface-Enhanced Raman Spectroscopy
DOI: https://doi.org/10.1016/B978-0-12-821121-2.00008-1

[4–6]. Therefore, it helps to reveal the genetic and epigenetic landscape of cancerous lesions and extract molecular information of the biomarkers.

Currently, the widely characterized circulating biomarkers from liquid biopsies can be categorized as circulating tumor cells (CTCs), circulating tumor nucleic acids, cancer cell–derived extracellular vesicles (EVs), and disease-associated proteins (Fig. 6.1, right) [7,8]. Each of them was discovered and treated as a promising biomarker due to their distinct features which provide characteristic information regarding tumor formation, metastasis, and progression. Specifically, CTCs were first identified in 1869 with a blood test from a metastatic cancer patient and defined as tumor cells in 1976 [9]. They shed off from the primary tumor or metastatic tumor into vasculature or the lymphatic system [9]. Similarly, fragmented circulating DNA was firstly found in 1948 freely floating in blood and outside cells. Its definition was refined as circulating tumor DNA in 1973 since a reduced amount of circulating tumor DNA from patients was noticed after passing through chemotherapy [10,11]. Subsequently, another nucleic acid–RNA, predominantly microRNA (miRNA), was discovered in 1993 from *Caenorhabditis elegans*. Cell-free RNA has been found in a wide range of body fluids since 2008 [12,13]. Biolipid-enwrapped vesicles from cell culture media were unexpectedly discovered by transmission electron microscopy (TEM) in 1981. The term "exosomes" was initially coined to describe such vesicles in a size range of 40–1000 nm [14]. In the subsequent three decades, this nomenclature was gradually adopted to describe the vesicles with a varying size range

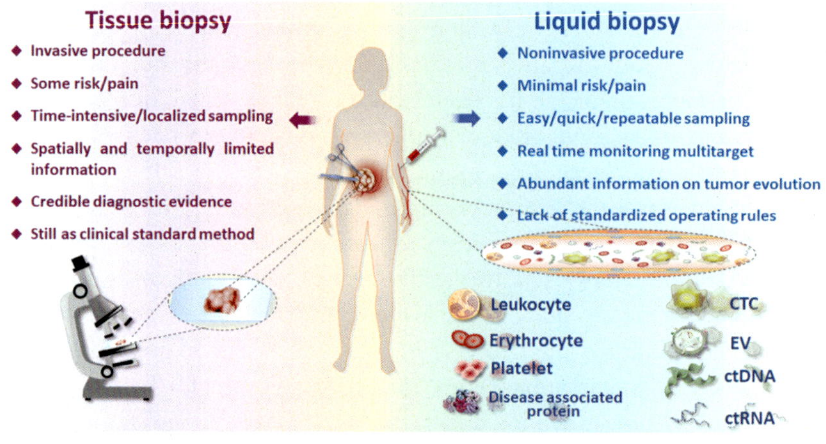

Figure 6.1 Schematic illustration of tissue biopsy and liquid biopsy.

from 30 to 200 nm and circulating in different body fluids. EVs were paid particular attention because their composition and amount uniquely reflected their parental cell origin, and physiological and pathological states. EVs are also relevant to the environmental conditions and/or stimuli, implicating a potential to indicate disease occurrence [15]. Disease-associated proteins have also been identified but generally are not as explicit identifiers of disease or its state as the above four types of biomarkers, which specifically depend on the target disease types because they contain more information [7,8].

The recent interest in biomarkers from liquid biopsies, has made it clear that traditional analysis techniques, which usually aim at either the protein or nucleic acid, are not sufficient to characterize the more complex particles which contain detailed diagnostic information as noted above. The development of novel sensing strategies, for example, by integrating optical or electrochemical techniques with immunological design, statistical analysis, and microfluidics is showing potential to characterize the complexity of the multibiocomponent assemblies present in liquid biopsies. These sensing approaches are expected to fit the demands of ultrasensitive and highly specific identification and quantification toward target biomarkers in a biological environment. Among the approaches being developed, optical sensing assays based on surface-enhanced Raman scattering/spectroscopy (SERS) currently show the most potential for diagnostic molecular fingerprints.

SERS is well known for its significant enhancement of Raman scattering via the effect from both chemical and electromagnetic mechanisms, as discussed in Chapter 1, *Principles of surface-enhanced Raman spectroscopy*. Two primary tactics are widely applied in SERS-based sensing platforms: so-called label-free (direct) SERS and labeled (indirect) SERS. The former gives rich fingerprint information of target molecule as summarized in Chapter 4, *Label-free surface-enhanced Raman scattering for clinical applications*, while the SERS-label approach using the designed multiple SERS nanotags is presented in Chapter 5, *Surface-enhanced Raman scattering nanotags design and synthesis*.

The main purpose of this chapter is to introduce the SERS-based specific applications for circulating biomarkers detection in clinical diagnosis. A general introduction regarding sample preparation and detection methods of liquid biopsy biomarkers is given, followed by four sections each of which emphasizes one aforementioned circulating biomarkers in liquid biopsy (i.e., CTCs, circulating EVs, circulating tumor DNA, circulating tumor RNA, and disease-associated proteins). Each section starts with the biomarker features and clinical significance, followed by the discussion of SERS-based assays for detection in a clinical setting. The chapter ends with an evaluation of the different systems, current challenges, and future perspectives for SERS applications in liquid biopsy.

6.2 Sample preparation and detection methods

Since circulating biomarkers naturally coexist with much larger amounts of nontarget biomaterials (e.g., hemocytes), enrichment and purification of the biological samples are essential for generating accurate analysis results. Typically, sample preparation methods can be categorized as affinity- and physical property—based approaches. In terms of affinity-based preparation, since CTCs, circulating EVs, and disease-associated proteins all have proteins, the corresponding antibodies can be conjugated onto a substrate [e.g., a magnetic bead (MB)] to identify and interact with these proteins on the target via immunoaffinity. The substrate/target complex can then be isolated by a method designed for the substrate, which in the case of MBs is a strong magnet. The main limitation of immunoaffinity-based methods is the cost of ligands (e.g., antibodies, peptides, or aptamers). The physical property—based collection is performed by using the size, density, etc. of the target biomarkers. For instance, ultracentrifugation is widely employed in EVs isolation by using a very high-speed centrifugation force to precipitate the fraction containing EVs. Sometimes, affinity- and physical property—based isolation methods are combined to obtain target biomarkers with higher purity and high concentration. More details on sample preparation are discussed in some particular cases in the following subsections.

In liquid biopsy applications, two main SERS measurement strategies are so-called label-free SERS and labeled SERS. For label-free SERS, the purified biomarkers are directly incubated with SERS substrate, such as metallic nanoparticles (as discussed in Chapter 2, *Nanoplasmoinc materials for surface-enhanced Raman scattering*), followed by a SERS measurement. The generated SERS response comes from the interaction of the target biomarker itself with the SERS substrate. Thus, the SERS spectrum can be used to analyze the fingerprint signature of the target biomarker (Fig. 6.2A). The biomarker origins can be classified by processing of the data with statistical tools. In labeled SERS, a SERS nanotag is required, typically consisting of a Raman reporter molecule and a ligand decorated nanoscale substrate (Fig. 6.2B and C). The

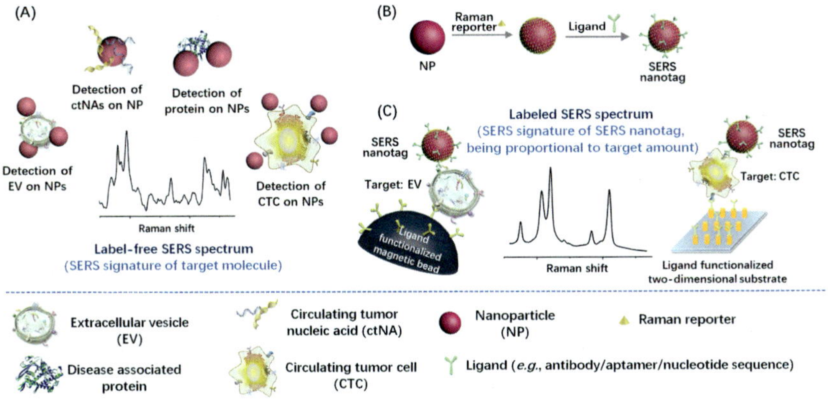

Figure 6.2 Schematic diagram (A) label-free SERS strategy for liquid biopsy biomarker detection; (B) SERS-nanotag preparation; and (C) typical labeled SERS strategy for liquid biopsy biomarker detection in MB- and two-dimensional substrate–based assay.

characteristic SERS spectrum of the Raman reporter molecule is used for signal output, while the ligand (e.g., antibody, aptamer, and peptide) is applied to bind target biomarkers specifically. Therefore, the signal intensity or shift of SERS nanotag in the labeled SERS measurement can be used to monitor the target biomarker proportionally.

The abovementioned SERS strategies have been employed for biomarker quantification and characterization (e.g., phenotyping of biomarkers on CTCs and circulating EVs, genotyping of circulating tumor nucleic acids). Phenotyping of biomarkers includes the distribution, expression level and their response to treatment [16–19]. Genotyping reflects the gene type, genetic change (e.g., mutation), and epigenetic modification (e.g., methylation) of target nucleic acids [10,20–22].

6.3 Circulating tumor cells

6.3.1 Features and current techniques for circulating tumor cells analysis

CTCs are the tumor cells that have passively detached from lesion sites into the bloodstream. They are intrinsically equipped with the pathological characteristics of the solid parental tumor [23–25]. Although less than 0.01% of CTCs hold high metastatic potential to seed, proliferate, and colonize distant tissue [26–30], CTCs are indicative of metastasis risk [31,32].

The clinical value of CTCs in diagnosis has been confirmed by numerous studies across different cancers by dynamically monitoring CTCs amounts and their genetic/proteomic profile. However, their lifetime are less than 30 min [28] and they are uncommon relative to the background of peripheral erythrocytes and leukocytes: 1–100 CTCs among 5 billion erythrocytes [33] and 10 million leukocytes per 10 mL blood [34–36], posing the challenge in CTC studies. Currently, CTC purification methods have been developed based on either physical properties (e.g., size, density) or immunoaffinity [28,37]. However, the heterogeneity of CTC size and other features as cells undergo different growth stages or apoptosis may result in insufficient purity and low specificity so more work is still required in improving CTCs enrichment [38,39].

Immunoaffinity-based CTCs isolation includes two subcategories: positive enrichment and negative enrichment. The positive enrichment approach relies on specific binding to tumor-associated antigens on CTCs to collect CTCs. A typical approach is CellSearch (Janssen Diagnostics), the first and the only system approved by the Food and Drug Administration (FDA) for identification, isolation, and enumeration of CTCs [7,28]. CellSearch and other positive enrichment strategies then identify CTCs with a fluorescent stain containing of 4′,6-diamidino-2-phenylindole (DAPI) + /CD45−/cytokeratin + : DAPI stains nuclei so identifies cells, fluorescence-labeled anti-CD45 antibody excludes white blood cells, and fluorescence-labeled anticytokeratin antibody is used to identify epithelial cells [37,40].

Negative enrichment, by way of contrast, targets collecting the background cells (e.g., leukocyte) to obtain a CTC-enriched sample. However, the biased isolation posed by excluding subpopulation of CTCs due to the preset surface expression, as well as the labor-intensive operation are the limitation of the immunoaffinity-based CTCs enrichment [41,42].

6.3.2 SERS strategy for CTCs analysis

6.3.2.1 Quantification of circulating tumor cells

Since SERS has a sensitive response for the molecules close to a SERS substrate (usually a metallic nanoparticle), it becomes a promising candidate to detect low level CTCs against a background of blood cells and biomolecules in circulation. As outlined above and discussed in Section 6.2, approaches have been established for CTCs enrichment. Then, indirect labeling of CTCs by SERS nanotags (composed of a metallic nanoparticle, a Raman reporter molecule, and a specific linker ligand targeting a specific CTC via a surface molecule) make quantification possible. Many studies have applied the indirect SERS strategy to quantify CTCs in body fluids to achieve sensitive detection [43–46]. For example, Tang et al. constructed SERS nanotags by conjugating a Raman reporter molecule (4-mercaptobenzoic acid, MBA) and luteinizing hormone-releasing hormone onto AuNPs to recognize LHRH receptors expressed on the HeLa cells, which were spiked in mouse blood [43]. This assay achieved sensitive and specific detection for the target CTCs with a detection limit of 5 CTCs mL^{-1} in blood.

6.3.2.2 Characterization of circulating tumor cells surface biomarkers

SERS can be used for characterizing CTCs surface biomarkers via indirectly labeling CTCs surface biomarkers with SERS nanotags (indirect SERS measurement) or directly reading the SERS signature of CTCs surface biomarkers (direct SERS measurement). In terms of indirect SERS measurement, a SERS nanotag is used to bind specifically with a target biomarker and to induce a SERS signal in a reporter molecule that is proportionally to the amount of the labeled biomarker. For example, Nima et al. have reported simultaneously labeling four SERS nanotags to target four different biomarkers in a sample composed of human blood spiked with breast cancer cells [47]. This work achieved the visual evaluation of multibiomarker expression on CTCs by applying SERS mapping to showcase the distribution of these four biomarkers.

The direct SERS approach relies on the biomolecular composition of CTCs being evident from the SERS response of all CTC molecular components that are near the surface and hence accessed by a metallic SERS substrate that induces a plasmonic enhancement effect. For example, Kaminska and coworkers presented a diagnostic assay based on directly obtaining the SERS signature of CTCs and analyzing the difference between CTCs and normal cells by a multivariate statistical method (principal component analysis, PCA) [48]. Prostate cancer cell line (PC3) and cervical carcinoma cell line (HeLa) cells were enriched on a SERS-active membrane which contained pores of appropriate size to capture preferentially the larger cancerous cells. As a result, the assay demonstrates a simple approach to integrate CTCs isolation, enrichment, and detection in one step, reaching a diagnostic sensitivity of 98% to statistically differentiate PC3, HeLa, and normal cells in a blood sample. The SERS signature also can reflect the molecular information of CTCs, for example, the vibration modes of nucleic acids were observed at 785 and 1093 cm^{-1}, the peaks at 1268 and 1660 cm^{-1} were from amide I and amide III bands, respectively.

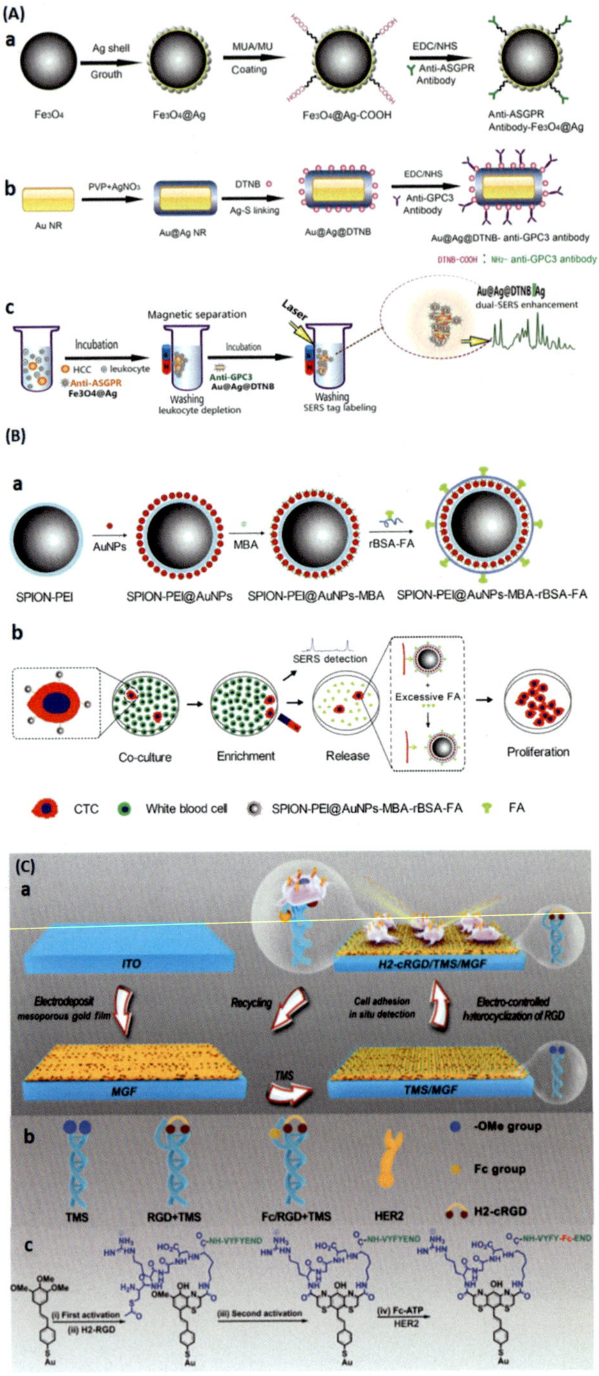

(A)
a
Ag shell
Grouth
MUA/MU
Coating
EDC/NHS
Anti-ASGPR Antibody
Fe3O4
Fe3O4@Ag
Fe3O4@Ag-COOH
Anti-ASGPR Antibody-Fe3O4@Ag
b
PVP+AgNO3
DTNB
Ag-S linking
EDC/NHS
Anti-GPC3 Antibody
Au NR
Au@Ag NR
Au@Ag@DTNB
Au@Ag@DTNB- anti-GPC3 antibody
DTNB-COOH : NH2- anti-GPC3 antibody
c
Incubation
HCC
leukocyte
Anti-ASGPR Fe3O4@Ag
Magnetic separation
Washing
leukocyte depletion
Incubation
Anti-GPC3 Au@Ag@DTNB
Laser
Washing
SERS tag labeling
Au@Ag@DTNB Ag
dual-SERS enhancement
(B)
a
AuNPs
MBA
rBSA-FA
SPION-PEI
SPION-PEI@AuNPs
SPION-PEI@AuNPs-MBA
SPION-PEI@AuNPs-MBA-rBSA-FA
b
SERS detection
Excessive FA
Co-culture
Enrichment
Release
Proliferation
CTC
White blood cell
SPION-PEI@AuNPs-MBA-rBSA-FA
FA
(C)
a
ITO
H2-cRGD/TMS/MGF
Electrodeposit mesoporous gold film
Recycling
Cell adhesion in situ detection
Electro-controlled heterocyclization of RGD
MGF
TMS
TMS/MGF
b
TMS
RGD+TMS
Fc/RGD+TMS
HER2
-OMe group
Fc group
H2-cRGD
c
(i) First activation
(ii) H2-RGD
(iii) Second activation
(iv) Fc-ATP
HER2

6.3.3 SERS-based assays for CTCs analysis in clinical samples

SERS has been applied to clinical CTC samples, often utilizing magnetic separation to achieve positive enrichment of CTCs. Pang et al. [49], in 2018, set up an indirect sandwich immunoassay by first collecting hepatocellular carcinoma CTCs using anti—asialoglyco protein receptor antibodies (anti-ASGPR), which are highly expressed on the surface of human hepatocytes (Fig. 6.3A). Then, the obtained CTCs were labeled with SERS-active Ag/Au nanorods consisting of Raman reporter molecules and antiglypican-3 antibodies. Glypican-3 is specifically overexpressed on hepatocellular carcinoma cell line, so the assay achieved a high specificity toward hepatocellular carcinoma cells in human blood samples with a limit of detection (LOD) of 1 cell mL^{-1} and an analytical range from 1 to 100 cells mL^{-1}. Its quantification capability in the clinical setting was demonstrated by screening blood samples from eight hepatocellular carcinoma patients with controls of three breast cancer patients and three healthy samples. The results indicate that hepatocellular carcinoma CTCs were only detected from the eight hepatocellular carcinoma patients' blood samples, proving the potential for clinical hepatocellular carcinoma diagnosis.

Xue et al. [50] presented a sensing strategy in 2019 (Fig. 6.3B) which utilized AuNPs, a Raman reporter, and folic acid functionalized MBs. They first isolate HeLa cells from blood samples via the specific interaction

◀ **Figure 6.3** (A) Schematic illustration for the conjugation of (a) anti-ASGPR antibody on monodispersed silver-coated magnetic nanoparticles and (b) antiGPC3 antibody on core—shell plasmonic nanoparticles (Au@Ag@DTNB); (c) the operating principle for hepatocellular carcinoma CTCs detection [49]. (B) Workflow for (a) synthesis of the magnetic SERS nanotag (SPION-PEI@AuNPs-MBA-rBSA-FA) and (b) SERS nanotag—based CTCs analysis [50]. (C) (a) Schematic working principle of H2-cRGD peptide and *trans*-30,40,50-trimethoxy-4-mercaptostillbene (TMS) decorated mesoporous gold film (MGF) platform for cell capture and in situ SERS detection of HER2; (b) mechanism of HER2 and peptide via stepwise electrochemical activation and surface heterocyclization reactions [51]. *Reproduced with permission from (A) Y. Pang, C. Wang, R. Xiao, Z. Sun, Dual-selective and dual-enhanced SERS nanoprobes strategy for circulating hepatocellular carcinoma cells setection, Chemistry 24 (2018) 7060—7067, Copyright 2018, Wiley-VCH; (B) T. Xue, S. Wang, G. Ou, Y. Li, H. Ruan, Z. Li, et al., Detection of circulating tumor cells based on improved SERS-active magnetic nanoparticles, Anal. Methods 11 (2019), 2918—2928, Copyright 2019, The Royal Society of Chemistry; (C) H. Dong, D. Yao, Q. Zhou, L. Zhang, Y. Tian, An integrated platform for the capture of circulating tumor cells and in situ SERS profiling of membrane proteins through rational spatial organization of multi-functional cyclic RGD nanopatterns, Chem. Commun. 55 (2019) 1730—1733, Copyright 2019, The Royal Society of Chemistry.*

between folic acid on functionalized MBs and the folate receptor on HeLa cells, showing ~90% capture efficiency and ~97% purity of HeLa cells. The LOD was calculated as one cell per mL of blood. Following the analysis of the purification strategy, the HeLa cells were released from MBs by adding excess folic acid to the culture media, and then the released cells can be used for downstream applications. The potential of this approach for clinical applications was explored by inspecting patient samples of first-stage cervical cancer (which also have folate receptors on the cell surface) and the blood of healthy people. The results showed that detectable CTCs only existed in cancer patient blood samples.

As an alternative to magnetic isolation, Dong et al. developed a two-dimensional (2D) chip composed of a mesoporous gold film coated indium tin oxide platform as the SERS substrate with a designed peptide, DNEYFYV(H2)−NH−CQO−CRGDKC-acetyl (denoted as H2-RGD) for capturing CTCs prior to analysis of the protein HER2 level which is overexpressed on CTCs [Fig. 6.3C(a)] [51]. The RGD sequence (CRGDKC) is a common peptide for capturing CTCs, while H2 (DNEYFYV) is used as a specific peptide to bind with HER2 on CTCs. After the capture of CTCs by the RGD peptide, HER2 on the CTCs transferred a γ-phosphate ferrocene group to the tyrosine residues of the H2 peptide, causing a new peak observed at 320 cm^{-1} in the SERS spectra which proportionally increased with the increasing amount of HER2. Monitoring the HER2 induced spectral intensity change, gave an assay with an excellent linear range from 5 to 100 fg mL^{-1} HER2 with a low LOD of 1.1 fg mL^{-1}. The in situ profiled HER2 level on CTCs collected from patient samples showed good alignment with the results from an enzyme-linked immunosorbent assay (ELISA) kit.

The competitive advantage of SERS over other optical techniques, such as fluorescence spectroscopy, is its multiplexing capability which enables analysis of multiple components simultaneously within one sample. This is possible due to customized SERS nanotags and the high spectral resolution from narrow characteristic peaks of Raman reporter molecules which enable the distribution and expression levels of surface biomarkers on CTCs to be measured in a single run. Tsao et al. [52] established a multiplex SERS-nanotag assay for characterizing surface biomarkers on melanoma cells. They used it to monitor the phenotypic evolution of the CTCs from melanoma patients during treatment Fig. 6.4A. The assay displays a sensitive quantification of ≈10 cells from 10 mL blood. Four types of customized SERS nanotags separately targeted

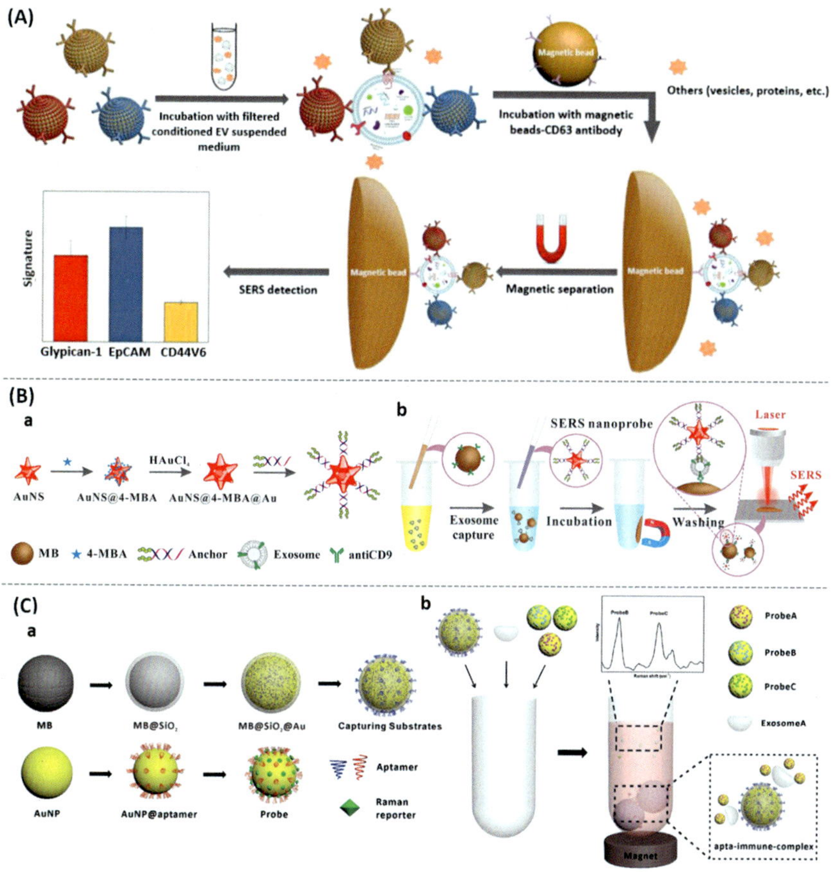

Figure 6.4 (A): Evaluation of expression levels of EpCAM, CD44V6, and Glypican-1 on magnetically enriched EVs based on SERS-nanotag labeling [78]. (B): (a) Fabrication of SERS nanoprobes (AuNS@4-MBA@Auanchor); (b) SERS sensing strategy for EVs detection [81]. (C): (a) Schematic illustration for the fabrication of capturing substrate and SERS nanotags; (b) the working principle for EVs detection by SERS-based assay [82]. *Reproduced with permission from (A) W. Zhang, L. Jiang, R.J. Diefenbach, D.H. Campbell, B.J. Walsh, N.H. Packer, et al., Enabling sensitive phenotypic profiling of cancer-derived small extracellular vesicles using surface-enhanced Raman spectroscopy nanotags, ACS Sens. 5 (2020) 764–771, Copyright 2020, The American Chemical Society; (B) Y. Tian, C. Ning, F. He, B. Yin, B. Ye, Highly sensitive detection of exosomes by SERS using gold nanostar@Raman reporter@nanoshell structures modified with a bivalent cholesterol-labeled DNA anchor, Analyst 143 (2018) 4915–4922, Copyright 2018, The Royal Society of Chemistry; (C) Z. Wang, S. Zong, Y. Wang, N. Li, L. Li, J. Lu, et al., Screening and multiple detection of cancer exosomes using an SERS-based method, Nanoscale 10 (2018) 9053–9062, Copyright 2018, The Royal Society of Chemistry.*

melanoma surface markers, including melanoma chondroitin sulfate proteoglycan (MCSP), melanoma cell adhesion molecule, erythroblastic leukemia viral oncogene homolog 3 (ErbB3), and low-affinity nerve growth factor receptor. Characterization of the distribution and relative amounts of these four biomarkers on three melanoma cell lines was achieved. The practicability of the assay was confirmed by monitoring the biomarker-pattern evolution of 10 stage-IV melanoma patients before and after immunological and molecular targeted therapies. The results show that ErbB3 (an important factor in resistance and metastasis development) was upregulated in some patients who showed tumor progression while the MCSP expression level was reduced in some patients after treatment.

Ultrasensitivity is another critical feature of SERS, offering a single molecule level detection ability and rendering SERS a good candidate for quantifying rare CTCs. Wang et al. [53] fabricated a SERS nanotag including a Raman reporter molecule, polyethylene glycol, and the epidermal growth factor peptide to identify CTCs in peripheral blood samples of 19 patients with squamous cell carcinoma of the head and neck (SCCHN). Since more than 90% of SCCHN cells overexpress EGFR, the EGF peptide played the role of recognition ligand to label SCCHN CTCs specifically. A sensing range of 1–720 CTCs per mL of whole blood was obtained from measurements on blood samples from 19 patients' blood.

6.3.4 Insights on SERS-based CTCs analysis in a clinical setting

In summary, SERS has been applied to CTCs quantification and characterization of clinical patients' samples. Its advantages include the followings. First, the sensitivity of SERS meets the demand of probing CTCs in early disease diagnosis with a desirable LOD of 1 CTC per mL blood [49,50,52,53]. Second, the multiplexing capability of SERS enables the evaluation of multiple biomarkers on a CTC surface [52]. This offers rich information on the surface bio-components evolution of CTCs, helping with the understanding of the patient's CTCs response to clinical therapy. Third, the fingerprint information from SERS spectra of specific biomarkers provides a simple and straightforward way to study the molecular change of target biomarker on CTCs (e.g., HER2 in Dong et al.'s work [51]).

Based on the above summary, what we suggest for the next frontier of SERS-based CTC study focuses on (1) more profound and comprehensive study for interior components of CTCs with genomic/transcriptomic/epigenetic/proteomic analyses by the improved CTCs enrichment and SERS technique; (2) fingerprint information of multiple surface biomarkers on CTCs based on novel plasmonic nanomaterials which enable direct SERS detection; (3) release of CTCs from the SERS capture unit (e.g., MBs) for further downstream study; (4) development of more SERS-based CTC assays to serve as routine clinical diagnosis and monitoring of treatment and prognosis.

6.4 SERS analysis of extracellular vesicles

6.4.1 Biological roles and current analysis techniques of extracellular vesicles

EVs are lipid bilayer-surrounded vesicles released by cells or shed from cell membranes into the extracellular environment. Typically, EVs serve as biological messengers for intercellular communication [54,55], as well as having varied biogeneses, range in size. The three most described subcategories of EVs include exosomes, microvesicles, and apoptotic bodies [56—58]. EVs are very attractive clinical diagnosis targets as they always carry a multitude of biomaterials on their surface. Briefly, EVs membrane includes functional proteins (e.g., CD63), lipids (e.g., flotillin-1), and antigen-presenting molecules (e.g., MHC I and MHC II) [55,59].

When an EV circulates in the body, it can activate/inhibit the function of a recipient cell via interacting with the surface receptors of recipient cell or merging with the recipient cell membrane [60]. The aqueous-soluble contents of an EV can be transferred into the recipient cell. The delivered biomaterials can influence or even phenotypically reprogram surrounding and distant cells [58]. EVs that come from cancerous cells can thus initiate oncogenic processes such as proliferation, migration, and immunosuppression [57—59,61]. It has been shown that tumor-derived EVs comprise more at least 10-fold higher content of potentially active molecules compared to the EVs secreted by normal cells [62—64], making them attractive diagnostic targets and have potential as therapeutic delivery vehicles.

Many methods have been utilized to isolate and collect EVs from different body fluids [64—66]. As with CTCs, the methods generally rely on EV physical properties (e.g., size or density: ultracentrifuge, ultrafiltration, and polymeric precipitation) [65], or are based on bio-affinity of EVs surface components [67,68]. Ultracentrifugation is the most commonly used in the laboratory due to the simplicity of processing large sample volumes. However, centrifugation methods cannot eliminate contamination from aggregated proteins that share similar size or density with EVs. They are also time consuming and require a specialized high-cost instrument. Immunoaffinity approaches offer highly purity isolation of target EVs, but the high cost of affinity ligands (e.g., antibodies) and the relatively low sample volume limits their routine application usage [66]. To date, no single ideal isolation strategy provides satisfactory EV purity and yield with reasonable time, consumable and instrument costs. Once an EV sample has been obtained, much attention has been paid to its qualitative and quantitative analyses in terms of EV morphology, size distribution, concentration, surface markers/intrinsic biomaterials, and even internal biomarkers. The commonly used techniques in these downstream analyses include TEM/scanning electronic microscopy (SEM), nanoparticle tracking analysis, tunable resistive pulse sensing, dynamic light scattering (DLS), zeta potential, flow cytometry, Western blot, mass spectrometry, and spectroscopic methods [69]. Depending on the specific requirements, two or more techniques are frequently combined to quantify and characterize the enriched EVs. SERS has shown promise as a method that can at least in part avoid the need for EV isolation while providing multibiomolecule characterization within one test. The following section summarizes some applications of the SERS technique in EVs analysis to illustrate its future potential.

6.4.2 SERS strategies for EVs detection and characterization

6.4.2.1 Discrimination of EVs origins

In any biological sample, numerable EVs from different cells, mingle together and share a similar size range. Differentiating their parental sources thus poses a considerable challenge. In general, EVs from different sources (cell-type, disease stage, etc.) carry different molecular compositions. Screening the inherent molecular signatures of EV-components is the most straightforward way to identify EVs from different parental cells, even if the differences are small.

Direct SERS measurement combined with statistical analysis has been frequently employed to probe and discriminate EVs based on their native SERS fingerprint [70–77]. Yan et al., for example, established a sensing platform to detect and discriminate human serum EVs and cancerous lung cell–derived EVs (from HCC827 cells and H1975 cells, respectively) by direct SERS measurement [71]. A hybrid platform composed of single-layer graphene coated a quasi-periodic array of gold pyramids acted as the SERS enhancement substrate with an enhancement factor of 10^{12}. The isolated EVs were placed on the substrate for direct SERS signal collection which showed a series of peaks associated with EVs molecular information (e.g., the peaks at 1012, 1509, and 1613 cm^{-1} representing the vibrational mode of phenylalanine, the ring breathing mode in DNA bases and the vibration mode of tyrosine). The SERS spectra of EVs from different sources were analyzed with PCA which showed a statistically distinguishable classification with an overlap of less than 5% among different EVs sources and a sensitivity of more than 84%. These results prove the assay has the potential to be used for identifying EV origins for early diagnosis and tumor subtyping.

6.4.2.2 Profiling of EVs bio-composition

Since EVs partially inherit cell membranes when they are released to extracellular space, the surface composition heterogeneity of EVs has the potential to reflect the aspects of the parental cell membrane. SERS multiplexing capability is appropriate to decode the distribution of multibiomarker expression on EVs [78–80]. For example, Zhang and coworkers conducted a study to profile the expression level of three biomarkers on cancerous EVs via multiple SERS nanotags, as shown in Fig. 6.4A [78]. Specifically, protein CD63 is a widely detected protein on EVs with different origins. Its antibody was conjugated onto MBs to enrich EVs from three cancerous cell lines: Panc-1, SW480, and C3. Three specific SERS nanotags targeting different biomarkers (EpCAM, CD44V6, and Glypican-1) were applied to label the enriched EVs on the MBs. Consequently, a sensitive detection toward EVs as low as 2.3×10^6 particles mL^{-1} was achieved, and the distribution pattern of these three biomarkers was obtained on the spiked target EVs in human plasma. This work holds great potential for cancer diagnosis, subtyping, and therapy monitoring in real clinical application. Such strategies have been testified with clinical-source EVs as discussed in the following section.

6.4.3 SERS-based assay for EV analysis with clinical samples

The integration of immunoassays with a SERS readout is applied in EVs analysis by using various recognizing elements (e.g., antibody, aptamer). Tian et al. constructed a sensing platform by magnetically enriching EVs via using an antibody to bind with another common biomarker on EV surfaces—CD9 (Fig. 6.4B) [81]. Then, the enriched EVs were labeled with SERS nanotags which was constructed by a gold nanostar (AuNS) and a bivalent cholesterol-labeled DNA strand. The SERS nanotags labeled EVs were realized via the cholesterol inserting into the lipid bilayer of EVs. Then, the CD9 antibody modified MB was applied to the enriched SERS nanotag–labeled EV sample to form immunocomplexes via interacting with the CD9. This SERS nanotag has a concentrated local electromagnetic field for SERS enhancement, contributing to an ultrasensitive LOD of 27 EVs per microliter. The assay was implemented on serum from three liver cancer patients and three healthy individuals. The results show that the detected amount of EVs from cancerous samples is higher than that from healthy serum, which matches well with the results obtained from the same samples quantified with the qNano technique. This work illustrates that SERS-based sensing assay is able to quantify EVs sufficiently to differentiate EVs amount between samples from sick and healthy individuals.

Another SERS-based immunoassay with multiplexing detection focuses on multiple EVs types. Wang et al. synthesized a double layer (SiO_2@Au, in which "@" means a coating/wrapping layer)-coated MBs (MB@SiO_2@Au) to anchor a CD63 aptamer for capturing EVs (Fig. 6.4C) [82]. Three SERS nanotags were created by individually modified them with an additional unique aptamer to target specific biomarkers: HER2 protein in breast cancer cell SKBR3, membrane antigen PSMA in prostate cancer cell LNCaP, and carcinoembryonic antigen overexpressed in colorectal cancer cell T84. After concentrating the magnetic pellet which includes the captured EVs labeled with the SERS nanotags, the supernatant was subjected to SERS measurement and a reduced intensity signal (due to the nonspecific SERS nanotags left in solution) was observed. The detection toward each type of EVs shows a LOD of 32 SKBR3 EVs per mL, 203 LNCaP EVs per mL, and 73 T84 EVs per mL. The crosstalk among these three types of EVs labeled with SERS nanotags does not affect the multiplexing of the assay. Blood samples from breast cancer, prostate cancer, colorectal cancer patients, and a healthy blood sample were analyzed to determine the feasibility of this assay for clinical application. The relatively weaker SERS signals were observed in the cancer patient samples which contain the target EVs compared to the signal of healthy blood sample group.

Profiling surface-active components is a crucial aspect of EVs study. Kwizera et al. developed a 3D-printed platform modified with an Au thin film and an antibody array for EVs capture (Fig. 6.5A) [83]. Gold nanorods (AuNRs) were utilized as SERS nanotags to label the captured EVs on the substrate, by conjugating with an organic dye QSY21 as Raman reporter. The established platform was linked with an anti-CD63 antibody to detect

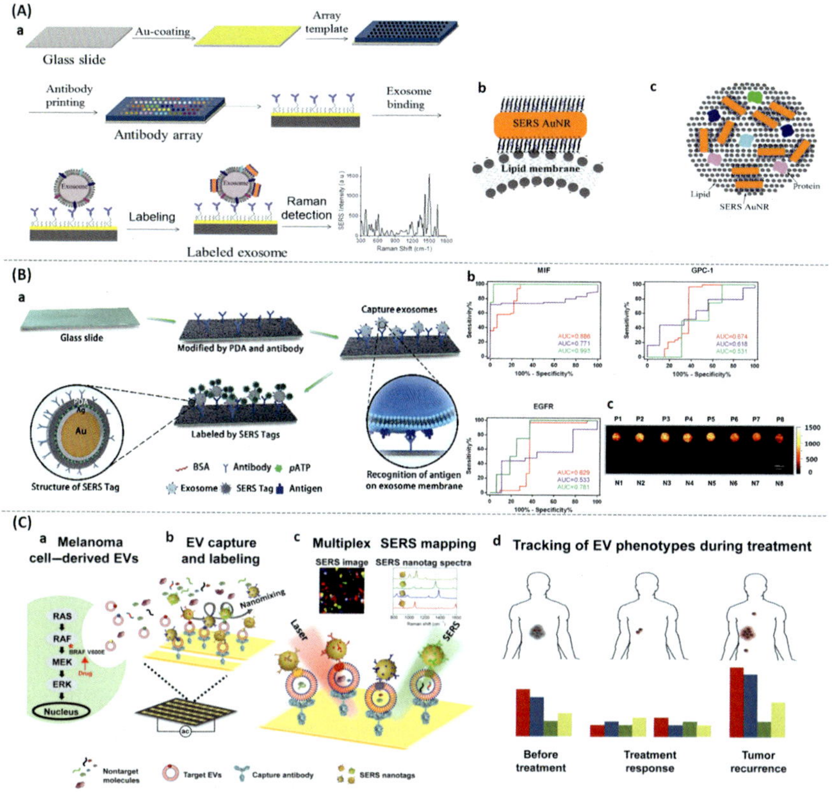

Figure 6.5 (A): (a) Schematic diagram for the fabrication of capture array used for exosome identification and detection; (b) side view and (c) top view of the interaction between lipid membrane of exosome and SERS AuNR [83]. (B): (a) A schematic view of the PDA chip and PEARL SERS tag-based exosome sensors; (b) receiver operating characteristic (ROC) curves were calculated for single exosome markers (MIF, GPC-1, and EGFR) (red: pancreatic cancer vs healthy controls; purple: metastasis vs nonmetastatic; and green: P1–2 vs P3). AUC stands for the area under the curve; (c) Raman imaging scanning results of sensing chip containing serum samples from pancreatic cancer patients (P1–8) and healthy individuals (N1–8) tested using the anti-MIF platform [84]. (C): Illustration for the working principle of EV phenotype analyzer chip (EPAC) and its application for tracking EV phenotyping during treatment [79]. *Reproduced with permission from (A) E.A. Kwizera, R. O'Connor, V. Vinduska, M. Williams, E.R. Butch, S.E. Snyder, et al., Molecular detection and analysis of exosomes using surface-enhanced Raman scattering gold nanorods and a miniaturized device, Theranostics 8 (2018) 2722–2738, Copyright 2018, Ivyspring International Publisher; (B) T. Li, R. Zhang, H. Chen, Z. Huang, X. Ye, H. Wang, et al., An ultrasensitive polydopamine bi-functionalized SERS immunoassay for exosome-based diagnosis and classification of pancreatic cancer, Chem. Sci. 9 (2018) 5372–5382, Copyright 2018, The Royal Society of Chemistry; (C) J. Wang, A. Wuethrich, A.A. Sina, R.E. Lane, L.L. Lin, Y. Wang, et al., Tracking extracellular vesicle phenotypic changes enables treatment monitoring in melanoma, Sci. Adv. 6 (2020) 3223–3237, Copyright 2020, The American Association for the Advancement of Science.*

EVs released by breast cancer cell MDA-MB-231. A LOD of 2×10^6 EVs per mL and a working range of $10^6 - 10^8$ EVs per mL were obtained. Eight surface proteins on EVs with three categories were studied: epithelial marker EpCAM, breast cancer markers CD44, HER2, EGFR, IGFR, and EVs markers CD81, CD63, CD9. The SERS measurement displayed much higher sensitivities for CD44 and the three EVs biomarkers (CD81, CD63, and CD9) than the others. This pattern is in accord with the characteristic expression of MDA-MB-231 cells obtained from ELISA results—overexpressed CD44 with low expression of HER2, EGFR, and EpCAM. In addition, the surface biomarker profiling on different cells–derived EVs was also explored. By analyzing EVs released from breast cancer cells MDA-MB-468, SKBR3, and normal breast cell MCF12A with both a SERS assay and flow cytometry, the protein expression patterns from the three types of EVs were shown to be entirely different. The results from the two techniques (SERS and flow cytometry) have good agreement with each other. To evaluate the potential of SERS for clinical application, EVs from 10 HER2-positive breast cancer patient samples from stages I, II, and III, as well as five healthy donors' plasma samples were tested. The results indicate that this assay could identify HER2 and EpCAM biomarkers on EVs to diagnose HER2-positive breast cancer patients. This work confirms that the SERS-based assay could profile EVs surface biomarker expression levels among different cell types and detect surface protein markers of clinical samples for diagnosis purposes.

Similarly, Li et al. performed a study on a substrate embedded with polydopamine and an antibody for EVs enrichment (Fig. 6.5B) [84]. An ultrathin polydopamine-encapsulated antibody–reporter–Ag(shell)–Au (core) multilayer SERS nanotag was designed for signal readout. Antibodies for migration inhibitory factor and glypican-1 were chosen to capture PANC-1 EVs for quantification due to their high expression on PANC-1 EVs surface. The sensitivity was determined with a linear detection range of $5.44 \times 10^2 - 2.72 \times 10^4$ EVs per mL and a LOD of 9×10^{-19} mol L^{-1}. Pancreatic cancer patients' samples ($n = 71$) and healthy individuals' samples ($n = 32$), were discriminated by the SERS assay. In addition, the assay was able to distinguish metastatic tumors from metastasis-free tumors, and in particular tumor-node metastasis from P1–P2 stages were differentiated from P3 stage matastastasis with a high discriminatory sensitivity of 95.7% (Fig. 6.5B(b)). Further investigation of eight pancreatic cancer patients' samples and eight healthy samples by image scanning shows that this assay can also offer a more straightforward and visual way to indicate and discriminate pancreatic cancer patients from healthy individuals (Fig. 6.5B(c)).

Another design from Wang and coworkers integrates EV enrichment, SERS nanotags labeling, and SERS measurement within one microchip (Fig. 6.5C), named as EV phenotype analyzer chip [79]. This chip was applied to probe four biomarkers carried on melanoma EVs by collecting the corresponding SERS signal of each biomarker. The levels of these biomarkers are associated with melanoma progression and treatment efficiency and the approach was successfully employed to monitor EVs molecular variations during cancer treatment. The SERS data from the chip correctly differentiated EVs from patients ($n = 11$) and healthy donors ($n = 12$) and tracked the phenotyping changes of EVs from patients ($n = 8$) who underwent targeted therapy with a detection limit of 10^5 EVs mL^{-1} in 40 min, which is sufficiently sensitive for clinical application while being faster (within 40 min) than conventional analyses (e.g., ELISA, >2 h).

Label-free SERS detection can directly reflect the intrinsic molecular fingerprint of target analytes which is attractive for EVs profiling and identifying cancerous EVs. For example, Park et al. established a SERS-based analysis for lung cancer diagnosis (Fig. 6.6) [85]. Briefly, EVs were

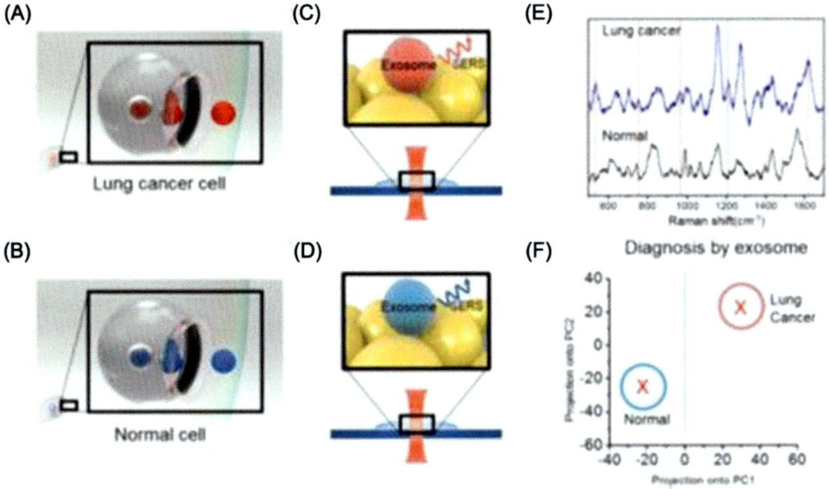

Figure 6.6 (A and B) EVs released by lung cancer cell and normal cell into the extracellular environment; (C and D) SERS detection of lung cancer cell and normal cell–derived EVs; (E) SERS spectra obtained from lung cancer cell and normal cell–derived EVs. The red lines indicate specific peaks of lung cancer-derived EVs. (F) EVs classification by PCA of SERS spectra [85]. *Reproduced with permission from J. Park, M. Hwang, B. Choi, H. Jeong, J.H. Jung, H.K. Kim, et al., Exosome classification by pattern analysis of surface-enhanced Raman spectroscopy data for lung cancer diagnosis, Anal. Chem. 89 (2017) 6695–6701, Copyright 2017, The American Chemical Society.*

placed on the gold nanoparticles coated glass slide for direct SERS measurement. The SERS spectra from both lung cancer cell-released EVs and normal cell–derived EVs were analyzed by PCA, which showed the peak at 732 cm^{-1} is highly relevant to lung cancer cell–derived EVs and the peak at 1150 cm^{-1} is associated with normal cell–derived EVs. By applying the first and second principal components (PC1 and PC2), both cancerous EVs and normal EVs were found to form their population within distinct boundaries, indicating SERS/PCA analysis can be used to differentiate cancerous from normal EVs. However, the assay did not show well-defined clusters for EVs in patients' blood samples and healthy samples. These results suggest that real blood sample contains much more complicated biological materials. Therefore, further study on EV sample purification and more robust statistical methods are required to differentiate the small SERS spectral variances reliably. Following this, Carmicheal et al. combined PCA with differential function analysis (DFA) for classifying cancerous EVs and normal EVs from real clinical samples [74]. As with the work of Park et al., another assay could differentiate spectra of EVs derived from cancerous cell lines and controls with high sensitivity and specificity. However, when the PCA-DFA was applied to classify 10 early pancreatic cancer patients' samples and 10 healthy controls, the results exhibit a range of predictive accuracies: characterization efficiency of pancreatic cancer patients' samples varied from 30% to 90%, while that of healthy control samples ranges from 20% to 87%. A possible explanation for this result could be that serum from cancer patients, especially from those with early-stage disease, inherently has a larger portion of EVs arising from normal epithelium rather than tumor cells. This adds to the complexity of the SERS spectra, leading to the significantly increased difficulty for algorithmic determination.

6.4.4 Insights on SERS-based EVs analysis with clinical setting

To sum up, SERS-based assays for EVs analysis have mainly focused on EVs quantification, bio-composition characterization, and the differentiation between target tumor-derived EVs and healthy control EVs using highly specific molecular targeting via antibodies or aptamers [81–84]. In this way, target EVs can be enriched and labeled. Approaches include (1) using specific targeting both to capture and SERS nanotag [82,84] and (2) only utilizing such recognizing elements in the capture and applying other interactions, such as electrostatic force and hydrophobic affinity, to attach SERS nanotags

[81,83]. Whatever combination is designed for a particular assay, the molecular pattern of EV surface components are probed, both for common EV proteins (e.g., CD63 and CD81) or specific disease-associated EV biomarkers (e.g., MIF and GPC-1 which have high expression levels on pancreatic cancer-derived EVs) [81,83,84]. This enables early diagnosis in clinical patient samples by applying SERS-based assays. Label-free SERS integrated with statistical analysis has been used successfully for EV analysis and is extremely attractive as simple sample preparation and data collection compared to indirect SERS. However, more work is required as there are examples of incorrect discrimination of EVs in patient samples [74,85]. The challenge can be ascribed to the heterogeneity of tumor released EV cargos in size, components, and quantity, affected at least in part by disease stage which points to additional diagnostic possibilities in the future. The considerable background of EVs from normal cells conceals the unique tumor-EV signatures at very early disease stages. Hence, further work on purification and enrichment or for analysis selectivity for tumor-EVs from real clinical sample is necessary. The required work may include more comprehensive training cohort for both normal EVs and specific target EVs and more sophisticated statistical discriminant methods.

6.5 SERS analysis of circulating tumor-derived nucleic acids

6.5.1 Biological significance and current analysis techniques for ctNAs

The bloodstream and other body fluids contain circulating nucleic acids which are fragments of genomic and mitochondrial DNA, RNA, and miRNA [86]. They are spontaneously released by live cells as part of a homeostatic regulation mechanism or generated from necrotic/apoptotic cells into the extracellular space [87]. They play roles in or are signals for physiological and pathological processes, endowing them the potential for clinical application, such as early diagnosis, tumor heterogeneity evaluation, and drug-resistance mechanism identification [88,89]. As potential liquid biopsy biomarkers, the most extensively investigated circulating tumor-derived nucleic acids (ctNAs) include circulating tumor-derived DNA (ctDNA) and circulating miRNA (ctmiRNA).

ctDNA released by tumor cells and tumor-derived EVs is a subset of circulating cell-free DNA (cfDNA). It has a commonly observed length of 160–200 base pairs [87,90] and typically has been eliminated by the liver, kidney, or spleen. It has a short half-life varying from 15 min to a few hours

[91,92]. A growing body of evidence has shown that cancer patients generally carry ctDNA ranging from 0.01% to more than 50% of the cfDNA [92]. In addition, up to 1000 ng mL^{-1} cfDNA can be found in cancer patients while approximately 10 times less cfDNA (100 ng mL^{-1}) is detectable for healthy individuals [93,94]. The variation in level can be ascribed to a set of influencing factors, such as cancer stage, tumor burden, tumor heterogeneity, response to therapy, and clearance mechanism [87,93,95].

Tumor-associated genetic/epigenetic alterations are coded into ctDNA (e.g., mutations and copy number variations). ctDNA thus has great potential as a diagnostic tool for solid tumors providing information about disease stage, type and genomic/epigenomic status of the parental tissues [10,87]. Currently, both level and characteristics of ctDNA are treated as potential indicators for use in early diagnosis, dynamic tracking of disease evolution, and to develop personalized therapies. Among these, ctDNA mutation and anomalous methylation status have received the most attention in research and clinical applications. Taking abnormal methylation level as an example, the hypermethylated promoter regions of RASSF1A gene, FHIT gene, and the APC gene discovered in plasma ctDNA show great potential to be used as biomarker for diagnosis of early stage renal cancer with a sensitivity of 56.8% and specificity of 96.7% [96].

MicroRNA (miRNA) is used to refer to endogenously noncoding RNA with a length of 19–23 nucleotides (nt). The majority of miRNAs remain intracellular while a small portion is released into circulation. miRNAs take the pivotal roles of up/down-regulators of translation, altering the expression of thousands of genes [97]. They also influence other biological processes such as proliferation, differentiation, and apoptosis. Apparently, dysregulation of miRNAs is responsible for some pathologies such as cardiovascular diseases, metastasis, and oncogenesis [89]. Circulating miRNAs are detected in almost all body fluids and are subsequently uptaken by recipient cells where they implement their functions [98–100]. Therefore, circulating miRNA, especially circulating tumor-associated miRNA (ctmiRNA), can deliver and spread undesirable outcomes when they are internalized by a receipt cell or seized by cell surface receptors. For example, miRNA-21 is abundantly expressed in mammalian cells and its upregulation is associated with many types of cancers including glioblastoma, breast, lung, and thyroid cancer [101]. Conversely, there is some evidence demonstrating that oncogenic ctmiRNA decreases after tumor resection [89]. The remarkable stability of miRNA is exhibited by its resistance to high temperature, a wide range of pH variations, multiple freeze-thaw cycles, and prolonged storage [102]. Therefore, ctmiRNA has the potential to be a promising noninvasive biomarker for clinical diagnosis.

Given the clinical significance of ctDNAs and ctmiRNA, much work has been undertaken to develop a range of quantitative and qualitative analysis approaches. Due to their heterogeneity in physical feature and structure, fluctuating concentrations relative to the abundant nontumor cfDNA background and multiple types of genetic/epigenetic variances and detection methods are required with high sensitivity and specificity. The most widely applied ctDNA and ctmiRNA analyses are sequencing and polymerase chain reaction (PCR)-based strategies. Sequencing has dramatically improved in speed, accuracy, throughput, and cost-saving since the first sequencing method called Sanger sequencing emerged in the 1970s [103,104]. Nowadays, next-generation sequencing methods allow parallel sequencing for millions of fragments in a single run and with customized designs for particular targets (e.g., single-nucleotide polymorphisms and methylation patterns) [105].

Although many researchers have demonstrated that next-generation sequencing can be effectively used for ctDNA/ctmiRNA determination with desirable sensitivity and specificity [104,106], there still remains some distance to clinical utility with standardized workflows because of (1) limited sensitivity for detecting rare mutations with allele frequencies $<1\%$; (2) insufficient accuracy due to the errors generated from nonspecific interactions between primers of multiplexed PCR during library setup; (3) the very complicated bioinformatic interpretation and workflow; (4) the lack of validation studies; and (5) its expense [107–109].

In terms of PCR-based tactics, real-time PCR (also named quantitative PCR, qPCR), droplet digital PCR (ddPCR), and reverse transcription-quantitative PCR (RT-qPCR, particularly for RNA-associated detection) are the primary techniques for ctNAs detection. They are capable of quantifying specific targets, but with some restrictions, including strictly controlled temperature (e.g., within $\pm 0.3°C$ required by ddPCR with the required temperatures varying from sequence to sequence) [110], prior knowledge about the target sequence of interest, limited throughput, and failed amplification due to environmental PCR inhibitors [111].

In general, it has become apparent that the ctNAs sample handling greatly influences the sensitivity and accuracy of the analysis outcomes, including the origin of the specimen (serum or plasma), postvenipuncture processing, pretreatment, and storage for blood samples [112]. In addition, the pretreatment of methylated ctNAs usually requires a bisulfite conversion treatment. Specifically, cytosine is deaminated and converted into uracil, while methylated cytosine remains mainly unaffected. This harsh chemical processing is conducted with prolonged incubation under high thermal conditions and low pH media, resulting in severe sample degradation and nonspecific conversion from methylated cytosine [113,114]. As a result, there is in needs for new robust, reliable, and accurate approaches for measuring ctNAs in body fluids and screening their genotypes for further personalized therapy. In the following section, the SERS-based strategies for ctNAs analysis have been summarized.

6.5.2 SERS strategies for ctNAs analysis

Although sequencing and PCR approaches dominate ctNA analysis, there are a number of interesting applications of SERS-based methods used for analysis of ctNA samples.

6.5.2.1 Selectively labeling ctNA regions of interest

SERS nanotag—enabled specific labeling has been utilized for ctNAs analysis to achieve sensitive multiplex detection with potential high-throughput screening. The identification and enrichment of target ctNAs are typically implemented via complementary oligonucleotides anchored on a SERS nanotag and/or a capture substrate such as a MB. For example, Zhou and coworkers [115] immobilized two thiolated DNA fragments on Ag microspheres (the capture substrate) and functionalized SERS nanotags, respectively (Fig. 6.7A). Each fragment included a region to complementarily hybridize with a target miRNA. Therefore, the target miRNA plays a role of linking the Ag microsphere and SERS nanotag together. The assay was applied to detect simultaneously three types of target miRNAs (miRNA-21, miRNA-122, and miRNA-223) which are specific to hepatocellular carcinoma (Fig. 6.7A). This approach achieved a detection limit of 10 fM for simultaneously analyzing three ctmiRNAs, indicating a highly sensitive and specific candidate for liver cancer clinical diagnosis.

6.5.2.2 Direct readout of ctNA molecular information

By pretreating the sample and functionalizing the SERS substrate, it is possible to use SERS to observe changes in the vibrational frequencies of nucleotide bases and the DNA backbones of target ctNAs which can indicate a change in the ctNA structure. Although NAs have a low affinity toward the bare metal surface, by functionalizing SERS substrate with a

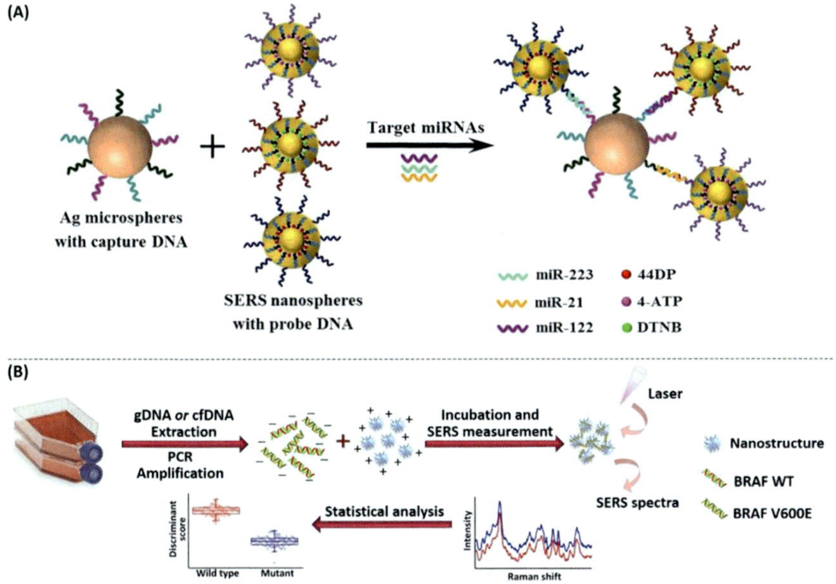

Figure 6.7 (A) Schematic workflow of the multiplex SERS assay for triple-target miRNA detection [115]. (B) Direct SERS readout and differentiation of DNA mutation by the assist of the statistical method [116]. *Reproduced with permission from (A) W. Zhou, Y. Tian, B. Yin, B. Ye, Simultaneous surface-enhanced Raman spectroscopy detection of multiplexed microRNA biomarkers, Anal. Chem. 89 (2017) 6120–6128, Copyright 2017, American Chemical Society; (B) Y. Liu, N. Lyu, V.K. Rajendran, J. Piper, A. Rodger, Y. Wang, Sensitive and direct DNA mutation detection by surface-enhanced Raman spectroscopy using rational designed and tunable plasmonic nanostructures, Anal. Chem. 92 (2020) 5708–5716, Copyright 2020, American Chemical Society.*

positively charged surface it is possible to actively enrich negatively charged ctNAs via electrostatic forces. Coupling such measurement with a PCR-based amplification simultaneously increases the amount of the target ctNAs and purifies the complex matrix during ctNAs extraction and PCR amplification, paving the road for generating reliable SERS signatures from target ctNAs themselves.

The direct readout approach is illustrated by the DNA mutation detection approach of (Fig. 6.7B) from Liu et al. They engineered four types of positively charged nanostructures with different morphologies (nanostar, nanoflower, nanoshell, and nanosphere) which were utilized as SERS substrates to measure the SERS response of negatively charged DNA [116]. The positively charged nanostar exhibited the highest SERS enhancement with an enhancement factor of 1.49×10^6. By performing

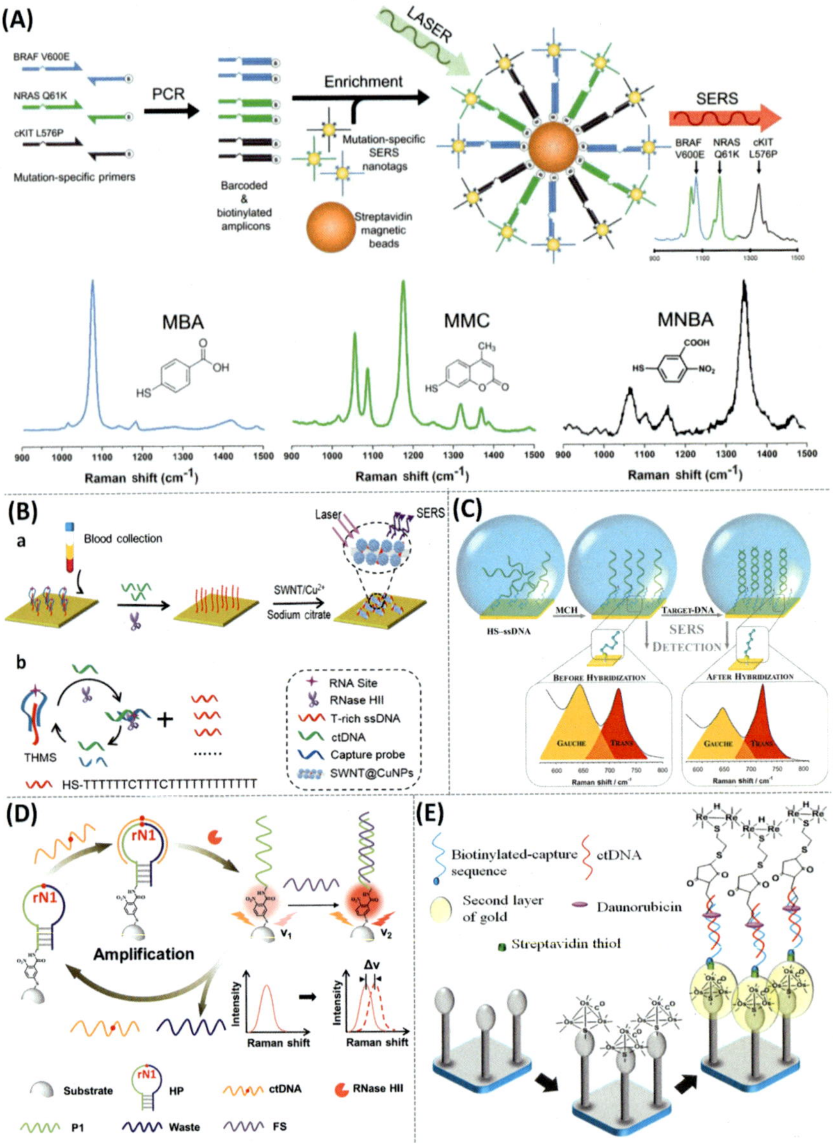

Figure 6.8 (A) Schematic illustration of PCR/SERS assay for multiplex DNA mutation detection via magnetic enrichment for specifically amplified and labeled mutant DNA fragments [117]. (B): (a) Working principle for SWNT-based SERS assay coupling with RNase HII-assisted amplification for highly sensitive detection of ctDNA

direct SERS measurements with nanostars, the SERS signatures of the BRAF gene and its one-base mismatched mutant BRAF V600E were acquired. Then, utilizing a statistical analysis via the combination of PCA and linear discriminant analysis (LDA), the mutant BRAF V600E was distinguished from wild type (WT) both for whole-genome DNA samples and cell-free DNA samples. The assay feasibility was validated by the correct classification of the gene type in a blind human plasma test. The assay also showed a 10-fold more sensitive detection limit than conventional method (gel electrophoresis) and comparable sensitivity with qPCR. In addition, the shifts in vibrational frequencies give information about structural changes in target gene sequences since the SERS spectrum is from

◀ mutation. The enlarged image illustrates T-rich DNA−mediated CuNPs growth to enhance the SERS signal of SWNTs; (b) mechanism for the formation of T-rich ssDNA from THMS structure in the presence of RNase HII and target ctDNA [118]. (C): Schematic diagram of the detection for BRAF V600E mutation in ctDNA via monitoring SERS intensity change induced by the conformation change of an alkanethiol linker conjugated on capture DNA before and after binding with target ctDNA [119]. (D) Working principle of frequency-shifted SERS sensing assay triggered by target ctDNA and assisted via RNase HII enzyme [120]. (E): (a) Experimental workflow for the fabrication of Au−osmium carbonyl−Au functionalized SERS substrate used for ctDNA detection; (b) SERS spectra of osmium carbonyl and the rhenium carbonyl and daunorubicin [121]. *Reproduced with permission from (A) E.J. Wee, Y. Wang, S.C. Tsao, M. Trau, Simple, sensitive and accurate multiplex detection of clinically important melanoma DNA mutations in circulating tumour DNA with SERS nanotags, Theranostics 6 (2016) 1506−1513, Copyright 2016, Ivyspring International Publisher; (B) Q. Zhou, J. Zheng, Z. Qing, M. Zheng, J. Yang, S. Yang, et al., Detection of circulating tumor dna in human blood via DNA-mediated surface-enhanced Raman spectroscopy of single-walled carbon nanotubes, Anal. Chem. 88 (2016) 4759−4765, Copyright 2016, American Chemical Society; (C) A. Kowalczyk, J. Krajczewski, A. Kowalik, J.L. Weyher, I. Dzięcielewski, M. Chłopek, et al., New strategy for the gene mutation identification using surface enhanced Raman spectroscopy (SERS), Biosens. Bioelectron. 132 (2019) 326−332, Copyright 2019, Elsevier B.V.; (D) J. Zhang, Y. Dong, W. Zhu, D. Xie, Y. Zhao, D. Yang, et al., Ultrasensitive detection of circulating tumor DNA of lung cancer via an enzymatically amplified SERS-based frequency shift assay, ACS Appl. Mater. Interfaces 11 (2019) 18145−18152, Copyright 2019, American Chemical Society; (E) D. Lin, T. Gong, Z.Y. Hong, S. Qiu, J. Pan, C.Y. Tseng, et al., Metal carbonyls for the biointerference-free ratiometric surface-enhanced Raman spectroscopy-based assay for cell-free circulating DNA of Epstein-Barr virus in blood, Anal. Chem. 90 (2018) 7139−7147, Copyright 2018, American Chemical Society.*

the DNA itself and reflects the molecular vibration modes of DNA (e.g., the peak at 732 cm^{-1} from adenine).

Hence, SERS is capable of ctNAs detection and analysis. Both indirect and direct SERS facilitated by clinical ctNAs examinations have been established and summarized in the next section.

6.5.3 ctDNA analysis by SERS

6.5.3.1 Mutant ctDNA detection

Point mutations are frequently observed in cancer patients making them an attractive diagnostic target. Since SERS can show an ultrasensitive response, even to single molecule level, it has been utilized to detect point mutations in ctDNA. For example, Wee et al. aimed at multiple mutation detection on ctDNA in serum samples from melanoma patients [117]. As shown in Fig. 6.8A, the extracted ctDNAs were amplified and labeled with biotin-end, followed by a hybridization with the complementary sequence on SERS nanotags. The mixture was enriched and isolated by streptavidin modified MBs for further SERS measurement. It was proved that the assay has excellent specificity for simultaneously detecting multi-mutant DNAs in a test system and high sensitivity for determining 0.1% mutant DNA against the WT DNA background. The sensitivity is 10-fold higher than commercial PCR-based assays (1%). In addition, this platform can accomplish genotyping screening for the extracted DNAs from both cell lines and serum samples, which were validated by using droplet digital PCR as a standard method. In conclusion, this work demonstrates that a SERS-based sensing technique could achieve specific and sensitive analysis for multiple mutations on ctDNA from clinical samples with high accuracy. Lyu and coworkers applied the same sensing scheme to detect the KRAS gene with three different mutant types with similar good results [122].

In another work, Li et al. designed a sensing workflow focusing on six types of mutation detection in plasma samples from 49 colorectal cancer patients [123]. The gene fragments enclosing target mutation loci from plasma DNA were obtained via PCR amplification. The amplicons were then mixed and annealed with complementary fluorescence-labeled probe sequences, while WT and nontargeted mutations were not labeled. After removing the unbound probe sequences, the probe-labeled target segments were then extracted and incubated with Ag colloids to generate a SERS spectrum under laser illumination. Multiple linear regression was applied to decompose the SERS spectra of multiple tags to indicate the signal from each component, identifying the mutation types included in the sample. Fisher's exact test was employed to decipher the correlation between mutation type and patients' characters (e.g., gender, age, tumor position, and cancer stages). The results illustrate that only BRAF V600E and PIK3CA E542K mutation are positively correlated with right-side colon cancer. Furthermore, combining with hierarchical clustering analysis, a visual dendrogram was created to support Fisher's exact test results. Consequently, this work demonstrated that PCR/SERS together with statistical analysis could achieve multiplex mutation detection with a LOD of 10^{-11} M. It was then possible to explore the relationship between mutation type and a patients' clinical characteristics.

In addition to the usage of SERS-active dyes and small molecules as signal indicators, some other nanomaterials [e.g., single-walled carbon nanotubes (SWNTs)] have also been used for SERS signal readout in ctDNA analysis. For instance, Zhou and coworkers proposed a strategy for mutant ctDNA determination by tracking the SERS signal of SWNTs [118]. As displayed in Fig. 6.8B, this work designed a triple-helix molecular switch structure (THMS) as a recognition unit, including a thymine-rich single-stranded DNA (named as T-rich ssDNA) as the signal transduction probe and an RNA site—embedded capture probe. When the target ctDNA was specifically bound by the capture probe in THMS structure, the T-rich ssDNA was released. An enzyme used to catalyze the hydrolysis of the RNA site in the formed complex of ctDNA and capture probe, the ctDNA was released from the complex which was bound by the capture probe in THMS again to retrigger another cycle and generate T-rich ssDNAs. Next, the released T-rich ssDNAs were absorbed on SWNTs surface due to $\pi-\pi$ stacking interactions to form SWNT/ssDNA complexes. Since thymine in T-rich ssDNA has a high affinity to Cu^{2+}, the SWNT/ssDNA complex was employed as a template to facilitate the in situ growth of CuNPs. The SERS signal of the radial breathing mode and tangential mode (G-band) from SWNTs was enhanced due to the extensive local electromagnetic field between SWNTs and CuNPs. This sensing platform exhibited ultrasensitive detection with a LOD of 0.3 fM and was applied for ctDNA screening enclosing KRAS G12D mutant type in plasma samples from colorectal cancer patients.

By monitoring the conformation change (*gauche* to *trans*) of an alkanethiol ligand, Kowalczyk et al. detected the BRAF V600E mutant ctDNA

(Fig. 6.8C) [119]. The whole assay was constructed on a sputtered gold layer covered gallium nitride (GaN) substrate. When the target ctDNAs were complemented by a capture probe, the SERS intensity change of ν (C—S) bond changed due to the conformation change associated with the duplex formation. Thus, the ratio of the *gauche* and *trans* conformers indicates the target ctDNA level. The assay was applied to samples from a thyroid patient's tissue and plasma, giving results well aligned with ddPCR results, implying high accuracy for clinical ctDNA sample analysis.

By examining a Raman shift change, Zhang and coworkers used a RNA site—embedded hairpin DNA on an Ag nanoparticle film (AgNF) to detect the KARS G12D mutation in ctDNA [120]. When the mutant ctDNA was hybridized with the complementary part of hairpin DNA, the hairpin DNA structure was cut at the RNA site in the presence of RNase HII enzyme shown in Fig. 6.8D and the ctDNA was release to the next cycle for intensifying the signal. The residual DNA of the previous hairpin structure (standing on the AgNF) was complementarily bound with a foreign sequence (FS), leading to a Raman shift change of the Raman reporter molecule on AgNF before and after the binding with FS. Therefore, these successive reactions enabled a highly sensitive subfemtomolar-level detection of KARS G12D in ctDNA. To demonstrate the assay feasibility, the assay was performed with serum samples from lung cancer patients. The results indicated significantly different expression levels of mutant ctDNA from primary and nonprimary lung adenocarcinoma.

6.5.3.2 Aberrant methylated ctDNA detection

SERS can be used to directly identify the presence of functional groups, so it has proved useful for directly tracking the methylation level of ctDNA in clinical samples. The chemical modification of cytosine with a methyl group ($-CH_3$) in ctNA sequences (known as methylated cytosine, 5mC) is not preserved in standard PCR amplification. A bisulfite pretreatment is usually applied to convert unmethylated cytosines into uracil, however, methylated cytosines cannot be converted. Following the bisulfite treatment, all cytosines in PCR amplification are correlated to the original methylated cytosines which can be used for methylation study. Recently, Li et al. established a bisulfite treatment-based SERS/PCR method to analyze the methylation level of three specific genes (p16, MGMT, and RASSFI) in clinical plasma samples [124]. The SERS spectra of bisulfite-treated ctDNA amplicons were obtained after mixing with AgNPs, which directly gave a signature for the methylated target.

The methylation level was calculated by processing spectra via multiple linear regressions to determine the ratio of cytosine-guanine and adenine-thymine (CG/AT). Its correlation with clinical characteristics was investigated using Fisher's exact test. A classification-regression tree was employed to evaluate the diagnostic value of methylation levels of these three genes. The results illustrate that at least one of the three genes can be discovered among 87.5% of patients, while this frequency drops to 11.8% for the control noncancerous group. They also found that the methylation level of the p16 gene was significantly elevated in nonsmall-cell lung cancer patients who had smoked heavily for many years.

Moisoiu and colleagues [125] developed an alternative approach by using direct SERS measurement of DNA methylation in acute myeloid leukemia patients without bisulfite treatment. A reduction in the peak intensity at 1005 cm^{-1} for methylated cytosine was observed, indicating consistency with the well-described hypomethylation in acute myeloid leukemia patients. Based on the statistical models of PCA-LDA and support vector machine, the classification of patients' and control samples was achieved with high accuracy.

6.5.3.3 Other ctDNA detection

SERS has also been implemented for other types of cancer-associated ctDNA analysis. For example, Lin et al. reported a ratiometric assay to detect ctDNA from Epstein-Barr viruses (considered a potential biomarker for diagnosing nasopharyngeal carcinoma) in the blood [121]. They fabricated a gold shell coated osmium carbonyl nanopillar with a capture sequences for enriching target ctDNAs via hybridization. The complex of capture sequence/ctDNA were labeled by the conjugation of rhenium carbonyl and daunorubicin for SERS signal readout (Fig. 6.8E). Since osmium carbonyl was protected in the gold layer on the top of the nanopillar structure, its SERS vibrational mode (2025 cm^{-1}) was used as an internal reference. Therefore the SERS intensity ratio of rhenium carbonyl (2113 cm^{-1}) and osmium carbonyl (2025 cm^{-1}) was calculated for the quantification study. To demonstrate its feasibility in clinical application, it was used to determine ctDNAs from 15 cancer patient samples. The results indicate the target ctDNA level in nasopharyngeal carcinoma patients' blood was higher than that in blood from healthy donors, consistent with PCR results, confirming the potential of this work in clinical nasopharyngeal carcinoma diagnosis.

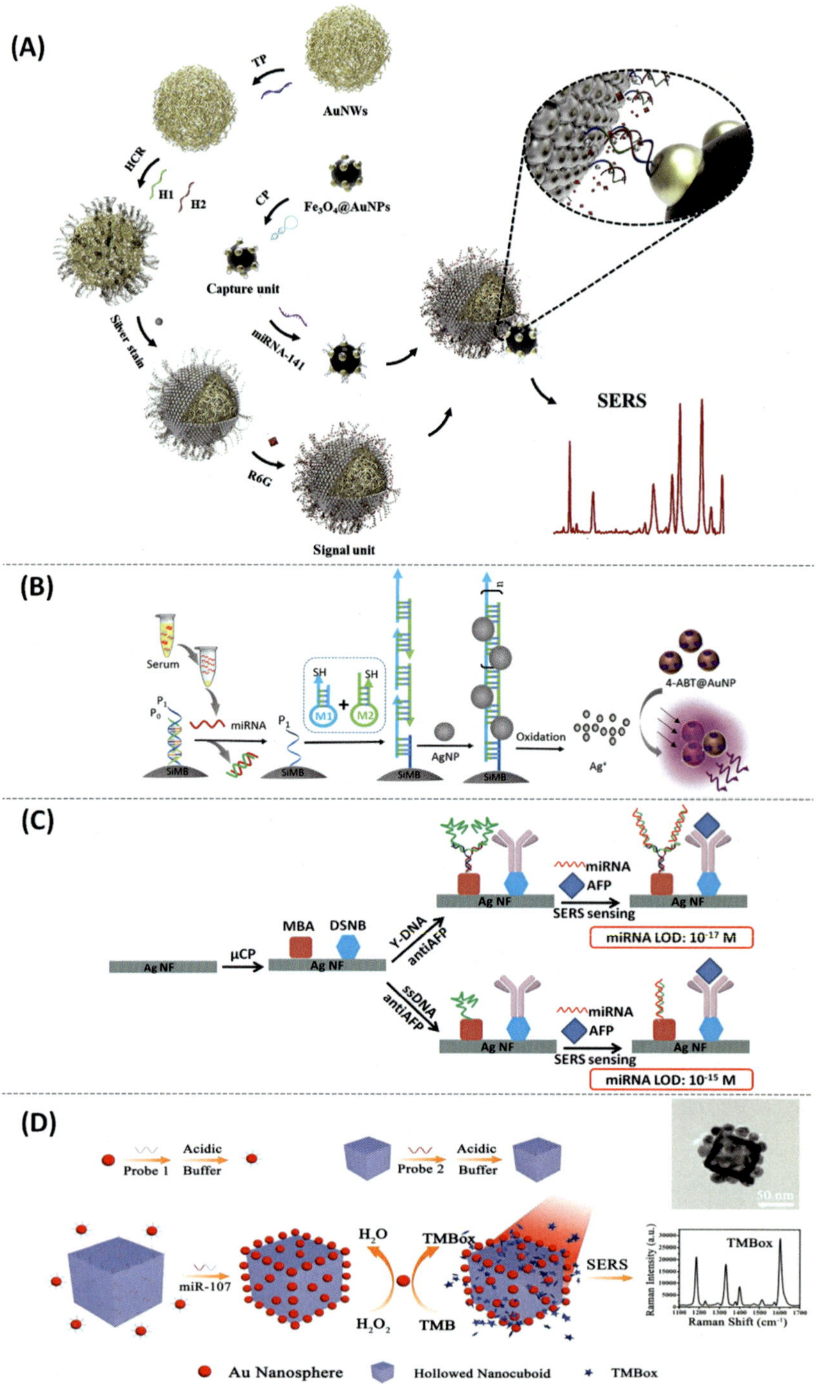
(A)
TP
AuNWs
HCR
H1
H2
CP
Fe3O4@AuNPs
Capture unit
Silver stain
miRNA-141
SERS
R6G
Signal unit
(B)
Serum
P1
P0
miRNA
SiMB
SH
M1
M2
AgNP
Oxidation
Ag+
4-ABT@AuNP
(C)
Ag NF
μCP
MBA
DSNB
Y-DNA
antiAFP
ssDNA
miRNA
AFP
SERS sensing
miRNA LOD: 10-17 M
miRNA LOD: 10-15 M
(D)
Probe 1
Acidic
Buffer
Probe 2
50 nm
miR-107
H2O
H2O2
TMBox
TMB
SERS
Raman Intensity (a.u.)
Raman Shift (cm-1)
Au Nanosphere
Hollowed Nanocuboid

6.5.4 ctRNA analysis by SERS

6.5.4.1 ctmiRNA detection

Enzyme-free hybridization chain reaction (HCR) has been used in ctmiRNA detection, in which an initiator triggers hybridization and leads to the polymerization of oligonucleotides into long nicked dsDNA polymers under mild conditions [126]. Shao et al. developed a sandwich assay via enzyme-free hybridization to enrich miRNA141 by a functionalized capture unit and a signal unit to quantify the target (Fig. 6.9A) [126]. In the assembly of the signal unit, a dsDNA (with a tail for interacting with capture unit later) was formed from two hairpin DNA sequences on gold nanowires (AuNWs) through HCR. Next, when $AgNO_3$ solution was added, Ag^+ ion was adsorbed on AuNWs and inserted into the dsDNA skeleton via electrostatic force, followed by the reduction into AgNPs by using hydroquinone. The AgNPs loaded AuNWs as a signal unit was used to improve the SERS signal of subsequently added Raman reporter molecule—rhodamine 6 G (R6G). In the capture unit the target miRNA141 was immobilized by hybridizing with a capture sequence on AuNPs decorated MBs to generate an end-residue sequence for coupling

Figure 6.9 (A) Graphical illustration of the fabrication of SERS assay for miRNA141 detection, which includes assembling of signal unit based on gold nanowire vesicles (AuNWs) and AuNPs decorated magnetic capture unit [126]. (B) Scheme design for enzyme-free quadratic SERS signal amplification for miRNA-21 detection in human serum via miRNA-initiated hybridization chain reaction (HCR) and Ag^+-modulated cascade amplification [127]. (C) Schematic construction of SERS sensing assay for concurrent analysis of hepatocellular carcinoma—associated biomarker miRNA-223 and α-fetoprotein [128]. (D) Schematic illustration for delicate usage of target miRNA as self-assembly of SERS substrate for miRNA-107 detection [129]. *Reproduced with permission from (A) H. Shao, H. Lin, Z. Guo, J. Lu, Y. Jia, M. Ye, et al., A multiple signal amplification sandwich-type SERS biosensor for femtomolar detection of miRNA, Biosens. Bioelectron. 143 (2019) 111616, Copyright 2019, Elsevier B.V.; (B) J. Zheng, D. Ma, M. Shi, J. Bai, Y. Li, J. Yang, et al., A new enzyme-free quadratic SERS signal amplification approach for circulating microRNA detection in human serum, Chem. Commun. 51 (2015) 16271—16274, Copyright 2015, The Royal Society of Chemistry; (C) L. Cheng, Z. Zhang, D. Zuo, W. Zhu, J. Zhang, Q. Zeng, et al., Ultrasensitive detection of serum microRNA using branched DNA-based SERS platform combining simultaneous detection of alpha-fetoprotein for early diagnosis of liver cancer, ACS Appl. Mater. Interfaces 10 (2018) 34869—34877, Copyright 2018, American Chemical Society; (D) J. Li, M.K. Koo, Y. Wang, M. Trau, Native microRNA targets trigger self-assembly of nanozyme-patterned hollowed nanocuboids with optimal interparticle gaps for plasmonic-activated cancer detection, Small 15 (2019) 1904689—1904698, Copyright 2019, Wiley-VCH.*

with the tail sequence of the signal unit. Thus, this sandwich assay was formed for SERS signal readout. The sensing platform was validated for specifically detecting miRNA141 with a LOD of 0.03 fM. By detecting prostate cancer patients' and healthy volunteers' serum samples, the assay showed a higher expression of miRNA141 in patient samples than that in healthy donor samples in accord with a parallel examination by the standard RT-PCR method.

In another work constructed by Zheng et al., an enzyme-free HCR design was triggered by a target miRNA-21 to form a long DNA polymer chain with sulfhydryl labels (Fig. 6.9B) [127]. AgNPs were loaded onto the DNA polymer chain via Ag-S interaction, and oxidized into Ag^+ to induce the aggregation of Raman report molecule modified AuNPs, which were served as SERS probes for signal readout. The assay performed well to analyze miRNA-21 with a LOD of 0.3 fM. The assay was then implemented for both chronic lymphocytic leukemia patients' and healthy donors' serum samples and a higher expression of miRNA-21 was found in chronic lymphocytic leukemia patients' serum at stage 3 than in the serum of patients at stage 1 and healthy controls.

Except the SERS intensity change used to indicate the target amount, the Raman shift in SERS measurement was also applied to evaluate the amount of target. A work published by Cheng et al., displayed in Fig. 6.9C, employed target concentration change accompanied Raman shift variation to simultaneously determine potential liver cancer–associated biomarker miRNA-223 and α-fetoprotein [128]. The AgNPs coated films were functionalized with Raman reporter molecules *p*-methoxybenzoic acid (MBA) and 5,5′-dithiobis(succinimidyl-2-nitrobenzoate) (DSNB) as SERS substrate, which was further modified with capture elements for enriching miRNA-223 and α-fetoprotein. The characteristic modes of DSNB (1335 cm^{-1}) and MBA (1580 cm^{-1}) corresponding to miRNA-223 and α-fetoprotein, respectively, both showed frequency shifts that were used to quantify the level of miRNA-223 and α-fetoprotein. A LOD of 10^{-17} M for both miRNA-223 and α-fetoprotein in fetal bovine serum was achieved. This platform was further demonstrated by analyzing these two potential biomarkers in serum from hepatocellular carcinoma patients and healthy individuals. The accuracy for α-fetoprotein detection was confirmed by an electrochemiluminescence immunoassay, the most widely adopted clinical technology. In conclusion, this sensing assay holds great potential to be applied for instantaneously analyzing miRNA-223 and α-fetoprotein in hepatocellular carcinoma related clinical applications.

Li and coworkers utilized target miRNA-107 as a trigger for a self-assembling nanosubstrate which acts as a "nanozyme" [130] to catalytically generate a SERS response toward target miRNA-107 for prostate cancer diagnosis [129]. As shown in Fig. 6.9D, the assembly of AuNPs patterned hollow nanocubodies was implemented by two half-complementary probes toward miRNA-107. Due to the hotspots created from the small gap between AuNPs and nanocubodies, the assay reached a low detection limit of 79 fM. The assay feasibility was verified by testing patient urine samples. The result showed a significantly higher level of miRNA-107 from patients than healthy donors.

6.5.4.2 Other disease–associated RNA detection

Other types of circulating RNAs as potential disease biomarkers have also been investigated by SERS-based strategies. Koo et al. integrated SERS with RT-recombinase polymerase amplification (RPA) to synchronously measure five prostate cancer–associated RNAs (as shown in Fig. 6.10A) [131]. The extracted RNAs were specifically amplified by RT-RPA rather than RT-PCR which requires thermal cycling. The formed RNA amplicons–SERS nanotag complex was magnetically enriched for SERS measurement by employing the linkage with complementary sequences on their own unique SERS nanotags. The sensitivity study suggested that the assay can detect the target RNAs with a LOD of 100 copies and a five-plexed assay on RNA extracted from patient tissues and urine samples was possible with performance similar to the commonly used qRT-PCR.

Intrinsic SERS detection has also been used to assess the level of potential RNA biomarker in body fluids and to predict risk for developing cancer by incorporating data collection with statistical tools. Wang and coworkers carried out SERS measurements on the mixture of prostate cancer specific RNA biomarker T1E4 amplicons (amplified by RT-RPA) by using the positively charged AgNPs as SERS substrate to naturally exploit spectral fingerprints of target RNA. The implication in the clinic of SERS results was analyzed with the aid of the PCA-LDA model [133]. The assay achieved a specificity of 93%, a sensitivity of 95%, and an accuracy of 94.2% in the analysis of 43 patients' urinary samples, exhibiting excellent feasibility and reliability. Following study was then undertaken from the same group to apply this strategy for analysis of two prostate cancer specific RNA biomarkers (T2: ERG and PCA3) and one prostate cell-enriched RNA (KLK2) in large-scale clinical samples (Fig. 6.10B) [132]. A training cohort ($n = 80$) is used to determine cut-off limit for prostate cancer detection and high-risk

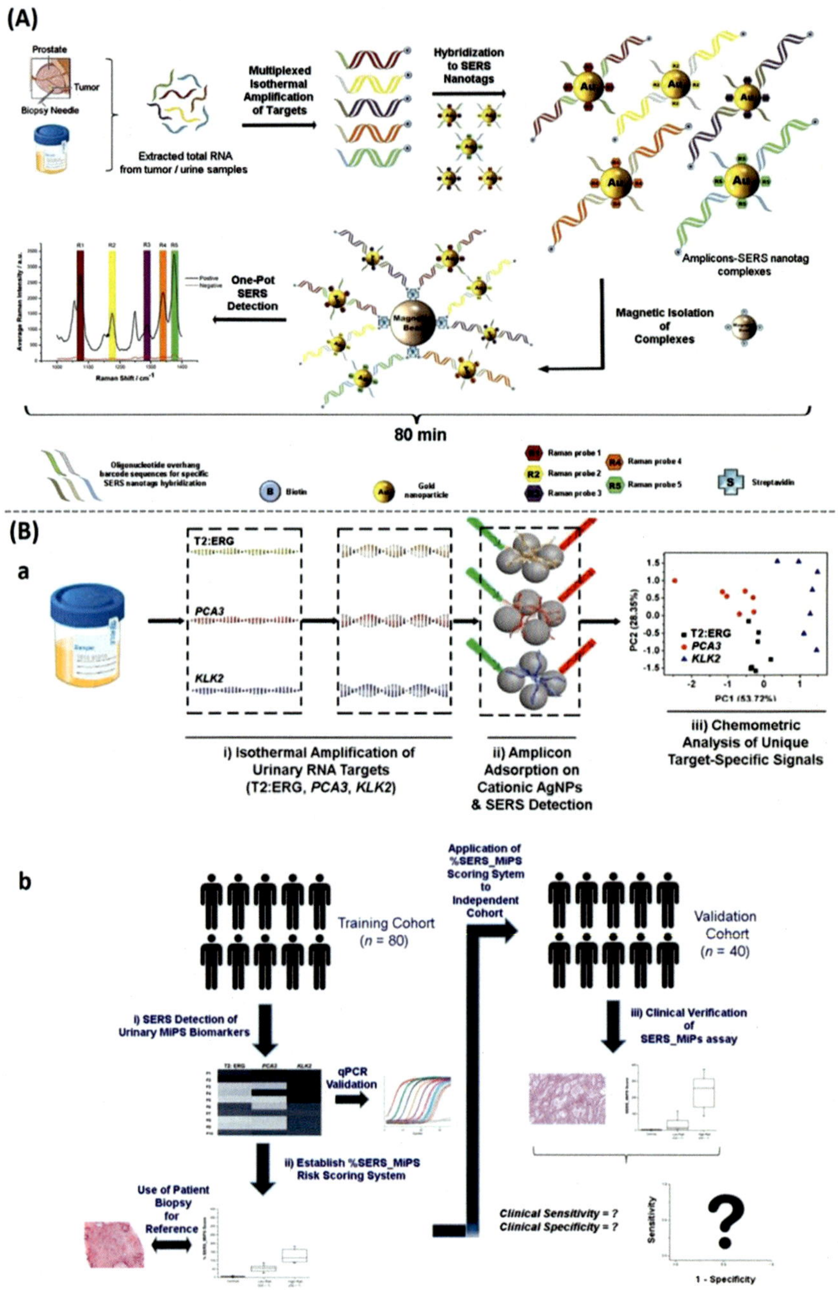

(A)
Prostate
Tumor
Biopsy Needle
Extracted total RNA
from tumor / urine samples
Multiplexed
Isothermal
Amplification
of Targets
Hybridization
to SERS
Nanotags
Amplicons-SERS nanotag
complexes
Magnetic Isolation
of
Complexes
One-Pot
SERS
Detection
Raman Shift / cm-1
80 min
Oligonucleotide overhang barcode sequences for specific SERS nanotags hybridization
Biotin
Gold nanoparticle
Raman probe 1
Raman probe 2
Raman probe 3
Raman probe 4
Raman probe 5
Streptavidin
(B)
a
T2:ERG
PCA3
KLK2
PC2 (28.35%)
PC1 (53.72%)
i) Isothermal Amplification of
Urinary RNA Targets
(T2:ERG, PCA3, KLK2)
ii) Amplicon
Adsorption on
Cationic AgNPs
& SERS Detection
iii) Chemometric
Analysis of Unique
Target-Specific Signals
b
Training Cohort
(n = 80)
Application of
%SERS_MiPS
Scoring Sytem
to
Independent
Cohort
Validation
Cohort
(n = 40)
i) SERS Detection of
Urinary MiPS Biomarkers
qPCR
Validation
ii) Establish %SERS_MiPS
Risk Scoring System
Use of Patient
Biopsy
for
Reference
iii) Clinical Verification
of
SERS_MiPs assay
Clinical Sensitivity = ?
Clinical Specificity = ?
Sensitivity
1 - Specificity
?

discrimination, while a validation cohort ($n = 40$) was employed to examine the performance of above cut-off limit. By resolving SERS spectra of these three target amplicons from training cohort and validation cohort of prostate cancer patients, a risk stratification model was established on the outcome correlation between SERS measurement and clinical validated Mi-Prostate Score platform. The stratification model showcased risk scores to suggest to patients the possibility of developing cancer with high sensitivity and accuracy. Therefore this assessing strategy in the clinical application has been well-characterized, being expected to play roles in early diagnosis indeed.

6.5.5 Insights into SERS-based ctNAs analysis with clinical samples

In general, SERS-based assays for ctNAs analysis can be used to give quantification [117,118,126], discrimination [124,125,133] and genotyping pattern profiling [123,132]. First of all, the sensitivity of quantitative detection toward target NAs is comparable with or even more sensitive than current mainstream techniques (e.g., the identification of 0.1% mutant DNA against its WT DNA background is 10-fold more sensitive than the 1% of commercial PCR-based assays) [117]. Second, due to the extremely sensitive SERS response of target molecule, tiny spectral variations either in intensity [118] or Raman shift [128] may be utilized to identify the differences between tumor-derived and healthy DNAs (e.g., point mutations) [123] or chemical modifications (e.g., abnormal methylation) [124,125]. Furthermore, the multiplexing capability from multiple SERS nanotags has the potential for analyzing several biologically relevant targets simultaneously [117,123,124,132]. The SERS methods are not limited to a single type of NAs [128], and are therefore superior to droplet digital PCR in which is currently limited to two channels. SERS-based approaches may be used to exploit the associations between genotypes and clinical characteristics

◀ **Figure 6.10** (A) Schematic workflow of SERS-based five-plexed assay for concurrently analyzing prostate cancer (PCa)–associated RNAs [131]. (B): (a) Graphical representation of direct SERS combined with statistical analysis for urinary RNA analysis; (b) clinical verification through above SERS-based assay for risk prediction toward PCa patients [132]. *Reproduced with permission from K.M. Koo, E.J. Wee, P.N. Mainwaring, Y. Wang, M. Trau, Toward precision medicine: a cancer molecular subtyping nano-strategy for RNA biomarkers in tumor and urine, Small 12 (45) (2016) 6233–6242, Copyright 2016, Wiley-VCH.; (B) K.M. Koo, J. Wang, R.S. Richards, A. Farrell, J.W. Yaxley, H. Samaratunga, et al., Design and clinical verification of surface-enhanced Raman spectroscopy diagnostic technology for individual cancer risk prediction. ACS Nano 12 (2018) 8362–8371, Copyright 2018, American Chemical Society.*

by dint of statistical processes to help with clinical decision-making [123,124,132]. Using a range of PCR techniques [117,124,131] have made available a number of ingenious PCR/SERS platforms which offer more possibilities in assay design and clinical translation.

SERS-based ctNAs analysis can now focus on the development and validation of some assays for specific ctNAs from clinical samples including more accuracy in indirect SERS—based approaches for methylation evaluation, direct SERS—based methods for mutation detection, and both direct/indirect-SERS measurement for analyzing methylation level on ctRNAs.

6.6 Tumor-associated proteins

6.6.1 Clinical significance and current analysis techniques of circulating proteins

Proteins are biomacromolecules constructed from repeated subunits (amino acid) with a molecular weight varying from thousands to nearly 1 million daltons [134]. They take vital roles in maintaining living organisms and promoting/inhibiting biological activities and changes in structure, amount, or sequence of proteins in the blood, urine, saliva, cerebrospinal fluid, sweat, or cystic fluid, are treated as promising biomarkers for clinical diagnosis and therapy monitoring [135]. For instances, amyloid-β sheet domain is associated with the occurrence of Alzheimer's disease; [136] the glycoproteins prostate-specific antigen (PSA), is considered to be a diagnostic biomarker for prostate diseases when present at concentrations above 4 ng mL^{-1} [137].

Assays to identify or quantify circulating proteins mainly focus on ligand-binding, Western blot and ELISA, and mass spectrometry—based strategies. Each approach has advantages and disadvantages. Western blots and ELISA assays are antibody based and can be specific, however, operational errors may be easily caused by lengthy workflows, the challenges of running reproducible gels and consistent results may be affected/limited by antibody quality [134]. Mass spectrometry—based approaches can resolve proteins at the sequence level, thus providing greater information and not requiring appropriate antibodies [138]; however, mass spectrometry—based analysis suffers from lower sensitivity and throughput and the requirement of extremely expensive instrumentation [138,139]. Sample preparation for mass spectrometry—based analysis is often lengthy and identifying a biomarker against the background of abundant endogenous proteins [140].

Therefore, new fast, selective (or specific) and simple approaches are required for routine clinical purposes and SERS has been targeted for this purpose. In the following section, SERS-based circulating protein analyses are categorized and discussed as three subclasses: (1) magnetic enrichment–based immunoassay, (2) immunoassay on 2D substrate, and (3) assay with unique signal-output design.

6.6.2 SERS-based strategy for protein analysis

6.6.2.1 Magnetic enrichment–based immunoassay

Magnetically isolating/enriching target proteins incorporated with SERS nanotags as signal indicators is a common way to determine proteins of interest in patients' body fluids. Lin et al. performed a rapid and straightforward detection of a specific colorectal cancer biomarker—carcinoembryonic antigen with a typical sandwich design [141]. In the presence of carcinoembryonic antigen, Au shell enclosed MBs and Raman reporter modified SERS nanotags were assembled via the immunoaffinity of specific antibodies toward carcinoembryonic antigen. After magnetic separation, the assembled sandwich complex was examined under laser illumination. The assay exhibited a LOD of 0.1 ng mL^{-1}, which is sensitive enough for evaluating carcinoembryonic antigen in human serum since its concentration is higher than 20 ng mL^{-1} suggesting a malignant tumor. Clinical serum samples from patients with a benign tumor, colorectal cancer patients, and healthy donors were assessed, and the results are in line with those from the validated method—electrochemiluminescence immunoassay—commonly used in hospitals.

Customized SERS nanotags enabled the simultaneous detection of multiple targets in one test showing potential for high-throughput simultaneous screening of more than one disease with small sample volume and saving-cost. Chon et al. synchronously detected carcinoembryonic antigen and α-fetoprotein for diagnosing colorectal cancer and liver cancer in one test [142]. The MBs were conjugated with specific antibodies to capture carcinoembryonic antigen and α-fetoprotein and, subsequently tagged with the hollow gold nanospheres (HGNs)-based SERS nanotags through their antibodies which were also anchored on SERS nanotags. The sensing performance of this platform was checked by testing clinical patients' blood samples and comparing results with the accepted commercial chemiluminescent microparticle immunoassay (CMIA). The results present high accuracy and concordance with the CMIA analysis, indicating promising capability for the simultaneous diagnosis of multi-disease. Work from the same research group contributed by Cheng et al. focused on the precise diagnosis of prostate cancer by sensing two PSAs: free-PSA (f-PSA) and complex-PSA (c-PSA) [143]. Increased level of total PSA (t-PSA) is the conventional way to diagnose prostate cancer; however, both prostate cancer and noncancerous factors [e.g., benign prostatic hyperplasia (BPH), prostatitis] can cause its elevated level. A lower level of f-PSA has been found in men with prostate cancer than in those with benign conditions. Thus, the free to total PSA ratio (f/t-PSA) is considered a diagnostic basis to discriminate prostate cancer and BPH. This guides clinical decisions for patients whose t-PSA level falls in the "diagnostic gray zone" with a 4.0–10.0 ng mL^{-1} concentration. In this work, the t-PSA antibody was conjugated onto a MB to capture both c-PSA and f-PSA, followed by their own specific SERS nanotags labeling to indicate the concentration of c-PSA and f-PSA simultaneously and independently. After demonstrating good sensing performances (a LOD of 0.012 ng mL^{-1} for f-PSA and 0.15 mg mL^{-1} for c-PSA) with negligible cross-reactivity, the assay was applied to examine the amount of t-PSA and f/t-PSA ratio in 30 clinical samples. The proposed method especially gives better precision based on f/t-PSA ratio evaluation for 13 samples at gray zone than the validated method.

Another similar example was performed by Bai et al., in which functionalized MBs and the specific SERS nanotags were used to seize and mark α-fetoprotein, carcinoembryonic antigen, and ferritin at one time to predict the liver cancer risk [144]. The assay was utilized for screening 39 suspected liver cancer patients. This SERS immunoassay gave statistically consistent results with evident advantages over hospital tests due to easy operation and rapid and exact measurement. The technical highlight is that three Raman active molecules with single narrow Raman bands in the bio-silent region at 2105, 2159, and 2227 cm^{-1} were employed and encoded with SERS nanotags as a signal indicator, thus eliminating signals from the biological matrix.

6.6.2.2 Assays with unique signal-output design

In most cases, SERS signal intensity is in proportion to the amount of target protein, yet a few works are designed to use SERS "turn-off" sensing assay in a reverse manner. Feng et al. established an assembling structure consisting of gold nanorods (AuNRs) as the core and with multiple AgNPs as satellites, in which two DNA segments were separately anchored on AuNRs and AgNPs to link them together by hybridization (Fig. 6.11A) [145]. The sequence of above DNA linkers on AgNPs has a highly specific affinity to the target protein mucin-1, resulting in the Raman reporter coated AgNPs can be released from AuNRs core in the presence of mucin-1 via the interaction between this high-affinity DNA linker and mucin-1. Thus, the SERS intensity from functional AgNPs linked AuNRs decreased along with the increasing mucin-1. By monitoring the signal change, the assay achieved a LOD of 4.3 aM in analyzing mucin-1 from 0.005 to 1 fM, and a high specificity toward mucin-1 in the presence of six other cancer biomarkers. The feasibility study was conducted by measuring clinical serum samples, giving a result consistent with the standard clinical diagnostic assay.

Enzyme catalysis-induced SERS signal changes have been used as a signal generation system to monitor PSA and adrenal stimulant ractopamine (Rac) indirectly by Liang and coworkers (Fig. 6.11B) [146]. The principle is based on the reduction reaction of hydrogen peroxide in the presence of catalase. Subsequently, the decreased SERS signal from the complex of AgNPs-Raman reporter (4-mercaptobenzoic acid, 4-MBA) was generated via the degradation of AgNPs by residual hydrogen peroxide. In this way, a sandwich assay with signal "turn-on" behavior and a competitive assay with signal "turn-off" were designed for detecting PSA and Rac, respectively. Both these platforms showed a LOD of 10^{-9} ng mL^{-1} for PSA and of 10^{-6} ng mL^{-1} for Rac. The detection of PSA in suspected prostatic cancer patients gave the same results as the validation method (time-resolved fluorescent immunoassay).

Assessing a target protein amount through the change in Raman shift can also be utilized as an indicator. As displayed in Fig. 6.11C, Tang et al. monitored the frequency movements of a protein complex (Raman active molecule-capture antibody) anchored on a substrate in the presence of α-fetoprotein and glypican-3, which are diagnostic biomarkers for hepatocellular carcinoma [147]. The dense patterned of AgNPs on the substrate (approximately 1 nm apart) ensured highly sensitive detection toward α-fetoprotein and glypican-3 down to subpicomolar level. Compared to the detection sensitivity of conventional ELISA approach, this work showed 2-orders of magnitude more sensitive detection for glypican-3 to discriminate the nonhepatocellular carcinoma patient and hepatocellular carcinoma-positive patient. It was also demonstrated to be a promising clinical diagnostic candidate by testifying α-fetoprotein and glypican-3 in a hepatocellular carcinoma patient serum with high accuracy.

6.6.2.3 Immunoassays on 2D arrays

In some cases, functionalized 2D solid substrate–based protein enrichment facilitates SERS measurements with high throughput for replicates by using multiindividual sensing wells in parallel. For instance,

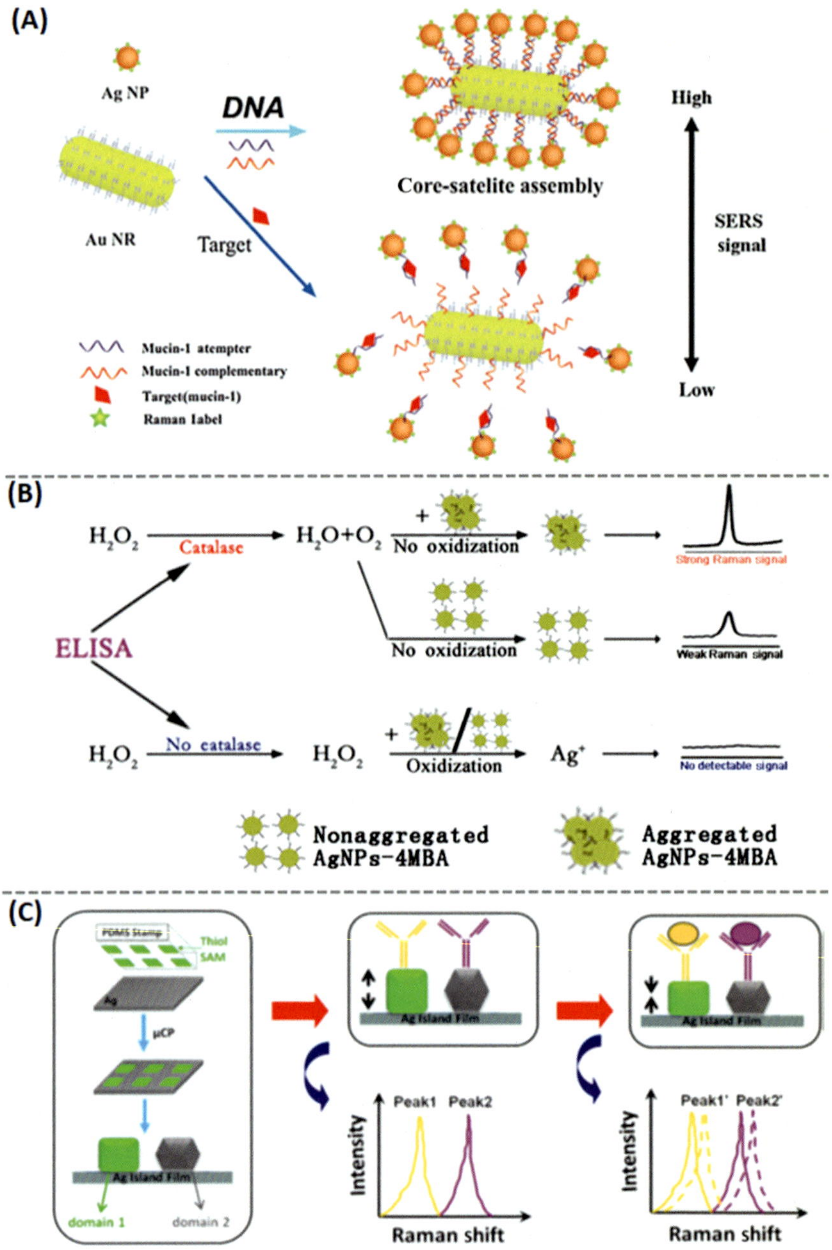
(A)
Ag NP
DNA
Au NR
Target
Core-satelite assembly
High
SERS signal
Low
Mucin-1 atempter
Mucin-1 complementary
Target(mucin-1)
Raman label
(B)
H2O2
Catalase
H2O+O2
No oxidization
Strong Raman signal
ELISA
No oxidization
Weak Raman signal
H2O2
No catalase
H2O2
Oxidization
Ag+
No detectable signal
Nonaggregated AgNPs-4MBA
Aggregated AgNPs-4MBA
(C)
PDMS Stamp
Thiol SAM
Ag
μCP
Ag Island Film
domain 1
domain 2
Peak1
Peak2
Intensity
Raman shift
Peak1'
Peak2'

Banaei et al. conducted SERS-based multiplex detection for three pancreatic cancer biomarkers on a gold nanoshell coated silica substrate with a 3 × 15 array of functionalized wells (Fig. 6.12A) [148]. CA19-9 antigen is a traditional clinical pancreatic cancer diagnostic biomarker; however, it cannot effectively distinguish pancreatic cancer from several benign liver diseases. Therefore, combining CA19-9 with two other biomarkers, MUC4 and MMP7, resulted in a detection sensitivity of under 2 ng mL^{-1} for each of protein that was used to screen five pancreatic cancer patients, five nonpancreatic cancer patients with other types of pancreatitis, and five healthy individuals. The results suggest that CA19-9 levels are highest in pancreatic cancer patients and high in pancreatitis patients, while the amounts of MUC4 and MMP7 are significantly higher in pancreatic cancer patients than in the pancreatitis group. The expression of these three biomarkers in pancreatic cancer patients is heterogeneous depending on cancer stage. Therefore this SERS-based multiplex analysis exhibits a more precise pancreatic cancer diagnostic potential than a single CA19-9 test.

Another similar contribution from Krasnoslobodtsev et al. involved antibody conjugated mica to enrich MUC4, and subsequently, AuNP-based SERS nanotags were used for signal readout (Fig. 6.12B) [149]. The assay was utilized to screen pancreatic cancer patients and accurately discriminated them from healthy individuals. Instead of using bare AuNPs as SERS probe-substrate, in Chang and coworkers' paper, the SERS

◀ **Figure 6.11** (A) Schematic construction of core-satellite structure–based SERS signal "turn-off" assay [145]. (B) Graphical working principle of using SERS signal as the ELISA signal generation system based on catalase induced hydrogen peroxide reduction and AgNPs degradation by residual hydrogen peroxide [146]. (C) Assay principle of Raman frequency shift-based detection on dense AgNPs coated substrate for duplex diagnostic biomarkers of hepatocellular carcinoma [147]. *Reproduced with permission from (A) J. Feng, X. Wu, W. Ma, H. Kuang, L. Xu, C. Xu, A SERS active bimetallic core–satellite nanostructure for the ultrasensitive detection of Mucin-1, Chem. Commun. 51 (2015) 14761–14763, Copyright 2016, American Chemical Society; (B) J. Liang, H. Liu, C. Huang, C. Yao, Q. Fu, X. Li, et al., Aggregated silver nanoparticles based surface-enhanced Raman scattering enzyme-linked immunosorbent assay for ultrasensitive detection of protein biomarkers and small molecules, Anal. Chem. 87 (2015) 5790–5796, Copyright 2015, American Chemical Society; (C) B. Tang, J. Wang, J.A. Hutchison, L. Ma, N. Zhang, H. Guo, et al., Ultrasensitive, multiplex Raman frequency shift immunoassay of liver cancer biomarkers in physiological media, ACS Nano 10 (2016) 871–879, Copyright 2016, American Chemical Society.*

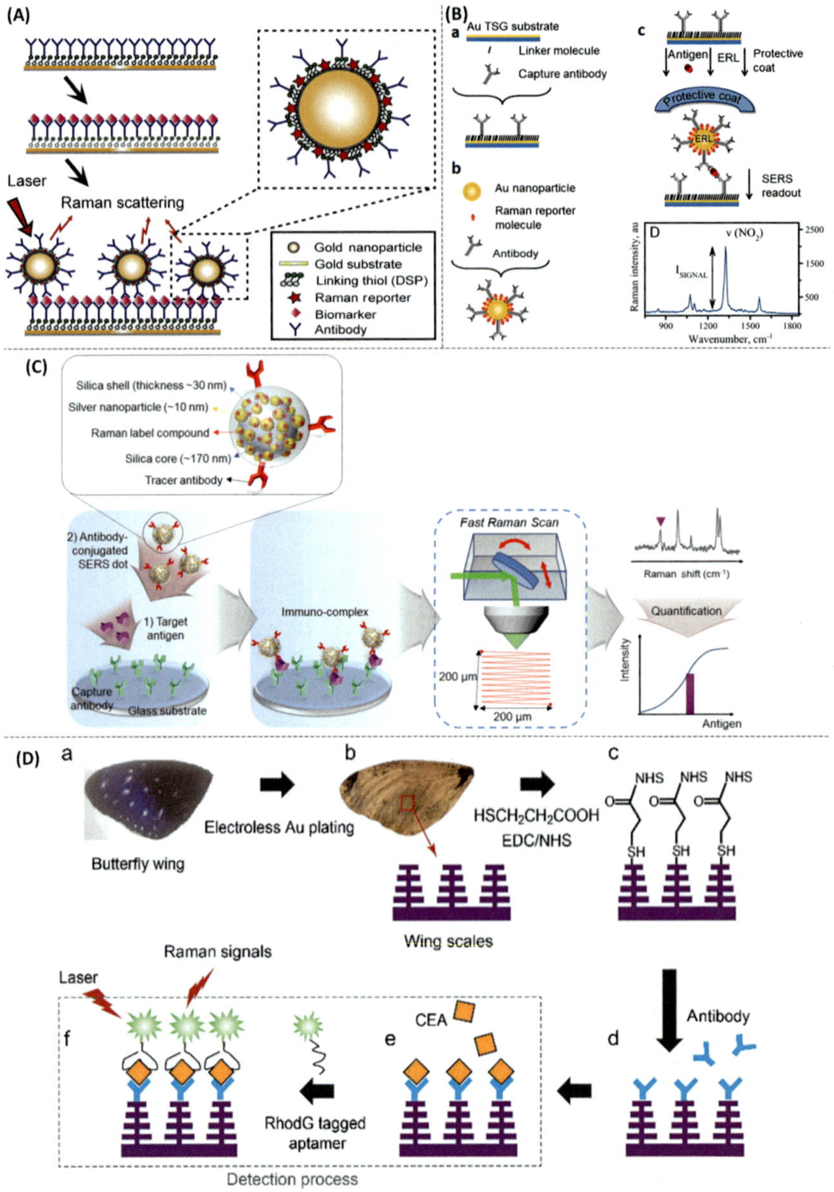

Figure 6.12 (A) The workflow of SERS-based sandwich immunoassay for detecting pancreatic cancer biomarker on a functionalized gold substrate [148]. (B) Graphical fabrication of (a) capture substrate and (b) SERS probes, and (c) immunoassay

substrate was made up of an AgNPs coated silica core and an outer silica shell used for signal readout (Fig. 6.12C) [150]. This design relieved NPs aggregation and improved signal reproducibility, achieving better linearity and lower LOD (0.11 pg mL^{-1}) for PSA analysis than those from commercial immunoradiometric assay and chemiluminescence immunoassay, though the results were good alignment.

3D hierarchical substrates have been developed to enhance assay sensitivity. For example, Song et al. fabricated Au butterfly wing-like (Au BWs) surfaces as the primary substrate to analyze carcinoembryonic antigen (Fig. 6.12D) [151]. The "rib" and "strut" structures in Au BWs contribute to a dense distribution of hotspots and carcinoembryonic antigen was enriched by the antibodies attached to the Au BWs enabling an enhanced SERS signal on Au BWs surface. Comparing with commercial Q-SERS substrates composed of AuNPs (~70 nm), this Au BWs substrate gains one magnitude lower LOD (10 ng mL^{-1}) for detecting carcinoembryonic antigen. This platform has acceptable accuracy in test with clinical samples compared with the commercial Abbott carcinoembryonic antigen reagent kit analysis. Li et al. also synthesized a 3D hierarchical Au nanostar coupled with an Au nanotriangle array as a plasmonic hotspot

---

◀ workflow for detecting pancreatic cancer biomarker MUC4 [149]. (C) Schematic illustration of (a) construction for SERS-based immunoassay on glass chip by using multilayer encapsulated SERS dots as signal generator, (b) characterization of single SERS dot on SERS intensity map overlaid with the corresponding SEM image, (c) the plotting results of determining PSA at a concentration of 0–1000 ng mL^{-1} by established assay [150]. (D) Graphical establishment of Au butterfly wings-based SERS sensing platform for carcinoembryonic antigen detection [151]. *(A) Reproduced with permission from N. Banaei, A. Foley, J.M. Houghton, Y. Sun, B. Kim, Multiplex detection of pancreatic cancer biomarkers using a SERS-based immunoassay, Nanotechnology 28 (2017) 455101–455110, Copyright 2017, IOP Publishing; (B) A.V. Krasnoslobodtsev, M.P. Torres, S. Kaur, I.V. Vlassiouk, R.J. Lipert, M. Jain, et al., Nano-immunoassay with improved performance for detection of cancer biomarkers, Nanomedicine 11 (2015) 167–173, Copyright 2015, Elsevier B.V.; (C) H. Chang, H. Kang, E. Ko, B.H. Jun, H.Y. Lee, Y.S. Lee, et al., PSA detection with femtomolar sensitivity and a broad dynamic range using SERS nanoprobes and an area-scanning method, ACS Sens. 1 (2016) 645–649, Copyright 2016, American Chemical Society; (D) G. Song, H. Zhou, J. Gu, Q. Liu, W. Zhang, H. Su, et al., Tumor marker detection using surface enhanced Raman spectroscopy on 3D Au butterfly wings, J. Mater. Chem. B 5 (2017) 1594–1600, Copyright 2017, The Royal Society of Chemistry.*

supplier to produce highly sensitive SERS readouts [152]. This array outperformed spherical AuNPs coated Au nanotriangle array, in detecting vascular endothelial growth factor, giving results consistent with an ELISA assay in clinical plasma samples. This sensitivity can be ascribed to the dense hotspots from the localized surface plasmon resonance coupling of nanostar spikes and nanotriangle tips.

6.6.3 Insights on SERS-based assays for disease-associated protein detection

Overall, disease-associated protein has been extensively assessed by SERS-based sensing strategies, often with sandwich substrates [141–144] and sometimes combined with novel signal generation methods [145–147]. The ultrasensitive detection and high accuracy of SERS have the potential to meet clinical diagnostic requirement with less sample volume and faster assay times than current methods [143]. Simultaneous detection of multiprotein targets within one sample volume facilitates better precise diagnosis or following of disease or treatment progression [143,148]. Direct SERS also has the potential to directly give information about protein structure changes which could be correlated with pathological phenomenon. Combining SERS with other structure analysis techniques, such as Raman optical activity spectroscopy, circular dichroism spectroscopy, and molecular imprinted technology [153,154], provides complementary information to support this application.

6.7 Conclusions and perspectives

With decades of development, SERS has become a widely accredited technique for qualitative characterization, trace level quantification, specimen identification, and imaging [155]. This chapter focuses on its roles and potential in liquid biopsy for screening circulating biomarkers in clinical samples. A wide range of target enrichment strategies, SERS signal generation approaches and SERS enhancement by plasmonic nanomaterials have been developed to analyze liquid biopsy biomarkers in clinical samples. In comparison with current conventional hospital tests, SERS assays exhibit the following advantages: (1) sensitive and specific identification of trace analytes in an abundant nontarget background; (2) multianalyte evaluation; (3) small sample volume; (4) fast measurement; (5) signal generation without interferences from water and biological autofluorescence; and (6) acquisition of inherent vibrational signatures of the analyte itself.

By combining the SERS data with increasingly sophisticated statistical analyses, it is possible to extract information regarding cancer progression, discrimination of cancerous status from noncancerous status, and treatment efficacy as well as answering questions on phenotyping and genotyping of analytes—all of which contribute to improved clinical decision-making. Moreover, as a vibrational technique, direct SERS analysis of analytes can give molecular information (such as lipids and proteins carried on target) which can contribute to understanding fundamental molecular mechanisms of disease occurrence.

Despite the attractions of SERS assays, some practical barriers currently hinder the transformation of these proposed strategies into real clinical usage. The key issues are that assays often (1) exhibit lack of reproducible spectra both for indirect SERS (based on SERS nanotags) or direct analyte SERS usually due to the nanoscale substrate structure; (2) require large sample cohorts for validation; (3) have spectra that are difficult to interpret for multiple-component samples (e.g., EVs and CTCs); (4) are specific to individual laboratories or even workers—standardized analysis procedure are the common issue for all sensing techniques-based platforms which straightly determines the application in large scale with reliable outcomes. The challenge of producing reproducible and stable functionalized substrates is most important requiring reliable particle size, shape and molecular information with controlled antibody conjugation and modification of surface chemical functional groups to minimize batch-to-batch variation [156—158]. Complementary techniques including electronic spectroscopy, DLS, Zeta potential, UV-visible spectroscopy, infrared spectroscopy, differential centrifugal sedimentation, and microscopy are all required to provide specific characterization of morphology, size distribution, surface charge, and chemical composition of nanosubstrate for SERS [158].

References

[1] N. Hawkes, Cancer survival data emphasise importance of early diagnosis, Br. Med. J. 364 (2019) l408.

[2] C. Foster, L. Calman, A. Richardson, H. Pimperton, R. Nash, Improving the lives of people living with and beyond cancer: generating the evidence needed to inform policy and practice, J. Cancer Policy 15 (2018) 92—95.

[3] R.A. Mathai, R.V.S. Vidya, B.S. Reddy, L. Thomas, K. Udupa, J. Kolesar, et al., Potential utility of liquid biopsy as a diagnostic and prognostic tool for the assessment of solid tumors: implications in the precision oncology, J. Clin. Med. 8 (2019) 373—390.

[4] K. Pantel, C. Alix-Panabieres, Liquid biopsy and minimal residual disease—latest advances and implications for cure, Nat. Rev. Clin. Oncol. 16 (2019) 409—424.

[5] W. Li, H. Wang, Z. Zhao, H. Gao, C. Liu, L. Zhu, et al., Emerging nanotechnologies for liquid biopsy: the detection of circulating tumor cells and extracellular vesicles, Adv. Mater. 31 (2019) 1805344–1805369.
[6] E. Heitzer, I.S. Haque, C.E.S. Roberts, M.R. Speicher, Current and future perspectives of liquid biopsies in genomics-driven oncology, Nat. Rev. Genet. 20 (2019) 71–88.
[7] Y. Zhang, X. Mi, X. Tan, R. Xiang, Recent progress on liquid biopsy analysis using surface-enhanced Raman spectroscopy, Theranostics 9 (2019) 491–525.
[8] J. Wang, K.M. Koo, Y. Wang, M. Trau, Engineering state-of-the-art plasmonic nanomaterials for SERS-based clinical liquid biopsy applications, Adv. Sci. 6 (2019) 1900730–1900765.
[9] R. Vaidyanathan, R.H. Soon, P. Zhang, K. Jiang, C.T. Lim, Cancer diagnosis: from tumor to liquid biopsy and beyond, Lab Chip 19 (2018) 11–34.
[10] L.A. Diaz, A. Bardelli, Liquid biopsies: genotyping circulating tumor DNA, J. Clin. Oncol. 32 (2014) 579–586.
[11] E. Crowley, F. Di Nicolantonio, F. Loupakis, A. Bardelli, Liquid biopsy: monitoring cancer-genetics in the blood, Nat. Rev. Clin. Oncol. 10 (2013) 472–484.
[12] C. Sole, E. Arnaiz, L. Manterola, D. Otaegui, C.H. Lawrie, The circulating transcriptome as a source of cancer liquid biopsy biomarkers, Semin. Cancer Biol. 58 (2019) 100–108.
[13] M. Cui, H. Wang, X. Yao, D. Zhang, Y. Xie, R. Cui, et al., Circulating microRNAs in cancer: potential and challenge, Front. Genet. 10 (2019) 626.
[14] E.G. Trams, C.J. Lauter, N. Salem, U. Heine, Exfoliation of membrane ecto enzymes in the form of micro vesicles, Biochim. Biophys. Acta 645 (1981) 63–70.
[15] Y.L. Tai, P.Y. Chu, B.H. Lee, K.C. Chen, C.Y. Yang, W.H. Kuo, et al., Basics and applications of tumor-derived extracellular vesicles, J. Biomed. Sci. 26 (2019) 35–52.
[16] M.S. Panagopoulou, A.W. Wark, D.J.S. Birch, C.D. Gregory, Phenotypic analysis of extracellular vesicles: a review on the applications of fluorescence, J. Extracell. Vesicles 9 (1) (2020) 1710020.
[17] A.S. McDaniel, R. Ferraldeschi, R. Krupa, M. Landers, R. Graf, J. Louw, et al., Phenotypic diversity of circulating tumour cells in patients with metastatic castration-resistant prostate cancer, BJU Int. 120 (2017) E30–E44.
[18] M. Wallwiener, A.D. Hartkopf, S. Riethdorf, J. Nees, M.R. Sprick, B. Schonfisch, et al., The impact of HER2 phenotype of circulating tumor cells in metastatic breast cancer: a retrospective study in 107 patients, BMC Cancer 15 (2015) 403–409.
[19] F.A. Coumans, S.T. Ligthart, J.W. Uhr, L.W. Terstappen, Challenges in the enumeration and phenotyping of CTC, Clin. Cancer Res. 18 (2012) 5711–5718.
[20] H. Liu, C. Yang, X. Zhao, J. Le, G. Wu, J. Wei, et al., Genotyping on ctDNA identifies shifts in mutation spectrum between newly diagnosed and relapse/refractory DLBCL, OncoTargets Ther. 13 (2020) 10797–10806.
[21] N. Bubnoff, Liquid biopsy: approaches to dynamic genotyping in cancer, Oncol. Res. Treat. 40 (2017) 409–416.
[22] G. Siravegna, A. Bardelli, Blood circulating tumor DNA for non-invasive genotyping of colon cancer patients, Mol. Oncol. 10 (2016) 475–480.
[23] M. Giuliano, A. Shaikh, H.C. Lo, G. Arpino, S. De Placido, X.H. Zhang, et al., Perspective on circulating tumor cell clusters: why it takes a village to metastasize, Cancer Res. 78 (2018) 845–852.
[24] A. Balakrishnan, D. Koppaka, A. Anand, B. Deb, G. Grenci, V. Viasnoff, et al., Circulating tumor cell cluster phenotype allows monitoring response to treatment and predicts survival, Sci. Rep. 9 (2019) 7933.
[25] A. Kulasinghe, J. Zhou, L. Kenny, I. Papautsky, C. Punyadeera, Capture of circulating tumour cell clusters using straight microfluidic chips, Cancers 11 (2019) 89–100.

[26] S. Valastyan, R.A. Weinberg, Tumor metastasis: molecular insights and evolving paradigms, Cell 147 (2) (2011) 275–292.
[27] V. Plaks, C.D. Koopman, Z. Werb, Circulating tumor cells, Science 341 (2013) 1186–1188.
[28] D.S. Micalizzi, S. Maheswaran, D.A. Haber, A conduit to metastasis circulating tumor cell biology, Genes Dev. 31 (2017) 1827–1840.
[29] W.C. Wang, X.F. Zhang, J. Peng, X.F. Li, A.L. Wang, Y.Q. Bie, et al., Survival mechanisms and influence factors of circulating tumor cells, Biomed. Res. Int. 2018 (2018) 6304701–6304710.
[30] A. Kowalik, M. Kowalewska, S. Gozdz, Current approaches for avoiding the limitations of circulating tumor cells detection methods-implications for diagnosis and treatment of patients with solid tumors, Transl. Res. 185 (2017) 58–84.
[31] C. Zhang, Y. Guan, Y. Sun, D. Ai, Q. Guo, Tumor heterogeneity and circulating tumor cells, Cancer Lett. 374 (2016) 216–223.
[32] T. Mamdouhi, J.D. Twomey, K.M. McSweeney, B. Zhang, Fugitives on the run: circulating tumor cells (CTCs) in metastatic diseases, Cancer Metastasis Rev. 38 (2019) 297–305.
[33] M. Hashimoto, F. Tanaka, K. Yoneda, T. Takuwa, A. Kuroda, S. Matsumoto, et al., The clinical value of circulating tumour cells (CTCs) in patients undergoing pulmonary metastasectomy for metastatic colorectal cancer, J. Thorac. Dis. 10 (2018) 1569–1577.
[34] Z. Shen, A. Wu, X. Chen, Current detection technologies for circulating tumor cells, Chem. Soc. Rev. 46 (2017) 2038–2056.
[35] F.S. Iliescu, W.J. Sim, H. Heidari, D.P. Poenar, J. Miao, H.K. Taylor, et al., Highlighting the uniqueness in dielectrophoretic enrichment of circulating tumor cells, Electrophoresis 40 (2019) 1457–1477.
[36] T. Li, N. Li, Y. Ma, Y.J. Bai, C.M. Xing, Y.K. Gong, A blood cell repelling and tumor cell capturing surface for high-purity enrichment of circulating tumor cells, J. Mater. Chem. B 7 (2019) 6087–6098.
[37] C. Alix-Panabieres, K. Pantel, Circulating tumor cells: liquid biopsy of cancer, Clin. Chem. 59 (2013) 110–118.
[38] M.M. Ferreira, V.C. Ramani, S.S. Jeffrey, Circulating tumor cell technologies, Mol. Oncol. 10 (2016) 374–394.
[39] W.J. Allard, J. Matera, M.C. Miller, M. Repollet, M.C. Connelly, R. Chandra, et al., Tumor cells circulate in the peripheral blood of all major carcinomas but not in healthy subjects or patients with nonmalignant diseases, Clin. Cancer Res. 10 (2004) 6897–6904.
[40] D.L. Adams, R.K. Alpaugh, S. Tsai, C.M. Tang, S. Stefansson, Multi-phenotypic subtyping of circulating tumor cells using sequential fluorescent quenching and restaining, Sci. Rep. 6 (2016) 33488–33497.
[41] C.H. Chu, R. Liu, T. Ozkaya-Ahmadov, M. Boya, B.E. Swain, J.M. Owens, et al., Hybrid negative enrichment of circulating tumor cells from whole blood in a 3D-printed monolithic device, Lab Chip 19 (2019) 3427–3437.
[42] L. Chen, A.M. Bode, Z. Dong, Circulating tumor cells: Moving biological insights into detection, Theranostics 7 (2017) 2606–2619.
[43] R. Tang, R. Hu, X. Jiang, F. Lu, LHRH-targeting surface-enhanced Raman scattering tags for the rapid detection of circulating tumor cells, Sens. Actuators B: Chem. 284 (2019) 468–474.
[44] T.M. Morgan, X. Wang, X. Qian, J.M. Switchenko, S. Nie, K.R. Patel, et al., Measurement of circulating tumor cells in squamous cell carcinoma of the head and neck and patient outcomes, Clin. Transl. Oncol. 21 (2019) 342–347.

[45] Y. Zhang, P. Yang, M.A. Habeeb Muhammed, S.K. Alsaiari, B. Moosa, A. Almalik, et al., Tunable and linker free nanogaps in core-shell plasmonic nanorods for selective and quantitative detection of circulating tumor cells by SERS, ACS Appl. Mater. Interfaces 9 (2017) 37597–37605.
[46] X. Wu, L. Luo, S. Yang, X. Ma, Y. Li, C. Dong, et al., Improved SERS nanoparticles for direct detection of circulating tumor cells in the blood, ACS Appl. Mater. Interfaces 7 (2015) 9965–9971.
[47] Z.A. Nima, M. Mahmood, Y. Xu, T. Mustafa, F. Watanabe, D.A. Nedosekin, et al., Circulating tumor cell identification by functionalized silver-gold nanorods with multicolor, super-enhanced SERS and photothermal resonances, Sci. Rep. 4 (2014) 4752–4760.
[48] A. Kaminska, T. Szymborski, E. Witkowska, E. Kijenska-Gawronska, W. Swieszkowski, K. Nicinski, et al., Detection of circulating tumor cells using membrane-based SERS platform: a new diagnostic approach for 'liquid biopsy', Nanomaterials 9 (2019) 366–381.
[49] Y. Pang, C. Wang, R. Xiao, Z. Sun, Dual-selective and dual-enhanced SERS nanoprobes strategy for circulating hepatocellular carcinoma cells selection, Chemistry 24 (2018) 7060–7067.
[50] T. Xue, S. Wang, G. Ou, Y. Li, H. Ruan, Z. Li, et al., Detection of circulating tumor cells based on improved SERS-active magnetic nanoparticles, Anal. Methods 11 (2019) 2918–2928.
[51] H. Dong, D. Yao, Q. Zhou, L. Zhang, Y. Tian, An integrated platform for the capture of circulating tumor cells and in situ SERS profiling of membrane proteins through rational spatial organization of multi-functional cyclic RGD nanopatterns, Chem. Commun. 55 (2019) 1730–1733.
[52] S.C. Tsao, J. Wang, Y. Wang, A. Behren, J. Cebon, M. Trau, Characterising the phenotypic evolution of circulating tumour cells during treatment, Nat. Commun. 9 (2018) 1482–1492.
[53] X. Wang, X. Qian, J.J. Beitler, Z.G. Chen, F.R. Khuri, M.M. Lewis, et al., Detection of circulating tumor cells in human peripheral blood using surface-enhanced Raman scattering nanoparticles, Cancer Res. 71 (2011) 1526–1532.
[54] N.M. Namee, L. O'Driscoll, Extracellular vesicles and anti-cancer drug resistance, Biochim. Biophys. Acta 2018 (1870) 123–136.
[55] M.C. Cufaro, D. Pieragostino, P. Lanuti, C. Rossi, I. Cicalini, L. Federici, et al., Extracellular vesicles and their potential use in monitoring cancer progression and therapy: the contribution of proteomics, J. Oncol. 2019 (2019) 1639854–1639873.
[56] G. Raposo, W. Stoorvogel, Extracellular vesicles: exosomes, microvesicles, and friends, J. Cell Biol. 200 (2013) 373–383.
[57] P. Penfornis, K.C. Vallabhaneni, J. Whitt, R. Pochampally, Extracellular vesicles as carriers of microRNA, proteins and lipids in tumor microenvironment, Int. J. Cancer 138 (2016) 14–21.
[58] C.P. O'Neill, K.E. Gilligan, R.M. Dwyer, Role of extracellular vesicles (EVs) in cell stress response and resistance to cancer therapy, Cancers 11 (2019) 136–150.
[59] M. Malloci, L. Perdomo, M. Veerasamy, R. Andriantsitohaina, G. Simard, M.C. Martinez, Extracellular vesicles: mechanisms in human health and disease, Antioxid. Redox Signal. 30 (2019) 813–856.
[60] J.M. Pitt, G. Kroemer, L. Zitvogel, Extracellular vesicles: masters of intercellular communication and potential clinical interventions, J. Clin. Invest. 126 (2016) 1139–1143.
[61] R. Shah, T. Patel, J.E. Freedman, Circulating extracellular vesicles in human disease, N. Engl. J. Med. 379 (2018) 958–966.

[62] W. Li, C. Li, T. Zhou, X. Liu, X. Liu, X. Li, et al., Role of exosomal proteins in cancer diagnosis, Mol. Cancer 16 (2017) 145–157.
[63] M.P. Bebelman, M.J. Smit, D.M. Pegtel, S.R. Baglio, Biogenesis and function of extracellular vesicles in cancer, Pharmacol. Ther. 188 (2018) 1–11.
[64] M. Venturella, F.M. Carpi, D. Zocco, Standardization of blood collection and processing for the diagnostic use of extracellular vesicles, Curr. Pathobiol. Rep. 7 (2019) 1–8.
[65] M.Y. Konoshenko, E.A. Lekchnov, A.V. Vlassov, P.P. Laktionov, Isolation of extracellular vesicles: general methodologies and latest trends, Biomed. Res. Int. 2018 (2018) 8545347–8545374.
[66] M.I. Ramirez, M.G. Amorim, C. Gadelha, I. Milic, J.A. Welsh, V.M. Freitas, et al., Technical challenges of working with extracellular vesicles, Nanoscale 10 (2018) 881–906.
[67] E. Pariset, V. Agache, A. Millet, Extracellular vesicles: isolation methods, Adv. Biosyst. 1 (2017) 1700040–1700052.
[68] S. Gholizadeh, M. Shehata Draz, M. Zarghooni, A. Sanati-Nezhad, S. Ghavami, H. Shafiee, et al., Microfluidic approaches for isolation, detection, and characterization of extracellular vesicles: current status and future directions, Biosens. Bioelectron. 91 (2017) 588–605.
[69] H. Shao, H. Im, C.M. Castro, X. Breakefield, R. Weissleder, H. Lee, New technologies for analysis of extracellular vesicles, Chem. Rev. 118 (2018) 1917–1950.
[70] A. Pramanik, J. Mayer, S. Patibandla, K. Gates, Y. Gao, D. Davis, et al., Mixed-dimensional heterostructure material-based SERS for trace level identification of breast cancer-derived exosomes, ACS Omega 27 (2020) 16602–16611.
[71] Z. Yan, S. Dutta, Z. Liu, X. Yu, N. Mesgarzadeh, F. Ji, et al., A label-free platform for identification of exosomes from different sources, ACS Sens. 4 (2019) 488–497.
[72] A. Merdalimova, V. Chernyshev, D. Nozdriukhin, P. Rudakovskaya, D. Gorin, A. Yashchenok, Identification and analysis of exosomes by surface-enhanced Raman spectroscopy, Appl. Sci. 9 (2019) 1135–1156.
[73] J.C. Fraire, S. Stremersch, D. Bouckaert, T. Monteyne, T. De Beer, P. Wuytens, et al., Improved label-free identification of individual exosome-like vesicles with Au@Ag nanoparticles as SERS substrate, ACS Appl. Mater. Interfaces 11 (2019) 39424–39435.
[74] J. Carmicheal, C. Hayashi, X. Huang, L. Liu, Y. Lu, A. Krasnoslobodtsev, et al., Label-free characterization of exosome via surface enhanced Raman spectroscopy for the early detection of pancreatic cancer, Nanomedicine 16 (2019) 88–96.
[75] H. Shin, H. Jeong, J. Park, S. Hong, Y. Choi, Correlation between cancerous exosomes and protein markers based on surface-enhanced Raman spectroscopy (SERS) and principal component analysis (PCA), ACS Sens. 3 (2018) 2637–2643.
[76] K. Sivashanmugan, W.L. Huang, C.H. Lin, J.D. Liao, C.C. Lin, W.C. Su, et al., Bimetallic nanoplasmonic gap-mode SERS substrate for lung normal and cancer-derived exosomes detection, J. Taiwan Inst. Chem. Eng. 80 (2017) 149–155.
[77] S. Stremersch, M. Marro, B.E. Pinchasik, P. Baatsen, A. Hendrix, S.C. De Smedt, et al., Identification of individual exosome-like vesicles by surface enhanced Raman spectroscopy, Small 12 (2016) 3292–3301.
[78] W. Zhang, L. Jiang, R.J. Diefenbach, D.H. Campbell, B.J. Walsh, N.H. Packer, et al., Enabling sensitive phenotypic profiling of cancer-derived small extracellular vesicles using surface-enhanced Raman spectroscopy nanotags, ACS Sens. 5 (2020) 764–771.
[79] J. Wang, A. Wuethrich, A.A. Sina, R.E. Lane, L.L. Lin, Y. Wang, et al., Tracking extracellular vesicle phenotypic changes enables treatment monitoring in melanoma, Sci. Adv. 6 (2020) 3223–3237.

[80] C. Ning, L. Wang, Y. Tian, B. Yin, B. Ye, Multiple and sensitive SERS detection of cancer-related exosomes based on gold–silver bimetallic nanotrepangs, Analyst 145 (2020) 2795–2804.
[81] Y. Tian, C. Ning, F. He, B. Yin, B. Ye, Highly sensitive detection of exosomes by SERS using gold nanostar@Raman reporter@nanoshell structures modified with a bivalent cholesterol-labeled DNA anchor, Analyst 143 (2018) 4915–4922.
[82] Z. Wang, S. Zong, Y. Wang, N. Li, L. Li, J. Lu, et al., Screening and multiple detection of cancer exosomes using an SERS-based method, Nanoscale 10 (2018) 9053–9062.
[83] E.A. Kwizera, R. O'Connor, V. Vinduska, M. Williams, E.R. Butch, S.E. Snyder, et al., Molecular detection and analysis of exosomes using surface-enhanced Raman scattering gold nanorods and a miniaturized device, Theranostics 8 (2018) 2722–2738.
[84] T. Li, R. Zhang, H. Chen, Z. Huang, X. Ye, H. Wang, et al., An ultrasensitive polydopamine bi-functionalized SERS immunoassay for exosome-based diagnosis and classification of pancreatic cancer, Chem. Sci. 9 (2018) 5372–5382.
[85] J. Park, M. Hwang, B. Choi, H. Jeong, J.H. Jung, H.K. Kim, et al., Exosome classification by pattern analysis of surface-enhanced Raman spectroscopy data for lung cancer diagnosis, Anal. Chem. 89 (2017) 6695–6701.
[86] N. Bellassai, G. Spoto, Biosensors for liquid biopsy: circulating nucleic acids to diagnose and treat cancer, Anal. Bioanal. Chem. 408 (2016) 7255–7264.
[87] Z. Qin, V.A. Ljubimov, C. Zhou, Y. Tong, J. Liang, Cell-free circulating tumor DNA in cancer, Chin. J. Cancer 35 (2016) 36–45.
[88] D. Chu, B.H. Park, Liquid biopsy: unlocking the potentials of cell-free DNA, Virchows Arch. 471 (2017) 147–154.
[89] V. Armand-Labit, A. Pradines, Circulating cell-free microRNAs as clinical cancer biomarkers, Biomol. Concepts 8 (2017) 61–81.
[90] J. Donaldson, B.H. Park, Circulating tumor DNA: measurement and clinical utility, Annu. Rev. Med. 69 (2018) 223–234.
[91] S. Salvi, G. Gurioli, U. De Giorgi, V. Conteduca, G. Tedaldi, D. Calistri, et al., Cell-free DNA as a diagnostic marker for cancer: current insights, OncoTargets Ther. 9 (2016) 6549–6559.
[92] X. Han, J. Wang, Y. Sun, Circulating tumor DNA as biomarkers for cancer detection, Genom. Proteom. Bioinform. 15 (2017) 59–72.
[93] N.O. Tuaeva, L. Falzone, Y.B. Porozov, A.E. Nosyrev, V.M. Trukhan, L. Kovatsi, et al., Translational application of circulating DNA in oncology: review of the last decades achievements, Cells 8 (2019) 1251–1281.
[94] A. Kustanovich, R. Schwartz, T. Peretz, A. Grinshpun, Life and death of circulating cell-free DNA, Cancer Biol. Ther. 20 (2019) 1057–1067.
[95] L.B. Ahlborn, O. Østrup, Toward liquid biopsies in cancer treatment: application of circulating tumor DNA, APMIS: Acta Pathol. Microbiol. Immunol. Scand. 127 (2019) 329–336.
[96] M. Elazezy, S.A. Joosse, Techniques of using circulating tumor DNA as a liquid biopsy component in cancer management, Comput. Struct. Biotechnol. J. 16 (2018) 370–378.
[97] F. Urabe, J. Matsuzaki, Y. Yamamoto, T. Kimura, T. Hara, M. Ichikawa, et al., Large-scale circulating microRNA profiling for the liquid biopsy of prostate cancer, Clin. Cancer Res. 25 (2019) 3016–3025.
[98] L. Giannopoulou, M. Zavridou, S. Kasimir-Bauer, E.S. Lianidou, Liquid biopsy in ovarian cancer: the potential of circulating miRNAs and exosomes, Transl. Res. 205 (2019) 77–91.
[99] P.S. Mitchell, R.K. Parkin, E.M. Kroh, B.R. Fritz, S.K. Wyman, E.L. Pogosova-Agadjanyan, et al., Circulating microRNAs as stable blood-based markers for cancer detection, Proc. Natl. Acad. Sci. U.S.A. 105 (2008) 10513–10518.

[100] X. Wang, D. Kong, C. Wang, X. Ding, L. Zhang, M. Zhao, et al., Circulating microRNAs as novel potential diagnostic biomarkers for ovarian cancer: a systematic review and updated meta-analysis, J. Ovarian Res. 12 (2019) 1–12.
[101] J. Wang, J. Chen, S. Sen, MicroRNA as biomarkers and diagnostics, J. Cell. Physiol. 231 (2016) 25–30.
[102] R. Singh, B. Ramasubramanian, S. Kanji, A.R. Chakraborty, S.J. Haque, A. Chakravarti, Circulating microRNAs in cancer: hope or hype? Cancer Lett. 381 (2016) 113–121.
[103] S. Morganti, P. Tarantino, E. Ferraro, P. D'Amico, G. Viale, D. Trapani, et al., Complexity of genome sequencing and reporting: next generation sequencing (NGS) technologies and implementation of precision medicine in real life, Crit. Rev. Oncol. Hematol. 133 (2019) 171–182.
[104] G.J. Netto, L.K. Karen, Genomic Applications in Pathology, Springer, Switzerland, 2019.
[105] T.Y. Low, M.A. Mohtar, M.Y. Ang, R. Jamal, Connecting proteomics to next-generation sequencing: proteogenomics and its current applications in biology, Proteomics 19 (2019) 1–10.
[106] H. El-Achi, J.D. Khoury, S. Loghavi, Liquid biopsy by next-generation sequencing: a multimodality test for management of cancer, Curr. Hematol. Malig. Rep. 14 (2019) 358–367.
[107] S. Serrati, S. De Summa, B. Pilato, D. Petriella, R. Lacalamita, S. Tommasi, et al., Next-generation sequencing: advances and applications in cancer diagnosis, OncoTargets Ther. 9 (2016) 7355–7365.
[108] Y. Bai, Z. Wang, Z. Liu, G. Liang, W. Gu, Q. Ge, Technical progress in circulating tumor DNA analysis using next generation sequencing, Mol. Cell. Probes 49 (2020) 101480–101488.
[109] D. Qin, Next-generation sequencing and its clinical application, Cancer Biol. Med. 16 (2019) 4–10.
[110] X. Li, M. Ye, W. Zhang, D. Tan, N. Jaffrezic-Renault, X. Yang, et al., Liquid biopsy of circulating tumor DNA and biosensor applications, Biosens. Bioelectron. 126 (2019) 596–607.
[111] X. Mao, C. Liu, H. Tong, Y. Chen, K. Liu, Principles of digital PCR and its applications in current obstetrical and gynecological disease, J. Transl. Res. 11 (2019) 7209–7222.
[112] R.M. Trigg, L.J. Martinson, S. Parpart-Li, J.A. Shaw, Factors that influence quality and yield of circulating-free DNA: a systematic review of the methodology literature, Heliyon 4 (2018) e00699.
[113] T. Kurinomaru, R. Kurita, Bisulfite-free approaches for DNA methylation profiling, Anal. Methods 9 (2017) 1537–1549.
[114] D. Dietrich, Current status and future perspectives of circulating cell-free DNA methylation in clinical diagnostics, Lab. Med. 40 (2016) 335–343.
[115] W. Zhou, Y. Tian, B. Yin, B. Ye, Simultaneous surface-enhanced Raman spectroscopy detection of multiplexed microRNA biomarkers, Anal. Chem. 89 (2017) 6120–6128.
[116] Y. Liu, N. Lyu, V.K. Rajendran, J. Piper, A. Rodger, Y. Wang, Sensitive and direct DNA mutation detection by surface-enhanced Raman spectroscopy using rational designed and tunable plasmonic nanostructures, Anal. Chem. 92 (2020) 5708–5716.
[117] E.J. Wee, Y. Wang, S.C. Tsao, M. Trau, Simple, sensitive and accurate multiplex detection of clinically important melanoma DNA mutations in circulating tumour DNA with SERS nanotags, Theranostics 6 (2016) 1506–1513.
[118] Q. Zhou, J. Zheng, Z. Qing, M. Zheng, J. Yang, S. Yang, et al., Detection of circulating tumor dna in human blood via DNA-mediated surface-enhanced Raman spectroscopy of single-walled carbon nanotubes, Anal. Chem. 88 (2016) 4759–4765.

[119] A. Kowalczyk, J. Krajczewski, A. Kowalik, J.L. Weyher, I. Dzięcielewski, M. Chłopek, et al., New strategy for the gene mutation identification using surface enhanced Raman spectroscopy (SERS), Biosens. Bioelectron. 132 (2019) 326–332.
[120] J. Zhang, Y. Dong, W. Zhu, D. Xie, Y. Zhao, D. Yang, et al., Ultrasensitive detection of circulating tumor DNA of lung cancer via an enzymatically amplified SERS-based frequency shift assay, ACS Appl. Mater. Interfaces 11 (2019) 18145–18152.
[121] D. Lin, T. Gong, Z.Y. Hong, S. Qiu, J. Pan, C.Y. Tseng, et al., Metal carbonyls for the biointerference-free ratiometric surface-enhanced Raman spectroscopy-based assay for cell-free circulating DNA of Epstein-Barr virus in blood, Anal. Chem. 90 (2018) 7139–7147.
[122] N. Lyu, V.K. Rajendran, R.J. Diefenbach, K. Charles, S.J. Clarke, A. Engel, et al., Multiplex detection of ctDNA mutations in plasma of colorectal cancer patients by PCR/SERS assay, Nanotheranostics 4 (2020) 224–232.
[123] X. Li, T. Yang, C.S. Li, Y. Song, H. Lou, D. Guan, et al., Surface enhanced Raman spectroscopy (SERS) for the multiplex detection of Braf, Kras, and Pik3ca mutations in plasma of colorectal cancer patients, Theranostics 8 (2018) 1678–1689.
[124] X. Li, T. Yang, C.S. Li, Y. Song, D. Wang, L. Jin, et al., Polymerase chain reaction-surface-enhanced Raman spectroscopy (PCR-SERS) method for gene methylation level detection in plasma, Theranostics 10 (2020) 898–909.
[125] V. Moisoiu, A. Stefancu, S.D. Iancu, T. Moisoiu, L. Loga, L. Dican, et al., SERS assessment of the cancer-specific methylation pattern of genomic DNA: towards the detection of acute myeloid leukemia in patients undergoing hematopoietic stem cell transplantation, Anal. Bioanal. Chem. 411 (2019) 7907–7913.
[126] H. Shao, H. Lin, Z. Guo, J. Lu, Y. Jia, M. Ye, et al., A multiple signal amplification sandwich-type SERS biosensor for femtomolar detection of miRNA, Biosens. Bioelectron. 143 (2019) 111616.
[127] J. Zheng, D. Ma, M. Shi, J. Bai, Y. Li, J. Yang, et al., A new enzyme-free quadratic SERS signal amplification approach for circulating microRNA detection in human serum, Chem. Commun. 51 (2015) 16271–16274.
[128] L. Cheng, Z. Zhang, D. Zuo, W. Zhu, J. Zhang, Q. Zeng, et al., Ultrasensitive detection of serum microRNA using branched DNA-based SERS platform combining simultaneous detection of alpha-fetoprotein for early diagnosis of liver cancer, ACS Appl. Mater. Interfaces 10 (2018) 34869–34877.
[129] J. Li, M.K. Koo, Y. Wang, M. Trau, Native microRNA targets trigger self-assembly of nanozyme-patterned hollowed nanocuboids with optimal interparticle gaps for plasmonic-activated cancer detection, Small 15 (2019) 1904689–1904698.
[130] R. Zhang, K. Fan, X. Yan, Nanozymes: created by learning from nature, Sci. China Life Sci. 63 (2020) 1183–1200.
[131] K.M. Koo, E.J. Wee, P.N. Mainwaring, Y. Wang, M. Trau, Toward precision medicine: a cancer molecular subtyping nano-strategy for RNA biomarkers in tumor and urine, Small 12 (45) (2016) 6233–6242.
[132] K.M. Koo, J. Wang, R.S. Richards, A. Farrell, J.W. Yaxley, H. Samaratunga, et al., Design and clinical verification of surface-enhanced Raman spectroscopy diagnostic technology for individual cancer risk prediction, ACS Nano 12 (2018) 8362–8371.
[133] J. Wang, K.M. Koo, E.J. Wee, Y. Wang, M. Trau, A nanoplasmonic label-free surface-enhanced Raman scattering strategy for non-invasive cancer genetic subtyping in patient samples, Nanoscale 9 (2017) 3496–3503.
[134] N. Feliu, M. Hassan, E. Garcia Rico, D. Cui, W. Parak, R. Alvarez-Puebla, SERS quantification and characterization of proteins and other biomolecules, Langmuir 33 (2017) 9711–9730.

[135] B. Yang, S. Jin, Y. Wang, H. Bao, J. Sun, L. Chen, et al., Disease-related proteins determination based on surface-enhanced Raman spectroscopy, Appl. Spectrosc. Rev. 54 (2019) 856–872.
[136] C.L. Masters, R. Bateman, K. Blennow, C.C. Rowe, R.A. Sperling, J.L. Cummings, Alzheimer's disease, Nat. Rev. Dis. Primers 1 (2015) 1–18.
[137] H.B. Carter, Prostate-specific antigen (PSA) screening for prostate cancer: revisiting the evidence, J. Am. Med. Assoc. 319 (2018) 1866–1868.
[138] B. Sabbagh, S. Mindt, M. Neumaier, P. Findeisen, Clinical applications of MS-based protein quantification, Proteom. Clin. Appl. 10 (2016) 323–345.
[139] N.C. Merbel, Protein quantification by LC-MS a decade of progress through the pages of bioanalysis, Bioanalysis 11 (2019) 629–644.
[140] K. Merrell, K. Southwick, S.W. Graves, M.S. Esplin, N.E. Lewis, C.D. Thulin, Analysis of low-abundance, low-molecular-weight serum proteins using mass spectrometry, J. Biomol. Tech. 15 (2004) 238–248.
[141] Y. Lin, G. Xu, F. Wei, A. Zhang, J. Yang, Q. Hu, Detection of CEA in human serum using surface-enhanced Raman spectroscopy coupled with antibody-modified Au and gamma-Fe_2O_3@Au nanoparticles, J. Pharm. Biomed. Anal. 121 (2016) 135–140.
[142] H. Chon, S. Lee, S.-Y. Yoon, S.-I. Chang, D.W. Lim, J. Choo, Simultaneous immunoassay for the detection of two lung cancer markers using functionalized SERS nanoprobes, Chem. Commun. 47 (2011) 12515–12517.
[143] Z. Cheng, N. Choi, R. Wang, S. Lee, K.C. Moon, S.-Y. Yoon, et al., Simultaneous detection of dual prostate specific antigens using surface-enhanced Raman scattering-based immunoassay for accurate diagnosis of prostate cancer, ACS Nano 11 (2017) 4926–4933.
[144] X. Bai, L. Wang, J. Ren, X. Bai, L. Zeng, A. Shen, et al., Accurate clinical diagnosis of liver cancer based on simultaneous detection of ternary specific antigens by magnetic induced mixing surface-enhanced Raman scattering emissions, Anal. Chem. 91 (2019) 2955–2963.
[145] J. Feng, X. Wu, W. Ma, H. Kuang, L. Xu, C. Xu, A SERS active bimetallic core–satellite nanostructure for the ultrasensitive detection of Mucin-1, Chem. Commun. 51 (2015) 14761–14763.
[146] J. Liang, H. Liu, C. Huang, C. Yao, Q. Fu, X. Li, et al., Aggregated silver nanoparticles based surface-enhanced Raman scattering enzyme-linked immunosorbent assay for ultrasensitive detection of protein biomarkers and small molecules, Anal. Chem. 87 (2015) 5790–5796.
[147] B. Tang, J. Wang, J.A. Hutchison, L. Ma, N. Zhang, H. Guo, et al., Ultrasensitive, multiplex Raman frequency shift immunoassay of liver cancer biomarkers in physiological media, ACS Nano 10 (2016) 871–879.
[148] N. Banaei, A. Foley, J.M. Houghton, Y. Sun, B. Kim, Multiplex detection of pancreatic cancer biomarkers using a SERS-based immunoassay, Nanotechnology 28 (2017) 455101–455110.
[149] A.V. Krasnoslobodtsev, M.P. Torres, S. Kaur, I.V. Vlassiouk, R.J. Lipert, M. Jain, et al., Nano-immunoassay with improved performance for detection of cancer biomarkers, Nanomedicine 11 (2015) 167–173.
[150] H. Chang, H. Kang, E. Ko, B.H. Jun, H.Y. Lee, Y.S. Lee, et al., PSA detection with femtomolar sensitivity and a broad dynamic range using SERS nanoprobes and an area-scanning method, ACS Sens. 1 (2016) 645–649.
[151] G. Song, H. Zhou, J. Gu, Q. Liu, W. Zhang, H. Su, et al., Tumor marker detection using surface enhanced Raman spectroscopy on 3D Au butterfly wings, J. Mater. Chem. B 5 (2017) 1594–1600.
[152] M. Li, S.K. Cushing, J. Zhang, S. Suri, R. Evans, W.P. Petros, et al., Three-dimensional hierarchical plasmonic nano-architecture enhanced surface-enhanced

Raman scattering immunosensor for cancer biomarker detection in blood plasma, ACS Nano. 7 (2013) 4967–4976.

[153] S. Ansari, S. Masoum, Molecularly imprinted polymers for capturing and sensing proteins: current progress and future implications, Trends Anal. Chem. 114 (2019) 29–47.

[154] H.R. Culver, N.A. Peppas, Protein-imprinted polymers: the shape of things to come? Chem. Mater. 29 (2017) 5753–5761.

[155] J. Wang, D. Liang, Q. Jin, J. Feng, X. Tang, Bioorthogonal SERS nanotags as a precision theranostic platform for in vivo SERS imaging and cancer photothermal therapy, Bioconjug. Chem. 31 (2020) 182–193.

[156] X. Qian, S. Nie, Single-molecule and single-nanoparticle SERS: from fundamental mechanisms to biomedical applications, Chem. Soc. Rev. 37 (2008) 912–920.

[157] K.K. Maiti, U.S. Dinish, C.Y. Fu, J.J. Lee, K.S. Soh, S.W. Yun, et al., Development of biocompatible SERS nanotag with increased stability by chemisorption of reporter molecule for in vivo cancer detection, Biosens. Bioelectron. 26 (2010) 398–403.

[158] J. Wang, W. Anderson, J. Li, L.L. Lin, Y. Wang, M. Trau, A high-resolution study of in situ surface-enhanced Raman scattering nanotag behavior in biological systems, J. Colloid Interface Sci. 537 (2019) 536–546.

CHAPTER 7

Surface-enhanced Raman spectroscopy-based microfluidic devices for in vitro diagnostics

Anupam Das and Jaebum Choo
Department of Chemistry, Chung-Ang University, Seoul, Republic of Korea

7.1 Introduction

For timely treatment of different immune-related diseases, rapid and accurate quantification of specific disease biomarkers is essential [1]. Therefore biochemical researchers and medical scientists have invested considerable effort into developing reliable immunoassays or molecular diagnostic tools that enable highly sensitive detection of disease-associated biomarkers at very low concentration [2]. In the past few decades, immunoassay has been considered as one of the most reliable and efficient tools for detection of protein biomarkers [3–8]. To date, many different immunoassay approaches, such as radio immunoassay (RIA) [9], immune-chemiluminescence assay (ICMA) [10], and enzyme-linked immunosorbent assay (ELISA) [11–13] have been developed.

Among them, ELISA is the most extensively used immunoassay technique to detect the presence of a target antigen or its complementary antibody, and various ELISA kits have been commercialized for the detection of specific disease biomarkers. For ELISAs using a 96-well plate format, however, tedious manual dilution steps require repetitive pipetting. Moreover, this assay does not satisfy the requirements for accurate assessment of a specific biomarker at a low concentration. In ELISA, quantification of a biomarker protein is achieved by measuring the absorption intensity change caused by enzyme-mediated color changes, but a discernable color change cannot be induced at low concentrations of a target marker [14,15]. Consequently, an ELISA detection reader fails to distinguish the color change [16]. RIA also allows inexpensive and fast screening of a clinical biomarker. In a RIA, the amount of radiolabeled antigen that has bound to an antibody is determined

Principles and Clinical Diagnostic Applications of Surface-Enhanced Raman Spectroscopy
DOI: https://doi.org/10.1016/B978-0-12-821121-2.00006-8

by measuring the radioactivity of the antigen–antibody pellet, and then, the antigen concentration in the sample is estimated. In many cases, however, the use of a radioisotope (usually ^{125}I) restricts its application in practical detection of a biomarker in human serum. It is also reported that RIA and ELISA techniques are comparable in sensitivity and specificity for several test biomarkers [17,18]. ICMA is another immunoassay technique popularly used in clinical laboratories, but it does not satisfy the high sensitivity requirement similarly to RIA or ELISA, regardless of the many advantages, such as absence of interfering emissions, a wide dynamic range, reduced incubation time, and rapid acquisition time [19–21].

The use of liquid chromatography coupled with mass spectrometry (LC/MS) has also increased for more confident and accurate in vitro diagnostics, because it shows high sensitivity detection capability. For instance, LC/MS shows very low limit of detection (LOD) values, less than 10 pg/mL, for the p53 cancer biomarker and *N*, *N′*, *N″*-triacetylfusarinine C siderophore [22,23]. Despite this high sensitivity of detection, LC/MS is not suitable for use in routine clinical diagnostic applications because of several analytical disadvantages, including the need for complicated and expensive instrumentation and long sample preparation time. Therefore there is need for a new sensitive in vitro detection technique that can be used for routine clinical diagnostics.

In recent years, application of surface-enhanced Raman scattering (SERS)-based assay platforms has drawn a lot of attention owing to its high sensitivity and multiplex detection capability [24]. Raman scattering signals are greatly enhanced (reported to reach factors up to 10^{14} for certain systems) at SERS active junctions known as “hot spots” as a result of electromagnetic (EM) and chemical enhancement effects [25]. Such strong enhancement shows great promise in overcoming the low sensitivity problems inherent in fluorescence or luminescence detection in ELISA or ICMA techniques. Moreover, the SERS-based detection method does not require culturing or amplification steps that are essential in bacteria colony counting or PCR assays because of it its high sensitivity. To date, SERS-based immunoassays have been successfully used in the detection of a number of biomarkers, as listed in Table 7.1.

There has also been great progress in the development of microfluidics for use in in vitro diagnostics. The use of microfluidics in biological assays has several advantages, including reduced consumption of reagents and cost effectiveness [36], high sensitivity [37], precise regulation of experimental conditions [38], short reaction times, high analytical throughput

Table 7.1 List of biomarkers and corresponding LODs estimated by SERS-based immunoassay methods.

Biomarker	LOD	Reference
Tau protein in Alzheimer's disease	<25 fM	[26]
Botulinum toxins A and B	5.7 and 1.3 ng/mL	[27]
Carcinoembryonic antigen (CEA)	10 pg/mL	[28]
Feline calicivirus (FCV)	10^6 FCV/mL	[29]
Hepatitis B virus	0.5 μg/mL	[30]
Metanephrine	<10 μM	[31]
Mucin 4	33 ng/mL	[32]
Mycobacterium avium subspecies paratuberculosis	1000 MAP/mL	[33]
Orientia tsutsugamushi	49.5 (Titer)	[34]
Prostate-specific antigen	1 pg/mL (~30 fM)	[35]

and improved production conversion, over the macroscale analytical tools [39,40]. For application of microfluidic devices in assays, however, a sensitive detection tool is absolutely needed to monitor the progress of a biological reaction, such as an immunoreaction or DNA hybridization, because the sample volume used in a microfluidic device is extremely small (pL to nL). Since SERS is a highly sensitive detection modality, the integration of SERS with many different types of microfluidic devices is promising for the development of a new diagnostic platform for early disease diagnosis [41,42]. In this chapter, recent advances in the integration of a SERS modality with different types of microfluidic devices, such as polydimethylsiloxane (PDMS)-based microfluidic channels, paper-based-lateral flow assay (LFA) strips, and nano-patterned microarray-embedded channels, will be introduced.

7.2 Various surface-enhanced Raman spectroscopy-based microfluidic devices for in vitro diagnostics

To date, a number of microfluidic devices have been developed as in vitro diagnostics, and different types of readers combined with these microfluidic devices, are used for detection of various biological targets, such as, small molecules, nucleic acids, proteins, pathogens, and cells. Various recognition receptors, including antibodies, DNA aptamers, peptides, and DNA enzymes are also utilized to selectively capture different disease targets. In Fig. 7.1, various recognition receptors, microfluidic devices, and readout methods for the detection of target samples have

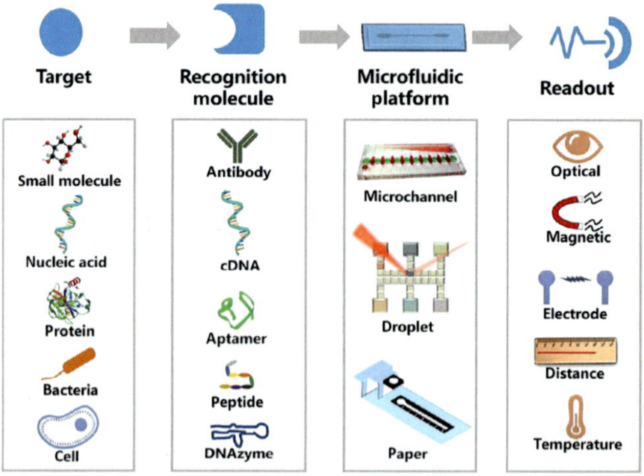

Figure 7.1 Schematic illustration of different types of target samples, recognition receptors, microfluidic devices, and readout methods. *Reprinted with permission from Y. Song, B. Lin, T. Tian, X. Xu, W. Wang, Q. Ruan, et al., Recent progress in microfluidics-based biosensing, Anal. Chem. 91 (2019) 388 – 404.*

been summarized [43]. In the following section, we will focus more on different types of microfluidic devices combined with an SERS detection system for in vitro diagnostic applications.

7.2.1 Application of paper-based microfluidics

The paper-based microfluidic device is also known as a LFA strip. This device is made of a hydrophilic cellulose or nitrocellulose fiber that helps a liquid flow from an inlet to a desired outlet through imbibition. In the test process, the mobile phase is pulled through the stationary phase by capillary force, and then, the mobile phase passes through a test line, where trapped antibody-conjugated gold nanoparticles, accumulate till they are visually detectable. This LFA technique has been extensively used in detecting many different types of disease biomarkers [44,45]. Some of the advantages of the paper-based assays include their portability, ease of use, short time to achieve test results, long-term stability, and low cost. However, the LFA strip has several drawbacks in terms of low sensitivity and limitations in quantitative analysis. Various optical readers with different detection labels have been used to resolve these problems. Among them, fluorescence readers have been the most widely utilized, with antibody-conjugated fluorescence labels being used for optical reading. Nonetheless, the fluorescence detection method suffers from lack of sensitivity, and a more sensitive detection method is needed. The SERS-based LFA technique is a promising analytical tool that can be used to overcome the sensitivity problem inherent in the fluorescence detection method.

Recently, Ren et al. reported a magnetic focus lateral flow sensor (mLFS) for the sensitive detection of cervical cancer biomarkers [46]. In this work, Fe_3O_4 core/Au shell nanoparticles with antibodies for target recognition were synthesized and utilized to investigate the magnetic focusing effect on SERS sensitivity. In this sensor, a low cost magnet was used to increase the residence time of the protein target, labeled with gold-encapsulated magnetic probes, at the detection zone. It was possible to control the movement of probe-labeled protein targets on the lateral flow strip by applying a magnetic field using an external magnet placed under the strip, as shown in Fig. 7.2. In this work, the valosin-containing protein, which is known as an important biomarker of cervical cancer,

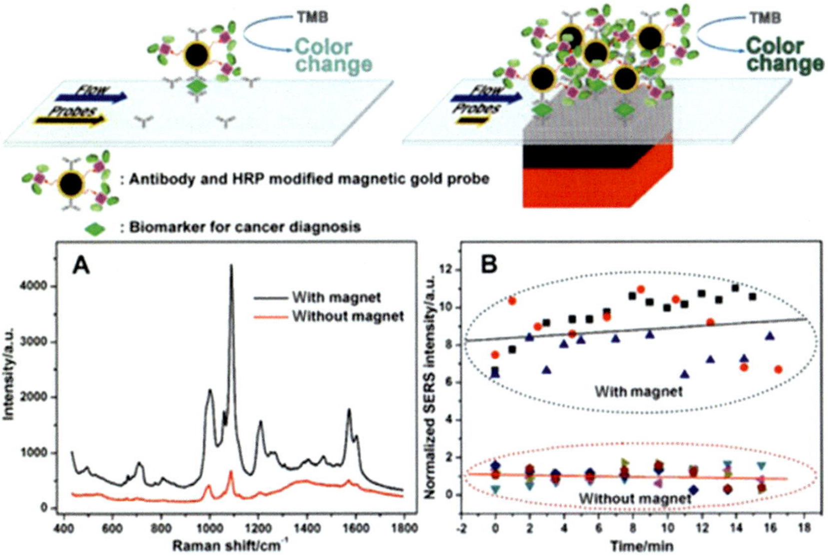

Figure 7.2 Schematic depicting the effect of magnetic focusing in a magnetic focus lateral flow sensor (mLFS). Without the magnet, the magnetic probe-labeled targets move along with the sample flow on the LF strip, resulting in low capture efficiency (top left). With the magnet, the probe-labeled targets are focused at the signal generation zone owing to magnetic focusing, thus increasing the capture efficiency of labeled targets (top right). SERS characterization of magnetic nanoparticles in the microchannel. (A) SERS signal after 15 min of sample addition; (B) normalized SERS intensity of the peak at 1091.5 cm^{-1} with respect to time from the application of solution with magnetic nanoparticles in the microchannel and the corresponding linear fit. Different symbols are from individual measurements. *Reprinted with permission from W. Ren, S.I. Mohammed, S. Wereley, J. Irudayaraj, Magnetic focus lateral flow sensor for detection of cervical cancer biomarkers, Anal. Chem. 91 (2019) 2876 – 2884.*

was tested as a target antigen, and the developed system could detect a concentration as low as 25 fg/mL with magnetic focusing to enhance target capture efficiency. The LOD of a target protein is 10^6-fold more sensitive than that of a conventional lateral flow (LF) system. A 6.0 × 0.5 cm sized lateral flow strip, assembled on plastic backing cards, was used for the experiment (Fig. 7.2). The strip was composed of four different segments: a nitrocellulose membrane, absorbent pad, sample pad, and conjugate pad. The nitrocellulose membrane was first attached to the plastic backing card, and then, the absorbent pad was attached at one end. The conjugate and sample pads were sequentially attached to the other side of the nitrocellulose membrane. For SERS detection, a microchannel was created using a glass capillary tube, and fixed on a glass slide to evaluate the behavior of SERS probes. The proposed magnetic focusing system showed improved sensitivity since a higher target concentration at the signal detection zone contributed to the increased capture efficiency.

Lee et al. developed a SERS-based lateral flow assay (SERS-LFA) strip for the accurate and rapid diagnosis of scrub typhus, which is an infectious disease caused by the Gram-negative intracellular bacterium *Orientia tsutsugamushi* (*O. tsutsugamushi*) [34]. *O. tsutsugamushi* is one of the most prevalent strains that cause human *Rickettsia* infection. This disease is mostly endemic across extensive parts of South-East Asia and the Pacific, with up to a 35% mortality rate. Currently, an immunofluorescence assay (IFA) is considered the gold standard serological assay technique for the diagnosis of *O. tsutsugamushi*. Here, specific antibodies in *O. tsutsugamushi* are detected based on increased titer concentration during the acute and convalescence phase of the disease. However, the lack of a numerical value as the cut-off point leads to a subjective decision of the exact titer concentration. Nested PCR replaces a serological assay in some cases, but it requires sophisticated instruments together with trained experts. To resolve this problem, a SERS-LFA diagnostic method was developed for the highly accurate quantitative analysis of IgG antibodies to *O. tsutsugamushi* (Fig. 7.3). Among various SERS-based assay platforms, the LFA is considered a cheap and easy handling point-of-care diagnostic technique. According to the experimental data [34], SERS-based assay results for 40 clinical samples (16 positive and 24 negative) demonstrate 100% specificity and 93.75% sensitivity. This result was comparable with the conventional IFA assay results, with 95.83% specificity and 93.75% sensitivity. SERS-LFA has many advantages over IFA, including a smaller sample volume, shorter assay time, simpler assay steps, and longer assay lifetime. The

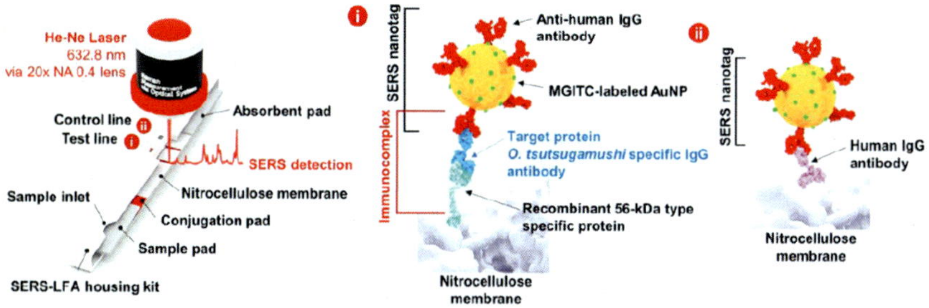

Figure 7.3 Schematic illustration of the SERS-LFA biosensor for the detection of *O. tsutsugamushi*-specific IgG antibodies. Recombinant proteins (56 kDa) are immobilized on the test line to capture IgG antibodies in human serum. (i) Sandwich immunocomplexes are formed when the antihuman IgG antibody-conjugated SERS nanotags bind to the target antibodies. (ii) Conversely, human IgG antibodies are immobilized on the control line and directly bind to the antibody-conjugated SERS nanotags. *Reprinted with permission from S.H. Lee, J. Hwang, K. Kim, J. Jeon, S. Lee, J. Ko, et al., Quantitative serodiagnosis of scrub typhus using surface-enhanced Raman scattering-based lateral flow assay platforms, Anal. Chem. 91 (2019) 12275 – 12282.*

proposed SERS-LFA technique meets the requirements of accurate quantification of IgG antibodies against *O. tsutsugamushi*.

In the SERS-LFA, selection of a SERS nanoprobe is important for sensitive detection of a specific biomarker. Particularly, it is well known that anisotropic geometries such as those of gold nanostars, provide stronger Raman activity than spherical geometries by producing a concentrated electromagnetic field at the tips of the anisotropic gold nanoparticles. Lin et al. successfully developed a highly sensitive SERS-LFA strip using gold nanostars (GNSc) instead of conventional gold nanospheres for the detection of bisphenol A (BPA) [47]. BPA is popularly used in food and customer product industries, but even a low-dose exposure to BPA adversely affects human health because it is a potent endocrine-disrupting chemical and a suspected carcinogen [48–50]. The authors used high-SERS-performance GNSc and 4-aminothiolphenol (4-ATP, the Raman reporter) to detect BPA. The HF102 type of nitrocellulose paper was selected as an optimal membrane because of its high sensitivity [51]. Because the molecular size of BPA is small, it only has a limited binding epitope. Consequently, the BPA LFA strip functions under a competitive format, as shown in Fig. 7.4. The LOD of this SERS-LFA strip for BPA was also demonstrated to be 20 and 205 times more sensitive than that of visual quantification and visual intensity, respectively. This BPA detection strip also has a wider detection range than commercially available LFA kits.

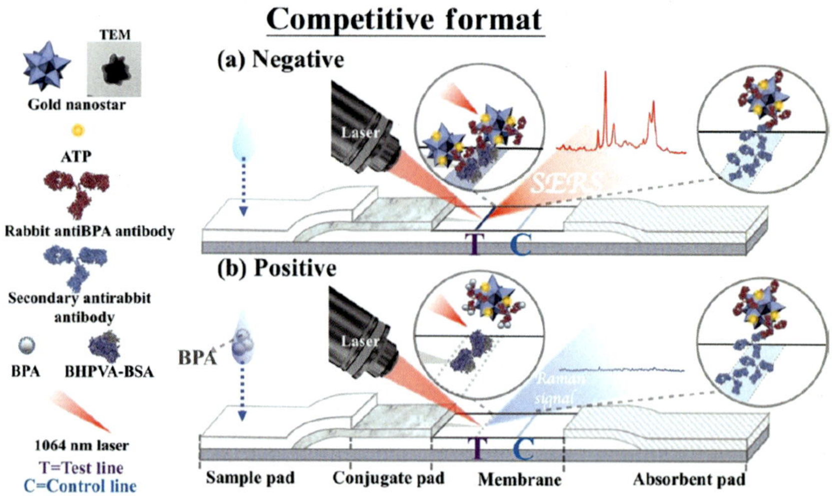

Figure 7.4 Schematic illustration of the bisphenol A (BPA) SERS-based lateral flow assay (SERS-LFA) test kit using a competitive format. (A) Negative response in the absence of BPA in the test sample leads to the appearance of color on the test (T) line. (B) Positive response in the presence of BPA in the test sample leads to the disappearance of color on the T line. *Reprinted with permission from L-K. Kin, L. A. Stanciu et al., Bisphenol A detection using gold nanostars in a SERS improved lateral flow immunochromatographic assay, Sens. Actuat. B-Chem. 276 (2018) 222-229.*

7.2.2 Magnetic particle-based microfluidics

Magnetic particles have been extensively used for immunoassays because of their ease of manipulation under a magnet [51]. The size of magnetic particles for immunoassays ranges from 5 to 100 nm, and they have a uniform distribution in aqueous solution. They can be easily concentrated by the use of an external magnetic field and transferred from one solution to another, which is why magnetic particles are widely used to isolate and concentrate various biomaterials such as bacteria, viruses, proteins, toxins, hormones, DNA, RNA, and even whole cells. Magnetic particles are also suitable as a substrate in optical detection because they have low optical signal background [52]. When magnetic particles are used as a substrate in fluorescence- or Raman-based optical assays, they provide several advantages compared to a two-dimensional substrate, such as a 96-well plate in ELISA. First, a three-dimensional immunoreaction pattern on the surface of a magnetic particle provides dramatically increased loading density of target molecules. Second, the reaction on a three-dimensional surface overcomes the slow reaction problem on a two-dimensional planar surface because of the diffusion-limited kinetic process. Finally, more reproducible quantification is possible because optical signals can be measured for an average nanoparticle ensemble in solution. For instance, quantitative evaluation of a target in SERS-based immunoassays could be determined by monitoring the characteristic Raman peak intensity of SERS nanotags on the surface of magnetic particles.

Yap et al. integrated SERS detection with a simple microfluidic channel in an efficient and low cost manner using magnetic particles (Fig. 7.5) [53]. For this purpose, a magnetic nanoparticle-based immunoassay system was designed and fabricated. Here, gold nanoparticles act as soluble SERS-reporting particles, whereas magnetic particles promote micromixing in a microfluidic chip. Using this SERS immunosubstrate, in conjunction with a simple microfluidic system, the assay time could be reduced from 4 h to 80 min, and detection specificity was enhanced by ~70%, compared to that in a nonmicrofluidic immunoassay. Compared to previous microfluidic SERS systems, the strategy in this work offers a simple microfluidic chip design with only one well for mixing, washing, and detection. Under optimized conditions, it was possible to attain an LOD of rabbit IgG as low as 1 pg/mL.

A wash-free magnetic immunoassay technique for the detection of prostate-specific antigen (PSA) using a SERS-based microdroplet sensor was developed by Gao et al. [54]. In this work, a magnetic bar, embedded in the droplet-based microfluidic system, segregates the bound and free

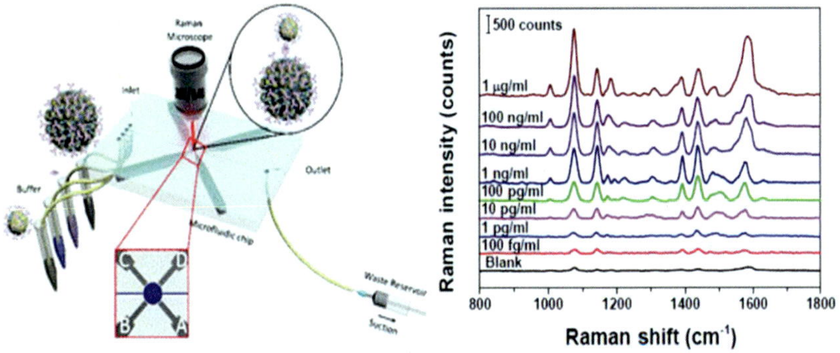

Figure 7.5 Schematic illustration of SERS-based immunoassay using a microfluidic chip (*left*). Microfluidic SERS immunoassay for rabbit IgG detection in the ~100 fg/mL − 1 μg/mL range (*right*). *Reprinted with permission from L.W. Yap, H. Chen, Y. Gao, K. Petkovic, Y. Liang, K.-J. Si, et al., Bifunctional plasmonic-magnetic particles for an enhanced microfluidic SERS immunoassay, Nanoscale 9 (2017) 7822–7829.*

SERS tags by splitting each droplet into two smaller parts (Fig. 7.6). SERS nanotags form immunocomplexes with PSA targets in one droplet, and fewer SERS tags remain in another supernatant solution droplet. Therefore SERS signal measurement enables quantitative evaluation of PSA markers. This approach provides a sensitive and rapid assay that is applicable for PSA cancer markers in clinical serum, without washing. For the quantitative evaluation of PSA markers, average SERS signals were measured at 174 droplets per minute. The LOD, determined by this approach, was estimated to be 0.1 ng/mL, which was significantly below the previously reported clinical cut-off value for the diagnosis of prostate

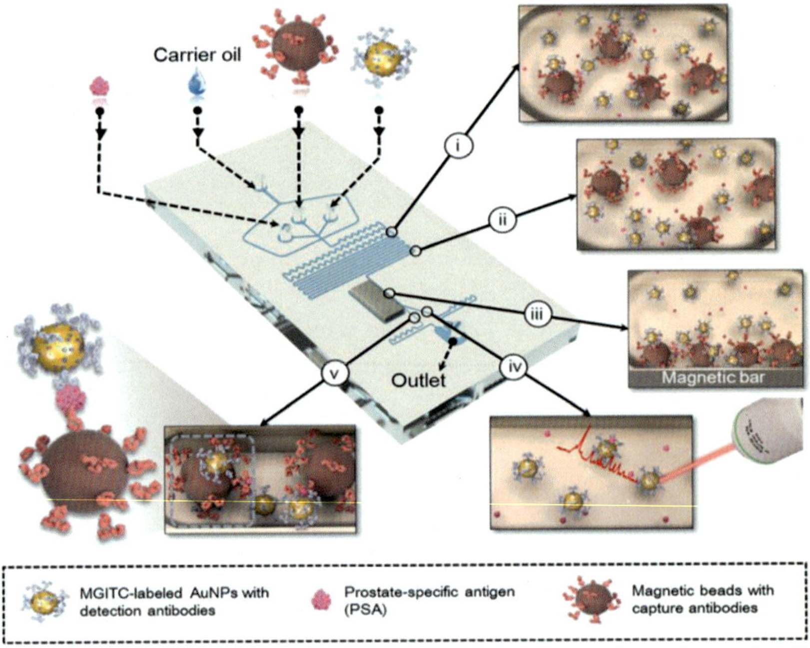

Figure 7.6 Schematic illustration of the SERS-based microdroplet sensor for the wash-free magnetic immunoassay. The sensor is composed of five compartments with the following functions. (i) Droplet generation and reagent mixing. (ii) Formation of magnetic immunocomplexes. (iii) Magnetic bar-mediated isolation of immunocomplexes (iv) Generation of larger droplets containing the supernatant for SERS detection. (v) Generation of smaller droplets containing magnetic immunocomplexes. MGITC refers to malachite green isothiocyanate; AuNPs refers to gold nanoparticles. *Reprinted with permission from R. Gao, Z. Cheng, A.J. deMello, J. Choo, Wash-free magnetic immunoassay of the PSA cancer marker using SERS and droplet microfluidics, Lab. Chip 16 (2016) 1022–1029.*

cancer. This SERS-based microdroplet sensor is a useful clinical tool for the early diagnosis of prostate cancer.

Xiong et al. reported a magnetic nanochain- (MagChain) integrated microfluidic chip for a broad range of applications [55]. In this study, the MagChain functioned as a nanoscale stir bar for rapid liquid mixing and as a capturing agent for a specific bioseparation, as shown in Fig. 7.7. Using these magnetic nanostructures, it was possible to simultaneously achieve highly efficient concerted liquid mixing and bioseparation in the MagChain-integrated microchip (MiChip). This also enabled a simple planar design of the MiChip, consisting of flat microchannels free of any built-in components. The use of the MagChain stir bar under continuous mixing conditions overcame the problem associated with diffusion-limited reaction kinetics inherent in conventional microfluidic systems. This approach, combined with SERS nanoprobes for signal transduction, allows ultrasensitive detection and quantification of cancer protein biomarkers and bacterial species with 1 μL of fluid in 8 min.

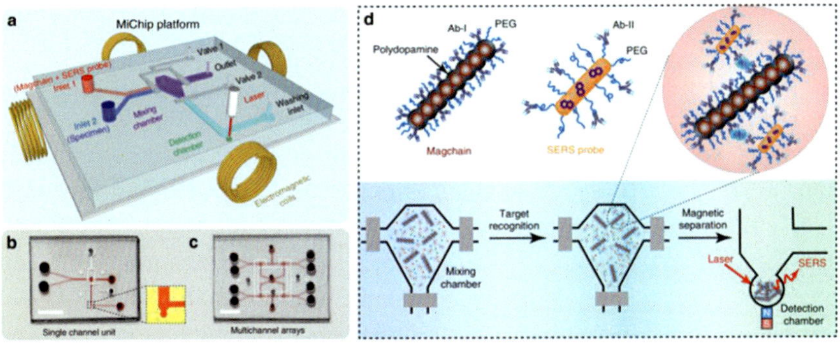

Figure 7.7 Design of the magnetic nanochain-integrated MiChip. (A) Schematic illustration of the MiChip assay platform. (B) and (C) Photographs of the MiChip: single channel unit (B) and multichannel arrays (C). The microchannel is filled with a red dye for better visualization. Scale bar: 0.5 cm. (D) The MiChip assay for the detection of biomarkers. The specimen, antibody-conjugated magnetic nanochains (Magchain), and SERS-encoded probes (SERS probe) are mixed in the mixing chamber. The targets of interest in the specimen are recognized by the antibodies on the Magchains and the SERS probes to form sandwich immune complexes. The immune complexes are then isolated in the detection chamber and subjected to Raman spectroscopic detection. *Reprinted with permission from Q. Xiong, C.Y. Lim, J. Ren, J. Zhou, K. Pu, M.B. Chan-Park, et al., Magnetic nanochain integrated microfluidic biochips, Nat. Commun. 9 (2018) 1743–1753.*

7.2.3 Gold-patterned microarray-embedded microfluidic platforms

A gold-patterned microarray chip has been widely used as a SERS-based immunoassay platform [56—58]. Here, a hybrid microarray pattern, composed of hydrophilic wells and other hydrophobic areas, was used as a sensing platform for SERS-based immunoassays. This plasmonics-based approach does not require an expensive array reader, which is essential for the conventional microarray-based assay because a hydrophilic target sample and SERS nanotags are automatically arranged on the surface of gold-patterned microwells. In addition, the LOD and dynamic range for detectable concentrations are greatly improved compared with those by the conventional ELISA. Nonetheless, this SERS-gold-patterned microarray chip-based immunoassay has some technical problems. First, it is hard to optimize uniform distribution of target markers in gold-patterned wells because all the sequential antibody/antigen immobilization and washing steps are difficult to control manually using a micropipette. Second, the reagents should be immobilized and reacted on the surface of the gold substrates in air. In many cases, however, the exposure of proteins to air seriously reduces their biological activity. Finally, the repetition of washing steps to remove nonbinding species makes this assay technique inconvenient. Thus this manually controlled process makes the SERS-based gold-patterned microarray platform less attractive.

To overcome these problems, Lee and coworkers [57] developed a fully automatic and programmable gold array-embedded gradient microfluidic chip, as shown in Fig. 7.8. In this work, gold-patterned microarray wells were embedded in a gradient microfluidic device. All sample distribution and washing steps could be automatically controlled using the gradient microfluidic technique. Moreover, a highly sensitive SERS analysis was possible if continuous flow and homogeneous mixing conditions were maintained in a microfluidic channel. Through this gradient approach, the tedious manual dilution process of repetitive pipetting and inaccurate dilution could be eliminated because various concentrations of target antigen were automatically generated by microfluidic gradient generators with N cascade-mixing stages. The total assay time from serial dilution to SERS detection took less than 60 min because all experimental conditions for the formation and detection of immunocomplexes could be automatically controlled inside the exquisitely designed microfluidic channel. This SERS-based microfluidic assay technique is expected to be a powerful clinical tool for fast and sensitive cancer marker detection.

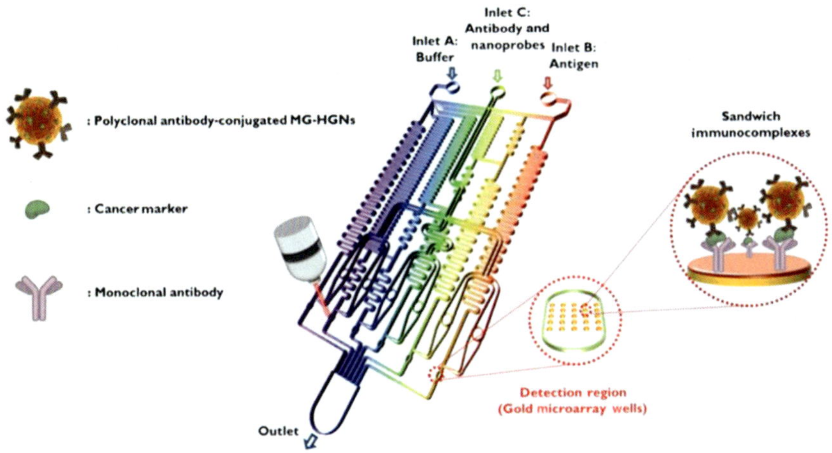

Figure 7.8 Layout of a gold array-embedded gradient chip for the SERS-based immunoassay. The illustrations in the enlarged circles represent the formation of sandwich immunocomplexes on the surface of 5 × 5 round gold wells embedded in the gradient channel. *Reprinted with permission from M. Lee, K. Lee, K.H. Kim, K.W. Oh, J. Choo, SERS-based immunoassay using a gold array embedded gradient microfluidic chip, Lab. Chip 12 (2012) 3720–3727.*

Wang and coworkers reported a SERS-based mapping technique for sensitive and reproducible analysis of multiple mycotoxins [58]. In many cases, the reproducibility issue caused by heterogeneous distribution of hot spots of a plasmonic substrate impeded the application of the SERS technique for quantitative analysis of target molecules. In the case of a two-dimensional plasmonic substrate, the Raman intensity value for quantitative evaluation was determined by averaging several randomly selected detection points, but the standard deviation was out of acceptable range in many cases. To resolve this problem, a Raman mapping technique was developed for the reliable quantification of mycotoxin molecules on a three-dimensional plasmonic substrate. For this purpose, three-dimensional (3D) gold nanopillar substrates were fabricated by thermal evaporation methods, and their reproducibility performance was tested. The high uniformity of the densely packed 3D nanopillars over the substrate minimizes spot-to-spot variations in the Raman signal intensity in a scanning area when Raman mapping is performed. According to the average Raman mapping data for 1368 pixel points, the LODs were estimated to be 5.09, 5.11, and 6.07 pg/mL for ochratoxin A, fumonisin B, and aflatoxin B1, and these values were approximately two orders of magnitude more sensitive than those determined using ELISAs.

7.2.4 Continuous-flow microfluidics

The category of microfluidics can be classified into two types; one is continuous flow microfluidics and the other is droplet-based microfluidics. Continuous-flow microfluidics depends on the control of a steady state liquid flow through micro-channels by accelerating or hindering flow in capillary elements. Commonly, the actuation of liquid flow is implemented by external pressure sources, such as mechanical pumps, integrated micropumps, or by combinations of capillary forces and electrokinetic mechanisms [59]. Continuous-flow microfluidic devices are adequate for well-defined applications, but they are not suitable for tasks requiring a high degree of fluid manipulation because it is difficult to scale, integrate, and synchronize all the parameters, in many cases. For example, the parameters that govern the field of flow (electric field, pressure, or fluid resistance), vary along the flow path and thus make the fluid flow of a particular location dependent on the properties of the entire system. Nevertheless, a continuous-flow microfluidic system is a main stream approach because it is easy to implement and less sensitive to protein fouling problems [59].

Rodríguez-Lorenzo et al. demonstrated the potential of SERS combined with continuous-flow microfluidics for the detection and differentiation of foodborne pathogens [60]. Monoclonal antibody-conjugated gold nanostar SERS tags were used for the detection of *Listeria monocytogenes*. An enhanced SERS signal corresponding to antibody-conjugated SERS nanotags was observed in the presence of *L. monocytogenes* in real time. This method also enabled the discrimination of *L. monocytogenes* and *L. innocua* in 100 s under continuous flow. In this work, it was demonstrated that living organisms could be successfully detected by SERS under continuous flowing conditions, as shown in Fig. 7.9. This in-flow SERS-based optofluidic setup also showed a strong potential for high-throughput screening of specific bacterial pathogens. Total assay time was estimated to be less than 30 min, including bacterial incubation with SERS nanotags, whereas it takes several days with the traditional bacterial counting method.

Kamińska et al. performed multiplex immunoassays using SERS nanotags incorporated into a continuous-flow microfluidic device, as shown in Fig. 7.10. This SERS-based microfluidic technique was used for rapid and sensitive monitoring of selected type 1 cytokine (interleukins: IL-6, IL-8, and IL-18) levels in blood plasma [61]. In this work, molecules such as

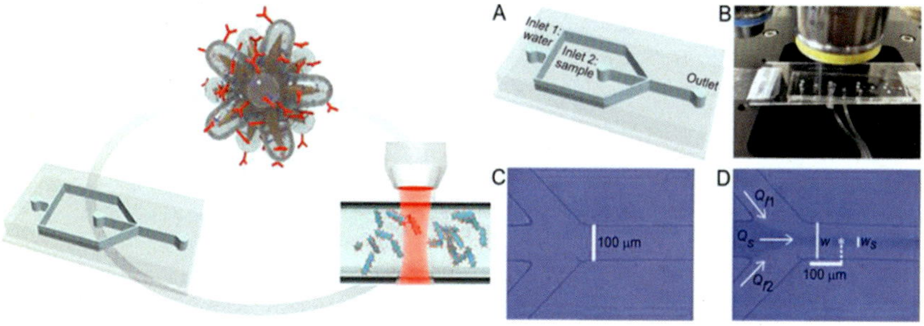

Figure 7.9 (A) Schematic illustration of the flow-focusing microfluidic device for the detection of SERS-targeted bacteria in-flow. (B) Experimental setup for the in-flow Raman signal acquisition in the microfluidic device. (C) Bright-field image of the T-junction of the device having a geometry of 100 × 80 μm (*w* × *h*). (D) Bright-field image of a blue dye to show the hydrodynamic flow focusing of the sample and the experimental focusing width, W_S. The white arrows, with Q_{f1} and Q_{f2}, indicate the flow direction (50 μL h^{-1} each) being the partial focusing flow rates and Q_s the sample flow rate (50 μL h^{-1}). The asterisk indicates where the laser was focused for Raman measurements. *Reprinted with permission from L. Rodríguez-Lorenzo, A. Garrido-Maestu, A.K. Bhunia, B. Espina, M. Prado, L. Diéguez, et al., Gold nanostars for the detection of foodborne pathogens via surface-enhanced Raman scattering combined with microfluidics, ACS Appl. Nano Mater. 2 (2019) 6081–6086.*

fuchsin (FC), p-mercatpobenzoic acid (p-MBA), and 5,5′-dithio-bis(2-nitro-benzoic acid) (DTNB) were used as Raman reporters, and principal component analysis was utilized for segregation of three immunocomplexes encoded by different Raman reporters during a simultaneous multiplex detection approach. The LODs were estimated to be 3.8, 7.5, and 5.2 pg/mL in a simultaneous multiplex detection for IL-6, IL-8, and IL-18, respectively. These LODs were greatly improved in comparison to those obtained by the standard ELISA methods.

7.2.5 Surface-enhanced Raman spectroscopy assays using droplet-based microfluidics

Continuous-flow microfluidics have been widely used for biological analysis, but this single-phase flow regime suffered from the difficulties associated with memory effects caused by aggregated nanoparticles on channel surfaces, implementation of substrates for assays in a microfluidic channel and washing of unreacted species. To resolve the problems in the continuous-flow regime, a droplet-based microfluidic system has been developed. Droplet-based microfluidics manipulates discrete volumes of fluids in immiscible phases with low Reynolds number and laminar flow

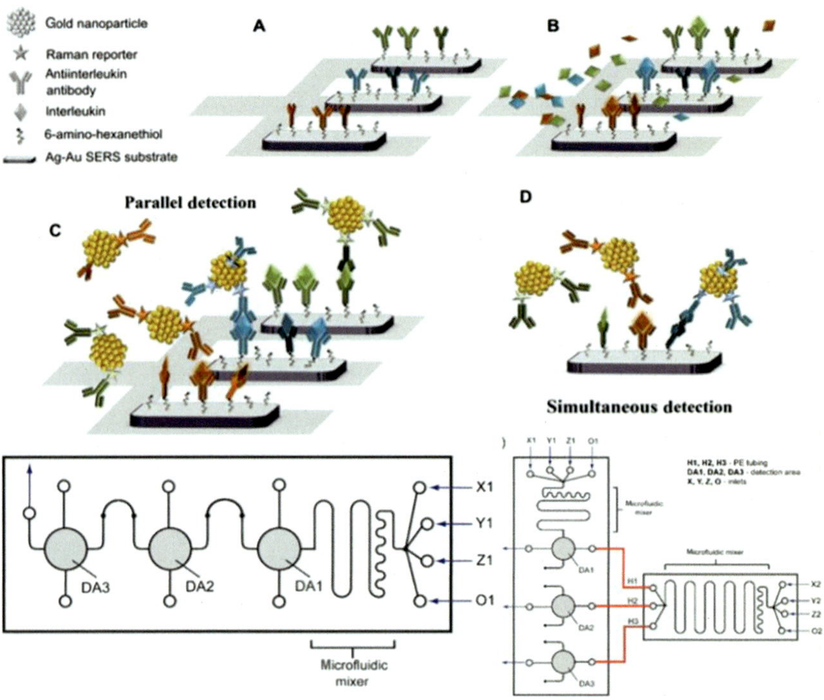

Figure 7.10 Sequential steps for SERS-based multiplex immunoassay. (A) Capturing substrate. (B) Mixture of interleukins (IL-6, IL-8, and IL-18) in human blood plasma injection. (C) Parallel approach and (D) simultaneous multiplex configuration (bottom). A schematic view of the microfluidic chip for simultaneous detection of three different biomarkers. *Reprinted with permission from A. Kamińska, K. Winkler, A. Kowalska, E. Witkowska, T. Szymborski, A. Janeczek, et al., SERS-based immunoassay in a microfluidic system for the multiplexed recognition of interleukins from blood plasma: towards picogram detection, Sci. Rep. 7 (2017) 10656–10666.*

regimes. Microdroplet devices allow for precise handling of miniature volumes (μL to fL) of fluids, and also provide high-throughput experiments through droplet generation, transport, mixing, merging, and splitting modules [59]. The advantages of droplet-based microfluidics include improved detection sensitivity and selectivity, reduction of reaction time, and lower false-positive or negative rates in biochemical reactions.

Sun et al. developed a rapid and ultrasensitive SERS-based microdroplet chip for the simultaneous detection of interleukin-8 (IL-8) and vascular endothelial growth factor (VEGF) secreted by a single cell (Fig. 7.11) [62]. A cross-typed water-in-oil microfluidic chip, generating high-throughput

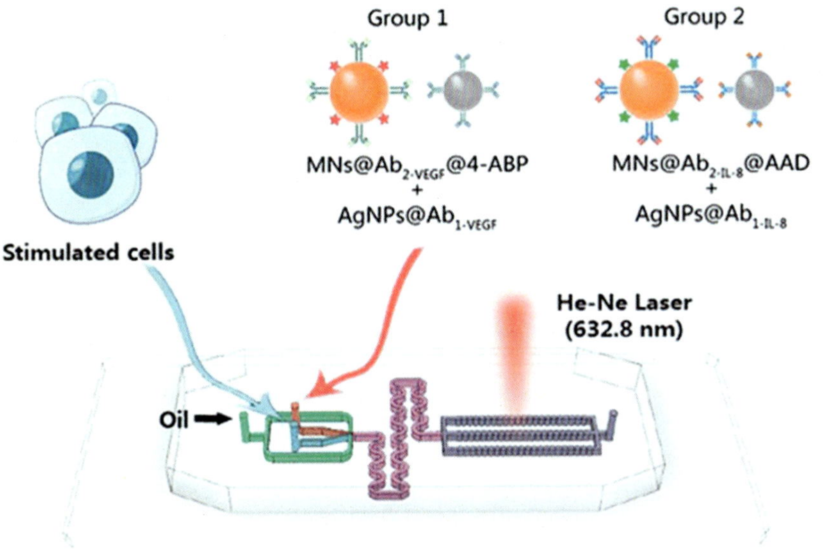

Figure 7.11 Workflow of the droplet-based microfluidics for single-cell encapsulation and SERS detection of vascular endothelial growth factor (VEGF) and interleukin-8 (IL-8) secreted by one cell. *Reprinted with permission from D. Sun, F. Cao, W. Xu, Q. Chen, W. Shi, S. Xu, Ultrasensitive and simultaneous detection of two cytokines secreted by single cell in microfluidic droplets via magnetic-field amplified SERS, Anal. Chem. 91 (2019) 2551–2558.*

water-in-oil droplets in a high-throughput manner, has been used to capture single cells. Then, the biomarkers, IL-8 and VEGF, secreted from the single cell were detected by SERS-based assays. Encapsulation of cytokines in a droplet allows an accumulation effect of target biomarkers with time, and the LOD for each biomarker was estimated to be 1.0 fg/mL in one droplet. A comparison of VEGF and IL-8 secretions of one to four cells in one droplet demonstrates that cell – cell interactions may promote angiogenesis of cancer cells through up-regulation of VEGF and IL-8.

In SERS-based assays using a continuous flow regime, the deposition of nanoparticle aggregates on channel surfaces induces the "memory effect," affecting both sensitivity and reproducibility. To resolve this problem, Jeon and coworkers developed a SERS-based gradient droplet system by combining a gradient microfluidic channel with a SERS detection platform (Fig. 7.12) [63]. Here, a droplet gradient chip was fabricated for high-throughput detection of various concentrations of a reagent. This chip was composed of two PDMS layers: one was a panel for serial

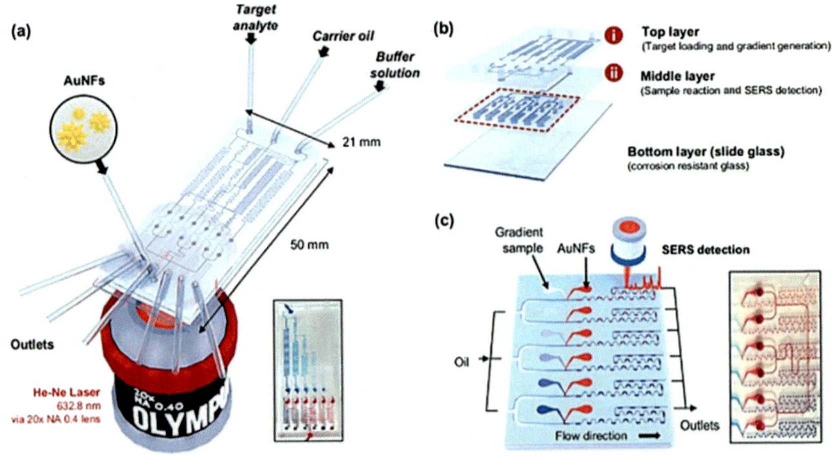

Figure 7.12 (A) Schematic design of a gradient droplet microfluidic chip for SERS-based high-throughput gradient analysis. Gold nanoflowers (AuNFs), targeting the analyte to be diluted, buffer solution, and carrier oil are injected into the channel inlets, and the SERS signals are measured using a He–Ne laser for multiple droplets containing various concentrations of a target reagent. (B) The system consists of two parallel layers: (i) target loading and serial dilution occur in the top panel, (ii) nanoparticle distribution, droplet generation, and SERS detection occur in the middle panel. (C) Image (*left*) and photo (*right*) of the middle layer showing the droplet generation and SERS detection. *Reprinted with permission from J. Jeon, N. Choi, H. Chen, J.-I., Moon, L. Chen, J. Choo, SERS-based droplet microfluidics for high-throughput gradient analysis, Lab. Chip 19 (2019) 674–681.*

dilution of a reagent using continuous, single-phase gradient channels, and the other was a panel for distribution of gold nanoflowers (SERS nanotags) and droplet generation. First, serial dilution of a reagent was obtained using microfluidic concentration gradient generators. Then, various concentrations of a reagent generated in different channels were simultaneously trapped into the tiny volume of droplets by injecting an oil mixture into the channel. Compared to the single-phase regime, this two-phase liquid/liquid segmented flow regime allows minimization of resident time distributions of reagents through localization of reagents in encapsulated droplets. Consequently, this droplet system greatly reduces the sample aggregation problem. This SERS-based gradient droplet system is expected to be a promising method of significant utility in simultaneously monitoring chemical or biological reactions for various concentrations of a reagent.

7.3 Summary

In recent decades, SERS-based immunoassay techniques, in combination with microfluidics, have emerged as potential clinical tools in in vitro diagnostics because of its high sensitivity, specificity, and multiplex detection capability. SERS has an ultrasensitive detection capability, and microfluidics enable accurate fluidic control and high-throughput analysis. Therefore the integration of SERS with microfluidic platforms offers a wide range of applications in biological detection or medical diagnosis. In this chapter, we discussed five different types of SERS-based microfluidic platforms: paper-based microfluidics, magnetic particle-based microfluidics, gold-patterned microwell-based microfluidics, continuous-flow microfluidics, and droplet-based microfluidics. These SERS-based microfluidic platforms provide many advantages such as low reagent consumption, short incubation time, wider concentration dynamic range, and cost savings over the conventional immunoassay techniques such as ELISA, RIA, and ICMA. We believe that these SERS-based microfluidic platforms have strong potential to be used as powerful high-throughput analytical tools for in vitro diagnostics, including immunoassays and molecular diagnostics.

Acknowledgment

The work was supported by the National Research Foundation of Korea (Grants 2019R1A2C3004375 and 2020R1A5A1018052).

References

[1] D.M. Bourdon, L. Meng, M. Alohali, A. Freeman, L. Le, J. Messina, Development of a powerful next-generation immunoassay platform for low level detection of serum based analytes, J. Immunol. 198 (2017).
[2] C.K. Dixit, R.M. Twyman, in: Encyclopedia of Analytical Science, third ed., Elsevier, Amsterdam, 2019.
[3] E. Engvall, P. Perlmann, Enzyme-linked immunosorbent assay, ELISA III. Quantitation of specific antibodies by enzyme-labeled anti-immunoglobulin in antigen-coated tubes, J. Immunol. 109 (1972) 129–135.
[4] J.W. Findlay, W.C. Smith, J.W. Lee, G.D. Nordblom, I. Das, B.S. DeSilva, et al., Validation of immunoassays for bioanalysis: a pharmaceutical industry perspective, J. Pharm. Biomed. Anal. 21 (2000) 1249–1273.
[5] M.F. Elsha, J.P. McCoy, Multiplex bead array assays: performance evaluation and comparison of sensitivity to ELISA, Methods 38 (2006) 317–323.
[6] I.A. Darwish, Immunoassay methods and their applications in pharmaceutical analysis: basic methodology and recent advances, Int. J. Biomed. Sci. 2 (2006) 217–235.
[7] J.L. Born, in: R. Kellner, J.-M. Mermet, M. Otto, H.M. Widmer (Eds.), Analytical Chemistry, Wiley-VCH, New York, 1998.

[8] N.T.K. Thanh, Magnetic Nanoparticles From Fabrication to Clinical Applications, CRC Press, Taylor and Francis Group, Boca Raton, 2012.
[9] E.A. Shirtcliff, D.A. Granger, E.B. Schwartz, M.J. Curran, A. Booth, W.H. Overman, Assessing estradiol in biobehavioral studies using saliva and blood spots: simple radioimmunoassay protocols, reliability, and comparative validity, Horm. Behav. 38 (2000) 137–147.
[10] T.B. Xin, H. Chen, Z. Lin, S.X. Liang, J.M. Lin, A secondary antibody format chemiluminescence immunoassay for the determination of estradiol in human serum, Talanta 82 (2010) 1472–1477.
[11] K.D. Elgert, Immunology: Understanding the Immune System, John Wiley & Sons, Hoboken, 2009.
[12] X. Tan, A. David, J. Day, H. Tang, H. Tang, E.R. Dixon, et al., Rapid mouse follicle stimulating hormone quantification and estrus cycle analysis using an automated microfluidic chemiluminescent ELISA system, ACS Sens. 3 (2018) 2327–2334.
[13] T. Tsumuraya, T. Sato, M. Hirama, I. Fujii, Highly sensitive and practical fluorescent sandwich ELISA for ciguatoxins, Anal. Chem. 90 (2018) 7318–7324.
[14] J. Jeon, N. Choi, J.-I. Moon, H. Chen, J. Choo, in: M. Tokeshi (Ed.), Applications of Microfluidic Systems in Biology and Medicine, vol. 7, Springer, New York, 2019.
[15] S. George, V. Chaudhery, M. Lu, M. Takagi, N. Amro, A. Pokhriyal, et al., Sensitive detection of protein and miRNA cancer biomarkers using silicon-based photonic crystals and a resonance coupling laser scanning platform, Lab. Chip 13 (2013) 4053–4064.
[16] S. Sakamoto, W. Putalun, S. Vimolmangkang, W. Phoolcharoen, Y. Shoyama, H. Tanaka, et al., Enzyme-linked immunosorbent assay for the quantitative/qualitative analysis of plant secondary metabolites, J. Nat. Med. 72 (2018) 32–42.
[17] J.C. Booth, G. Hannington, T.M.F. Bakir, H. Stern, H. Kangro, P.D. Griffiths, et al., Comparison of enzyme-linked immunosorbent assay, radioimmunoassay, complement fixation, anticomplement immunofluorescence and passive haemagglutination techniques for detecting cytomegalovirus IgG antibody, J. Clin. Pathol. 35 (1982) 1345–1348.
[18] R.D. Grange, J.P. Thompson, D.G. Lambert, Radioimmunoassay, enzyme and non-enzyme-based immunoassays, Br. J. Anaesth. 112 (2014) 213–216.
[19] D.G. Wild, The Immunoassay Handbook: Theory and Applications of Ligand Binding, ELISA and Related Techniques, Springer, Amsterdam, 2013.
[20] L. Cinquanta, D.E. Fontana, N. Bizzaro, Chemiluminescent immunoassay technology: what does it change in autoantibody detection? Auto. Immun. Highlights 8 (2017) 9.
[21] C.A. Spencer, Assay of Thyroid Hormones and Related Substancesin: K.R. Feingold, B. Anawalt, A. Boyce, et al. (Eds.), MDText.com, Inc., South Dartmouth, MA, 2000.
[22] J.F. Rusling, C.V. Kumar, J.S. Gutkind, V. Patel, Measurement of biomarker proteins for point-of-care early detection and monitoring of cancer, Analyst 135 (2010) 2496–2511.
[23] C.S. Carroll, L.N. Amankwa, L.J. Pinto, J.D. Fuller, M.M. Moo, Detection of a serum siderophore by LC-MS/MS as a potential biomarker of invasive aspergillosis, PLoS One 11 (2016) 0155451–0155467.
[24] J. Smolsky, S. Kaur, C. Hayashi, S.K. Batra, A.V. Krasnoslobodtsev, Surface-enhanced Raman scattering-based immunoassay technologies for detection of disease biomarkers, Biosensors 7 (2017) 7–27.
[25] K. Kneipp, H. Kneipp, I. Itzkan, R.R. Dasari, M.S. Feld, Ultrasensitive chemical analysis by Raman spectroscopy, Chem. Rev. 99 (1999) 2957–2976.

[26] A. Zengin, U. Tamer, T.A. Caykara, SERS-based sandwich assay for ultrasensitive and selective detection of alzheimer's tau protein, Biomacromolecules 14 (2013) 3001–3009.
[27] K. Kim, N. Choi, J.H. Jeon, G.E. Rhie, J. Choo, SERS-based immunoassays for the detection of botulinum toxins A and B using magnetic beads, Sensors 19 (2019) 4081–4088.
[28] M. Guo, J. Dong, W. Xie, L. Tao, W. Lu, Y. Wang, et al., SERS tags-based novel monodispersed hollow gold nanospheres for highly sensitive immunoassay of CEA, J. Mater. Sci. 50 (2015) 3329–3336.
[29] J.D. Driskell, K.M. Kwarta, R.J. Lipert, M.D. Porter, J.D. Neill, J.F. Ridpath, Low-level detection of viral pathogens by a surface-enhanced Raman scattering based immunoassay, Anal. Chem. 77 (2005) 6147–6154.
[30] S. Xu, X. Ji, W. Xu, X. Li, L. Wang, Y. Bai, et al., Immunoassay using probe-labelling immunogold nanoparticles with silver staining enhancement via surface-enhanced Raman scattering, Analyst 129 (2004) 63–68.
[31] S. Boca, C. Farcau, M. Baia, S. Astilean, Metanephrine neuroendocrine tumor marker detection by SERS using au nanoparticle/au film sandwich architecture, Biomed. Microdevices 18 (2016) 12–21.
[32] G. Wang, R.J. Lipert, M. Jain, S. Kaur, S. Chakraboty, M.P. Torres, et al., Detection of the potential pancreatic cancer marker muc4 in serum using surface-enhanced Raman scattering, Anal. Chem. 83 (2011) 2554–2561.
[33] B.J. Yakes, R.J. Lipert, J.P. Bannantine, M.D. Porter, Detection of mycobacterium avium subsp. paratuberculosis by a sonicate immunoassay based on surface-enhanced Raman scattering, Clin. Vaccine Immunol. 15 (2008) 227–234.
[34] S.H. Lee, J. Hwang, K. Kim, J. Jeon, S. Lee, J. Ko, et al., Quantitative serodiagnosis of scrub typhus using surface-enhanced Raman scattering-based lateral flow assay platforms, Anal. Chem. 91 (2019) 12275–12282.
[35] K.J. Yoon, H.K. Seo, H. Hwang, D. Pyo, I.Y. Eom, J.H. Hahn, et al., Bioanalytical application of SERS immunoassay for detection of prostate-specific antigen, Bull. Korean Chem. Soc. 31 (2010) 1215–1218.
[36] P. Lang, Y. Liu, Soft Matter at Aqueous Interfaces, Springer, Amsterdam, 2016.
[37] T.A. Duncombe, A.M. Tentori, A.E. Herr, Microfluidics: reframing biological enquiry, Nat. Rev. Mol. Cell Biol. 16 (2015) 554–567.
[38] D. Mark, S. Haeberle, G. Roth, F. von Stetten, R. Zengerle, Microfluidic lab-on-a-chip platforms: requirements, characteristics and applications, Chem. Soc. Rev. 39 (2010) 1153–1182.
[39] S. Kim, H.J. Kim, N.L. Jeon, Biological applications of microfluidic gradient devices, Integr. Biol. 2 (2010) 584–603.
[40] D.N. Breslauer, P.J. Lee, L.P. Lee, Microfluidics-based systems biology, Mol. BioSyst. 2 (2006) 97–112.
[41] G.T. Taylor, S.K. Sharma, K. Mohanan, Optimization of a flow injection sampling system for quantitative analysis of dilute aqueous solutions using combined resonance and surface-enhanced Raman spectroscopy (SERS), Appl. Spectrosc. 44 (1990) 635–640.
[42] J.J. Laserna, Combining fingerprinting capability with trace analytical detection: surface-enhanced Raman spectrometry, Anal. Chim. Acta 283 (1993) 607–622.
[43] Y. Song, B. Lin, T. Tian, X. Xu, W. Wang, Q. Ruan, et al., Recent progress in microfluidics-based biosensing, Anal. Chem. 91 (2019) 388–404.
[44] G.A. Posthuma-Trumpie, J. Korf, A. van Amerongen, Lateral flow (immuno)assay: its strengths, weaknesses, opportunities and threats. A literature survey, Anal. Bioanal. Chem. 393 (2009) 569–582.
[45] L. Zhan, S.Z. Guo, F. Song, Y. Gong, F. Xu, D.R. Boulware, et al., The role of nanoparticle design in determining analytical performance of lateral flow immunoassays, Nano Lett. 17 (2017) 7207–7212.

[46] W. Ren, S.I. Mohammed, S. Wereley, J. Irudayaraj, Magnetic focus lateral flow sensor for detection of cervical cancer biomarkers, Anal. Chem. 91 (2019) 2876–2884.
[47] L.-K. Lin, L.A. Stanciu, Bisphenol A detection using gold nanostars in a SERS improved lateral flow immunochromatographic assay, Sens. Actuat. B-Chem. 276 (2018) 222–229.
[48] D.D. Seachrist, K.W. Bonk, S.-M. Ho, G.S. Prins, A.M. Soto, R.A. Keri, et al., A review of the carcinogenic potential of Bisphenol A, Reprod. Toxicol. 59 (2016) 167–182.
[49] M. Wu, C. Pan, M. Yang, B. Xu, X. Lei, J. Ma, et al., Chemical analysis of fish bile extracts for monitoring endocrine disrupting chemical exposure in water: bisphenol A, alkylphenols, and norethindrone, Environ. Toxicol. Chem. 35 (2016) 182–190.
[50] D.D. Seachrist, K.W. Bonk, S.-M. Ho, G.S. Prins, A.M. Soto, R.A. Keri, A review of the carcinogenic potential of bisphenol A, Reprod. Toxicol. 59 (2016) 167–182.
[51] J. Li, D. McMillan, J. Macdonald, Enhancing the signal of lateral flow immunoassays by using different developing methods, Sensor. Mater. 27 (2015) 549–561.
[52] A.E. Urusov, A.V. Petrakova, A.V. Zherdev, B.B. Dzantiev, Application of magnetic nanoparticles in immunoassay, Nanotechnologies Russia 12 (2017) 471–479.
[53] L.W. Yap, H. Chen, Y. Gao, K. Petkovic, Y. Liang, K.-J. Si, et al., Bifunctional plasmonic-magnetic particles for an enhanced microfluidic SERS immunoassay, Nanoscale 9 (2017) 7822–7829.
[54] R. Gao, Z. Cheng, A.J. deMello, J. Choo, Wash-free magnetic immunoassay of the PSA cancer marker using SERS and droplet microfluidics, Lab. Chip 16 (2016) 1022–1029.
[55] Q. Xiong, C.Y. Lim, J. Ren, J. Zhou, K. Pu, M.B. Chan-Park, et al., Magnetic nanochain integrated microfluidic biochips, Nat. Commun. 9 (2018) 1743–1753.
[56] M. Lee, S. Lee, J.H. Lee, H. Lim, G.H. Seong, E.K. Lee, et al., Highly reproducible immunoassay of cancer markers on a gold-patterned microarray chip using surface-enhanced Raman scattering imaging, Biosens. Bioelectron. 26 (2011) 2135–2141.
[57] M. Lee, K. Lee, K.H. Kim, K.W. Oh, J. Choo, SERS-based immunoassay using a gold array embedded gradient microfluidic chip, Lab. Chip 12 (2012) 3720–3727.
[58] X. Wang, S.-G. Park, J. Ko, X. Xiao, V. Giannini, S.A. Maier, et al., Sensitive and reproducible immunoassay of multiple mycotoxins using surface-enhanced Raman scattering mapping on 3D plasmonic nanopillar arrays, Small 14 (2018) 1801623–1801633.
[59] L. Rodríguez-Lorenzo, A. Garrido-Maestu, A.K. Bhunia, B. Espina, M. Prado, L. Diéguez, et al., in: J. Zeng (Ed.), Design Automation Methods and Tools for Microfluidics-Based Biochips, Springer, Berlin, 2006.
[60] L. Rodríguez-Lorenzo, A. Garrido-Maestu, A.K. Bhunia, B. Espina, M. Prado, L. Diéguez, et al., Gold nanostars for the detection of foodborne pathogens via surface-enhanced Raman scattering combined with microfluidics, ACS Appl. Nano Mater. 2 (2019) 6081–6086.
[61] A. Kamińska, K. Winkler, A. Kowalska, E. Witkowska, T. Szymborski, A. Janeczek, et al., SERS-based immunoassay in a microfluidic system for the multiplexed recognition of interleukins from blood plasma: towards picogram detection, Sci. Rep. 7 (2017) 10656–10666.
[62] D. Sun, F. Cao, W. Xu, Q. Chen, W. Shi, S. Xu, Ultrasensitive and simultaneous detection of two cytokines secreted by single cell in microfluidic droplets via magnetic-field amplified SERS, Anal. Chem. 91 (2019) 2551–2558.
[63] J. Jeon, N. Choi, H. Chen, J.-I. Moon, L. Chen, J. Choo, SERS-based droplet microfluidics for high-throughput gradient analysis, Lab. Chip 19 (2019) 674–681.

CHAPTER 8

SERS for sensing and imaging in live cells

Janina Kneipp
Department of Chemistry, Humboldt-Universität zu Berlin, Berlin, Germany

8.1 Recent trends in SERS from animal cells: probe of cellular biochemistry

Surface-enhanced Raman scattering (SERS) in the absence of labels, tags or reporters is a powerful tool to obtain comprehensive and diverse information about the composition and structure of biomolecular samples [1]. While SERS has become widely used in analytical chemistry [2] specifically in labeling, in imaging or in quantitative tasks, the possibility to apply it for the characterization of molecular composition and interaction for a broad range of samples, for example, of analytes in a complex matrix or mixtures remains a challenge [3].

Nevertheless, as illustrated by the developments of the last years, the particular challenges in SERS experiments must be named and critically addressed, in order to create biodiagnostic applications that are reliable. The success of recent work is mainly based on novel, controllable SERS substrates and the appreciation of the fact that SERS relies on the *interaction* of the sample with a plasmonic substrate. This led to well-controlled and characterized plasmonic probes, ranging from the manipulation of nanoparticles [4–8] over nanoelectrodes or needles [9–14] to plasmonic surfaces [15,16].

An important topic that must be addressed are the strong signal fluctuations that are inherent to the method at low molecule concentration and when high enhancement regimes are used [17,18]. More importantly, the variation of non-resonant Raman cross sections for different types of molecules by orders of magnitude, and of selective molecule–metal interactions with a particular SERS substrate render some "weak scatterers" in multicomponent systems undetected. Recently, suitable molecular models including lipids [19], proteins [20], and specific drugs [21] have been proposed to

Principles and Clinical Diagnostic Applications of Surface-Enhanced Raman Spectroscopy
DOI: https://doi.org/10.1016/B978-0-12-821121-2.00009-3

better understand and interpret the SERS spectra of cells and tissues where different molecular species co-occur and interact.

The applicability of SERS probes with relevance for theranostic applications was shown to greatly rely on specific plasmonic but also on other optical properties, enabling their fast screening by microscopic imaging using linear and nonlinear effects, such as Rayleigh scattering [22] or second-harmonic generation [23]. In this context, tuning the optical properties and constructing multifunctional plasmonic-composite nanostructures is paving the way for fast and efficient multimodal imaging that employs SERS as one modality.

Different plasmon-enhanced vibrational probing at different excitation wavelength, for example, on- and off-resonance with electronic transitions in chromophores of specific proteins can lead to different information in SERS experiments with such samples in vivo, hemoglobin being one of the most popular examples for this type of "multimodal" probing by SERS and SERRS [24]. In a similar way, one and the same type of plasmonic nanoprobe can enhance SERS, as well as its two-photon excited analog, surface enhanced hyper Raman scattering (SEHRS) that was proposed very early after the discovery of SERS [25,26], and that delivers vibrational information which is complementary to that of SERS [27]. Effective SEHRS cross sections up to the order of 10^{-45} cm^4 s that were reported are similar or higher than those of two-photon excited fluorescence and led to a number of studies of cultured cells in vivo with a range of different excitation wavelengths up into the near infrared [23,28–32].

In the following, important aspects that have determined recent work in SERS probing of cells will be highlighted, in particular further developments of intracellular nanoprobes, the possibilities to study molecules with low Raman cross sections or in low abundance, with relevance to applications of SERS in biotechnology and theranostics.

8.2 Biomolecular SERS from intracellular nanoprobes

The SERS obtained from nanoparticles inside cells and tissues is always an indicator of the immediate environment of the nanostructures. Often, nanoparticles that serve as SERS nanoprobes are delivered into the cells by adding them to the culture medium, by the process of receptor-mediated endocytosis. Inside the cells, the nanoparticles accumulate in endolysosomal vesicles over time and can be quantified precisely by mass spectrometric micromapping (Fig. 8.1) [33,34].

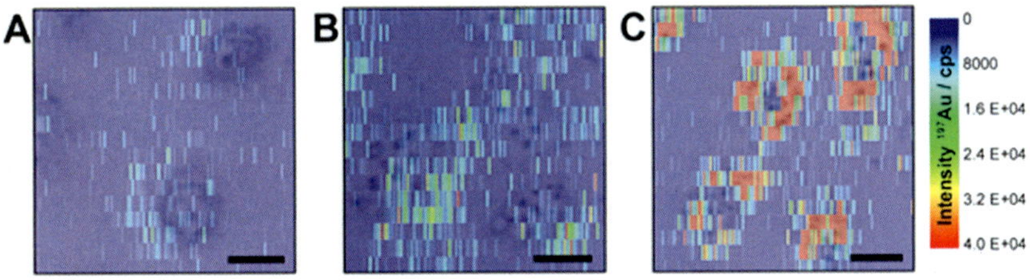

Figure 8.1 Intensity distribution of 197Au^{+} in individual 3T3 fibroblast cells after gold nanoparticle incubation for (A) 1 h, (B) 3 h and (C) 24 h measured by LA-ICP-Ms superimposed with corresponding brightfield images. With increasing incubation time, gold nanoparticles accumulate in the perinuclear region. Parameters: laser spot size 4 μm, line distance 6 μm, scan speed 5 $\mu m\ s^{-1}$, frequency 10 Hz, pixel size 6 × 1 μm, fluence 0.7 $J\ cm^{-2}$. *Reproduced with permission from D. Drescher, H. Traub, T. Büchner, N. Jakubowski, J. Kneipp, Properties of in situ generated gold nanoparticles in the cellular context, Nanoscale, 9 (32) (2017) 11647–11656.*

Inside the cell, the nanoparticles are surrounded by a corona of biomolecules, which is known to determine their fate inside the cell, rather than the pristine gold surface will do. From the theranostic perspective, understanding the composition and dynamics of the hard protein corona is very important, and therefore SERS can be very useful here to get a better understanding on the interaction of theranostic nanostructures with their cellular surroundings.

Using the SERS vibrational information obtained from those molecules that form the hard corona in vivo, it was shown that the composition of the hard protein corona and the induced cellular response depend strongly on the cell line and the culturing conditions [35]. This was shown for two different cell lines that were grown in two different culture media, through which gold nanoparticles were delivered to the cells. Specifically, using the SERS spectra, the secondary structure of the adsorbed proteins was shown to remain partially intact, but their interaction with the gold nanoparticles varied depending on the surface charge as an important property of the gold surface itself, independent of variations in physiological aspects of their uptake into the cells. The findings obtained by SERS were corroborated by data from mass spectrometry and the description of the cellular ultrastructure based on nanotomography [35].

In a similar fashion, the composition of the protein corona of gold nanostars was investigated using SERS (Fig. 8.2). There, a comparison of the biomolecular environment of the nanostars of slightly different geometry in the same cell line indicated differences in the structure and interaction

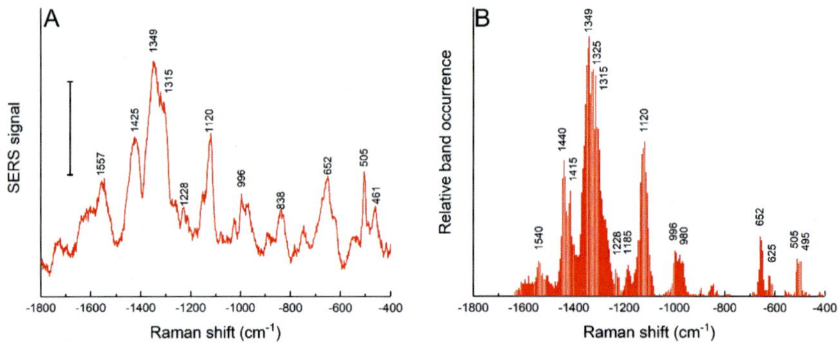

Figure 8.2 (A) SERS average spectrum and (B) relative band occurrence for HCT-116 cells incubated for 24 h with gold nanostars synthesized with 25 mM HEPES. Scale bar in (A): 10 cps. *Reproduced with permission from C. Spedalieri, G.P. Szekeres, S. Werner, P. Guttmann, J. Kneipp, Intracellular optical probing with gold nanostars, Nanoscale 13 (2) (2021) 968–979.*

of the proteins in the hard corona of the nanoparticles, and also a different interaction with membranous structures in the endolysosomal compartment [36]. Very similar SERS data are obtained in the cell culture medium with three different nanostar preparations, yet very different spectral patterns later in the endolysosomal system lead us to conclude that the specific molecular environment observed inside the cells is a consequence of postendocytotic processing depending on nanostar geometry [36].

Although the composition of the protein corona of gold nanoparticles inside the endo-/lysosomal system changes over time in the course of endocytic processing, SERS revealed that the proteins adsorbed to the immediate gold surface are strongly attached and become fragmented over time in the endolysosome while being part of the protein corona [20]. These results were obtained by comparing the protein SERS spectra of live J774 macrophage cells, containing gold nanoparticles in their active endolysosomal compartments to those of isolated cytoplasm devoid of active processing, unlike the endolysosomal environment. In the SERS spectra collected from the macrophages, a higher abundance of nonpolar group vibrations was observed, while the abundance of those vibrations that indicate an intact protein backbone decreased. To support the identification of the spectral features of potential enzymatic protein cleavage in the protein corona, the enzymatic cleavage of BSA by trypsin was studied in solution. Support of this conclusion was provided by an analysis of the low-mass peptides in the hard corona of the gold nanoparticles after isolation from the same cells by SDS-PAGE and HPLC-ESI-Q-TOF-Ms [20].

Different from the internalization of nanoparticles by receptor-mediated endocytosis, the biomolecular corona of nanoparticles will vary strongly for nanoparticles that are generated directly in the cells, in situ. In situ formation of gold nanoparticles was shown in cell cultures of several animal cell lines and in their supernatant after incubation with tetrachloroauric acid in buffer [37–40], and the SERS activity of the in situ generated nanoparticles was reported [38,40]. Interestingly, also hybrid structures with graphene oxide could be obtained in this way [39]. As was shown, SERS can probe the biomolecular species involved in the formation of the nanoparticles in situ [34]. In vitro, several biomolecular species were shown to be involved in gold nanoparticle formation and stabilization, including amino acids [41], protein side chains [42,43], glutathione [44], phospholipids [45], chitosan and chitin [46], bilirubin [47], and many more.

8.3 Probing lipid-rich environments in pathology

Lipids and an affected lipid metabolism play an important role in a vast range of pathologies, and lipids are a major constituent of the molecular environment of animal cells. Even after almost 20 years of in vivo SERS [48] from living cultured cells, obtaining selectively information from cellular lipids has been relatively rare. The lipids of cellular membranes do not produce prominent signals in the SERS spectra of cells, because of their low overall cross section [49] compared to those of other molecules such as labels, drugs, nucleotides, or aromatic amino acid side chains and because affinity of lipid molecules for the surface of gold nanoparticles in an aqueous environment is low [50]. There are several pathologies that are characterized by a high local abundance of lipids at the microscopic level, including metabolic disorders [51] as well as parasite infections [52] and certain models of neurodegenerative disease [53]. So far, probing of lipids in cells has not been in the focus of biodiagnostic SERS, and although lipid spectral contributions have also been discussed in a SERS analysis of exosomes [54], SERS characterization of these important organelles in cancer diagnostics mostly relies on protein biomarkers [55–57].

As discussed in this and also in the following subsection, the SERS spectra of intracellular lipid accumulations can be studied in lipid rich environments in certain pathologies by SERS.

In a macrophage model of Leishmania infection, SERS can be used to map the distinct distribution of cholesterol and its ester ergosterol, as well as of glycoinositol-phospholipids [52]. These molecules are secreted by the parasite in vacuoles inside the cells and are very specific to the infection (Fig. 8.3).

The vacuole that carries the eukaryotic parasites is easily visible by light microscopy [58] and can therefore be selectively studied by SERS mapping. Fig. 8.4 shows maps of typical bands associated with typical lipids, lipids specific to the parasite infection, and other molecular constituents such as proteins and carbohydrate compounds. The occurrence of the band at 420 cm^{-1} of cholesterol colocalizes with that of the band at 1140 cm^{-1}, indicating that often, although not always, cholesterol colocalizes with phospholipids in the infected cells. One of these exceptions is the region of the parasitophorous vacuole, where the cholesterol signal is mostly found around the vacuole membrane. It has been proposed that the parasites, which sit in the proximity to the vacuole membrane (Fig. 8.4, arrowheads) are surrounded by a cholesterol halo, and that cholesterol is sequestered into the parasitophorous vacuole by the parasites in infected cells [59]. The intensity of the band at 1602 cm^{-1}, attributed to the C = C double bond of ergosterol and/or ergosterol derivatives [60] is

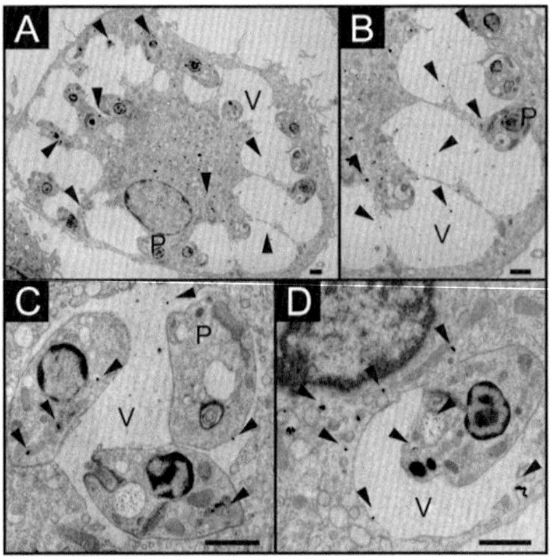

Figure 8.3 TEM micrographs of *Leishmania mexicana* infected macrophages showing localization of the gold nanoparticles in the parasitoporous vacuoles, parasites, and phagolysosomes. *Arrowheads* indicate gold nanoparticles (not all nanoparticles are labeled). P: parasites (amastigotes); V: parasitophorous vacuole. Scale bar: 1 μm. *Reproduced with permission from V. Živanović, G. Semini, M. Laue, D. Drescher, T. Aebischer, J. Kneipp, Chemical mapping of Leishmania infection in live cells by SERS microscopy, Anal. Chem. 90 (13) (2018) 8154–8161.*

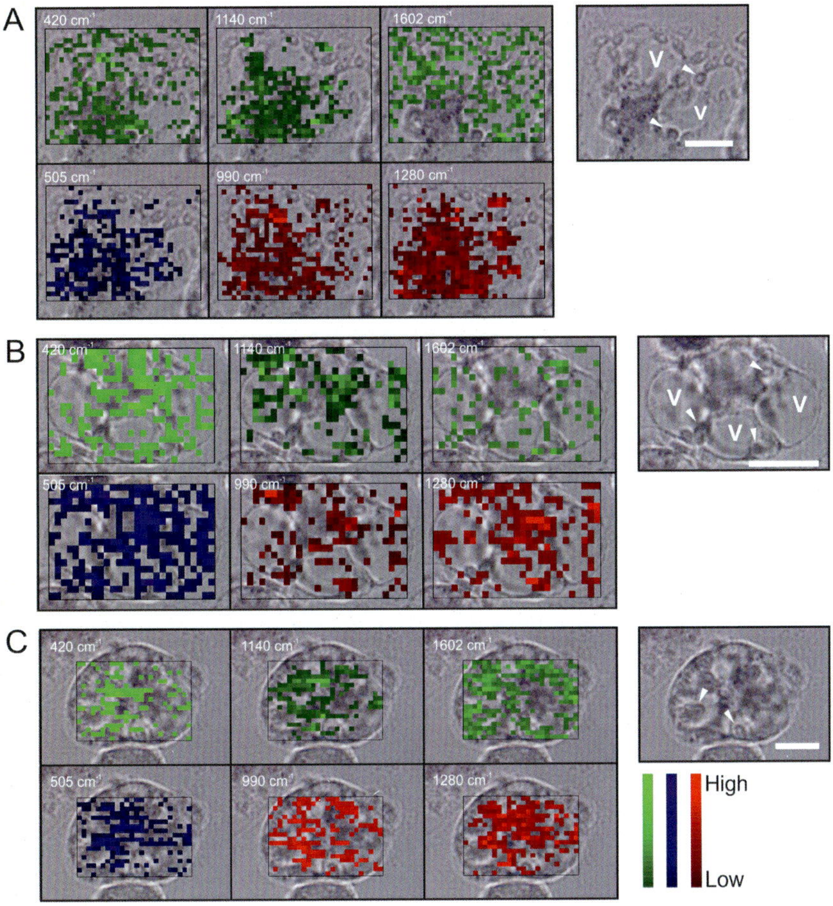

Figure 8.4 (A)–(C) Distribution of signals that are related to molecular species characteristic of the infection by *Leishmania mexicana*, representing lipids (*green*), proteins (*blue*), and phosphate (*red*) species together with bright field images of the three exemplary cells. Chemical images are generated by mapping the intensity of the bands at 420 cm^{-1} assigned to cholesterol, 505 cm^{-1} of disulfide in proteins, 990 cm^{-1} mainly assigned to phosphate in phosphoproteoglycans, 1140 cm^{-1} of phospholipid alkyl chains, 1278 cm^{-1} of phosphate in proteins and phopholipids, and 1602 cm^{-1} of ergosterol. *Arrowheads* in the brightfield images indicate individual parasites (amastigotes); V: parasitophorous vacuole. Scale bar: 10 μm. *Reproduced with permission from V. Živanović, G. Semini, M. Laue, D. Drescher, T. Aebischer, J. Kneipp, Chemical mapping of Leishmania infection in live cells by SERS microscopy, Anal. Chem. 90 (13) (2018) 8154–8161.*

high in the region of the vacuoles as well. While the sterol-specific filipin staining that is typically used in conventional microscopy of Leishmania infected cells cannot discriminate between cholesterol and ergosterol, here we see that these two molecules do not necessarily colocalize in the same focal volume. Furthermore, different from the signal of cholesterol, the ergosterol spectrum was not observed in the regions of the endolysosomes away from the vacuole regions. This confirms its very specific abundance in the parasite [61] and indicates that it is also abundant in the vacuole.

To characterize the interaction of lipid molecules with plasmonic nanostructures, as in the above disease model, SERS spectra of different lipids must be better understood. This can be achieved by specific lipid models, either pure lipids, often forming lipid vesicles in the typical experimental settings, with functionalized SERS substrates, or by analyzing hybrid bilayers [62–65].

SERS spectra obtained from liposomes that contain gold nanoparticles, as typically used for intra-endosomal probing, varied depending on vesicle composition and interaction. Specifically, liposomes composed of phosphatidylcholine, sphingomyelin, and cholesterol in different proportions reveal that very small changes in the lipid composition can alter the contact between liposomes and gold nanostructures [19]. In addition to the detection of the different types of lipids and the interaction of phospholipids with cholesterol by SERS, changes in membrane fluidity at the site of interaction of the lipids with the surface of gold nanoparticles were observed, and the influence of varying charge of the SERS substrate on the liposome–gold interaction was revealed [19]. As an example, negative charge can be introduced into the system either by excess citrate in the gold nanoparticles, or by the incorporation of phosphatidic acid, and influences packing of the lipids in the vesicles [66]. Understanding the interaction of lipid membranes and monitoring their integrity is also important in the development of new gold nanoparticle and gold nanoparticle-liposome-based drug delivery systems.

8.4 SERS for monitoring of drug action

In spite of the strong enhancement that can be observed in the endolysosomal system of cells or in specific regions of tissues and the great versatility with respect to SERS substrates that can be applied, the identification of distinct molecular components by state-of-the-art multivariate chemometric tools is often not possible in the real biological environment. In addition to the example of lipid molecules discussed in the previous section, the large spectral differences observed in the spectra of a particular protein at different concentration, or the dominating contributions of adenine and related compounds in many spectra of biomaterials prevent the observation of significant differences and pose a limit to the applicability of SERS as universal tool.

Tricyclic antidepressants (TCA) are drugs that have different modes of action and can be used for the therapy of a variety of diseases including tumors, since they act on enzymes in the *endo*-lysosomal system. A group of different TCA drugs, showing high SERS cross sections, albeit extreme differences in their interaction with plasmonic nanoparticles of gold and silver, were studied systematically regarding their interaction with gold and silver nanoparticles and in the presence and absence of other molecules and cell culture media [21]. The interaction of specific functional groups with the respective metal surfaces was systematically discussed, and the SERS data were also validated by other, complementary vibrational information, including two-photon excited SEHRS, as well as infrared spectroscopy [21]. It is possible to obtain the SERS spectra of these molecules with near-infrared excitation and with gold nanoparticles in cell culture media. TCA interact with silver nanostructures mainly via their ring moiety and less intensely with the alkyl chain [21,67]. In contrast, as revealed from the spectra obtained on biocompatible gold nanostructures, the methyl-aminopropyl side chain of the TCA plays a very important role in the interaction with the gold, along with parts of the ring system. This is very different from the interaction with the silver nanostructures. The NIR excited SERS spectra of the TCA-gold nanoparticles are greatly invariant with respect to changes in TCA concentration and size of the biocompatible gold nanostructures. They show remarkable stability in the presence of cell culture media and upon decrease of pH in the typical ranges of pH values in late endosomal structures.

Surprisingly, the detection and observation of TCA in living cells remains a great challenge, for reasons that are not well-established so far, most likely related to the altered interaction of the TCA with gold nanoparticles in the harsh conditions of the endolysosomal environment. As has been discussed for many molecules, different selectivity of the SERS enhancement when charge, pH or concentration change can cause a variation of relative intensities of spectral bands or the absence of vibrations of particular functional groups.

To study both the behavior of the drug itself as well as the potential effects it has in the cells, incubation experiments were conducted with the TCA desipramine, as well as two other TCA, imipramine and amitriptyline. They were incubated in different combinations with gold nanoparticles, before (Fig. 8.5A), after (Fig. 8.5B), or simultaneously (Fig. 8.5C). SERS spectra were collected from the live cells in phosphate buffer and analyzed by random forest (RF) machine learning.

The application of machine learning approaches to SERS data ranges back until about a decade ago, when both, unsupervised multivariate classification [69,70], as well as supervised machine learning tools, including artificial neural networks [71], support vector machines [72], and lately also RFs [73] were proposed for the classification of SERS spectra, for imaging as well as for diagnostic purpose. RF is a nonparametric machine learning method that is based on a large number of individual decision trees and that can be applied in versatile data analysis tasks, specifically to build classification and regression models. In addition to this application, RF can also be utilized to analyze the importance of individual variables and to select those that are important.

RF-based variable selection and relations analysis are valuable tools to elucidate the interaction of TCA with proteins and lipids based on a selection of spectral variables assigned to both the TCA and the respective biomolecules [68]. While other multivariate tools such as hierarchical cluster

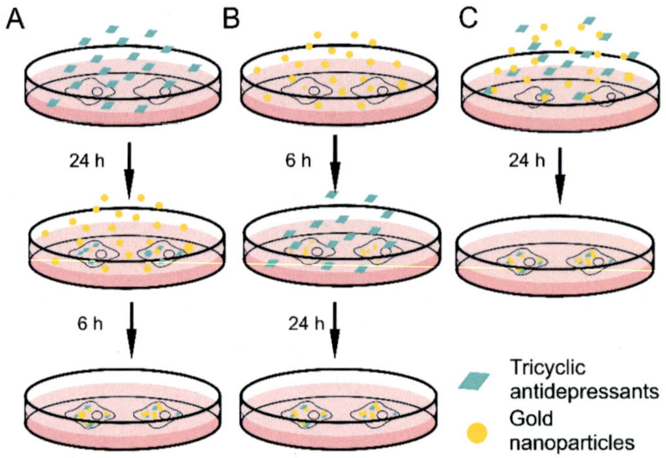

Figure 8.5 Representation of the incubation of the cells with the TCA and the gold nanoparticles that were used as SERS probes. (A) Cells were incubated with TCA molecules for 24 h prior to the incubation with gold nanoparticles for 6 h. (B) Cells were incubated with gold nanoparticles for 6 h prior to the incubation with TCA molecules for 24 h. (C) Cells were incubated with TCA molecules and gold nanoparticles for 24 h simultaneously. *Reproduced with permission from V. Zivanovic, S. Seifert, D. Drescher, P. Schrade, S. Werner, P. Guttmann, et al., Optical nanosensing of lipid accumulation due to enzyme inhibition in live cells, ACS Nano (2019).*

analysis and principal components analysis failed to identify differences between the drug-treated and untreated cells, RF helped to identify important variables in the large data sets that contained of thousands of spectra from different cellular regions obtained for the different incubation conditions (Fig. 8.5), as well as other variables that relate to them. Interestingly, the spectral bands assigned to the drugs were related to bands assigned to lipid or protein molecules, indicating their cooccurrence in the spectra. It must be pointed out here that the assignments of the variables in the RF analysis (and in any other pattern recognition-based tool) are only possible using known spectra of model compounds, such as those of the lipid vesicles [19] or the pure TCA drugs [21] discussed above.

8.5 Composite SERS probes for intracellular applications with different physical functions

The relatively facile synthesis of gold and silver nanostructures by wet-chemistry approaches in aqueous media enables different possibilities to combine both metals with other nanomaterials and to provide additional physical and chemical properties to them that can be beneficial in biotechnological and theranostic applications. As very popular example, the combination of silver or gold nanostructures with nanoparticles that provide magnetic properties should be mentioned, as the magnetic component can enable manipulation of cells and tissues in magnetic fields [74–76]. The manipulation of whole eukaryotic cells using magnetic nanostructures is especially powerful in microfluidic structures, where local magnetic forces can be applied efficiently [77,78], and different cell types can be sorted [79,80].

SERS spectra of reporter or label molecules can be obtained from composite plasmonic nanostructures with magnetic properties [81–86]. Spectra of magnetic SERS labels were also employed to visualize their interaction with cells, for example, when particles with plasmonic and magnetic properties bind to cell membranes, [87–89], or when they enter a cell [90,91]. In bacteria, plasmonic–magnetic nanostructures were shown to provide SERS spectral information from the molecules in the cell walls of bacterial cells [92–94].

By using the intrinsic SERS signals from the cells, the composition of the nanoparticle surface of composite nanostructures of gold with magnetite (Au-Magnetite) and silver with magnetite (Ag-Magnetite) was characterized [5].

Furthermore, the uptake behavior and intracellular fate of such magnetic–plasmonic nanostructures were studied using cryo soft-XRT and quantitative LA-ICP-Ms mapping.

The SERS enhancement factors for the composite structures were estimated to be on the order of 10^3 to 10^4, in accord with the observation that many of the metal nanoparticles do not form aggregates but are kept separated from one another within their respective nanocomposite particles. As was revealed by cryo soft-XRT data, the composite nanoparticles are contained in endosomes inside the cells, indicating their endocytic uptake. Fibroblast cells incubated with Ag-Magnetite and Au-Magnetite for 24 h were brought into an external magnetic field. After detaching the cells from their substrate by trypsin treatment, the cell suspensions were mixed thoroughly for 2 min, and a magnet was placed in the suspensions. Fig. 8.6 displays the motion of fibroblast cells containing Ag-Magnetite (Fig. 8.6A) and Au-Magnetite (Fig. 8.6B). Under the experimental conditions chosen, the velocity of the cells is on the order of 10 $\mu m\ s^{-1}$.

Relatively high variations (of a factor of two) were found between individual cells due to the fact that fibroblast cells naturally adhere again to the surface and subsequently detach again. These results indicate that Ag-Magnetite and Au-Magnetite can be utilized for applications in magnetic cell separations, as well as in microfluidics.

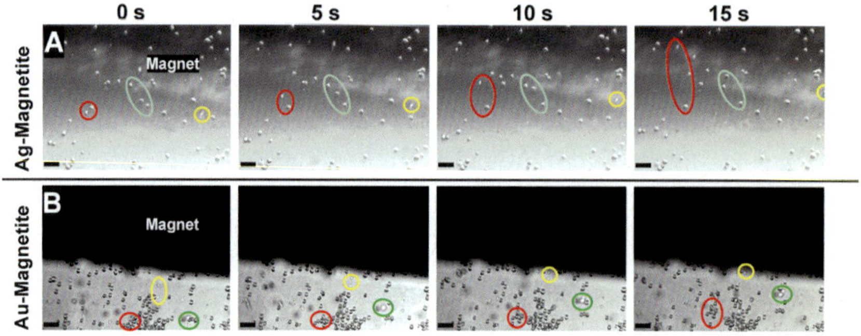

Figure 8.6 Microscopic images of suspended fibroblast cells after 24 h-exposure to (A) Ag-Magnetite and (B) Au-Magnetite. The series of micrographs is taken from a time span of 15 s. The motion of cell groups towards the magnet in the magnetic field is indicated by green, red, and yellow marks. Scale bar: 100 μm. For both composite structures, displacement in the magnetic field is mainly observed for single cells (*yellow labels*) but also for groups of cells (*green* and *red labels*). *Reproduced with permission from T. Buchner, D. Drescher, V. Merk, H. Traub, P. Guttmann, S. Werner, et al., Biomolecular environment, quantification, and intracellular interaction of multifunctional magnetic SERS nanoprobes, Analyst 141 (17) (2016) 5096–5106.*

Interestingly, the uptake efficiency of the magnetite composites is approximately two to three times higher than for the pure gold and silver nanoparticles, respectively [5], which was interpreted to reveal an altered uptake mechanism due to the presence of the magnetite [95,96]. A statistical analysis of the SERS spectra obtained with the composite nanostructures after different times of incubation (Fig. 8.7) showed that, while many bands resemble the cell spectra observed with pure silver nanoparticles, after longer incubation of 24 h (Fig. 8.7), several spectra contain also contributions from intrinsic cellular molecules, for example, of amino acid mixtures and proteins. The spectral fingerprints support that the composition of the biomolecular corona of the composite nanostructures is different from that of silver nanoparticles [5,97], which is suggested to be caused by the magnetite component that preferably interacts with other biomolecules than the silver parts would do in the absence of magnetite [98].

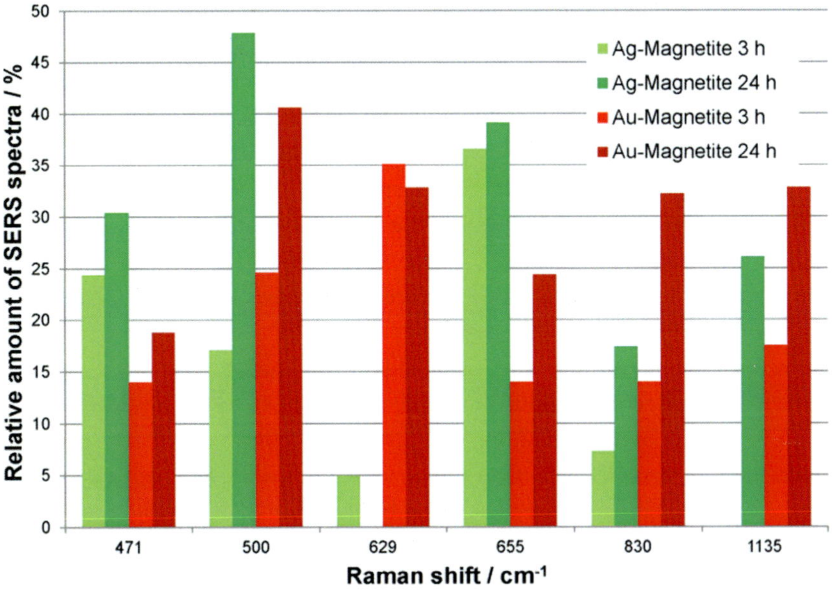

Figure 8.7 Relative amount of spectra containing SERS signals that occur in high abundance in fibroblast cells after incubation with Ag-Magnetite for 3 h (light green) and 24 h (dark green) and with Au-Magnetite for 3 h (light red) and 24 h (dark red). *Reproduced with permission from T. Buchner, D. Drescher, V. Merk, H. Traub, P. Guttmann, S. Werner, et al., Biomolecular environment, quantification, and intracellular interaction of multifunctional magnetic SERS nanoprobes, Analyst 141 (17) (2016) 5096–5106.*

The frequency of typical signals that occur in many of the SERS spectra obtained with both Ag-Magnetite (Fig. 8.7, green bars) and Au-Magnetite (Fig. 8.7, red bars) was analyzed. Especially for the bands at 500 cm^{-1} and at 830 cm^{-1} the occurrence is approximately doubled when incubation time is prolonged from 3 to 24 h (compare the light with dark green and bright with dark red bars, respectively, in Fig. 8.7). Even though all bands shown in Fig. 8.7 are found with both the Ag-Magnetite and Au-Magnetite nanoprobes, their frequency differs between the silver and gold composite structures. Interestingly, when comparing the bands that occur frequently in the composite Au-Magnetite nanoprobes with spectra measured with silica-coated gold nanoparticles reported previously [99], we find many similarities in the spectra, indicating that both silica and magnetite must share parts of their biomolecular surface composition and this composition differs from the surface of gold nanoparticles in a very similar fashion.

Different from a magnetic moiety, a silica shell can also add additional functions to labeled and unlabeled intracellular SERS probes. Specifically, the advantages of these coated SERS probes include physical robustness, stable signals, and immunity to their biological and chemical environment [88,100,101]. Core-shell structures with a plasmonic core that mimic the surface of silica nanoparticles, termed BrightSilica, have different molecules interacting with their surface depending on the cell type and the associated route of uptake [99]. At the same time, the stability of a silica shell in the harsh environment of the endosomes may be compromised, but porosity can be tuned and influenced, therefore, use of such structures in drug delivery has been discussed.

In addition to altered chemical and surface properties, SERS nanoprobes can also become multifunctional with respect to the optical properties inherent to part of an inorganic part of the composite that can enable much more versatile detection by a variety of different optical signals. While the plasmonic moiety itself can be used for SERS imaging and also provides large scattering signals that can be observed in dark field microscopy [102], some inorganic crystals provide very advantageous nonlinear optical properties [103] that make a combined imaging and prescreening of probes possible. Nonlinear processes are attractive in microscopy and spectroscopy since they can be excited with light in the near-infrared, which offers several advantages. The latter include deep tissue penetration capability, reduced photodamage, and spatially well-confined probed volumes, which can result in an improved lateral resolution. While imaging based on SERS spectra gives detailed molecular structure, most nonlinear optical signals, such as second-harmonic generation (SHG), are less specific but lead to information about the microscopic morphology of a sample. SHG is a strong, coherent two-photon process, where a noncentrosymmetric macromolecule or nanostructure yields an effective combination of two photons into a single photon of twice the frequency of the incident beam. It has been frequently used for optical imaging of biological samples [103,104].

As shown recently Madzharova et al., it is possible to extend the capabilities of plasmonic nanoprobes that deliver chemical information through a SERS signature with the requirements of fast nonlinear imaging [23]. There, barium titanate with well-established use as harmonic probe in SHG biomicroscopy [103] was chosen as the core material and functionalized with gold or silver nanoparticles. The concept of this material is displayed in Fig. 8.8A: While the barium titanate core material delivers a strong coherent SHG signal, the plasmonic nanoparticles on the surface can enhance Raman scattering, as well as the nonlinear incoherent signal of hyper Raman scattering [23]. Both SERS and surface-enhanced hyper Raman scattering (SEHRS) [27] produce molecular vibrational information from the cellular environment [28,31,32]. A synthesis method was established that yielded gold nanoparticles of variable sizes on the surface of polycrystalline $BaTiO_3$ nanoparticles, as verified by elemental analysis in scanning transmission electron microscopy (STEM-EDS) (Fig. 8.8B).

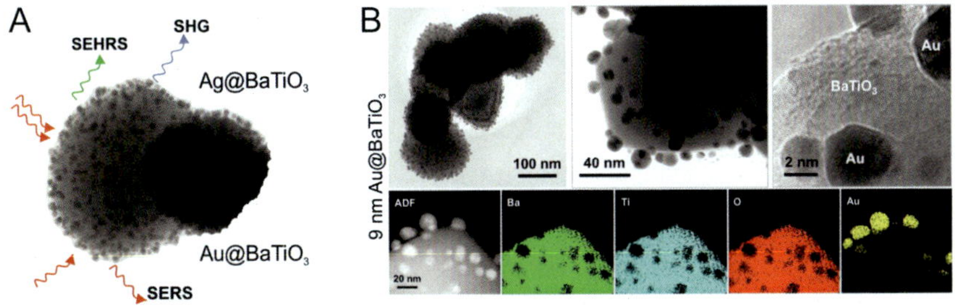

Figure 8.8 (A) Multimodal action of composite $Au@BaTiO_3$ and $Ag@BaTiO_3$ with the $BaTiO_3$ core delivering SHG signal and the plasmonic component enhancing two-photon excited SEHRS as well as SERS. (B) Transmission electron micrographs and STEM-EDS mapping of barium titanate particles with 9 ± 2 nm gold particles (9 nm $Au@BaTiO_3$). *Reproduced with permission from F. Madzharova, Á. Nodar, V. Živanović, M.R.S. Huang, C.T. Koch, R. Esteban, et al., Gold- and silver-coated barium titanate nanocomposites as probes for two-photon multimodal microspectroscopy, Adv. Funct. Mater. 29 (49) (2019) 1904289.*

Apart from these Au@$BaTiO_3$ nanostructures, also silver-$BaTiO_3$ nanocomposites (Ag@$BaTiO_3$) were obtained [23].

Fig. 8.9A shows SHG of the plasmonic barium titanate nanocomposite systems using pulsed laser excitation at 850 nm from aqueous solutions of the composite nanoparticles in backscattering collection geometry. Fig. 8.9A shows the quadratic dependence of the SHG signal on excitation intensity, and confirms the two-photon parametric process of SHG.

As application it was shown that the composite Au@$BaTiO_3$ and Ag@$BaTiO_3$ can be used for combined SHG and vibrational imaging in macrophage cells. For the SHG imaging experiments, J774 macrophage cells were used as a biological model system. In order to increase their biocompatibility and uptake by the cells, the nanoprobes were exposed to lipids [63], and then the macrophage cells were incubated with culture medium containing lipid coated composite nanoprobes for 3 h. SHG

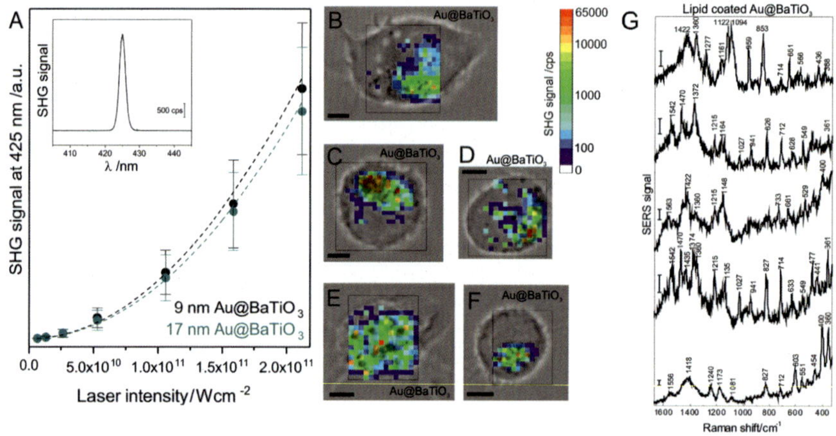

Figure 8.9 (A) SHG signal from 9 nm Au@$BaTiO_3$ (*black*) and 17 nm Au@$BaTiO_3$ (*green*) excited at 850 nm as a function of the peak laser intensity. The values are the mean of 30 measurements with 1 s acquisition time; the dotted lines are the fitted quadratic curves ($ax^2 + b$) to the data points. The inset shows an SHG spectrum of 17 nm Au@$BaTiO_3$ at 5×10^{10} W cm^{-2}. (B–F) Bright-field images of J774 macrophage cells containing lipid coated 9 and 17 nm Au@$BaTiO_3$ nanocomposites, overlaid with SHG maps. The color bar applies to all images B–F. Excitation: 850 nm; acquisition time: 1 s, peak laser intensity: 1×10^{11} W cm^{-2} (5 mW average power); step size: 1 μm; scale bars: 5 μm. (G) SERS spectra of the J774 cells containing Au@$BaTiO_3$, Excitation: 785 nm; acquisition time: 1 s, laser intensity: 8×10^5 W cm^{-2}. *Reproduced with permission from F. Madzharova, Á. Nodar, V. Živanović, M.R.S. Huang, C.T. Koch, R. Esteban, et al., Gold- and silver-coated barium titanate nanocomposites as probes for two-photon multimodal microspectroscopy, Adv. Funct. Mater. 29 (49) (2019) 1904289.*

maps of the cells containing Au@$BaTiO_3$ at 850 nm excitation are shown in Fig. 8.9B–F. The SHG maps show the spatial distribution of the nanoprobes inside the macrophage cells. Comparing the absolute SHG signals from the nanoprobes in the cells with those measured in solution, the former can be significantly higher at some positions in the cells, in agreement with the known formation of aggregate structures in endosomes [105]. Probing of intrinsic cellular structures without reporter molecules using the lipid coated Au@$BaTiO_3$ particles in the macrophage cells is presented in Fig. 8.9G. The SERS spectra show lipid bands due to the lipid coating of the nanoprobes or of cellular membranes, and also bands that are characteristic of other molecules in the cells. The SERS spectra of the intrinsic cellular molecules (Fig. 8.9G) inform us on the chemical environment of the nanoprobes, and thereby complement the morphological information obtained by SHG (Fig. 8.9B–F).

Acknowledgments

I am grateful to Tina Büchner, Daniela Drescher, Fani Madzharova, Virginia Merk, Vesna Živanović, Gergő Péter Szekeres, and Cecilia Spedalieri, who all have contributed to the research on label-free SERS of cultured cells and biomolecules in our group over the last decade and to the works discussed here, as well as to all collaborators and coauthors on the discussed papers. Funding for this research was provided by ERC grant 259432, by DFG GSC 1013 SALSA, and by Fonds der Chemischen Industrie.

References

[1] J. Langer, D.J. de Aberasturi, J. Aizpurua, R.A. Alvarez-Puebla, B. Auguié, J.J. Baumberg, et al., Present and future of surface-enhanced Raman scattering, ACS Nano 14 (1) (2020) 28–117.
[2] S.E.J. Bell, G. Charron, E. Cortés, J. Kneipp, M.L. de la Chapelle, J. Langer, et al., Towards reliable and quantitative surface-enhanced Raman scattering (SERS): from key parameters to good analytical practice, Angew. Chem. Int. (Ed.) 59 (14) (2020) 5454–5462.
[3] J. Baumberg, S. Bell, A. Bonifacio, R. Chikkaraddy, M. Chisanga, S. Corsetti, et al., SERS in biology/biomedical SERS: general discussion, Faraday Discuss. 205 (2017) 429–456.
[4] L.A. Austin, B. Kang, M.A. El-Sayed, Probing molecular cell event dynamics at the single-cell level with targeted plasmonic gold nanoparticles: a review, Nano Today 10 (5) (2015) 542–558.
[5] T. Buchner, D. Drescher, V. Merk, H. Traub, P. Guttmann, S. Werner, et al., Biomolecular environment, quantification, and intracellular interaction of multifunctional magnetic SERS nanoprobes, Analyst 141 (17) (2016) 5096–5106.
[6] X.X. Han, Y. Kitahama, T. Itoh, C.X. Wang, B. Zhao, Y. Ozaki, Protein-mediated sandwich strategy for surface-enhanced raman scattering: application to versatile protein detection, Anal. Chem. 81 (9) (2009) 3350–3355.

[7] I. Pavel, E. McCarney, A. Elkhaled, A. Morrill, K. Plaxco, M. Moskovits, Label-free SERS detection of small proteins modified to act as bifunctional linkers, J. Phys. Chem. C 112 (13) (2008) 4880–4883.
[8] S. Sloan-Dennison, Z.D. Schultz, Label-free plasmonic nanostar probes to illuminate: in vitro membrane receptor recognition, Chem. Sci. 10 (6) (2019) 1807–1815.
[9] V. Caprettini, J.A. Huang, F. Moia, A. Jacassi, C.A. Gonano, N. Maccaferri, et al., Angelis, enhanced Raman investigation of cell membrane and intracellular compounds by 3D plasmonic nanoelectrode arrays, Adv. Sci. 5 (12) (2018).
[10] E.A. Vitol, Z. Orynbayeva, M.J. Bouchard, J. Azizkhan-Clifford, G. Friedman, Y. Gogotsi, In situ intracellular spectroscopy with surface enhanced Raman spectroscopy (SERS)-enabled nanopipettes, ACS Nano 3 (11) (2009) 3529–3536.
[11] J. Dong, Q. Chen, C. Rong, D. Li, Y. Rao, Minimally invasive surface-enhanced Raman scattering detection with depth profiles based on a surface-enhanced Raman scattering-active acupuncture needle, Anal. Chem. 83 (16) (2011) 6191–6195.
[12] C. Yuen, Q. Liu, Towards in vivo intradermal surface enhanced Raman scattering (SERS) measurements: silver coated microneedle based SERS probe, J. Biophotonics 7 (9) (2014) 683–689.
[13] F. Lussier, T. Brulé, M. Vishwakarma, T. Das, J.P. Spatz, J.F. Masson, Dynamic-SERS optophysiology: a nanosensor for monitoring cell secretion events, Nano Lett. 16 (6) (2016) 3866–3871.
[14] P. Li, B. Zhou, X. Cao, X. Tang, L. Yang, L. Hu, et al., Functionalized acupuncture needle as surface-enhanced resonance Raman spectroscopy sensor for rapid and sensitive detection of dopamine in serum and cerebrospinal fluid, Chem. – A Eur. J. 23 (57) (2017) 14278–14285.
[15] M. Wang, X. Cao, W. Lu, L. Tao, H. Zhao, Y. Wang, et al., Surface-enhanced Raman spectroscopic detection and differentiation of lung cancer cell lines (A549, H1229) and normal cell line (AT II) based on gold nanostar substrates, RSC Adv. 4 (109) (2014) 64225–64234.
[16] Y.J. Zhang, Q.Y. Zeng, L.F. Li, M.N. Qi, Q.C. Qi, S.X. Li, et al., Label-free rapid identification of tumor cells and blood cells with silver film SERS substrate, Opt. Express 26 (25) (2018) 33044–33056.
[17] K. Kneipp, Y. Wang, H. Kneipp, L.T. Perelman, I. Itzkan, R.R. Dasari, et al., Single molecule detection using surface-enhanced Raman scattering (SERS), Phys. Rev. Lett. 78 (1997) 1667.
[18] H.X. Xu, J. Aizpurua, M. Kall, P. Apell, Electromagnetic contributions to single-molecule sensitivity in surface-enhanced Raman scattering, Phys. Rev. E 62 (3) (2000) 4318–4324.
[19] V. Živanović, Z. Kochovski, C. Arenz, Y. Lu, J. Kneipp, SERS and Cryo-EM directly reveal different liposome structures during interaction with gold nanoparticles, J. Phys. Chem. Lett. 9 (23) (2018) 6767–6772.
[20] G.P. Szekeres, M. Montes-Bayón, J. Bettmer, J. Kneipp, Fragmentation of proteins in the corona of gold nanoparticles as observed in live cell surface-enhanced Raman scattering, Anal. Chem. 92 (12) (2020) 8553–8560.
[21] V. Živanović, F. Madzharova, Z. Heiner, C. Arenz, J. Kneipp, Specific interaction of tricyclic antidepressants with gold and silver nanostructures as revealed by combined one- and two-photon vibrational spectroscopy, J. Phys. Chem. C. 121 (41) (2017) 22958–22968.
[22] L.A. Austin, B. Kang, C.-W. Yen, M.A. El-Sayed, Plasmonic imaging of human oral cancer cell communities during programmed cell death by nuclear-targeting silver nanoparticles, J. Am. Chem. Soc. 133 (44) (2011) 17594–17597.
[23] F. Madzharova, Á. Nodar, V. Živanović, M.R.S. Huang, C.T. Koch, R. Esteban, et al., Gold- and silver-coated barium titanate nanocomposites as probes for two-photon multimodal microspectroscopy, Adv. Funct. Mater. 29 (49) (2019) 1904289.

[24] D. Drescher, T. Buchner, D. McNaughton, J. Kneipp, SERS reveals the specific interaction of silver and gold nanoparticles with hemoglobin and red blood cell components, Phys. Chem. Chem. Phys. 15 (15) (2013) 5364–5373.
[25] A.V. Baranov, Y.S. Bobovich, Super-enhanced hyper-Raman scattering from dyes adsorbed on colloidal silver particles, JETP Lett. 36 (8) (1982) 339–343.
[26] D.V. Murphy, K.U. Vonraben, R.K. Chang, P.B. Dorain, Surface-enhanced hyper-Raman scattering from SO32- adsorbed on Ag powder, Chem. Phys. Lett. 85 (1) (1982) 43–47.
[27] F. Madzharova, Z. Heiner, J. Kneipp, Surface enhanced hyper Raman scattering (SEHRS) and its applications, Chem. Soc. Rev. 46 (13) (2017) 3980–3999.
[28] J. Kneipp, H. Kneipp, K. Kneipp, Two-photon vibrational spectroscopy for biosciences based on surface-enhanced hyper-Raman scattering, Proc. Natl. Acad. Sci. U S A. 103 (46) (2006) 17149–17153.
[29] J. Kneipp, H. Kneipp, B. Wittig, K. Kneipp, One- and two-photon excited optical pH probing for cells using surface-enhanced Raman and hyper-Raman nanosensors, Nano Lett. 7 (2007) 2819–2823.
[30] Z. Heiner, M. Gühlke, V. Živanović, F. Madzharova, J. Kneipp, Surface-enhanced hyper Raman hyperspectral imaging and probing in animal cells, Nanoscale 9 (23) (2017) 8024–8032.
[31] J. Kneipp, Interrogating cells, tissues, and live animals with new generations of surface-enhanced raman scattering probes and labels, ACS Nano 11 (2) (2017) 1136–1141.
[32] Z. Heiner, F. Madzharova, V. Živanović, J. Kneipp, Bio-probing with nonresonant surface-enhanced hyper-Raman scattering excited at 1550 nm, J. Raman Spectrosc. (2020).
[33] D. Drescher, C. Giesen, H. Traub, U. Panne, J. Kneipp, N. Jakubowski, Quantitative imaging of gold and silver nanoparticles in single eukaryotic cells by laser ablation ICP-MS, Anal. Chem. 84 (22) (2012) 9684–9688.
[34] D. Drescher, H. Traub, T. Büchner, N. Jakubowski, J. Kneipp, Properties of in situ generated gold nanoparticles in the cellular context, Nanoscale 9 (32) (2017) 11647–11656.
[35] G.P. Szekeres, S. Werner, P. Guttmann, C. Spedalieri, D. Drescher, V. Živanovic, et al., Relating the composition and interface interactions in the hard corona of gold nanoparticles to the induced response mechanisms in living cells, Nanoscale 12 (33) (2020) 17450–17461.
[36] C. Spedalieri, G.P. Szekeres, S. Werner, P. Guttmann, J. Kneipp, Intracellular optical probing with gold nanostars, Nanoscale 13 (2) (2021) 968–979.
[37] J.S. Anshup, C. Venkataraman, R.R. Subramaniam, S. Kumar, T.R.S. Priya, R.V. Kumar, et al., Growth of gold nanoparticles in human cells, Langmuir 21 (25) (2005) 11562–11567.
[38] H. Huang, W.W. Chen, J.J. Pan, Q.S. Chen, S.Y. Feng, Y. Yu, et al., SERS spectra of a single nasopharyngeal carcinoma cell based on intracellularly grown and passive uptake Au nanoparticles, Spectrosc. Biomed. Appl. 26 (3) (2011) 187–194.
[39] Z.M. Liu, C.F. Hu, S.X. Li, W. Zhang, Z.Y. Guo, Rapid intracellular growth of gold nanostructures assisted by functionalized graphene oxide and its application for surface-enhanced Raman spectroscopy, Anal. Chem. 84 (23) (2012) 10338–10344.
[40] A. Shamsaie, M. Jonczyk, J. Sturgis, J.P. Robinson, J. Irudayaraj, Intracellularly grown gold nanoparticles as potential surface-enhanced Raman scattering probes, J. Biomed. Opt. 12 (2) (2007).
[41] W.L. Fu, S.J. Zhen, C.Z. Huang, One-pot green synthesis of graphene oxide/gold nanocomposites as SERS substrates for malachite green detection, Analyst 138 (10) (2013) 3075–3081.

[42] S. Chattoraj, K. Bhattacharyya, Fluorescent gold nanocluster inside a live breast cell: etching and higher uptake in cancer cell, J. Phys. Chem. C 118 (38) (2014) 22339–22346.
[43] J.P. Xie, Y.G. Zheng, J.Y. Ying, Protein-directed synthesis of highly fluorescent gold nanoclusters, J. Am. Chem. Soc. 131 (3) (2009) 888.
[44] S. Chattoraj, M.A. Amin, S. Mohapatra, S. Ghosh, K. Bhattacharyya, Cancer cell imaging using in situ generated gold nanoclusters, Chemphyschem 17 (1) (2016) 61–68.
[45] P. He, X.Y. Zhu, Phospholipid-assisted synthesis of size-controlled gold nanoparticles, Mater. Res. Bull. 42 (7) (2007) 1310–1315.
[46] Z.D. Mu, X.W. Zhao, Z.Y. Xie, Y.J. Zhao, Q.F. Zhong, L. Bo, et al., In situ synthesis of gold nanoparticles (AuNPs) in butterfly wings for surface enhanced Raman spectroscopy (SERS), J. Mater. Chem. B 1 (11) (2013) 1607–1613.
[47] S.P. Shukla, M. Roy, P. Mukherjee, A.K. Tyagi, T. Mukherjee, S. Adhikari, Interaction of bilirubin with Ag and Au ions: green synthesis of bilirubin-stabilized nanoparticles, J. Nanopart. Res. 14 (7) (2012).
[48] K. Kneipp, A.S. Haka, H. Kneipp, K. Badizadegan, N. Yoshizawa, C. Boone, et al., Surface-enhanced Raman spectroscopy in single living cells using gold nanoparticles, Appl. Spectrosc. 56 (2) (2002) 150–154.
[49] R. Manoharan, J.J. Baraga, M.S. Feld, R.P. Rava, Quantitative histochemical analysis of human artery using Raman spectroscopy, J. Photochem. Photobiol. B Biol. 16 (2) (1992) 211–233.
[50] L. Wang, N. Malmstadt, Interactions between charged nanoparticles and giant vesicles fabricated from inverted-headgroup lipids, J. Phys. D Appl. Phys. 50 (41) (2017) 415402.
[51] B. Moody, C.M. Haslauer, E. Kirk, A. Kannan, E.G. Loboa, G.S. McCarty, In situ monitoring of adipogenesis with human-adipose-derived stem cells using surface-enhanced Raman spectroscopy, Appl. Spectrosc. 64 (11) (2010) 1227–1233.
[52] V. Živanović, G. Semini, M. Laue, D. Drescher, T. Aebischer, J. Kneipp, Chemical mapping of Leishmania infection in live cells by SERS microscopy, Anal. Chem. 90 (13) (2018) 8154–8161.
[53] A.H. Futerman, G. van Meer, The cell biology of lysosomal storage disorders, Nat. Rev. Mol. Cell Biol. 5 (7) (2004) 554–565.
[54] S. Stremersch, M. Marro, B.E. Pinchasik, P. Baatsen, A. Hendrix, S.C. De Smedt, et al., Identification of individual exosome-like vesicles by surface enhanced Raman spectroscopy, Small 12 (24) (2016) 3292–3301.
[55] H. Shin, H. Jeong, J. Park, S. Hong, Y. Choi, Correlation between cancerous exosomes and protein markers based on surface-enhanced Raman spectroscopy (SERS) and principal component analysis (PCA), ACS Sens. 3 (12) (2018) 2637–2643.
[56] J. Carmicheal, C. Hayashi, X. Huang, L. Liu, Y. Lu, A. Krasnoslobodtsev, et al., Label-free characterization of exosome via surface enhanced Raman spectroscopy for the early detection of pancreatic cancer, Nanomed. Nanotechnol. Biol. Med. 16 (2019) 88–96.
[57] J. Park, M. Hwang, B. Choi, H. Jeong, J.H. Jung, H.K. Kim, et al., Exosome classification by pattern analysis of surface-enhanced Raman spectroscopy data for lung cancer diagnosis, Anal. Chem. 89 (12) (2017) 6695–6701.
[58] G. Semini, T. Aebischer, Phagosome proteomics to study Leishmania's intracellular niche in macrophages, Int. J. Med. Microbiol. 308 (1) (2018) 68–76.
[59] G. Semini, D. Paape, A. Paterou, J. Schroeder, M. Barrios-Llerena, T. Aebischer, Changes to cholesterol trafficking in macrophages by Leishmania parasites infection, MicrobiologyOpen 6 (4) (2017).
[60] Ld Chiu, F. Hullin-Matsuda, T. Kobayashi, H. Torii, Ho Hamaguchi, On the origin of the 1602 cm^{-1} Raman band of yeasts; contribution of ergosterol, J. Biophotonics 5 (10) (2012) 724–728.

[61] L.J. Goad, G.G. Holz Jr, D.H. Beach, Sterols of ketoconazole-inhibited *Leishmania mexicana* mexicana promastigotes, Mol. Biochem. Parasitol. 15 (3) (1985) 257–279.
[62] M. Driver, Y. Li, J. Zheng, E. Decker, D. Julian McClements, L. He, Fabrication of lipophilic gold nanoparticles for studying lipids by surface enhanced Raman spectroscopy (SERS), Analyst 139 (13) (2014) 3352–3355.
[63] J.R. Matthews, C.M. Payne, J.H. Hafner, Analysis of phospholipid bilayers on gold nanorods by plasmon resonance sensing and surface-enhanced Raman scattering, Langmuir 31 (36) (2015) 9893–9900.
[64] K. Suga, T. Yoshida, H. Ishii, Y. Okamoto, D. Nagao, M. Konno, et al., Membrane surface-enhanced Raman spectroscopy for sensitive detection of molecular behavior of lipid assemblies, Anal. Chem. 87 (9) (2015) 4772–4780.
[65] R.W. Taylor, F. Benz, D.O. Sigle, R.W. Bowman, P. Bao, J.S. Roth, et al., Watching individual molecules flex within lipid membranes using SERS, Sci. Rep. 4 (2014) 5940.
[66] V. Živanović, J. Kneipp, Nano-bio interactions as characterized by SERS: the interaction of liposomes with gold nanostructures is highly dependent on lipid composition and charge, Proc. SPIE 10894, Plasmon. Biol. Med. XVI 10894 (2019) 1089404.
[67] A. Jaworska, K. Malek, A comparison between adsorption mechanism of tricyclic antidepressants on silver nanoparticles and binding modes on receptors. Surface-enhanced Raman Spectroscopic Studies, J. Colloid Interface Sci. 431 (2014) 117–124.
[68] V. Zivanovic, S. Seifert, D. Drescher, P. Schrade, S. Werner, P. Guttmann, et al., Optical nanosensing of lipid accumulation due to enzyme inhibition in live cells, ACS Nano (2019).
[69] A. Huefner, W.-L. Kuan, R.A. Barker, S. Mahajan, Intracellular SERS nanoprobes for distinction of different neuronal cell types, Nano Lett. 13 (6) (2013) 2463–2470.
[70] A. Matschulat, D. Drescher, J. Kneipp, Surface-enhanced Raman scattering hybrid nanoprobe multiplexing and imaging in biological systems, ACS Nano 4 (6) (2010) 3259–3269.
[71] S. Seifert, V. Merk, J. Kneipp, Identification of aqueous pollen extracts using surface enhanced Raman scattering (SERS) and pattern recognition methods, J. Biophotonics 9 (1–2) (2016) 181–189.
[72] A. Walter, A. März, W. Schumacher, P. Rösch, J. Popp, Towards a fast, high specific and reliable discrimination of bacteria on strain level by means of SERS in a microfluidic device, Lab Chip 11 (6) (2011) 1013–1021.
[73] K. Qian, Y. Wang, L. Hua, A. Chen, Y. Zhang, New method of lung cancer detection by saliva test using surface-enhanced Raman spectroscopy, Thorac. Cancer 9 (11) (2018) 1556–1561.
[74] K.E. McCloskey, J.J. Chalmers, M. Zborowski, Magnetic cell separation: characterization of magnetophoretic mobility, Anal. Chem. 75 (24) (2003) 6868–6874.
[75] B. Polyak, I. Fishbein, M. Chorny, I. Alferiev, D. Williams, B. Yellen, et al., High field gradient targeting of magnetic nanoparticle-loaded endothelial cells to the surfaces of steel stents, Proc. Natl. Acad. Sci. U S A 105 (2) (2008) 698–703.
[76] Y. Jing, N. Mal, P.S. Williams, M. Mayorga, M.S. Penn, J.J. Chalmers, et al., Quantitative intracellular magnetic nanoparticle uptake measured by live cell magnetophoresis, FASEB J 22 (12) (2008) 4239–4247.
[77] N. Xia, T. Hunt, B. Mayers, E. Alsberg, G. Whitesides, R. Westervelt, et al., Combined microfluidic-micromagnetic separation of living cells in continuous flow, Biomed. Microdevices 8 (4) (2006) 299–308.
[78] K. Kim, H.J. Jang, K.S. Shin, Ag nanostructures assembled on magnetic particles for ready SERS-based detection of dissolved chemical species, Analyst 134 (2) (2009) 308–313.

[79] N. Pamme, C. Wilhelm, Continuous sorting of magnetic cells via on-chip free-flow magnetophoresis, Lab Chip 6 (8) (2006) 974–980.
[80] J.D. Adams, U. Kim, H.T. Soh, Multitarget magnetic activated cell sorter, Proc. Natl. Acad. Sci. U S A 105 (47) (2008) 18165–18170.
[81] I.I.S. Lim, N.N. Peter, P. Hye-Young, W. Xin, W. Lingyan, M. Derrick, Z. Chuan-Jian, Gold and magnetic oxide/gold core/shell nanoparticles as bio-functional nanoprobes, Nanotechnology 19 (30) (2008) 305102.
[82] M. Spuch-Calvar, L. Rodríguez-Lorenzo, M.P. Morales, R.A. Álvarez-Puebla, L.M. Liz-Marzán, Bifunctional nanocomposites with long-term stability as SERS optical accumulators for ultrasensitive analysis, J. Phys. Chem. C 113 (9) (2009) 3373–3377.
[83] M. Gühlke, S. Selve, J. Kneipp, Magnetic separation and SERS observation of analyte molecules on bifunctional silver/iron oxide composite nanostructures, J. Raman Spectrosc. 43 (9) (2012) 1204–1207.
[84] X.X. Han, A.M. Schmidt, G. Marten, A. Fischer, I.M. Weidinger, P. Hildebrandt, Magnetic silver hybrid nanoparticles for surface-enhanced resonance Raman spectroscopic detection and decontamination of small toxic molecules, ACS Nano 7 (4) (2013) 3212–3220.
[85] T. Donnelly, W.E. Smith, K. Faulds, D. Graham, Silver and magnetic nanoparticles for sensitive DNA detection by SERS, Chem. Commun. 50 (85) (2014) 12907–12910.
[86] A. La Porta, A. Sanchez-Iglesias, T. Altantzis, S. Bals, M. Grzelczak, L.M. Liz-Marzan, Multifunctional self-assembled composite colloids and their application to SERS detection, Nanoscale 7 (23) (2015) 10377–10381.
[87] M.S. Noh, B.-H. Jun, S. Kim, H. Kang, M.-A. Woo, A. Minai-Tehrani, et al., Magnetic surface-enhanced Raman spectroscopic (M-SERS) dots for the identification of bronchioalveolar stem cells in normal and lung cancer mice, Biomaterials 30 (23–24) (2009) 3915–3925.
[88] B.-H. Jun, M.S. Noh, J. Kim, G. Kim, H. Kang, M.-S. Kim, et al., Multifunctional silver-embedded magnetic nanoparticles as SERS nanoprobes and their applications, Small 6 (1) (2010) 119–125.
[89] Y. Liu, Z. Chang, H. Yuan, A.M. Fales, T. Vo-Dinh, Quintuple-modality (SERS-MRI-CT-TPL-PTT) plasmonic nanoprobe for theranostics, Nanoscale 5 (24) (2013) 12126–12131.
[90] S. Charan, C.W. Kuo, Y.-W. Kuo, N. Singh, P. Drake, Y.-J. Lin, et al., Synthesis of surface enhanced Raman scattering active magnetic nanoparticles for cell labeling and sorting, J. Appl. Phys. 105 (7) (2009) 07B310.
[91] F. Bertorelle, M. Ceccarello, M. Pinto, G. Fracasso, D. Badocco, V. Amendola, et al., Efficient AuFeOx nanoclusters of laser-ablated nanoparticles in water for cells guiding and surface-enhanced resonance Raman scattering imaging, J. Phys. Chem. C 118 (26) (2014) 14534–14541.
[92] L. Zhang, J. Xu, L. Mi, H. Gong, S. Jiang, Q. Yu, Multifunctional magnetic–plasmonic nanoparticles for fast concentration and sensitive detection of bacteria using SERS, Biosens. Bioelectron. 31 (1) (2012) 130–136.
[93] Z. Fan, D. Senapati, S.A. Khan, A.K. Singh, A. Hamme, B. Yust, et al., Popcorn-shaped magnetic core–plasmonic shell multifunctional nanoparticles for the targeted magnetic separation and enrichment, label-free SERS imaging, and photothermal destruction of multidrug-resistant bacteria, Chem. Eur. J. 19 (8) (2013) 2839–2847.
[94] U. Tamer, D. Cetin, Z. Suludere, I.H. Boyaci, H.T. Temiz, H. Yegenoglu, et al., Gold-coated iron composite nanospheres targeted the detection of *Escherichia coli*, Int. J. Mol. Sci. 14 (3) (2013) 6223–6240.
[95] K.K. Comfort, E.I. Maurer, L.K. Braydich-Stolle, S.M. Hussain, Interference of silver, gold, and iron oxide nanoparticles on epidermal growth factor signal transduction in epithelial cells, ACS Nano 5 (12) (2011) 10000–10008.

[96] C. Wilhelm, F. Gazeau, J. Roger, J.N. Pons, J.C. Bacri, Interaction of anionic superparamagnetic nanoparticles with cells: kinetic analyses of membrane adsorption and subsequent internalization, Langmuir 18 (21) (2002) 8148–8155.
[97] D. Drescher, P. Guttmann, T. Buchner, S. Werner, G. Laube, A. Hornemann, et al., Specific biomolecule corona is associated with ring-shaped organization of silver nanoparticles in cells, Nanoscale 5 (19) (2013) 9193–9198.
[98] E. Casals, T. Pfaller, A. Duschl, G.J. Oostingh, V.F. Puntes, Hardening of the nanoparticle–protein corona in metal (Au, Ag) and oxide (Fe_3O_4, CoO, and CeO_2) nanoparticles, Small 7 (24) (2011) 3479–3486.
[99] D. Drescher, I. Zeise, H. Traub, P. Guttmann, S. Seifert, T. Buchner, et al., In situ characterization of SiO_2 nanoparticle biointeractions using brightsilica, Adv. Funct. Mater. 24 (24) (2014) 3765–3775.
[100] R.A. Jensen, J. Sherin, S.R. Emory, Single nanoparticle based optical pH probe, Appl. Spectrosc. 61 (8) (2007) 832–838.
[101] J.V. Jokerst, Z. Miao, C. Zavaleta, Z. Cheng, S.S. Gambhir, Affibody-functionalized gold-silica nanoparticles for Raman molecular imaging of the epidermal growth factor receptor, Small 7 (5) (2011) 625–633.
[102] B. Kang, L.A. Austin, M.A. El-Sayed, Observing real-time molecular event dynamics of apoptosis in living cancer cells using nuclear-targeted plasmonically enhanced Raman nanoprobes, ACS Nano 8 (5) (2014) 4883–4892.
[103] P. Pantazis, J. Maloney, D. Wu, S.E. Fraser, Second harmonic generating (SHG) nanoprobes for in vivo imaging, Proc. Natl. Acad. Sci. U S A 107 (33) (2010) 14535–14540.
[104] D. Staedler, T. Magouroux, R. Hadji, C. Joulaud, J. Extermann, S. Schwung, et al., Harmonic nanocrystals for biolabeling: a survey of optical properties and biocompatibility, ACS Nano 6 (3) (2012) 2542–2549.
[105] D. Drescher, T. Büchner, P. Guttmann, S. Werner, G. Schneider, J. Kneipp, X-ray tomography shows the varying three-dimensional morphology of gold nanoaggregates in the cellular ultrastructure, Nanoscale Adv. (2019).

CHAPTER 9

iSERS microscopy: point-of-care diagnosis and tissue imaging

Yuying Zhang[1,*], Vi. Tran[2,*], Mujo Adanalic[2] and Sebastian Schlücker[2]
[1]Medical School of Nankai University, Tianjin, P.R. China
[2]Department of Chemistry, CENIDE and ZMB, University of Duisburg-Essen, Essen, Germany

9.1 Point-of-care diagnosis

Point-of-care (POC) diagnosis is used for the on-site analysis close to the patient for the rapid detection of analytes with reducing costs and time in comparison to diagnostic testing in central laboratories. Early diagnosis, that is, diagnosis of a disease in its initial stages, can lead to proper early start of medical treatment resulting in improved health outcomes [1,2]. Characteristic features of POC testing (POCT) are: testing outside of a central laboratory, no need of sample preparations and pipetting steps, ready-to-use reagents, the use of special testing and portable devices, no qualified technicians or clinical personnel needed, and easy-to-use for end users [3].

The application field of membrane-based POCT ranges from clinical diagnostics to areas as agriculture, aquaculture, biowarfare, food safety, environmental health and safety, industrial testing, molecular diagnostics and theranostics on human samples as well as veterinary medicine [4]. Lateral flow assays (LFA, Fig. 9.1) are attractive and well-known for their user-friendly, low-cost format, and its simplicity since they typically do not require a readout device [4,6]. Samples are, for example, body fluids (e.g., whole blood, serum, urine, or saliva), milk, or water for detecting various molecules such as cancer markers, mycotoxins, pesticides, drugs, heavy metal ions, or even microorganisms [4,7]. The selectivity of the LFA is based on molecular recognition, that is, it exploits noncovalent interactions such as the selective recognition of an antigen by its antibody (key-lock principle/induced fit) or Watson-Crick based pairing between complementary oligonucleotide strands (hybridization) [7]. The first LFA, a urine

* These authors contributed equally.

Principles and Clinical Diagnostic Applications of Surface-Enhanced Raman Spectroscopy
DOI: https://doi.org/10.1016/B978-0-12-821121-2.00002-0

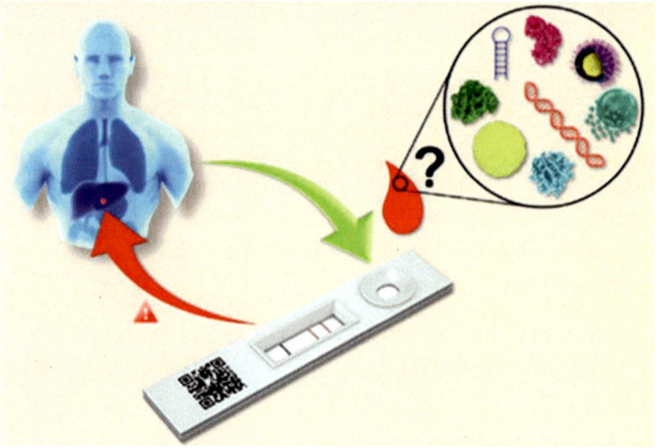

Figure 9.1 Point-of-care detection of diseases via lateral flow assay technology [5].

pregnancy test, was developed in 1976 for home use [4,7]. Since then the powerful technology has received a lot of popularity and is nowadays very widely used [4]. The central advantage of this technology is the fast outcome of a positive or negative visual test result as a yes/no answer within a few minutes (10–20 minutes) [8]. The qualitative test can therefore be visually evaluated shortly after application, usually without the need of additional devices. The simplicity of this membrane-based nitrocellulose (NC) assay is based on the fact that no trained person is required for carrying out the assay and that end users can perform the assay anywhere and at any time [7,9]. In addition, only small sample volumes are required for the assay [8]. The portable and compact test strip is based on various membranes and is very inexpensive to manufacture and also very reliable. The assay is characterized by a very high reproducibility of the obtained test results and by a high level of robustness, that is, a shelf life of around 12–24 months after production can be guaranteed for the test [8,10]. The manufacturing of the LFA strips is simple and inexpensive. This makes it possible to produce an assay in large numbers by easy scalability after the research and development phase [10]. The focus in the first years after LFA development was primarily on a cost-effective qualitative test and the development toward quantification has only taken place in recent years [4]. However, the improvement of the sensitivity and the quantification as well as multiplexing for parallel detection for this type of assay still remain a great challenge [4].

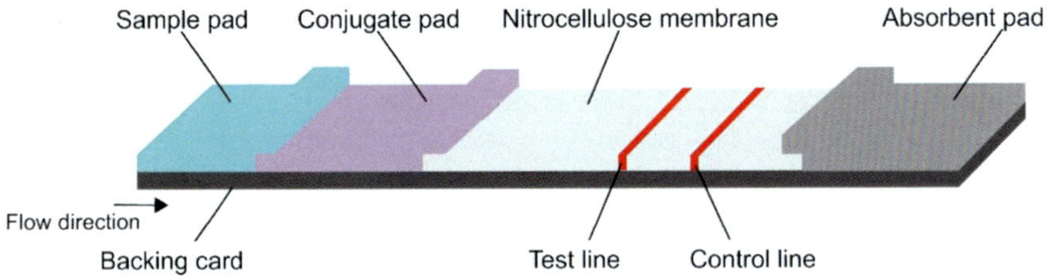

Figure 9.2 Compartments of a lateral flow assay test strip: sample pad, conjugate pad, nitrocellulose membrane, and absorbent pad. All materials are assembled onto a backing card. The sample probe is driven through the assay via capillary forces.

9.1.1 Principle of a lateral flow assay

The LFA test strip is composed of four parts: a sample pad, a conjugation pad, a reaction test membrane, and an absorbent pad (Fig. 9.2) [7].

The sample pad consists of cellulose or glass fiber, onto which the sample probe is applied. The sample pad overlaps with the conjugate pad. The conjugates—the labeled target-specific ligands such as antibody/gold nanoparticle conjugates—are embedded and dried in the conjugate pad, usually glass fiber. In the following, we will focus on antibody-based test as very widely used LFA. Here, the first immunochemical reaction is the recognition of the corresponding analyte (the antigen) by the antibodies of the conjugates in the conjugate pad. This immunocomplex is driven by capillary forces further into the direction of the test line (TL) and control line (CL). There, the second immunochemical reaction takes place: with the biorecognition elements at the control zone (for both positive and negative tests) and at the test zone (only in positive tests). The absorption pad (cellulose/glass fiber) stops the markers from passing back into the test membrane. For stability and better handling of the test strip, all components are overlapped on one another on a plastic backing card, which is made of polystyrene, polyester, or polyvinyl [4,7]. The test membrane typically consists of NC. This chromatographic system enables fluid transport. The sample flows through the membrane due to capillary forces. Onto the NC membrane, capture and control antibodies are immobilized on the corresponding TL and CL via contact or noncontact dispensing systems (tip or ink jet) [4]. The proteins interact electrostatically with the membrane via the strong dipoles of the nitrate ester of NC, for example, via hydrogen bonds and hydrophobic interactions. Therefore the NC has a high protein binding capacity. After the immobilization step the membrane is treated with a

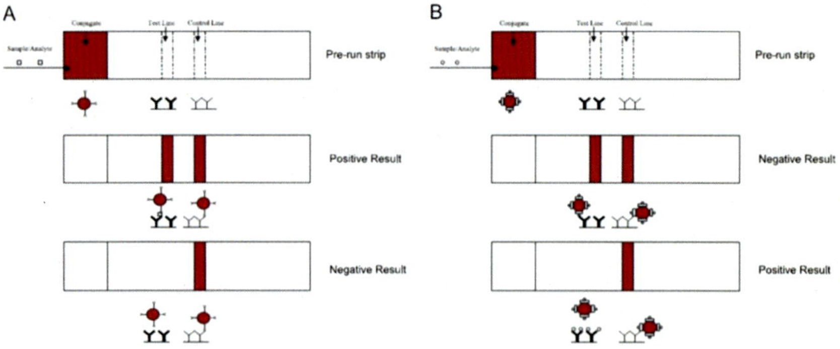

Figure 9.3 Two different types of lateral flow assay (LFA): sandwich LFA (A) and competitive LFA (B). In positive outcomes, two colored bands appear at test line and control line (CL) in sandwich assays, whereas only one colored band (CL) appears in competitive assays. The opposite case is observed for negative results [10].

blocking matrix (BSA, sucrose, casein, polymer, etc.) so that free potential protein binding sites are blocked to minimize nonspecific binding and background staining [4,7]. Two formats of the assay are available: sandwich assay and competitive assay (Fig. 9.3) [4,9,10].

Quasispherical colloidal gold nanoparticles (AuNP) with a size of 20–40 nm are routinely used labels in the LFA strips. A specific detection antibody (Ab) is conjugated covalently or traditionally electrostatically on the nanoparticle surface [10]. In a sandwich assay, three different antibodies are used. The detection Ab is conjugated to the labeling agents, for example, AuNP. This antibody binds specifically to an epitope of the target analyte molecule. The capture Ab is immobilized on the membrane on the TL, which specifically binds to the other epitope of the analyte. If the analyte is present, a sandwich immunocomplex is formed and the corresponding TL appears red-colored due to the gold conjugate. In the absence of the analyte, this TL remains unchanged since no gold conjugates can be bound via the complex. On the CL there are species-specific antiimmunoglobulin Abs (control antibodies) which form antibody–antibody complexes with the detection Abs on the AuNP, which indicates the functionality of the used antibodies and thus of the whole test validity. In a positive test two colored bands are expected, while in a negative test only one colored band at the CL occurs. However, if no CL appears at all, then the test is invalid. Competitive assays are used for detecting small molecules (e.g., drugs) with only one antigenic determinant, which are not feasible to bind two antibodies simultaneously. The positive result in this case is indicated by only one colored line at the CL [10].

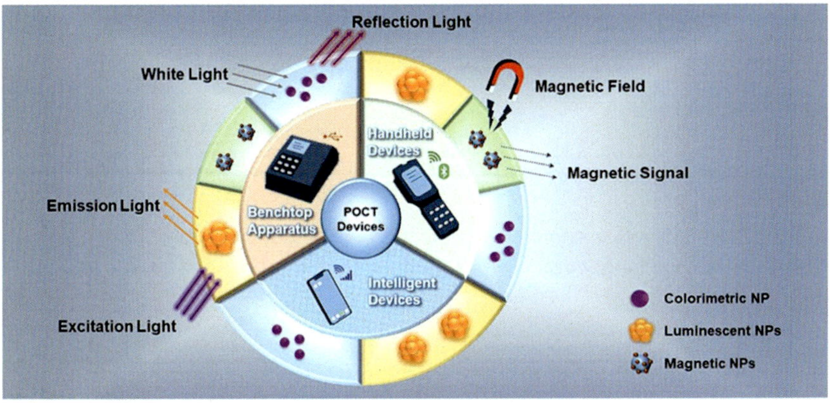

Figure 9.4 Point-of-care testing (POCT) devices: rapid POCT platforms with different size and performance: bench-top apparatuses, handheld devices, and detection devices from colorimetry, fluorescence to magnetism [11].

Different labels like colloidal nanoparticles, microbeads with dyes or magnetic beads are used so far. The readout of a LFA (see Fig. 9.4) can be colorimetric by naked eye, but also via fluorescent, magnetic, and electrochemical as well as optical detection methods including surface-enhanced Raman scattering (SERS) [4,11,12]. SERS implemented in the LFA technology is promising and very powerful concerning quantification, lowering the limit of detection (LOD) and therefore increasing the sensitivity of the assay and multiplexing. Multiplexing is possible by using a set of spectrally distinct SERS nanotag/antibody conjugates with different Raman reporter molecules on the gold surface: this enables the simultaneous spectral distinction of various targets with only one single test strip. Quantification is possible because the SERS signal intensity is proportional to the number of bound SERS nanotags in the sandwich assay which is directly related to the analyte concentration. In comparison to other methods the SERS technology can therefore solve current challenges and offers a very promising alternative analytical method. Complex samples like body fluids typically consist of various biomolecules and certain target molecules are typically present only in low concentrations [13]. This situation makes the biomarker detection very challenging. SERS as an ultrasensitive plasmon-assisted scattering technique enables the detection of molecules even in very low concentrations. Therefore the major advantage of SERS over the conventional colorimetric method is the detection in concentration ranges where a colored band is no longer visible (Fig. 9.5).

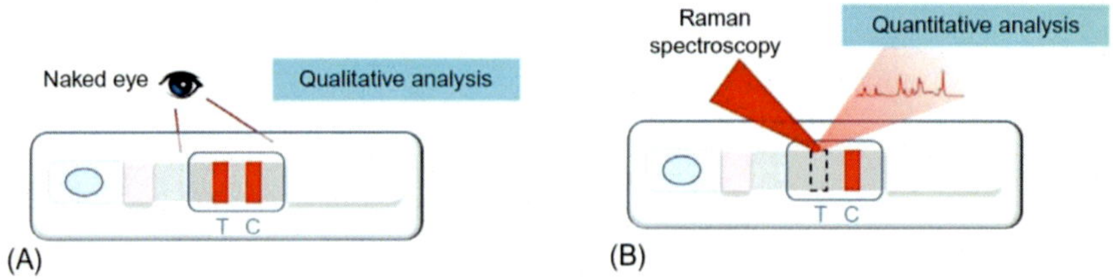

Figure 9.5 SERS readout enables the detection in low concentration ranges, where a colored band is no longer visible [14].

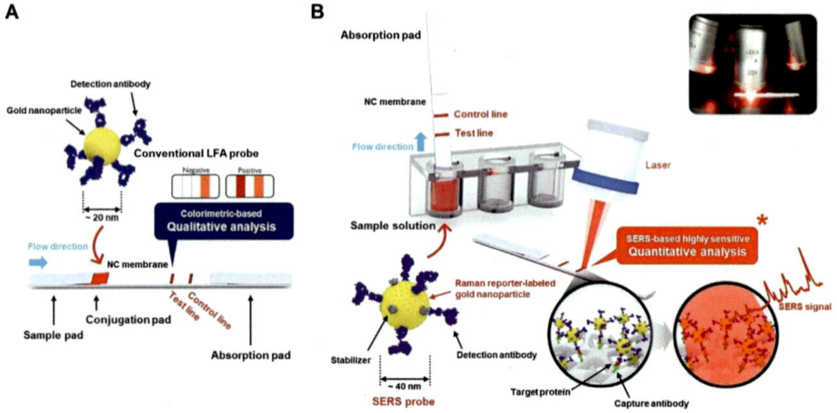

Figure 9.6 Principle of conventional lateral flow assays with gold colloids as tags for qualitative analysis via colorimetric optical detection (A) versus SERS-based LFA using SERS tags for quantification via SERS readout (B) [15].

9.1.2 SERS-based lateral flow assay

Early diagnosis heavily relies on detecting the target analyte at low concentrations. SERS in combination with a readout device enables detection with ultrahigh sensitivity as well as quantitative analysis of LFA using SERS conjugates (Fig. 9.6). The quantification is based on the proportionality of the SERS signal intensity to the analyte concentration. Since 2014 the applicability and feasibility of SERS nanotags as labeling agents in LFA toward specific and quantitative detection of a single target analyte with significant improved sensitivities have been shown by various research groups (see Table 9.1) [16,27–30]. Tang et al. developed a SERS-based LFA for hemoglobin with increased sensitivity compared to the conventional colorimetric test. Therefore flower-like AuNP with a rough Ag shell and the Raman reporter 4-mercapto benzoic acid

Table 9.1 Overview of SERS-based lateral/vertical flow assays.

Target analyte	Nanoparticle type	Raman reporter/dye	Sample	LOD	SERS readout	Ref.
Staphylococcal enterotoxin B	Hollow gold nanospheres	MGITC	In buffer	0.001 ng mL^{-1}	Raman microscope (3 mW; 633 nm)	[16]
hCG	Au/Au core/satellite particles	Thio-2-naphthol	Clinical serum	1.6 mIU mL^{-1}	Portable SERS reader (785 nm)	[17]
Bisphenol A	40 nm nanostars	4-ATP	In PBS/BSA buffer	0.073 ppb	FT-Raman spectrometer (0.1 W; 1064 nm)	[18]
Troponin I	Au shell, Au rods, Au spheresNonspherical gap-enhanced Raman tags (GERTs)	1,4-Nitrobenzene-thiole	In PBS/BSA buffer	0.1 ng mL^{-1}	Raman microscope (7 mW; 785 nm)	[19]
NeomycinCiprofloxacinEnrofloxacin	Gold nanoparticle	4-ATP	In BSA buffer/ milk	0.37 pg mL^{-1} 0.57 pg mL^{-1} 0.695 pg mL^{-1}	Confocal Raman microscope (5 mW; 633 nm)	[20]
Interleukin-6	Gold nanoparticleGold nanoshell/silica core	DTNB	Unprocessed whole blood/ PBS buffer	5 pg mL^{-1} (blood) 1 pg mL^{-1} (PBS)	Raman microscope (785 nm)	[21]
Stroke biomarker S100-β	Gold nanoparticleGold nanoshell/silica core	DTNB	In PBS buffer	0.14 pg mL^{-1}	Raman microscope (785 nm)	[22]
Listeria monocytogenes/ Salmonella typhimurium	Gold nanoparticle	4-MBA/DTNB	Milk	75 cfu mL^{-1} (extrapolated)	Raman microscope with high spectral resolution (785 nm)	[23]

(*Continued*)

Table 9.1 (Continued)

Target analyte	Nanoparticle type	Raman reporter/dye	Sample	LOD	SERS readout	Ref.
Myoglobin (Myo), cardiac troponin I (cTnI), creatine kinase-MB isoenzymes (CK-MB)	Gold nanoshell with silver core	Nile blue A	Clinical buffer	0.01 ng mL^{-1} 0.01 ng mL^{-1} 0.02 ng mL^{-1}	Raman microscope (532, 633, 785 nm)	[24]
∝-Fetoprotein/carcinoembryonic antigen (VFM)	Gold nanoshell with silver core	Nile blue A	Clinical by using Tween 20, BSA	0.27 pg mL^{-1} 0.96 pg mL^{-1}	Portable Raman instrument	[25]
Nuclear acid (RTI-pathogen)	Gold nanoshell with silver core	Nile blue A and methylene blue	Clinical	Nearby 0.03–0.04 pM	Confocal Raman microscope (785 nm)	[26]

(4-MBA) were used [28]. Choo and coworkers used spherical AuNP (30–40 nm) and malachite green isothiocyanate (MGITC) as Raman dye for the detection of HIV-1 DNA with an LOD of 0.24 pg mL^{-1} under optimal conditions, which resulted in a sensitivity increase by a factor of 1000 compared to the colorimetric selection [31]. With these types of SERS nanotags the detection limit of the thyroid stimulating hormone (TSH) in clinical samples could also be decreased by 2 orders of magnitude [14].

The Choo group also showed a quantitative SERS-based detection of *Staphylococcal enterotoxin B* using hollow gold nanospheres and MGITC (Fig. 9.7). The detection limit was estimated at approximated 0.001 ng mL^{-1}, which the authors classified as 3 orders of magnitude more sensitive by comparison with the usual ELISA method (LOD: 1 ng mL^{-1}) [16].

In 2018 the same group also designed a SERS-based LFA for four highly dangerous bacterial pathogens, which is 3–4 orders of magnitude more sensitive than conventional LFA kits [15]. Maneeprakorn et al. (2016) showed the applicability of gold nanostars in the assays as tags for the detection of *influenza A* nucleoprotein down to a concentration of 6.7 ng mL^{-1}. The sensitivity is increased by 37 and 300 times, respectively, compared to the fluorescence-based LFA and the conventional assay [32]. In the SEM images (Figs. 9.8 and 9.9) the presence of SERS

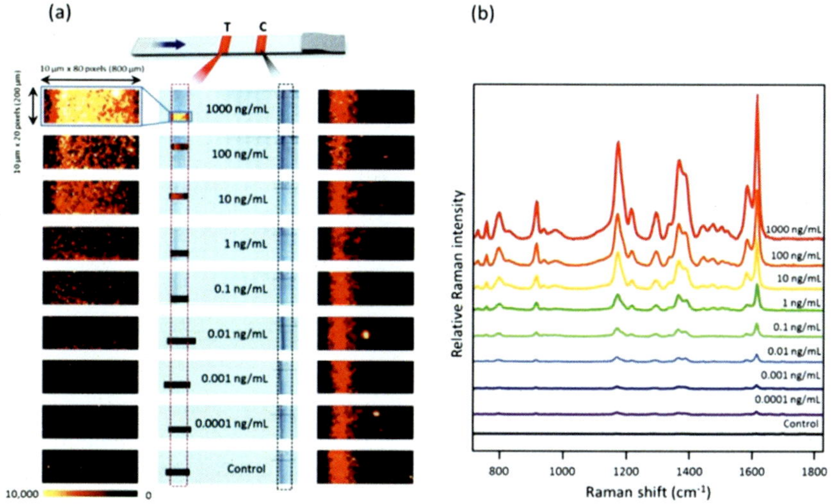

Figure 9.7 SERS readout: (A) SERS mapping for different SEB concentrations (0.1 pg mL^{-1}–1000 ng mL^{-1}). (B) Average SERS spectra of the SERS mapping zones for each concentration [16].

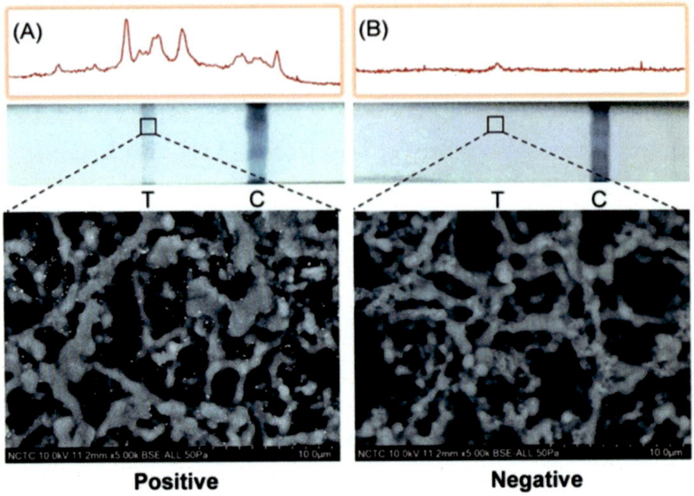

Figure 9.8 SERS spectra at the test line T (top), photographic images of LFA (middle), and SEM images of the test lines (bottom) of a positive (A) and a negative (B) test result. The SEM images show the presence of SERS nanotags in the positive case and the absence of the tags in the negative case. The presence of the tags at the test line is also correlated to the SERS signal intensity [32]. *LFA*, Lateral flow assay; *SEM*, scanning electron microscope.

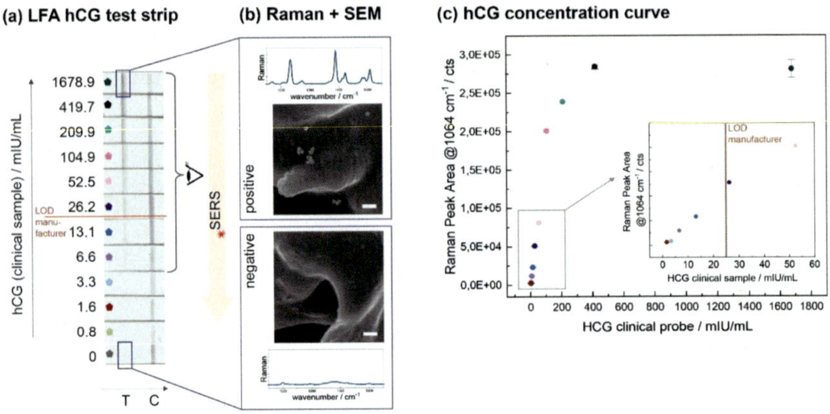

Figure 9.9 SERS readout of hCG lateral flow assay (LFA) using the portable SERS point of care (POC) reader from Fig. 9.18. (A) Test strips with different hCG concentrations in clinical serum samples. (B) Scanning electron microscope (SEM) images of the test line of a positive and negative case. (C) Quantitative SERS response (785 nm, 50 scans, 5 s) [17].

nanotags in the positive case and the absence of the tags in the negative case at the test zones are shown. The presence of the tags is also correlated to the SERS signal intensities measured at the TL [17,32].

Blanco-López, Graham and coworkers developed a *pneumolysin* assay using Au@Ag nanoshells and rhodamine-B isocyanate as SERS nanotags with a LOD of 1 pg mL^{-1}. They also showed a significant improvement in the sensitivity compared to the literature [29]. Deng et al. showed a LFA with flower-like Au/Ag core/shell particles with 4-MBA as Raman reporter for a sensitive detection of *β-adrenergic agonist bromobuterol* with a LOD of 0.15 pg mL^{-1} [33]. The Wu group presented a SERS-based lateral flow test with a detection limit of 0.86 ng mL^{-1} for *neuron-specific enolase* in blood plasma [34]. Lu's group compared the use of different SERS nanotags made of bimetallic (Au@Ag) and AuNP for the detection of cardiac troponin I in the assay. Au@Ag-Au nanoparticles comprising an Ag—Au core and an Au shell show the highest SERS sensitivity and are 50-fold more sensitive in the assay than the naked eye. They also predicted that the nanotag composed of a multishell and the use of more Raman molecules could result in an enormous increase in sensitivity [35]. To increase the sensitivity of the SERS nanotags, S. Wang et al. functionalized the Au/Ag nanoparticle with Raman-active molecules (DTNB) twice. For this purpose, a monolayer was formed on the AuNP, followed by another functionalization step with DTNB on the Ag shell. These particles have been successfully used for sensitive detection of *Mycoplasma pneumoniae* infections [36]. The applicability of these SERS nanotags has also been demonstrated for the detection of C-reactive protein in nonhuman primates [37]. The Schlücker group developed very bright Au/Au core/satellite nanoparticles, which show high SERS activities even at the single-particle level. The applicability of these nanotags for quantification was shown in conventional pregnancy tests as a model system (Fig. 9.9) [17]. Further work on SERS-based assays has also been recently published by Stanciu et al. and Tang et al. as well as others, demonstrating the feasibility of SERS-based assay [17,18,38].

9.1.3 SERS-based multiplex lateral flow assay

In addition to the detection of a single target, the demand for simultaneous detection of multiple analytes in the same sample in a single test is motivated by saving material cost, time and sample volume. Recent multiplex LFA are realized by using several separate test zones (dots or lines)

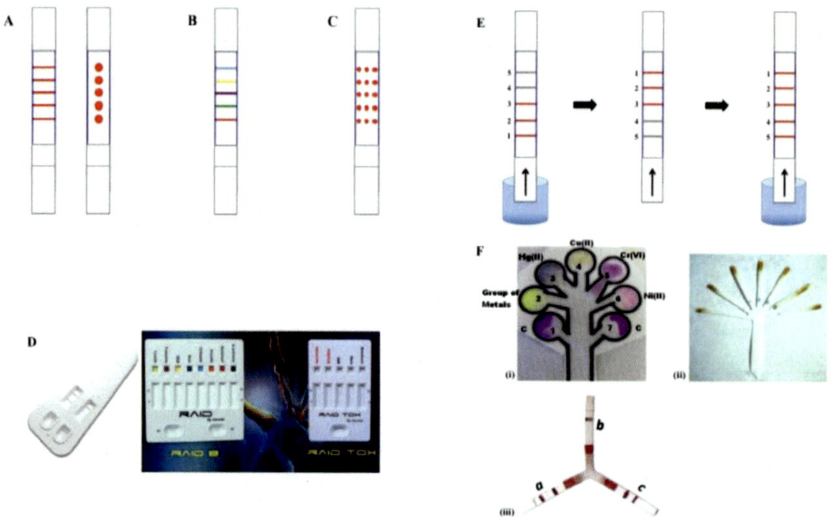

Figure 9.10 Increasing the detection capacity via multiple lines or dots on a single lateral flow assay (A), use of different colored tracers (B), display of a series of lines or dots in a format akin to a small microarray (C), combining multiple single lateral flow strips into one plastic cassette (D), a bidirectional lateral flow dipstick (E), and multidirectional lateral flow dipstick with different arms/reaction zones (F) [39].

on a single membrane (Fig. 9.10) [39]. Therefore different tags have been used, for example, colored polybeads for a qualitative assay [5].

In general, the use of different Raman reporter molecules for obtaining a set of spectrally distinct SERS nanotags is very important and beneficial for SERS-based multiplex LFA. For the proof of concept of a quantitative 6-plex cuvette experiment, Schlücker et al. employed six different Raman reporter molecules thio-2-naphthol (TN), 7-mercapto-4-methylcoumarin (MMC), mercapto-4-methyl-5-thiazoleacetic acid (MMTAA), 2,3,5,6-tetrafluoro-4-mercaptobenzoic acid (TFMBA), 4-mercaptonitrobenzoic acid (MNBA), and 2-bromo-4-mercaptobenzoic acid (BMBA) for obtaining six spectrally distinct SERS nanotags (Fig. 9.11). Furthermore, the use of different mixtures of the six nanotags and the spectral decomposition of their overall Raman signal into linear combinations of the known contributions of the individual tags via a least-square fit also demonstrated that quantitative experiments with six barcodes are possible.

The first working quantitative SERS-based multiplex LFA were developed, for example, by the groups of Hamad-Schifferli, Choo, Zhao, Wang, and Zhang [20,24,41–45]. The feasibility of a SERS-based LFA

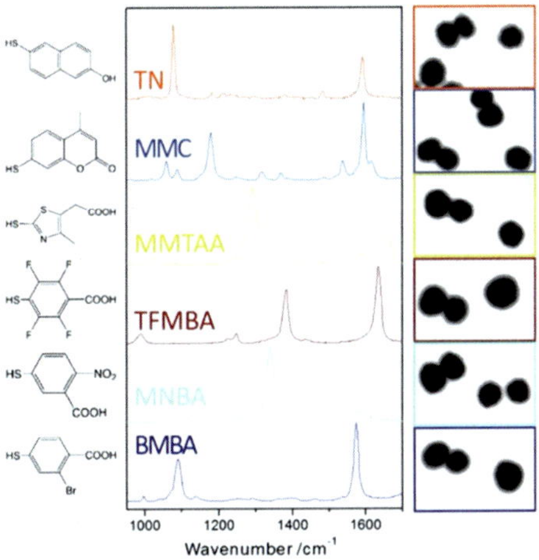

Figure 9.11 SERS tags with six different Raman reporter molecules (left), their corresponding SERS spectrum (middle), and a TEM image (right). The reporter molecules thio-2-naphthol (TN), 7-mercapto-4-methylcoumarin (MMC), mercapto-4-methyl-5-thiazoleacetic acid (MMTAA), 2,3,5,6-tetrafluoro-4-mercaptobenzoic acid (TFMBA), 4-mercaptonitrobenzoic acid (MNBA), and 2-bromo-4-mercaptobenzoic acid (BMBA) can be used for spectral differentiation in a multiplex assay [40].

for simultaneous detection was shown in the field of food safety to prevent food-borne infections caused by listeria (*Listeria monocytogenes*) and Salmonella (*Salmonella enterica*) by Wang and coworkers in 2017. To this end, Au/Ag nanoshells with 4-MBA were applied as SERS nanotags in a LFA with multiple separate TLs [45]. Choo and coworkers also developed a multiplex assay with multiple test zones for the simultaneous detection of *Kaposi's sarcoma herpes virus* (KSHV) and *bacillary angiomatosis* (BA). In this case, AuNPs in combination with the Raman dye MGITC were used as SERS nanotags. The group estimated a 10k-fold increase in sensitivity with LODs of 0.043 pM (KSHV) and 0.074 pM (BA) [42]. Besides, the Zhang group used a multiplex assay for the detection of the antibiotics neomycin (NEO) and quinolone antibiotics in food with an increase in sensitivity to LOD values of 0.37 pg mL^{-1} for NEO and 0.55 pg mL^{-1} for NOR. Spherical AuNP with the Raman reporter 4-NTP were used as tags in this multiplex LFA [20].

Hamad-Schifferli et al. were the first to use gold nanostars encoded by two different Raman reporters (BPE and 4-MBA) for the simultaneous detection of both Zika and dengue virus, in comparison to other groups using only one type of Raman reporter molecule (Fig. 9.12). They used

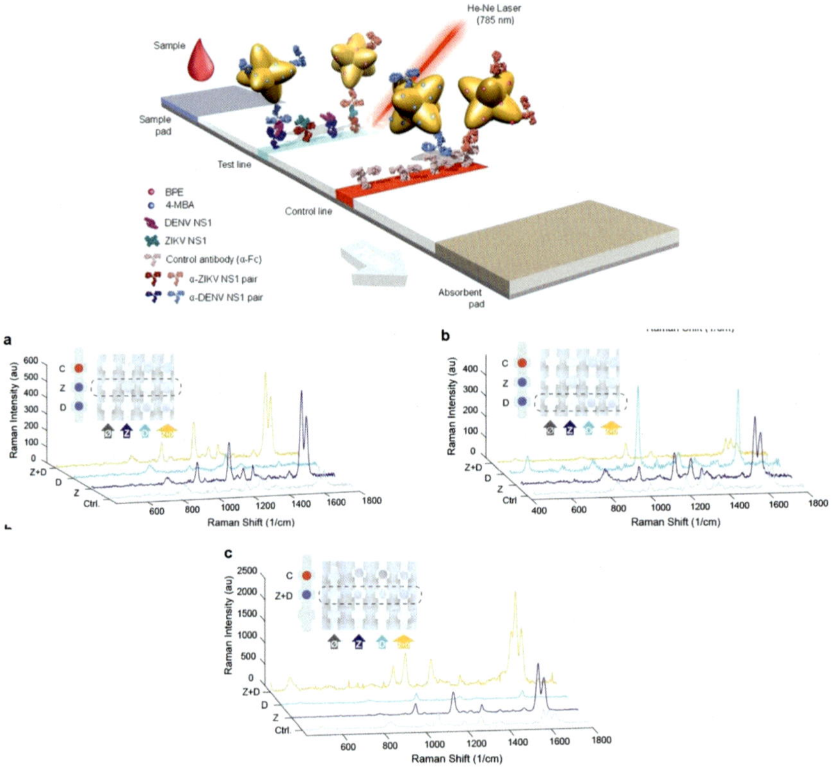

Figure 9.12 A multiplexed assay with two separate test zones for Zika and dengue virus detection: SERS spectra recorded the Zika (A) and dengue (B) test line. SERS spectra of multiplex assay with one single test zone for parallel detection of Zika and dengue virus (C). Samples included the control (0 ng mL^{-1} NS1) (*gray*), ZIKV NS1 (*blue*), DENV NS1 (*cyan*), and a mixture of ZIKV and DENV NS1 (*yellow*) [41].

two separate test zones for Zika and dengue as well as one test zone for their parallel detection. With their SERS-based multiplex dipstick assay, a detection limit of 0.72 ng mL^{-1} for Zika virus and 7.67 ng mL^{-1} for dengue virus could be achieved, which corresponds to a sensitivity increase of 15 and 7 times, respectively. Thus this assay leads to an earlier detection of the pathogens after infection [41].

A multiplex SERS LFA for three different biomarkers was first reported by Zhao et al. For the acute myocardial infarction an assay for the parallel detection of myoglobin (Myo), cardiac troponin I (cTnI), and creatine kinase-MB isoenzyme (CK-MB) was developed [24,43]. First, they designed a test strip with three TLs for detecting the three different biomarkers

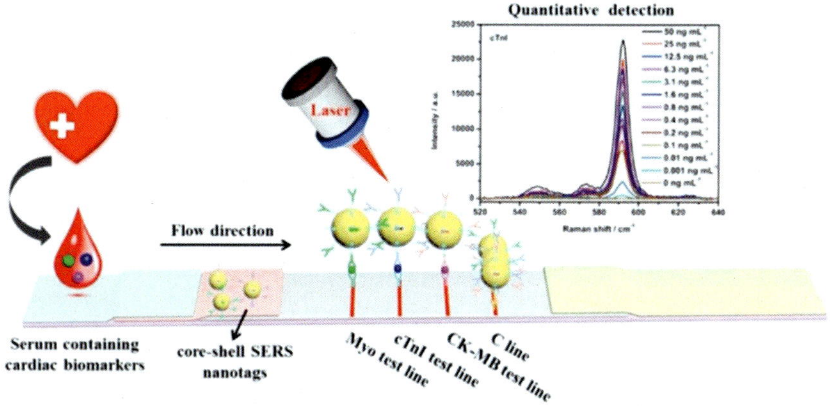

Figure 9.13 Multiplex LFA with three different test lines for simultaneous detection of the cardiac biomarkers Myo, cTnI, and CK-MB using Au/Ag nanoshells and the Raman dye Nile blue A [24].

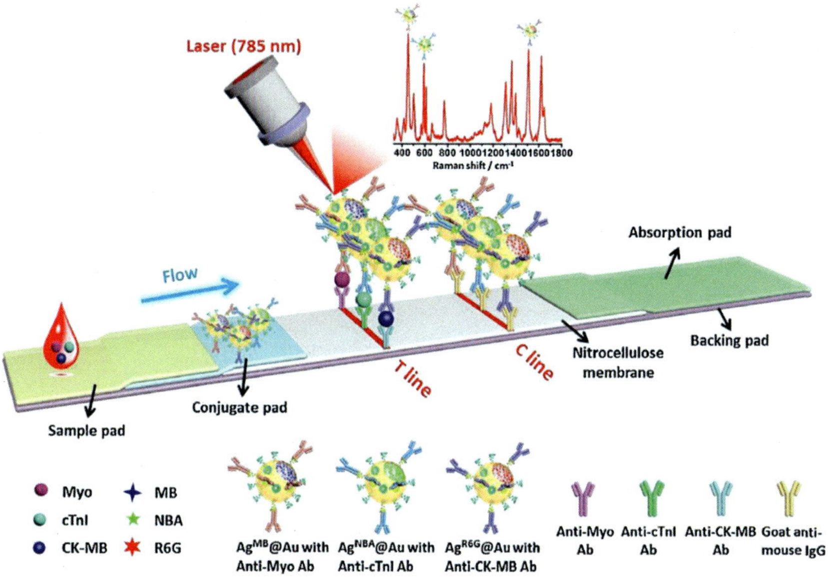

Figure 9.14 Multiplex LFA on a single test line for simultaneous detection of the three cardiac biomarkers Myo, cTnI, and CK-MB using different Raman dye-encoded core–shell nanotags [43].

(Fig. 9.13) using Nile blue A (NBA) encoded Ag/Au nanoshells. The LODs for Myo, cTnI, and CK-MB were determined as 3.2, 0.44, and 0.55 pg mL^{-1}, respectively, which are below the clinical cutoff values [24].

Based on this work, Zhao et al. (2018) continued to design a multiplex assay for the three cardiac biomarkers Myo, cTnI, and CK-MB together on a single TL (Fig. 9.14). Real serum samples were analyzed using three different encoded Ag/Au nanoshells employing the Raman dyes methylene blue (MB), Nile blue A (NBA), and Rhodamin 6 G (R6G). Compared to the readout time for a spatially separated assay (3xTL), the readout time could be reduced by 3 minutes with only one single TL, which is advantageous in the case of acute myocardial infarction in addition to the sensitivity of an assay, since earlier diagnosis is very important for initiating appropriate treatment [43]. With the SERS technology also more than three biomarkers for parallel quantitative detection with one single test strip are possible and feasible, for example, in the area of drug testing.

The number of testing zones sequentially located on one single test strip and therefore the number of analytes is limited, usually there are no more than five zones. For the multiplex approach two-dimensional immunochromatography, that is, an array of points with immunoreagents, is promising (Fig. 9.15). This assay combines advantages of immunochromatographic tests and immunochips [6].

In 2019 the Zhao group developed a multiplex 2×3 microarray on a SERS lateral flow strip encoding the nucleic acids of 11 common

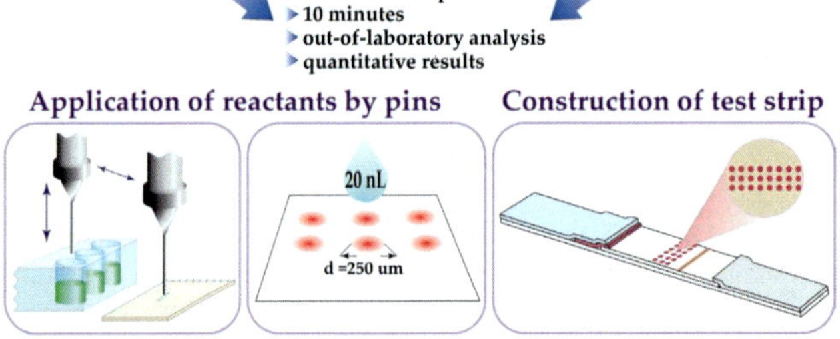

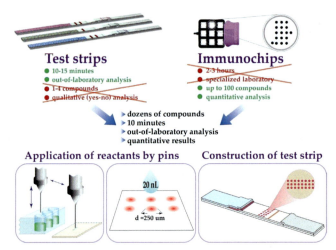

Figure 9.15 Two-dimensional immunochromatography increases the information content and reduces the consumption of reagents and materials for one analysis [6].

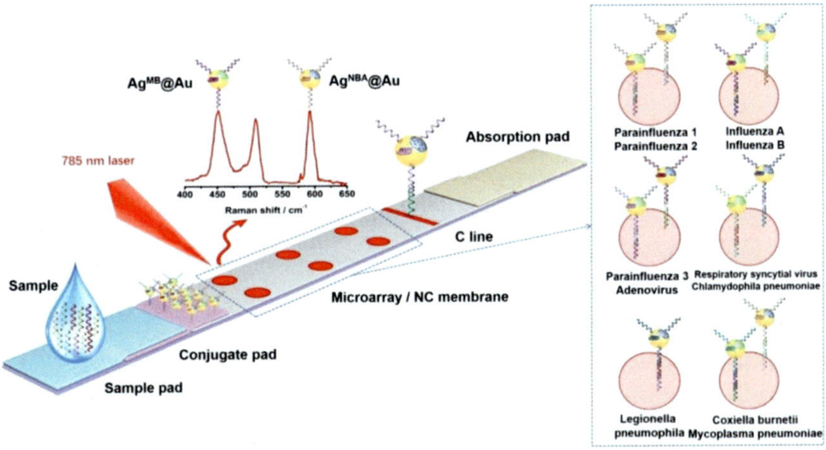

Figure 9.16 Schematic illustration of a multiplex SERS lateral flow assay (LFA) for the detection of 11 different RTI pathogens with RDs encoded core–shell SERS nanotags [26].

respiratory tract infection pathogens (Fig. 9.16). By using this microarray, they obtained ultra-high sensitivities using core shell SERS nanotags encoded with two Raman dyes (Nile blue A and methylene blue) as tags. They calculated the LODs for *influenza A*, *parainfluenza 1*, *parainfluenza 3*, *respiratory syncytial virus*, *coxiella burnetii*, *legionella pneumophila*, *influenza B*, *parainfluenza 2*, *adenovirus*, *chlamydophila pneumoniae*, and *mycoplasma pneumoniae* to be in the range between c.0.03 and 0.04 pM [26].

9.1.4 Portable Raman/SERS-POC reader

Portable, rapid, and user-friendly readout devices such as the known conventional optical POC readers are needed for the on-site analysis to avoid costs and precious time for evaluation in central laboratories. Miniaturization is needed to reduce consumption of reagents and readout time [1]. Generally, conventional confocal Raman microscopes have been used for the SERS readout in academic laboratories so far. To evaluate the test strips only point measurements or only a small area of the test zone via time-consuming Raman-mapping were carried out by Raman microscopy (Fig. 9.17). This is of course suboptimal, since time is a key parameter for POCT and the detection of all analyte molecules at the test zone is required to provide reliable quantitative information. Several Raman instrument manufacturers (e.g., InPhotonics, Ocean Optics, Thermo Fisher Scientific, Horiba) have developed portable devices toward miniaturization, handheld, or lightweight applications and high resolution. The lasers available in these portable Raman devices are usually

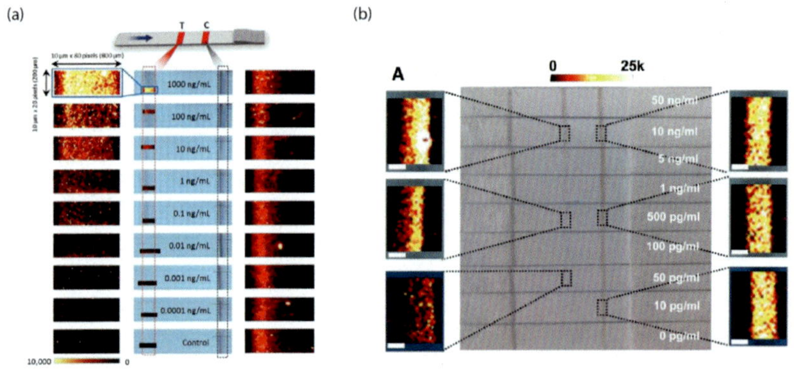

Figure 9.17 Raman mapping at the test lines of SERS-based LFA using conventional confocal Raman microscopes by the groups of (A) Choo and (B) Blanco-López, Graham, and coworkers [16,29]. Importantly, only a small part of the entire test line is analyzed by SERS.

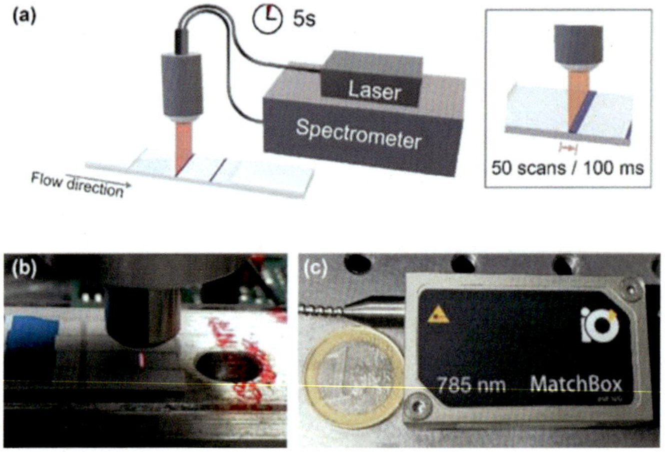

Figure 9.18 Portable Raman/SERS-based point of care (POC) reader with line illumination. (A) Schematic setup. (B) Custom-designed optical fiber probe with 4 mm width to cover the entire TL. (C) Compact 785 nm diode laser [17].

diode lasers [46]. However, these devices were designed for conventional on-site label-free Raman experiments and not for SERS-based LFA.

The Schlücker group recently developed a portable Raman/SERS-based POC reader with line illumination for rapid readout of the entire TL (Fig. 9.18). Fast SERS readout is achieved by covering the entire length of the TL (4 mm) using a custom-made fiber optical probe with a line focus (Fig. 9.18b), a motorized stage for rapid scanning along the flow direction across the entire width of the TL, and a powerful compact 785 nm diode laser (Fig. 9.18c) [17]. For proof of concept, a conventional pregnancy hCG test as model system, clinical specimens, and SERS nanotags comprising Au/Au core/satellite particles with the Raman reporter 2-naphthalenethiol were used. By using these SERS nanotags the sensitivity of the conventional hCG test could be increased 15 times in comparison to the manufacturer's LOD using conventional colloidal gold. For the SERS readout 50 scan positions with 100 ms per step were carried out to measure the whole the TL, resulting in an overall acquisition time of only 5 s, which is several orders of magnitude faster than conventional Raman point mapping of the entire TL.

A direct and fair comparison of the short readout times obtained with the SERS-POC reader with the much longer readout times reported in SERS LFA reported by other groups [16,29] via conventional Raman microscopy (Fig. 9.17) requires that for the latter the acquisition time for covering the entire TL area is considered.

Therefore the acquisition time required for the entire CL by using point illumination and mapping without spatial undersampling was extrapolated to be on the order of several days [17]. This readout time for the entire TL is obviously not appropriate for rapid on-site analysis. The readout time can be significantly reduced by several orders of magnitude by using a SERS reader with line illumination covering the whole TL without any spatial undersampling. For the sensitivity of the tests as well as lowering the LOD the Raman signal from the maximum possible number of SERS nanotags is needed for the test evaluation as well as for the reproducibility, so that ideally no compromises should be made in this respect [17].

The portable Raman/SERS-based POC-reader is a great improvement over all previously shown SERS-based LFA using conventional, expensive, and nonportable Raman microscopes, and it is a great progress regarding the use of SERS-based LFA for POCT.

9.2 Imaging

iSERS microscopy is a novel imaging technique that combines SERS tag-labeled antibodies with Raman microspectroscopy for protein localization in cells or tissue specimens (Fig. 9.19). The principle and staining process

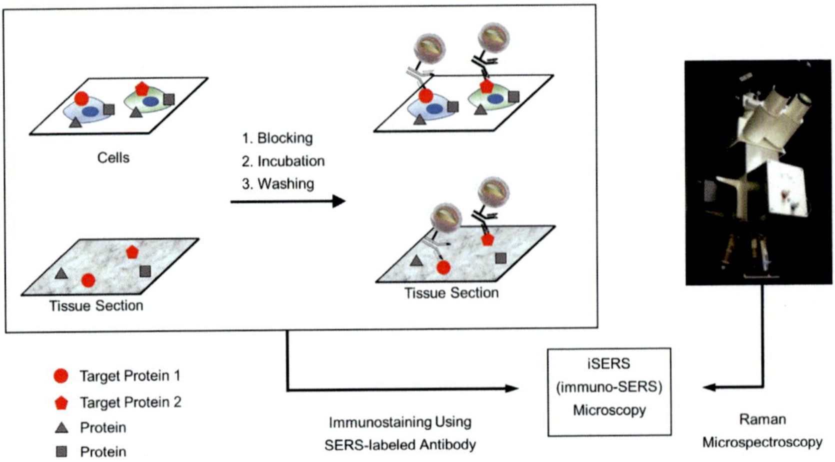

Figure 9.19 Principle of iSERS staining and iSERS microscopy.

is analogous to conventional immunohistochemistry (IHC) or immunofluorescence (IF), that is, the target molecule (antigen) is selectively recognized by the specific antibody and then visualized by the characteristic signals of the labels: either the color of the reaction product from the enzymatic conversion of a chromogen (IHC) or the characteristic fluorescence emission from fluorophores (IF). iSERS imaging offers the following advantages since it is based on the use of SERS nanotags as labels: (1) simultaneous localization of multiple biomarkers based on the narrow line width of Raman peaks; (2) quantification of protein expression using the spectral intensities of the corresponding SERS tags; (3) good image contrast due to minimal interference with disturbing autofluorescence from the biosamples by using red to near-infrared (NIR) excitation; (4) minimal photodamage of biological specimen by using only low laser powers due to the ultra-high sensitivity of SERS tags in combination with the option to use red to NIR excitation; (5) high photostability of the tags with no or only minimal photobleaching; (6) the need of only a single laser excitation to excite the characteristic spectral barcodes from all SERS nanotags. Thus in the last decade iSERS microscopy has been intensively explored in analyzing tumor markers in cell/tissue specimens ex vivo and even in detecting tumor lesions in murine models.

9.2.1 iSERS microscopy on cells

Cell surface receptors, such as epidermal growth factor receptor (EGFR), human epidermal growth receptor 2 (HER2), vascular endothelial growth factor receptor, insulin-like growth factor 1 receptor (IGFR), and folic acid receptor, have been reported to be overexpressed in human neoplasms and could be used as effective biomarkers for cancer diagnosis. Conjugation of SERS tags with recognition molecules (antibodies, aptamers, and small molecules) against these receptor molecules are capable of analyzing their locations and expression levels in different cells.

9.2.1.1 Detection of target proteins in cells using antibody-modified SERS tags

The first study employing antibody-modified SERS tags for cellular protein localization was reported by Kim et al. in 2006 [47]. They fabricated SERS tags by adsorbing Raman reporter molecules (4-mercaptotoluene [4-MT], 2-naphthalenethiol [2-NT], or thiophenol [TP]) on Ag NP-embedded silica spheres, and then antibodies against HER2 or CD10 were covalently conjugated on the sphere surface for target recognition. The antibody-modified SERS tags show linearity at their concentrations, and exhibited good specificity for the targets on cellular membranes. Soon later, SERS tags with various shapes and compositions were synthesized and applied for target localization on different cell membranes. Choo and coworkers prepared SERS tags by coating a silver layer over Raman reporter-labeled gold NPs and conjugated the tags with antiphospholipase Cγ1 (PLCγ1) antibodies [48]. The SERS labels were successfully utilized for highly sensitive imaging of PLCγ1 in live HEK293 cells (Fig. 9.20). In the following studies by the same group, crystal violet-labeled hollow Au nanospheres [49] and 4-mercaptopyridine (MP)-labeled Au nanorods [50] were conjugated with antibodies and used for targeted recognition of HER2 on MCF-7 cells, showing potential application of iSERS imaging in early cancer diagnosis. Nguyen et al. constructed SERS tags conjugated to human anti-CD19 antibodies for selectively imaging of chronic lymphocytic leukemia CLL cells. They examined the stability of the SERS labels under various conditions in common pathology protocols and found that the SERS signals remained strong 4 weeks after preparation and when incubated in standard histology stains such as Eosin, Hemotoxylin, and Giemsa, representing a significant advantage over fluorescent dyes [51].

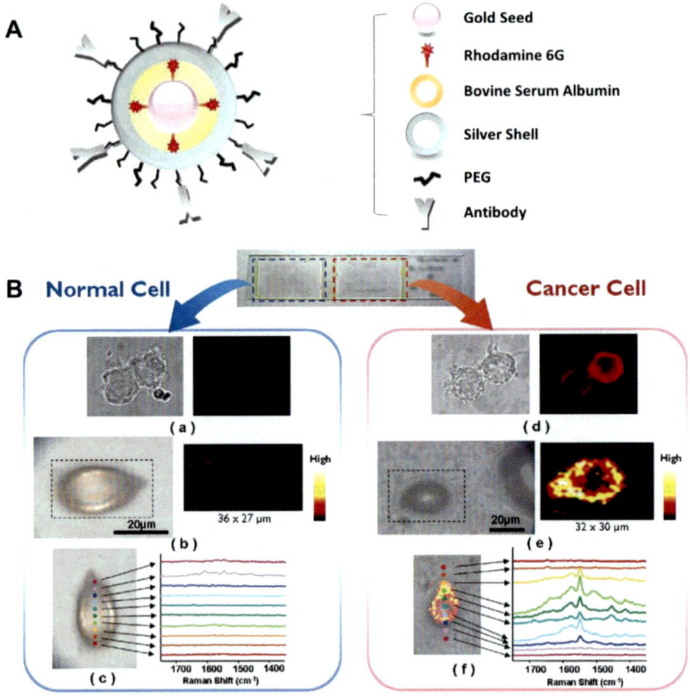

Figure 9.20 (A) Schematic of silver-coated gold nanotags and antibody bioconjugation. (B) Fluorescence images (a, d), SERS images (b, e), and SERS spectra (c, f) of normal HEK293 cells and PLCγ1-expressing HEK293 cells. *Adapted with permission from S. Lee, S. Kim, J. Choo, S.Y. Shin, Y.H. Lee, H.Y. Choi, et al., Biological imaging of HEK293 cells expressing PLCgamma1 using surface-enhanced Raman microscopy. Anal. Chem. 79 (3) (2007) 916–922, https://doi.org/10.1021/ac061246a. Copyright 2007 American Chemical Society.*

Co-localization of different biomarkers in the same sample is essential for study protein–protein interactions during biological processes and also important for precise characterization of tumor phenotypes. SERS tags offer tremendous multiplexing capacity, which allows simultaneous detection of a series of target molecules through conjugating with the corresponding antibodies. Kennedy and coworkers performed multiplexed imaging of b2-adrenergic receptor and caveolin-3 on the surface of a rat cardiomyocyte using 4-(mercaptomethyl) benzonitrile (MMBN) and d7-mercaptomethyl benzene (DMMB) functionalized silver NPs. The results show that 17% of the two receptors are colocalized, which was in accordance with previously reports [52]. Lee et al. synthesized targeting SERS

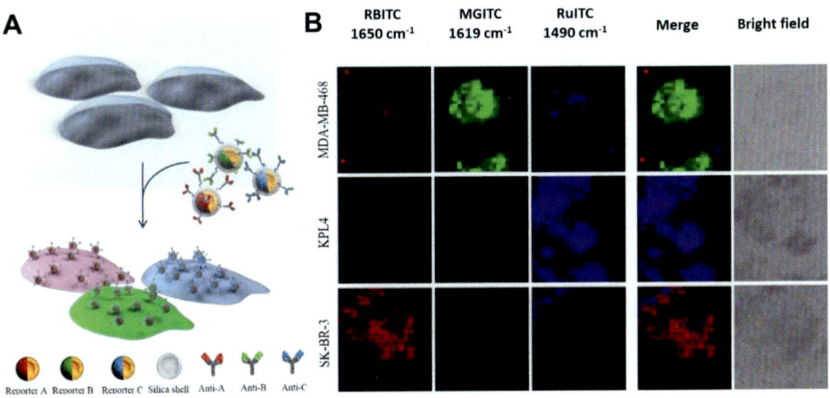

Figure 9.21 (A) Schematic illustration of rapid detection and quantification of growth factor receptors on breast cancer cell membranes using SERS tags. (B) SERS mapping images of three breast cancer cell lines measured at 1650 cm^{-1} (RBITC), 1619 cm^{-1} (MGITC), and 1490 cm^{-1} (RuITC), the merged SERS mapping images and the corresponding bright field images. *Adapted with permission from S. Lee, H. Chon, J. Lee, J. Ko, B.H. Chung, D.W. Lim, et al., Rapid and sensitive phenotypic marker detection on breast cancer cells using surface-enhanced Raman scattering (SERS) imaging. Biosens. Bioelectron. 51 (2014) 238–243, https://doi.org/10.1016/j.bios.2013.07.063. Copyright 2014 Elsevier.*

tags by conjugating Raman reporter-coded silica-encapsulated hollow gold nanospheres (SEHGNs) with specific antibodies against different receptor molecules (EGFR, ErbB2, and IGF-1R) [53]. Expression patterns of these receptor molecules in different human breast cancer cell lines (MDA-MB-468, KPL4, and SK-BR-3) were assessed by iSERS imaging (Fig. 9.21), showing similar results with western blot (WB), but with a simpler assay, a shorter detection time and at the single-cell level.

9.2.1.2 Detection of target protein in cells using aptamer-modified SERS tags

Aptamers are single-stranded DNA or RNA with unique tertiary structures that enable them to specifically bind with cognate molecular targets. As a class of molecular ligands, aptamers have some characteristic features compared with antibodies: (1) aptamers can be screened for a wide array of molecular targets; (2) the synthesis of aptamers is relatively easy, cost-effective, allows large-scale manufacture and chemical modification with low batch-to-batch variations; (3) their relatively simple chemical structures entail full conformational recovery even after thermal/chemical denaturation and therefore aptamers typically have a long shelf life [54].

AS1411 is a screened aptamer against nucleolin, which is overexpressed in a variety of tumor cells while retain a very low level in normal cells. Wu et al. fabricated AS1411 aptamer-modified Raman reporter-labeled Au nanoflowers for target recognition of nucleolin. They noticed that the Raman signals of previous SERS tags using fluorescent dyes (Indocyanine green, Crystal violet, RoseBengal, Rhodamine 6G, NIR dyes, etc.) as the Raman signatures are complicated and are overlapped with Raman signals originating from the cellular surrounding molecules (such as peptides, proteins, and cytochrome C) which may also be absorbed on the surface of metal NPs, while the bond vibrations of alkyne, azide, nitrile, and C—D moieties are located in a Raman-silent region of biological samples (approximately 1800 to 2800 cm^{-1}). Therefore they synthesized four different L-cysteine analogs containing alkyne, azide, nitrile, and diyne groups as reporter molecules. By coculture of MCF-7 and 3T3-L1 cells in the same plate and staining with the SERS tags, specific identification of MCF-7 cells was achieved, indicating the promising application of iSERS imaging in tumor cell screening [55].

In a recent study, Tan and coworkers fabricated SERS tags with stable and simple fingerprint spectrum through synthesis of isotopic cellular Raman-silent graphene-isolated-Au-nanocrystals (GIANs), and conjugated the tags with three phospholipid-polyethylene glycol-linked aptamers (AS1411 targeting nucleolin, S1.6 targeting mucin 1, and SYL3C targeting EpCAM) to identify proteins overexpressed on the cancer cell surface. The G100-AS1411, G050-S1.6, and G000-SYL3C GIAN SERS-encoders were successfully applied for pattern recognition of HepG2 and A549 cell lines (Fig. 9.22), showing high potential for cost-effective cancer cell identification with high sensitivity and low background interference [56]. To further broaden the multiplexing capacity of SERS tags, Zeng et al. developed a completely new readout technique, the so-called Click SERS. The basic principle is that the interparticle plasmonic coupling makes the controllable and artificial splicing of the SERS tags for emitting dynamic combinatorial signal output of single peaks. SERS tags encoded with two different reporter molecules (OPE 1 and MBN) but both functionalized with EGFR aptamers were incubated with HeLa cells at the same time. Owing to the double check, the location with both signals from OPE 1 and MBN demonstrated the true receptors' presence, while the remaining red and green areas were actually false positive results [57].

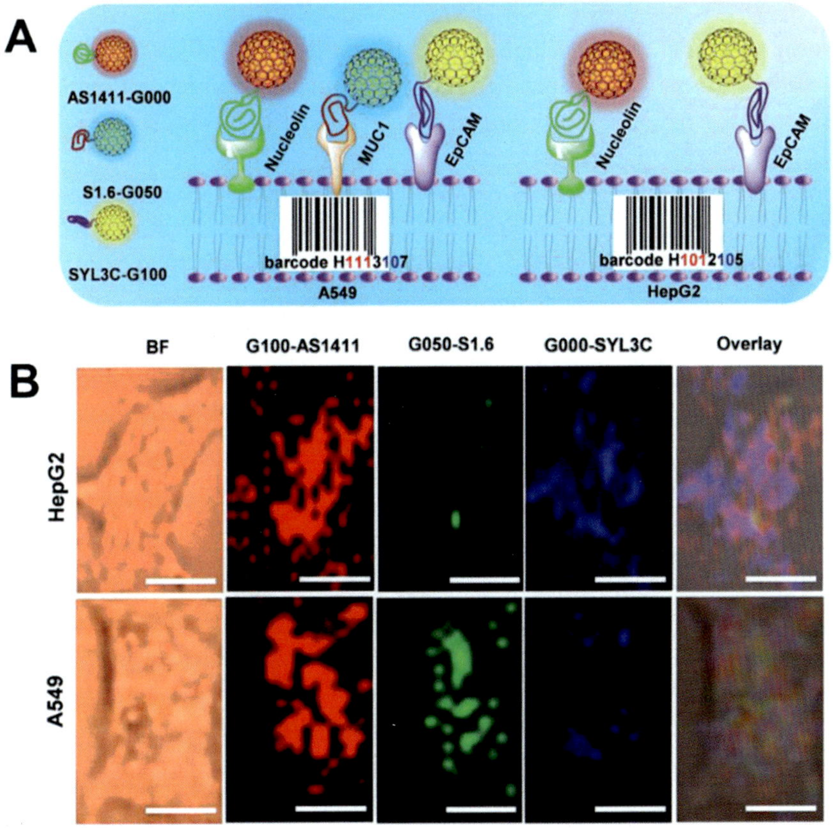

Figure 9.22 Cancer cell pattern recognition with aptamer-functionalized SERS tags. (A) Schematic illustration of pattern recognition and discrimination of cancer cell lines with multiplexed SERS tags. (B) SERS images of HepG2 and A549 cells, scale bar, 10 μm. *Adapted with permission from Y. Zou, S. Huang, Y. Liao, X. Zhu, Y. Chen, L. Chen, et al., Isotopic graphene-isolated-Au-nanocrystals with cellular Raman-silent signals for cancer cell pattern recognition. Chem. Sci. 9 (10) (2018) 2842–2849, https://doi.org/10.1039/c7sc05442d. Copyright 2018 The Royal Society of Chemistry.*

9.2.1.3 Detection of target proteins in cells using SERS tags modified with other recognition molecules

Affibody molecules are antibody mimetics which can bind to target proteins with high affinity. Compared to antibodies, they have lower molecular weight/size (7 kDa), higher stability, and are convenient to conjugate via the terminal cysteine. Gambhir and coworkers modified affibodies against EGFR onto SERS tags for molecular imaging on cells. They found that the targeting SERS tags reports EGFR-positive A431 tumor cells with a signal nearly 13-fold higher than EGFR-negative MDA-435S tumor cells, and 292-fold higher than the control tags [58].

In addition to antibodies, affibodies or aptamers which are rationally prepared for target recognition, some natural molecules such as folic acid (FA), transferrin that can selectively bind to their corresponding receptor molecules on cell membranes, and lectins that react with carbohydrates on cell surface, have also been utilized as targeting molecules for SERS tags modification. Folic acid is a basic component of cell metabolism and DNA synthesis. The expression levels of FA-binding proteins are usually elevated in malignant tumors compared to normal cells due to increased requirement of FA to maintain DNA synthesis in fast dividing cancer cells. Fasolato et al. fabricated FA-modified SERS tags to distinguish different cell populations based on measurement of the overall SERS signal intensity from a single cell. By treating incubated cells with the targeting SERS tags, they were able to distinguish two cancer cell lines (HeLa and PC-3) from one normal cell line (HaCaT), which have different expression levels of FA binding proteins [59]. Li et al. recently reported a SERS-based imaging approach to visualize castration-resistant prostate cancer cells using SERS tags modified with a urea-based small-molecule inhibitor of prostate-specific membrane antigen (PSMA). The authors demonstrated that the highly sensitive and specific SERS tags enable identification of a single prostate cancer cell, supporting further preclinical feasibility studies [60]. The elevated expression of sialic acid containing glycoproteins is symptomatic of disease and cancer progression, and therefore the SERS tags targeting sialic acid composition of mammalian cells have also been used to discriminate cancerous cells. Graham et al. designed lectin-functionalized SERS tags to investigate carbohydrate − lectin interactions at the cell surface. Through iSERS analysis, they identified an increased sialo-glycan expression of one prostate cell line (PC3) over another (PNT2A) [61]. Besides lectin, phenylboronic acid (PBA) also shows high binding affinity toward SA on the cell membrane through esterification. In a recent study reported by Liu and coworkers, PBA-modified alkyne-bridged AuNP dimers were fabricated and successfully applied for high-precision profiling of SA expression in cancer cells. The authors compared SERS mapping images constructed using the 1580 cm^{-1} (phenyl C = C bonds, green) peak with the one using the 2210 cm^{-1} (C≡C moieties, red) peak, showing a lower background interference in the cellular Raman-silent region [62]. For a multiplexed application, Chen et al. fabricated alkyne-modulated SERS tags (OPE1, OPE1, OPE2) and linked

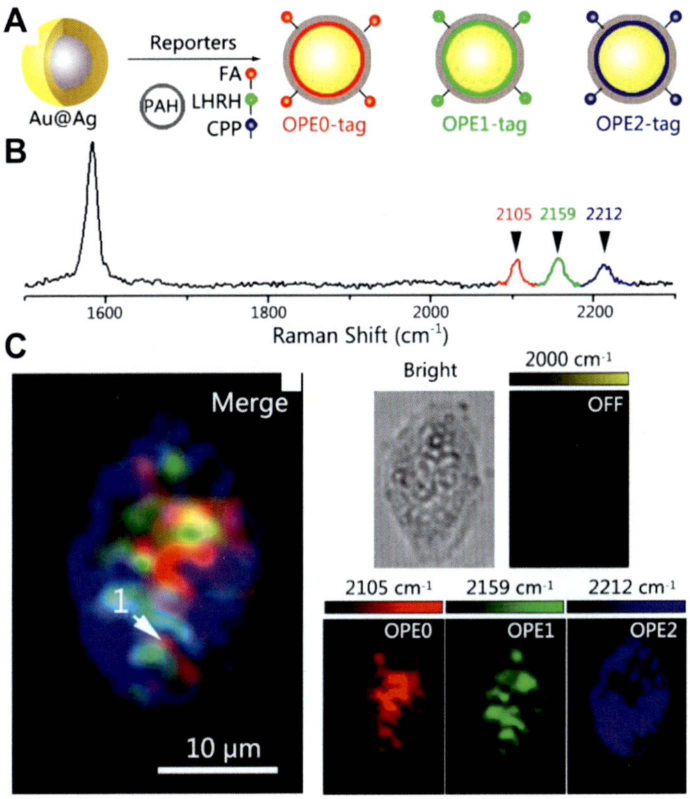

Figure 9.23 (A) Scheme of SERS tags coded with OPE0 (*red*), OPE1 (*green*), and OPE2 (*blue*) and conjugated with different ligands. (B) Normalized Raman spectra of the three SERS tags. (C) Three-color SERS imaging in a HeLa cell. *Adapted with permission from Y. Chen, J.-Q. Ren, X.-G. Zhang, D.-Y. Wu, A.-G. Shen, J.-M. Hu, Alkyne-modulated surface-enhanced Raman scattering-palette for optical interference-free and multiplex cellular imaging. Anal. Chem. 88 (12) (2016) 6115–6119, https://doi.org/10.1021/acs.analchem.6b01374. Copyright 2016 American Chemical Society.*

them with FA, luteinizing hormone-releasing hormone (LHRH), or CALNNR$_8$ peptide, respectively. With their characteristic peaks at 2105, 2158, and 2212 cm^{-1}, the SERS tags were successfully used for triplex cellular imaging of HeLa cells (Fig. 9.23) [63].

9.2.1.4 Monitoring receptor status on cellular membranes

Based on the quantitative analysis capability of SERS tags, iSERS microscopy has been utilized to monitor receptor status on cellular membranes after targeted therapy or extra stimulus. EGFR is an important prognostic

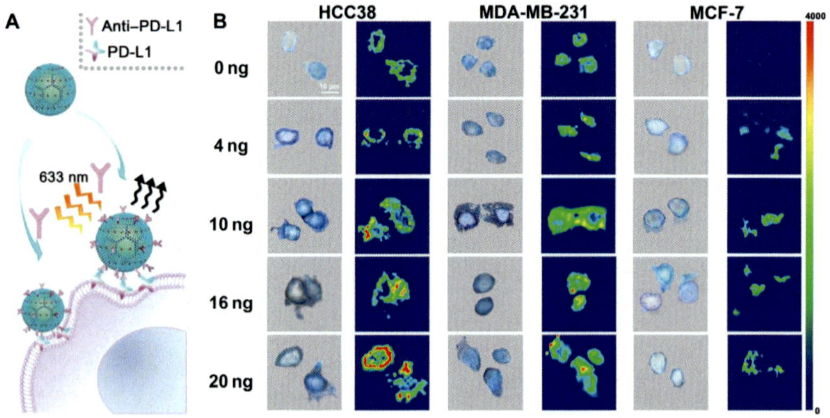

Figure 9.24 (A) Scheme showing specific recognition and detection of PD-L1 on triple-negative breast cancer cells. (B) SERS mapping images obtained from HCC38, MDA-MB-231, and MCF-7 cells incubated with different amount of IFN-γ for 48 h. *Adapted with permission from E. Feng, T. Zheng, X. He, J. Chen, Y. Tian, A novel ternary heterostructure with dramatic SERS activity for evaluation of PD-L1 expression at the single-cell level. Sci. Adv.4 (11) (2018) eaau3494, https://doi.org/10.1126/sciadv.aau3494. Copyright 2018 American Association for the Advancement of Science.*

marker expressed in cancer cells and cetuximab is the first monoclonal antibody drug that targets EGFR, which blocks EGFR-mediated tyrosine kinase activity and inhibits cellular proliferation. Chung and coworkers reported a strategy that fabricating EGF-conjugated SERS tags to quantify EGFR expression on MDA-MB-468 cancer cells and monitor its variation after treatment with cetuximab. The distribution of EGFR on cellular membrane was localized, the amount of EGFR uninhibited by cetuximab was quantified, and the whole process took less than 4 h, showing the great potential of iSERS imaging to improve prognostic efficacy of targeted therapy [64]. Programmed cell death receptor ligand 1 (PD-L1) is one kind of crucial immune checkpoint molecules which are overexpressed in many cancers but rarely expressed in regular epithelial tissues. Quantitative analysis of PD-L1 expression on the cell surface is important for guidance of PD-L1—based immunotherapy. In a recent study, Feng et al. synthesized a novel ternary heterostructure SERS substrate (Fe_3O_4@GO@TiO_2) with a significant enhancement factor of 8.08×10^6, then labeled the NPs with CuPc and conjugated them with anti(PD-L1) antibodies. The fabricated SERS conjugates were applied for in situ quantification of PD-L1 on triple-negative breast cancer cell surface at the single-cell level, and the expression variation and distribution of PD-L1 during drug treatment (Fig. 9.24) was successfully monitored [65].

9.2.1.5 Profiling of circulating tumor cells

Tumor cells that shed from solid tumor sites and enter the circulatory system are called circulating tumor cells (CTCs). Many studies have confirmed that CTCs play an important role in cancer metastasis, and the phenotypic information of CTCs represents the molecular characteristics of the corresponding tumor tissue. Therefore detection and characterization of CTCs help in early diagnosis of cancer metastasis and may have important prognostic and therapeutic implications [66]. Although a variety of technologies have been set up to separate and identify CTCs in the past decades, the time-consuming procedures and high instrument costs remain a major obstacle to hinder their use in clinical diagnostics, and the phenotypic information of CTCs can hardly be obtained.

Due to their high sensitivity and high multiplexing capacity, SERS tags have been modified with recognition molecules against CTCs biomarkers and employed in CTCs detection and profiling [67–69]. Nima et al. fabricated a series of SERS tags by labeling Au/Ag nanorods with four different Raman reporter molecules (4-mercaptobenzoic acid [4-MBA], p-aminothiophenol [PATP], p-nitrothiophenol [PNTP], and 4-(methylsulfanyl) thiophenol [4MSTP]). The SERS tags were conjugated with four kinds of antibodies against different cancer markers (antiEpCAM, antiCD44, antiKeratin 18, and antiinsulin-like growth factor antigen), respectively. Using a cocktail of these four SERS nanotag/antibody conjugates, the authors could identify single breast cancer cells in unprocessed human blood sample [70]. Recent studies have indicated that the molecular phenotype of CTCs evolves dynamically. For instance, Jordan et al. found that CTCs in women with HER2-negative breast cancer acquire a HER2-positive subpopulation after multiple courses of therapy [71]. Therefore real-time monitoring of phenotypic evolution of CTCs is crucial for understanding progression of cancer and beneficial for guiding clinical treatment. Tsao and coworkers developed a SERS-based technique to monitor phenotypic evolution of CTCs from clinical melanoma patients during molecular targeted therapies. Antibodies against melanoma-chondroitin sulfate proteoglycan (MCSP), melanoma cell adhesion molecule (MCAM), erythroblastic leukemia viral oncogene homolog 3 (ErbB3), and low-affinity nerve growth factor receptor (LNGFR) were conjugated with SERS tags labeled with 4-MBA, 2,3,5,6-tetrafluoro-4-mercaptobenzoic acid (TFMBA),

4-mercapto-3-nitro benzoic acid (MNBA), and 4-mercaptopyridine (MPY), respectively. The target-specific SERS tags were applied to monitor CTC signature changes in blood samples serially collected from ten stage-IV melanoma patients during therapy treatment, showing that CTC populations shifted after treatment with dabrafenib and trametinib for 40 days and a different cluster was formed on day 48 [72]. The SERS-based CTC detection technique is extremely sensitive (10 cells in 10 mL of blood), highly multiplexed (simultaneously monitoring expression profiles of several surface proteins) and simple (does not need initial enrichment of CTCs), and therefore holds great potential to be translated into clinical use.

Circulating cancer stem cells (CCSCs) as a rare type of CTCs provide important information for characterization of both cancer tissues and their metastatic derivatives. However, due to the scarcity of CCSCs among hematologic cells in the blood and the complexity of the phenotype confirmation process, CCSC analysis is extremely challenging. Cho and coworkers developed a new technique for selective isolation and noninvasive analysis of CCSCs and CTC subtypes based on a combination of SERS detection with a microfluidic chip. Antibodies against different biomarkers (HER2, CD133, EGFR, EpCAM, MUC1) were individually conjugated with SERS tags encoded by five different Raman reporter molecules (Nile Blue A [NBA], 2-quinolinethiol [QNT], 1-naphthalenethiol [NPT], 4-mercaptopyridine [MPD], and thiophenol [TP]), then applied for selective recognition and SERS imaging of the corresponding surface markers. The Raman intensity differences could be considered as barcode signals to represent ON and OFF values of each surface marker (Fig. 9.25). Accordingly, the characteristic information of detected CCSCs/CTCs were clearly distinguished based on the observed barcode signals. This approach was used to screen blood samples from xenograft models, showing that upon CCSC detection, all subjects exhibited liver metastasis. The authors concluded that this SERS-based method is especially powerful for predicting tumor's metastatic capabilities through efficient CCSC/CTC detection, therefore, can be tailored for precision medicine involving cancer metastasis prevention and the development of effective therapeutics against it [73].

9.2.2 iSERS microscopy on tissues

In most hospitals world-wide, histopathological cancer diagnosis is based on comprehensive analysis of formalin-fixed and paraffin-embedded

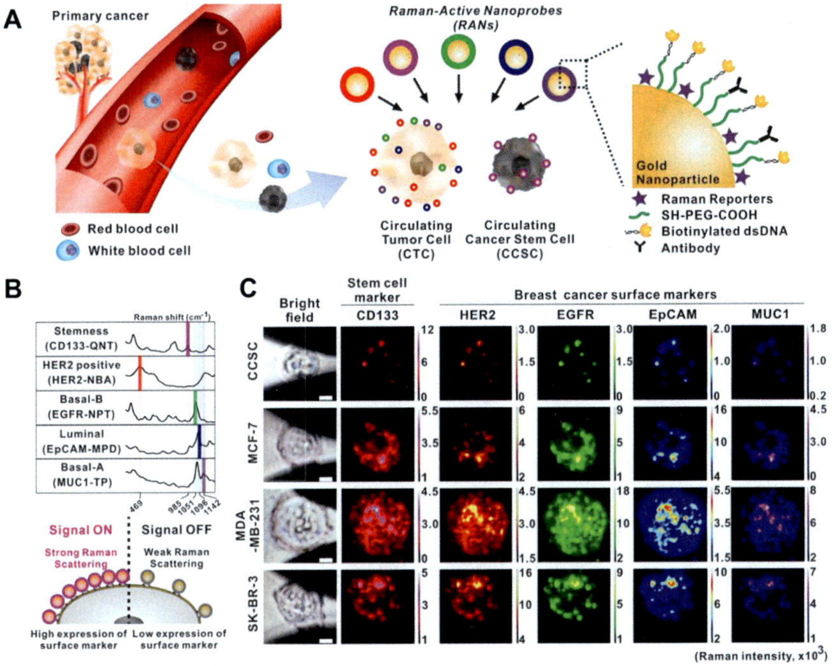

Figure 9.25 (A) Scheme showing the CCSCs and CTCs in blood sample are labeled with five types of SERS probes. (B) The representative SERS peaks of the five types of SERS probes and scheme of selective encoding of live CCSCs/CTCs based on expressed surface markers. (C) Optical microscope images and Raman mapping images obtained from detected CCSCs/CTCs. *CCSCs*, Circulating cancer stem cells; *CTCs*, circulating tumor cells. *Adapted with permission from H.-Y. Cho, M.K. Hossain, J.-H. Lee, J. Han, H.J. Lee, K.-J. Kim, et al., Selective isolation and noninvasive analysis of circulating cancer stem cells through Raman imaging. Biosens. Bioelectron. 102 (2018) 372–382, https://doi.org/10.1016/j.bios.2017.11.049. Copyright 2018 Elsevier.*

(FFPE) tissue specimens [74]. IHC is a powerful investigative tool that can provide supplemental information to the routine morphological assessment of tissues. Using IHC to study cellular markers provides important diagnostic, prognostic, and predictive information relative to disease status and biology. Especially, immunoanalysis capable of simultaneously detecting several biomarkers may offer greater insights into disease heterogeneity, conserve limited tissue samples, and improve diagnostic accuracy [75]. Based on their high sensitivity and high multiplexing capacity, SERS nanotag-antibody conjugates have been intensively investigated in applications of analyzing multiple biomarkers in tissue samples.

9.2.2.1 Detection of target protein in fixed tissue samples

The proof of concept study with antibody-modified SERS tags for tissue-based diagnostics was reported by Schlücker and coworkers in 2006 [76]. In this work, SERS tags were fabricated using gold/silver nanoshells as metal substrates and aromatic thiols as Raman reporter molecules and then conjugated with antiprostate specific antigen (PSA) antibodies. PSA was chosen as a target protein because of its selective histological abundance in the epithelium of the prostate gland. After incubation of prostate tissue sections with the SERS probes, iSERS microspectroscopy was

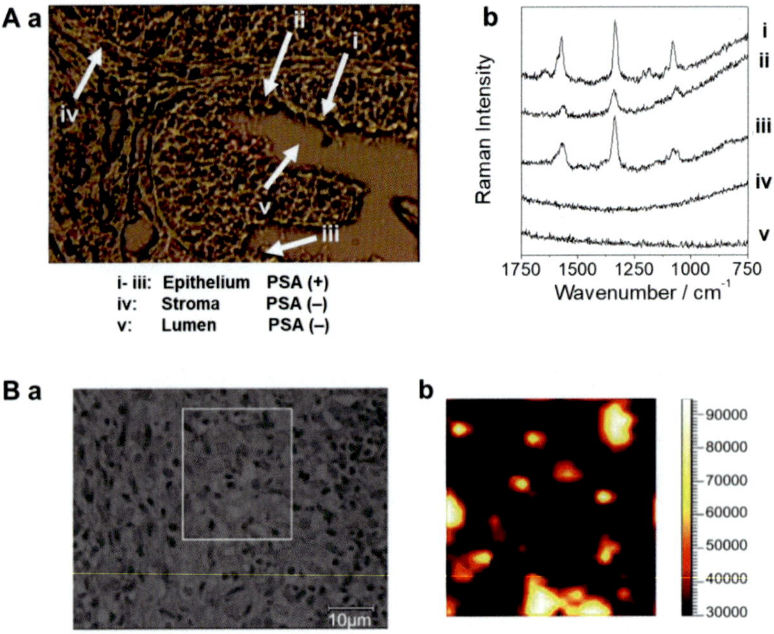

Figure 9.26 (A) Proof of concept study for iSERS microscopy using antibody-modified SERS tags for tissue diagnostics. (a) Bright field image of epithelial tissue from the prostate; (b) Raman spectra recorded at the locations indicated by arrows: (i–iii) epithelium, (iv) stroma, (v) lumen. (B) Bright field image (a) and SERS image (b) of a region from cancerous nasopharyngeal tissue section. *(Figure 9.26A) Adapted with permission from S. Schlücker, B. Küstner, A. Punge, R. Bonfig, A. Marx, P. Ströbel, Immuno-Raman microspectroscopy: in situ detection of antigens in tissue specimens by surface-enhanced Raman scattering. J. Raman Spectrosc. 37 (7) (2006) 719–721, https://doi.org/10.1002/jrs.1534. Copyright 2006 John Wiley & Sons, Ltd. (Figure 9.26B) Adapted with permission from Y. Chen, X. Zheng, G. Chen, C. He, W. Zhu, S. Feng, et al., Immunoassay for LMP1 in nasopharyngeal tissue based on surface-enhanced Raman scattering. Int. J. Nanomed. 7 (2012) 73–82, https://doi.org/10.2147/IJN.S26854. Copyright 2012 Dove Medical Press Ltd.*

performed. As shown in Fig. 9.26A, strong signals were detected only in the PSA(+) epithelium, but not in the PSA(−) stroma or lumen regions, demonstrating the specificity of the antibody-conjugated SERS tags.

In subsequent studies, again Au/Ag nanoshells [78] but also other metal nanoparticles such as Au/Ag core−shell particles [77], Au nanostars [79], composite organic-inorganic nanoparticles (COINs) [80−82], silica-encapsulated small clusters (dimers, trimers) of AuNPs [83], and polydopamine-coated AuNPs [84] have been explored as SERS tags to detect antigens in tissue specimens. Sun and coworkers fabricated COINs labeled with Basic Fushin (BFU) or acridine orange (AOH) individually, conjugated them with antiPSA antibodies, and simultaneous applied them to the tissue samples. The characteristic Raman signatures from both COINs were detected at almost every location in the epithelium, demonstrating that the steric hindrance of NPs did not influence their binding to target molecules [80]. P63 as a tumor suppressor protein is expressed in the basal cells of many epithelial organs, but is not detected in human prostate adenocarcinomas, and therefore may serve as a biomarker for prostate cancer diagnosis [85]. Schütz et al. conjugated SERS tags with antiP63 antibodies and incubated them with prostate biopsies. The iSERS imaging result show that p63 is only abundant in the basal epithelium of benign prostate tissue, but not in the secretory epithelium or the stroma/connective tissue [79,83]. Epstein-Barr virus (EBV)-encoded latent membrane protein 1 (LMP1) is closely associated with the occurrence and development of nasopharyngeal carcinoma and therefore can be used as a tumor marker for screening, predicting metastasis, and immune-targeted treatment of this disease. Chen et al. fabricated antiLMP1 antibodies-modified SERS tags and detected LMP1 expression in nasopharyngeal tissue samples from 34 cancer patients and 20 healthy controls (Fig. 9.26B). The results (sensitivity of 97.1% and specificity of 100%) were superior to those of conventional IHC staining and in excellent agreement with those of in situ hybridization for EBV-encoded small RNA, demonstrating the potential of iSERS imaging to be applied in differential diagnosis of nasopharyngeal carcinoma [77].

To develop multiplexed iSERS imaging as a clinical tool for tissue-based cancer diagnosis, many efforts have been made to improve this technology, like synthesis of bright and uniform SERS nanotags, elimination of false-positive signals produced by nonspecific adsorption of NPs onto the tissue samples, preventing uneven staining due to NPs aggregation, and accelerating the scanning speed, etc.

Knudsen et al. have compared the staining quality of COINs-tagged PSA antibody to that of Alexa fluorophore-labeled PSA antibody on adjacent tissue sections, and they found that COINs provided comparable signal intensities with fluorophores, but with a lower staining accuracy, which may be attributed to an elevated false-negative rate of COINs [81]. Schlücker et al. further investigated the influence of heterogeneous SERS intensity on staining quality. They compared SERS signals from a glass-coated monomer, dimer, and trimer of AuNPs by correlative SEM/SERS experiments. With 30 ms integration time, the glass-coated trimer (red) and dimer (blue) yielded detectable SERS signals, while the glass-coated monomer (black) did not. When antibodies are labeled with a mixture of monomers, dimers, and trimers and applied for tissue staining, the ones labeled with a dimer/trimer could be detected, while for the ones labeled with monomers, no SERS signal is detectable due to the poorer plasmonic enhancement of the single AuNP, and a false-negative result is obtained [83]. Therefore synthesis of SERS tags with uniform brightness is essential for obtaining reliable and precise staining results. Li et al. recently reported a universal strategy for the one-pot preparation of SERS tags by incorporating Raman reporters during dopamine polymerization on NP surface. Background-free SERS tags were prepared and conjugated with antibodies against different biomarkers (HER2, ER, PR, and EGFR) and then applied for multiplexed staining of clinical breast cancer biopsies (Fig. 9.27). Based on the easily differentiated single narrow bands in the cellular Raman-silent region, individual SERS probes can be conveniently discriminated [84]. This method enables easy preparation of a large library of SERS probes for multiplexed protein localization and promotes the clinical translation of iSERS imaging-based cancer diagnostics.

NPs are usually "sticky" and often bind nonspecifically to biospecimens especially tissue sections. The main causes are: (1) the increased size, weight, and surface area of NPs compared to traditional labels like dyes or molecular fluorophores; (2) conjugation of the targeting molecules onto NPs usually requires the introduction of reactive functional groups on the NPs, while excess or unreacted groups often lead to increased nonspecific protein interaction; (3) aggregation of NPs during the preparation process or in physiological media. To improve the stability of SERS tags and in the meantime maximize their coverage with Raman reporter molecules, Schlücker et al. coated Au/Ag nanoshells with a dual-SAM of ethylene glycol-modified Raman reporter molecules (a short monoethylene glycol

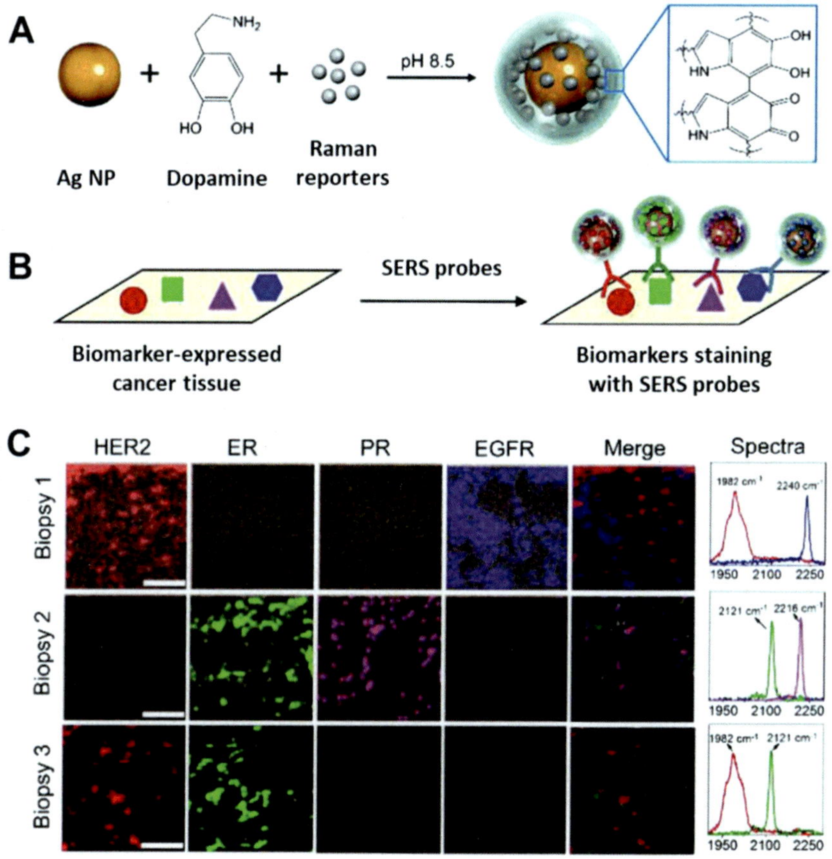

Figure 9.27 (A) Schematic illustration of the one-pot synthesis of SERS tags. (B) Schematic of the multiplexed biomarker staining. (C) Multiplexed iSERS mapping images of four biomarkers on breast cancer biopsies (HER2 in red, ER in green, PR in magenta, and EGFR in blue). Scale bar = 50 μm. *Adapted with permission from J. Li, H. Liu, P. Rong, W. Zhou, X. Gao, D. Liu, A universal strategy for the one-pot synthesis of SERS tags. Nanoscale 10 (17) (2018) 8292–8297, https://doi.org/10.1039/C8NR00564H. Copyright 2018 The Royal Society of Chemistry.*

[MEG-OH] and a longer triethylene glycol [TEG-COOH]) covalently bound to a same type of reporter molecule. This dual-SAM design provides complete coverage of the NP surface by Raman reporters for maximum signal intensity, improves water solubility of SERS tags, facilitates antibody bioconjugation via the longer TEG spacers with terminal COOH groups, and allows controlled bioconjugation via the stoichiometry of the two spacers [78]. The authors recently found that not only the properties of the NPs, but also the pretreatment manners of the tissue sample affect the nonspecific adsorption of SERS tags. They systematically investigated the influence of heat- and protease-induced epitope retrieval (HIER and PIER, respectively) on the immunostaining quality of PSA on human prostate tissue sections. The results showed that HIER pretreatment but not PIER led to nonspecific adsorption of the antibody-SERS tags conjugates onto epithelial cells. Therefore enzymatic treatment is recommended for future NP-based tissue immunostaining [86].

Unlike IHC or IF which is generally performed in a wide field configuration, iSERS is typically performed in the mapping mode for a spectrally resolved detection at each pixel. Although acquisition times per pixel have been shortened to the millisecond regime with the development of brighter SERS tags, it is time-consuming to scan a large tissue area. To improve the time efficiency of iSERS, Schlücker and coworkers recently proposed a strategy using wide field IF to guide local iSERS imaging of regions of interest [86b]. They fabricated fluorophore/SERS tag-dual labeled antibodies for localization of HER2/PSA on human breast cancer tissue sections/human prostate tissue sections. Correlative fluorescence-SERS images were recorded by an in-house integrated confocal Raman-fluorescence microscope. Fluorescence images were first recorded for global analysis, then smaller regions of interest were selected for spectral detection of each pixel using Raman mapping. The correlative imaging approach allows fast and highly multiplexed analysis of cancer biomarkers on the same tissue section.

9.2.2.2 Detection of target protein in fresh tissue samples

In addition to fixed tissue samples, biomarkers in fresh tissue samples can also be analyzed using target-specific SERS tag-antibody conjugates. Based on its specific recognition and multiplexed imaging capability, iSERS imaging has been explored as a new technique for intraoperative guidance in the last decade. The complete surgical resection is the ideal first-line treatment in surgical oncology, but is difficult to achieve when the surrounding normal tissues need to be maximally reserved. If postoperative pathology identifies residual tumor at the surgical margins, reexcision surgeries are often necessary. Therefore developing efficient intraoperative methods that can identify residual tumors and guide their complete removal during tumor-resection procedures are highly desired in the clinic.

Gambhir and coworkers fabricated SERS tags which can specifically recognize EGFR and incubated them with ex vivo tissues freshly resected from mice. iSERS imaging was then performed and reported EGFR-positive A431 tumors with a signal nearly 35-fold higher than EGFR-negative MDA-435S tumors. But the authors pointed out that some error sources might induce variability of SERS imaging, including particle-to-particle variation in size and intensity, tumor heterogeneity, inconsistencies with the optics, and nonspecific binding arises from both NP sources and ligand sources [58]. To improve the signal specificity and reliability, Liu et al. topically applied an equimolar mixture of targeted SERS tags and one untargeted SERS tag to the tissue specimen, and then constructed the SERS images using the ratio of signals from targeted tags to untargeted tags. This ratiometric biomarker detection assay allows an unambiguous assessment of the molecular expression of tumor markers, since the simultaneous scanning of targeted NPs and untargeted NPs eliminates nonspecific binding effects as well as fluctuations with the laser, CCD variation and focal plane [87]. For better understanding of nonspecific accumulation, diffusion, and chemical binding of SERS tags in complex tissue specimens, the authors froze and sectioned the tissue to image the depth-dependent accumulation of SERS tags, showing that larger

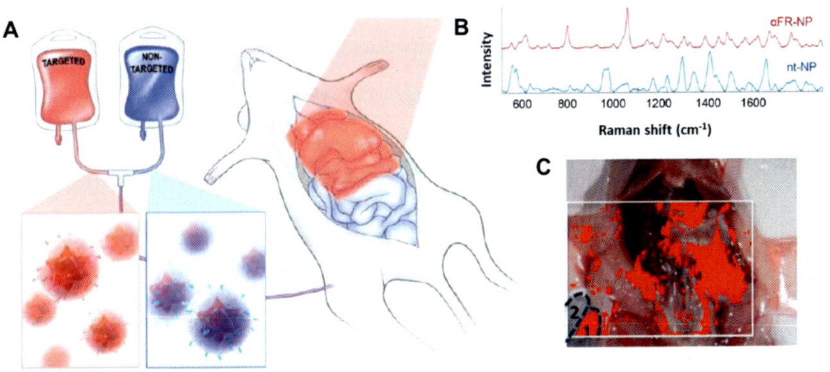

Figure 9.28 (A) Schematic illustration of the topically application of SERS tags in mouse peritoneal cavity. (B) Spectra from targeted αFR-NPs (*red*) and nontargeted nt-NPs (*blue*). (C) Ratiometric SERS image for surgical guidance, showing only regions with positive ratios in red. *Adapted with permission from A. Oseledchyk, C. Andreou, M. A. Wall, M.F. Kircher, Folate-targeted surface-enhanced resonance Raman scattering nanoprobe ratiometry for detection of microscopic ovarian cancer. ACS Nano 11 (2) (2017) 1488–1497, https://doi.org/10.1021/acsnano.6b06796. Copyright 2016 American Chemical Society.*

NPs would exhibit less diffusion below tissue surfaces, and therefore enabling higher targeted-to-untargeted NP ratios [88]. Ovarian cancer spreads locally within the peritoneal cavity, but it is currently impossible for surgeons to visualize microscopic implants, impeding their removal and leading to tumor recurrences in most patients. Kircher and coworkers fabricated folate receptor-targeted SERS tags to recognize these lesions since folate receptors are typically overexpressed in ovarian cancer. Tumors as small as 370 μm were detected (Fig. 9.28), indicating that ratiometric iSERS imaging holds promise for intraoperative detection of microscopic residual tumors in ovarian cancer and other diseases with peritoneal spread [89].

To promote the convection of the SERS tags at fresh tissue surfaces, Liu and coworkers developed a dipping and mechanical vibration (DMV) method to accelerate the binding of SERS tags to their respective biomarker targets. By utilizing a custom-developed device for automated DMV staining, specific binding of SERS tags to biomarkers on fresh human breast tissues was achieved 5 min after topical application (Fig. 9.29) [90]. Simultaneous imaging of a diverse panel of disease-related biomarkers that determines "quantitative molecular phenotype (QMP)" of the tissue may greatly improve the diagnostic accuracy [91]. In a clinical study recently

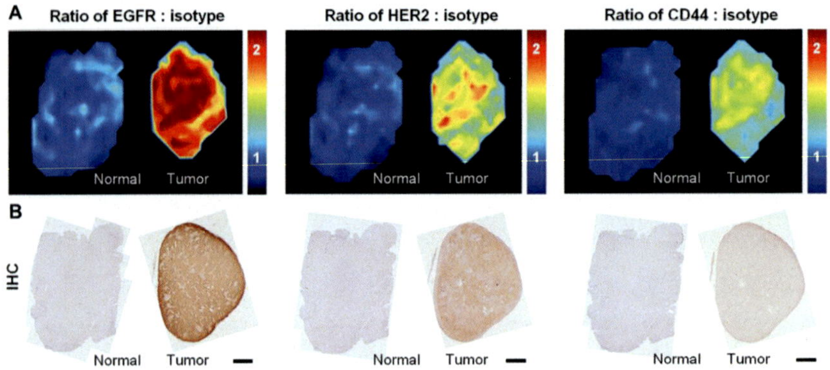

Figure 9.29 (A) Multiplexed molecular imaging of an A431 tumor xenograft through automated DMV staining with a four-flavor SERS tag mixture (EGFR-NPs, HER2-NPs, CD44-NPs, and isotype-NPs). (B) IHC validation data for EGFR, HER2, and CD44. Scale bars, 2 mm. *Adapted with permission from Y.W. Wang, J.D. Doerksen, S. Kang, D. Walsh, Q. Yang, D. Hong, et al., Multiplexed molecular imaging of fresh tissue surfaces enabled by convection-enhanced topical staining with SERS-coded nanoparticles. Small 12 (40) (2016) 5612–5621, https://doi.org/10.1002/smll.201601829. Copyright 2016 Wiley-VCH.*

performed by the same group, mixed SERS probes were applied on 57 freshly excised tissue specimens for simultaneous quantification of four biomarkers (HER2, ER, EGFR, and CD44), enabling a highly sensitive (89.3%) and specific (92.1%) detection of breast carcinoma [92]. The specimens were stained topically with a mixture of untargeted NPs and targeted NPs for 5 min, followed by a 10 s rinse step in PBS and SERS imaging. The entire staining and imaging procedure was completed within 15 minutes, demonstrating the great potential of iSERS imaging for intraoperative guidance to reduce the rate of reexcision in cancer patients.

9.3 Summary and perspectives

The SERS technology implemented in LFA overcomes the current limitations of the LFA technology by (1) providing several orders of magnitude higher sensitivity for early detection/diagnosis, (2) quantification (so far only semiquantitative), and (3) multiplexing for the detection of different analytes simultaneously. In combination with a portable SERS-POC-reader and a short readout time within seconds the SERS-LFA technology is capable to provide quantitative information by analyzing the whole test zone. Furthermore, by using multiplex SERS-LFA on one single test strip the required sample volume, material costs, reagent consumption as well as the application and operation time for assay preparation and test evaluation have been significantly reduced. Due to the universality of the SERS-LFA method it can be used and adopted for many application areas, for example, human and veterinary diagnostics, food analysis as well as the detection of drugs and biowarfare agents. Since the first application of iSERS microscopy in cellular samples [47] and tissue specimens [76] in 2006, abundant studies have been performed to improve the technique and promote its clinic translation. Progress in this field includes: (1) fabrication of bright, stable, and uniform SERS nanotags for bioconjugation to antibodies or equivalent target-specific recognition elements; (2) optimizing the immunostaining procedures such as the incubation process and biosample pretreatment approaches; and (3) improving the quality and time-efficiency of iSERS imaging. With the recently developed SERS probes, imaging, and analysis methods, efficient profiling of multiple biomarkers on CTCs [72] or fresh tissue samples [92] from clinical patients was demonstrated. However, there is still plenty of room for improvements, including aspects such as improving the uniformity and stability of SERS nanotags as well as the development of custom-made instruments (rapid scanning and high-resolution). Considering that advances in the field have progressed rapidly in the past years, we are optimistic that clinical applications of iSERS microcopy especially for cancer diagnostics will be boosted in the future.

References

[1] A.J. Tudos, G.J. Besselink, R.B. Schasfoort, Trends in miniaturized total analysis systems for point-of-care testing in clinical chemistry, Lab. Chip 1 (2) (2001) 83–95. Available from: https://doi.org/10.1039/b106958f.

[2] S.K. Vashist, Point-of-care diagnostics: recent advances and trends, Biosensors (Basel) 7 (4) (2017). Available from: https://doi.org/10.3390/bios7040062.

[3] P. Luppa, POCT—Patientennahe Labordiagnostik, 2nd (ed.), Springer-Verlag, Berlin, Heidelberg, 2012.

[4] R. Wong, H. Tse, Lateral Flow Immunoassay, Humana Press, Totowa, NJ, 2009.

[5] T. Mahmoudi, M. La Guardia, B. de Baradaran, Lateral flow assays towards point-of-care cancer detection: A review of current progress and future trends, Trends Anal. Chem. 125 (2020) 115842. Available from: https://doi.org/10.1016/j.trac.2020.115842.

[6] A.V. Zherdev, B.B. Dzantiev, Ways to reach lower detection limits of lateral flow immunoassays, in: L. Anfossi (Ed.), Rapid Test—Advances in Design, Format and Diagnostic Applications, InTech, 2018.

[7] E.B. Bahadır, M.K. Sezgintürk, Lateral flow assays: principles, designs and labels, Trends Anal. Chem. 82 (2016) 286–306. Available from: https://doi.org/10.1016/j.trac.2016.06.006.

[8] G.A. Posthuma-Trumpie, J. Korf, A. van Amerongen, Lateral flow (immuno)assay: its strengths, weaknesses, opportunities and threats. A literature survey, Anal. Bioanal. Chem. 393 (2) (2009) 569–582. Available from: https://doi.org/10.1007/s00216-008-2287-2.

[9] M. Sajid, A.-N. Kawde, M. Daud, Designs, formats and applications of lateral flow assay: a literature review, J. Saudi Chem. Soc. 19 (6) (2015) 689–705. Available from: https://doi.org/10.1016/j.jscs.2014.09.001.

[10] B. O'Farrell, Lateral flow technology for field-based applications-basics and advanced developments, Top. Compan. An. Med. 30 (4) (2015) 139–147. Available from: https://doi.org/10.1053/j.tcam.2015.12.003.

[11] J. Yang, K. Wang, H. Xu, W. Yan, Q. Jin, D. Cui, Detection platforms for point-of-care testing based on colorimetric, luminescent and magnetic assays: a review, Talanta 202: (2019) 96–110. Available from: https://doi.org/10.1016/j.talanta.2019.04.054.

[12] H. Puig, I. de, Bosch, L. Gehrke, K. Hamad-Schifferli, Challenges of the nano-bio interface in lateral flow and dipstick immunoassays, Trends Biotechnol. 35 (12) (2017) 1169–1180. Available from: https://doi.org/10.1016/j.tibtech.2017.09.001.

[13] J. Smolsky, S. Kaur, C. Hayashi, S.K. Batra, A.V. Krasnoslobodtsev, Surface-enhanced raman scattering-based immunoassay technologies for detection of disease biomarkers, Biosensors (Basel) 7 (1) (2017) 7. Available from: https://doi.org/10.3390/bios7010007.

[14] S. Choi, J. Hwang, S. Lee, D.W. Lim, H. Joo, J. Choo, Quantitative analysis of thyroid-stimulating hormone (TSH) using SERS-based lateral flow immunoassay, Sens. Actuator B Chem. 240 (2017) 358–364. Available from: https://doi.org/10.1016/j.snb.2016.08.178.

[15] R. Wang, K. Kim, N. Choi, X. Wang, J. Lee, J.H. Jeon, et al., Highly sensitive detection of high-risk bacterial pathogens using SERS-based lateral flow assay strips, Sens. Actuator B Chem. 270 (2018) 72–79. Available from: https://doi.org/10.1016/j.snb.2018.04.162.

[16] J. Hwang, S. Lee, J. Choo, Application of a SERS-based lateral flow immunoassay strip for the rapid and sensitive detection of staphylococcal enterotoxin B, Nanoscale 8 (22) (2016) 11418–11425. Available from: https://doi.org/10.1039/c5nr07243c.

[17] V. Tran, B. Walkenfort, M. König, M. Salehi, S. Schlücker, Rapid, quantitative, and ultrasensitive point-of-care testing: a portable SERS reader for lateral flow assays in clinical chemistry, Angew. Chem. Int. (Ed.) 58 (2) (2019) 442–446. Available from: https://doi.org/10.1002/anie.201980261.

[18] L.-K. Lin, L.A. Stanciu, Sens. Actuator B Chem. 276 (2018) 222–229. Available from: https://doi.org/10.1016/j.snb.2018.08.068.

[19] B.N. Khlebtsov, D.N. Bratashov, N.A. Byzova, B.B. Dzantiev, N.G. Khlebtsov, SERS-based lateral flow immunoassay of troponin I by using gap-enhanced Raman tags, Nano Res. 12 (2) (2019) 413–420. Available from: https://doi.org/10.1007/s12274-018-2232-4.

[20] Q. Shi, J. Huang, Y. Sun, R. Deng, M. Teng, Q. Li, et al., A SERS-based multiple immuno-nanoprobe for ultrasensitive detection of neomycin and quinolone antibiotics via a lateral flow assay, Microchim. Acta 185 (2) (2018) 84. Available from: https://doi.org/10.1007/s00604-017-2556-x.

[21] Y. Wang, J. Sun, Y. Hou, C. Zhang, D. Li, H. Li, et al., A SERS-based lateral flow assay biosensor for quantitative and ultrasensitive detection of interleukin-6 in unprocessed whole blood, Biosens. Bioelectron. 141 (2019) 111432. Available from: https://doi.org/10.1016/j.bios.2019.111432.

[22] Y. Wang, Y. Hou, H. Li, M. Yang, P. Zhao, B. Sun, A SERS-based lateral flow assay for the stroke biomarker S100-β, Microchim. Acta 186 (8) (2019) 548. Available from: https://doi.org/10.1007/s00604-019-3634-z.

[23] Z. Wu, Simultaneous detection of *Listeria monocytogenes* and *Salmonella typhimurium* by a SERS-based lateral flow immunochromatographic assay, Food Anal. Methods 12 (5) (2019) 1086–1091. Available from: https://doi.org/10.1007/s12161-019-01444-4.

[24] Di Zhang, L. Huang, B. Liu, H. Ni, L. Sun, E. Su, et al., Quantitative and ultrasensitive detection of multiplex cardiac biomarkers in lateral flow assay with core-shell SERS nanotags, Biosens. Bioelectron. 106 (2018) 204–211. Available from: https://doi.org/10.1016/j.bios.2018.01.062.

[25] Di Zhang, L. Huang, B. Liu, Q. Ge, J. Dong, X. Zhao, A vertical flow microarray chip based on SERS nanotags for rapid and ultrasensitive quantification of α-fetoprotein and carcinoembryonic antigen, Microchim. Acta 186 (11) (2019) 699. Available from: https://doi.org/10.1007/s00604-019-3792-z.

[26] Di Zhang, L. Huang, B. Liu, Q. Ge, J. Dong, X. Zhao, Rapid and ultrasensitive quantification of multiplex respiratory tract infection pathogen via lateral flow microarray based on SERS nanotags, Theranostics 9 (17) (2019) 4849–4859. Available from: https://doi.org/10.7150/thno.35824.

[27] N.R. Stambach, S.A. Carr, C.R. Cox, K.J. Voorhees, Rapid detection of listeria by bacteriophage amplification and SERS-lateral flow immunochromatography, Viruses 7 (12) (2015) 6631–6641. Available from: https://doi.org/10.3390/v7122962.

[28] Q. Fu, H.L. Liu, Z. Wu, A. Liu, C. Yao, X. Li, et al., Rough surface Au@Ag core-shell nanoparticles to fabricating high sensitivity SERS immunochromatographic sensors, J. Nanobiotechnol. 13 (2015) 81. Available from: https://doi.org/10.1186/s12951-015-0142-0.

[29] L. Blanco-Covián, V. Montes-García, A. Girard, M.T. Fernández-Abedul, J. Pérez-Juste, I. Pastoriza-Santos, et al., Au@Ag SERRS tags coupled to a lateral flow immunoassay for the sensitive detection of pneumolysin, Nanoscale 9 (5) (2017) 2051–2058. Available from: https://doi.org/10.1039/c6nr08432j.
[30] M. Li, H. Yang, S. Li, K. Zhao, J. Li, D. Jiang, et al., Ultrasensitive and quantitative detection of a new β-agonist phenylethanolamine A by a novel immunochromatographic assay based on surface-enhanced Raman scattering (SERS), J. Agric. Food Chem. 62 (45) (2014) 10896–10902. Available from: https://doi.org/10.1021/jf503599x.
[31] X. Fu, Z. Cheng, J. Yu, P. Choo, L. Chen, J. Choo, A SERS-based lateral flow assay biosensor for highly sensitive detection of HIV-1 DNA, Biosens. Bioelectron. 78 (2016) 530–537. Available from: https://doi.org/10.1016/j.bios.2015.11.099.
[32] W. Maneeprakorn, S. Bamrungsap, C. Apiwat, N. Wiriyachaiporn, Surface-enhanced Raman scattering based lateral flow immunochromatographic assay for sensitive influenza detection, RSC Adv. 6 (113) (2016) 112079–112085. Available from: https://doi.org/10.1039/c6ra24418a.
[33] X. Fu, Y. Chu, K. Zhao, J. Li, A. Deng, Ultrasensitive detection of the β-adrenergic agonist brombuterol by a SERS-based lateral flow immunochromatographic assay using flower-like gold-silver core-shell nanoparticles, Microchim. Acta 184 (6) (2017) 1711–1719. Available from: https://doi.org/10.1007/s00604-017-2178-3.
[34] X. Gao, P. Zheng, S. Kasani, S. Wu, F. Yang, S. Lewis, et al., Paper-based surface-enhanced Raman scattering lateral flow strip for detection of neuron-specific enolase in blood plasma, Anal. Chem. 89 (18) (2017) 10104–10110. Available from: https://doi.org/10.1021/acs.analchem.7b03015.
[35] T. Bai, M. Wang, M. Cao, J. Zhang, K. Zhang, P. Zhou, et al., Functionalized Au@Ag-Au nanoparticles as an optical and SERS dual probe for lateral flow sensing, Anal. Bioanal. Chem. 410 (9) (2018) 2291–2303. Available from: https://doi.org/10.1007/s00216-018-0850-z.
[36] X. Jia, C. Wang, Z. Rong, J. Li, K. Wang, Z. Qie, et al., Dual dye-loaded Au@Ag coupled to a lateral flow immunoassay for the accurate and sensitive detection of Mycoplasma pneumoniae infection, RSC Adv. 8 (38) (2018) 21243–21251. Available from: https://doi.org/10.1039/c8ra03323d.
[37] Z. Rong, R. Xiao, S. Xing, G. Xiong, Z. Yu, L. Wang, et al., SERS-based lateral flow assay for quantitative detection of C-reactive protein as an early bio-indicator of a radiation-induced inflammatory response in nonhuman primates, Analyst 143 (9) (2018) 2115–2121. Available from: https://doi.org/10.1039/c8an00160j.
[38] H. Shen, K. Xie, L. Huang, L. Wang, J. Ye, M. Xiao, et al., A novel SERS-based lateral flow assay for differential diagnosis of wild-type pseudorabies virus and gE-deleted vaccine, Sens. Actuator B Chem. 282 (2019) 152–157. Available from: https://doi.org/10.1016/j.snb.2018.11.065.
[39] J. Li, J. Macdonald, Multiplexed lateral flow biosensors: technological advances for radically improving point-of-care diagnoses, Biosens. Bioelectron. 83 (2016) 177–192. Available from: https://doi.org/10.1016/j.bios.2016.04.021.
[40] M. Schütz, S. Schlücker, Towards quantitative multi-color nanodiagnostics: spectral multiplexing with six silica-encapsulated SERS labels, J. Raman Spectrosc. 47 (9) (2016) 1012–1016. Available from: https://doi.org/10.1002/jrs.4913.
[41] M. Sánchez-Purrà, M. Carré-Camps, H. Puig, I. de, Bosch, L. Gehrke, K. Hamad-Schifferli, Surface-enhanced Raman spectroscopy-based sandwich immunoassays for multiplexed detection of Zika and dengue viral biomarkers, ACS Infect. Dis. 3 (10) (2017) 767–776. Available from: https://doi.org/10.1021/acsinfecdis.7b00110.

[42] X. Wang, N. Choi, Z. Cheng, J. Ko, L. Chen, J. Choo, Simultaneous detection of dual nucleic acids using a SERS-based lateral flow assay biosensor, Anal. Chem. 89 (2) (2017) 1163–1169. Available from: https://doi.org/10.1021/acs.analchem.6b03536.
[43] Di Zhang, L. Huang, B. Liu, E. Su, H.-Y. Chen, Z. Gu, et al., Quantitative detection of multiplex cardiac biomarkers with encoded SERS nanotags on a single T line in lateral flow assay, Sens. Actuator B Chem. 277 (2018) 502–509. Available from: https://doi.org/10.1016/j.snb.2018.09.044.
[44] B. Küstner, Wirkstoff-Substrat-Charakterisierung und Protein-Lokalisierung mittels Raman-Streuung, Julius-Maximilians-Universität Würzburg, 2009.
[45] H.-B. Liu, X.-J. Du, Y.-X. Zang, P. Li, S. Wang, SERS-based lateral flow strip biosensor for simultaneous detection of *Listeria monocytogenes* and *Salmonella enterica* serotype Enteritidis, J. Agric. Food Chem. 65 (47) (2017) 10290–10299. Available from: https://doi.org/10.1021/acs.jafc.7b03957.
[46] J. Sun, L. Gong, W. Wang, Z. Gong, D. Wang, M. Fan, Surface-enhanced Raman spectroscopy for on-site analysis: a review of recent developments, Luminescence (2020). Available from: https://doi.org/10.1002/bio.3796.
[47] J.-H. Kim, J.-S. Kim, H. Choi, S.-M. Lee, B.-H. Jun, K.-N. Yu, et al., Nanoparticle probes with surface enhanced Raman spectroscopic tags for cellular cancer targeting, Anal. Chem. 78 (19) (2006) 6967–6973. Available from: https://doi.org/10.1021/ac0607663.
[48] S. Lee, S. Kim, J. Choo, S.Y. Shin, Y.H. Lee, H.Y. Choi, et al., Biological imaging of HEK293 cells expressing PLC gamma1 using surface-enhanced Raman microscopy, Anal. Chem. 79 (3) (2007) 916–922. Available from: https://doi.org/10.1021/ac061246a.
[49] S. Lee, H. Chon, M. Lee, J. Choo, S.Y. Shin, Y.H. Lee, et al., Surface-enhanced Raman scattering imaging of HER2 cancer markers overexpressed in single MCF7 cells using antibody conjugated hollow gold nanospheres, Biosens. Bioelectron. 24 (7) (2009) 2260–2263. Available from: https://doi.org/10.1016/j.bios.2008.10.018.
[50] H. Park, S. Lee, L. Chen, E.K. Lee, S.Y. Shin, Y.H. Lee, et al., SERS imaging of HER2-overexpressed MCF7 cells using antibody-conjugated gold nanorods, Phys. Chem. Chem. Phys. 11 (34) (2009) 7444–7449. Available from: https://doi.org/10.1039/B904592A.
[51] C.T. Nguyen, J.T. Nguyen, S. Rutledge, J. Zhang, C. Wang, G.C. Walker, Detection of chronic lymphocytic leukemia cell surface markers using surface enhanced Raman scattering gold nanoparticles, Cancer Lett. 292 (1) (2010) 91–97. Available from: https://doi.org/10.1016/j.canlet.2009.11.011.
[52] D.C. Kennedy, K.A. Hoop, L.-L. Tay, J.P. Pezacki, Development of nanoparticle probes for multiplex SERS imaging of cell surface proteins, Nanoscale 2 (8) (2010) 1413–1416. Available from: https://doi.org/10.1039/c0nr00122h.
[53] S. Lee, H. Chon, J. Lee, J. Ko, B.H. Chung, D.W. Lim, et al., Rapid and sensitive phenotypic marker detection on breast cancer cells using surface-enhanced Raman scattering (SERS) imaging, Biosens. Bioelectron. 51 (2014) 238–243. Available from: https://doi.org/10.1016/j.bios.2013.07.063.
[54] G. Zhu, X. Chen, Aptamer-based targeted therapy, Adv. Drug. Deliv. Rev. 134 (2018) 65–78. Available from: https://doi.org/10.1016/j.addr.2018.08.005.
[55] J. Wu, D. Liang, Q. Jin, J. Liu, M. Zheng, X. Duan, et al., Bioorthogonal SERS nanoprobes for mulitplex spectroscopic detection, tumor cell targeting, and tissue imaging, Chemistry 21 (37) (2015) 12914–12918. Available from: https://doi.org/10.1002/chem.201501942.
[56] Y. Zou, S. Huang, Y. Liao, X. Zhu, Y. Chen, L. Chen, et al., Isotopic graphene-isolated-Au-nanocrystals with cellular Raman-silent signals for cancer cell pattern

recognition, Chem. Sci. 9 (10) (2018) 2842–2849. Available from: https://doi.org/10.1039/c7sc05442d.
[57] Y. Zeng, J.-Q. Ren, A.-G. Shen, J.-M. Hu, Splicing nanoparticles-based "click" SERS could aid multiplex liquid biopsy and accurate cellular imaging, J. Am. Chem. Soc. 140 (34) (2018) 10649–10652. Available from: https://doi.org/10.1021/jacs.8b04892.
[58] J.V. Jokerst, Z. Miao, C. Zavaleta, Z. Cheng, S.S. Gambhir, Affibody-functionalized gold-silica nanoparticles for Raman molecular imaging of the epidermal growth factor receptor, Small 7 (5) (2011) 625–633. Available from: https://doi.org/10.1002/smll.201002291.
[59] C. Fasolato, S. Giantulli, I. Silvestri, F. Mazzarda, Y. Toumia, F. Ripanti, et al., Folate-based single cell screening using surface enhanced Raman microimaging, Nanoscale 8 (39) (2016) 17304–17313. Available from: https://doi.org/10.1039/c6nr05057c.
[60] M. Li, S.R. Banerjee, C. Zheng, M.G. Pomper, I. Barman, Ultrahigh affinity Raman probe for targeted live cell imaging of prostate cancer, Chem. Sci. 7 (11) (2016) 6779–6785. Available from: https://doi.org/10.1039/c6sc01739h.
[61] D. Craig, S. McAughtrie, J. Simpson, C. McCraw, K. Faulds, D. Graham, Confocal SERS mapping of glycan expression for the identification of cancerous cells, Anal. Chem. 86 (10) (2014) 4775–4782. Available from: https://doi.org/10.1021/ac4038762.
[62] H. Di, H. Liu, M. Li, J. Li, D. Liu, High-precision profiling of sialic acid expression in cancer cells and tissues using background-free surface-enhanced Raman scattering tags, Anal. Chem. 89 (11) (2017) 5874–5881. Available from: https://doi.org/10.1021/acs.analchem.7b00199.
[63] Y. Chen, J.-Q. Ren, X.-G. Zhang, D.-Y. Wu, A.-G. Shen, J.-M. Hu, Alkyne-modulated surface-enhanced Raman scattering-palette for optical interference-free and multiplex cellular imaging, Anal. Chem. 88 (12) (2016) 6115–6119. Available from: https://doi.org/10.1021/acs.analchem.6b01374.
[64] E. Chung, J. Lee, J. Yu, S. Lee, J.H. Kang, I.Y. Chung, et al., Use of surface-enhanced Raman scattering to quantify EGFR markers uninhibited by cetuximab antibodies, Biosens. Bioelectron. 60 (2014) 358–365. Available from: https://doi.org/10.1016/j.bios.2014.04.041.
[65] E. Feng, T. Zheng, X. He, J. Chen, Y. Tian, A novel ternary heterostructure with dramatic SERS activity for evaluation of PD-L1 expression at the single-cell level, Sci. Adv. 4 (11) (2018) eaau3494. Available from: https://doi.org/10.1126/sciadv.aau3494.
[66] Y. Zhang, X. Mi, X. Tan, R. Xiang, Recent progress on liquid biopsy analysis using surface-enhanced Raman spectroscopy, Theranostics 9 (2) (2019) 491–525. Available from: https://doi.org/10.7150/thno.29875.
[67] M.Y. Sha, H. Xu, M.J. Natan, R. Cromer, Surface-enhanced Raman scattering tags for rapid and homogeneous detection of circulating tumor cells in the presence of human whole blood, J. Am. Chem. Soc. 130 (51) (2008) 17214–17215. Available from: https://doi.org/10.1021/ja804494m.
[68] X. Wang, X. Qian, J.J. Beitler, Z.G. Chen, F.R. Khuri, M.M. Lewis, et al., Detection of circulating tumor cells in human peripheral blood using surface-enhanced Raman scattering nanoparticles, Cancer Res. 71 (5) (2011) 1526–1532. Available from: https://doi.org/10.1158/0008-5472.CAN-10-3069.
[69] X. Wu, Y. Xia, Y. Huang, J. Li, H. Ruan, T. Chen, et al., Improved SERS-active nanoparticles with various shapes for CTC detection without enrichment process with supersensitivity and high specificity, ACS Appl. Mater. Interfaces 8 (31) (2016) 19928–19938. Available from: https://doi.org/10.1021/acsami.6b07205.

[70] Z.A. Nima, M. Mahmood, Y. Xu, T. Mustafa, F. Watanabe, D.A. Nedosekin, et al., Circulating tumor cell identification by functionalized silver-gold nanorods with multicolor, super-enhanced SERS and photothermal resonances, Sci. Rep. 4 (2014) 4752. Available from: https://doi.org/10.1038/srep04752.
[71] N.V. Jordan, A. Bardia, B.S. Wittner, C. Benes, M. Ligorio, Y. Zheng, et al., HER2 expression identifies dynamic functional states within circulating breast cancer cells, Nature 537 (7618) (2016) 102–106. Available from: https://doi.org/10.1038/nature19328.
[72] S.C.-H. Tsao, J. Wang, Y. Wang, A. Behren, J. Cebon, M. Trau, Characterising the phenotypic evolution of circulating tumour cells during treatment, Nat. Commun. 9 (1) (2018) 1482. Available from: https://doi.org/10.1038/s41467-018-03725-8.
[73] H.-Y. Cho, M.K. Hossain, J.-H. Lee, J. Han, H.J. Lee, K.-J. Kim, et al., Selective isolation and noninvasive analysis of circulating cancer stem cells through Raman imaging, Biosens. Bioelectron. 102 (2018) 372–382. Available from: https://doi.org/10.1016/j.bios.2017.11.049.
[74] D. Berg, S. Hipp, K. Malinowsky, C. Böllner, K.-F. Becker, Molecular profiling of signalling pathways in formalin-fixed and paraffin-embedded cancer tissues, Eur. J. Cancer 46 (1) (2010) 47–55. Available from: https://doi.org/10.1016/j.ejc.2009.10.016.
[75] E.C. Stack, C. Wang, K.A. Roman, C.C. Hoyt, Multiplexed immunohistochemistry, imaging, and quantitation: a review, with an assessment of Tyramide signal amplification, multispectral imaging and multiplex analysis, Methods 70 (1) (2014) 46–58. Available from: https://doi.org/10.1016/j.ymeth.2014.08.016.
[76] S. Schlücker, B. Küstner, A. Punge, R. Bonfig, A. Marx, P. Ströbel, Immuno-Raman microspectroscopy: in situ detection of antigens in tissue specimens by surface-enhanced Raman scattering, J. Raman Spectrosc. 37 (7) (2006) 719–721. Available from: https://doi.org/10.1002/jrs.1534.
[77] Y. Chen, X. Zheng, G. Chen, C. He, W. Zhu, S. Feng, et al., Immunoassay for LMP1 in nasopharyngeal tissue based on surface-enhanced Raman scattering, Int. J. Nanomed. 7 (2012) 73–82. Available from: https://doi.org/10.2147/IJN.S26854.
[78] C. Jehn, B. Küstner, P. Adam, A. Marx, P. Ströbel, C. Schmuck, et al., Water soluble SERS labels comprising a SAM with dual spacers for controlled bioconjugation, Phys. Chem. Chem. Phys. 11 (34) (2009) 7499–7504. Available from: https://doi.org/10.1039/B905092B.
[79] M. Schütz, D. Steinigeweg, M. Salehi, K. Kömpe, S. Schlücker, Hydrophilically stabilized gold nanostars as SERS labels for tissue imaging of the tumor suppressor p63 by immuno-SERS microscopy, Chem. Comm. 47 (14) (2011) 4216–4218. Available from: https://doi.org/10.1039/C0CC05229A.
[80] L. Sun, K.-B. Sung, C. Dentinger, B. Lutz, L. Nguyen, J. Zhang, et al., Composite organic-inorganic nanoparticles as Raman labels for tissue analysis, Nano Lett. 7 (2) (2007) 351–356. Available from: https://doi.org/10.1021/nl062453t.
[81] B. Lutz, C. Dentinger, L. Sun, L. Nguyen, J. Zhang, A. Chmura, et al., Raman nanoparticle probes for antibody-based protein detection in tissues, J. Histochem. Cytochem. 56 (4) (2008) 371–379. Available from: https://doi.org/10.1369/jhc.7A7313.2007.
[82] B.R. Lutz, C.E. Dentinger, L.N. Nguyen, L. Sun, J. Zhang, A.N. Allen, et al., Spectral analysis of multiplex Raman probe signatures, ACS Nano 2 (11) (2008) 2306–2314. Available from: https://doi.org/10.1021/nn800243g.
[83] M. Salehi, D. Steinigeweg, P. Ströbel, A. Marx, J. Packeisen, S. Schlücker, Rapid immuno-SERS microscopy for tissue imaging with single-nanoparticle sensitivity, J. Biophotonics 6 (10) (2013) 785–792. Available from: https://doi.org/10.1002/jbio.201200148.

[84] J. Li, H. Liu, P. Rong, W. Zhou, X. Gao, D. Liu, A universal strategy for the one-pot synthesis of SERS tags, Nanoscale 10 (17) (2018) 8292–8297. Available from: https://doi.org/10.1039/C8NR00564H.
[85] S. Signoretti, D. Waltregny, J. Dilks, B. Isaac, D. Lin, L. Garraway, et al., p63 is a prostate basal cell marker and is required for prostate development, Am. J. Pathol. 157 (6) (2000) 1769–1775. Available from: https://doi.org/10.1016/S0002-9440(10)64814-6.
[86] (a) Y. Zhang, X.-P. Wang, S. Perner, A. Bankfalvi, S. Schlücker, Effect of antigen retrieval methods on nonspecific binding of antibody-metal nanoparticle conjugates on formalin-fixed paraffin-embedded tissue, Anal. Chem. 90 (1) (2018) 760–768. Available from: https://doi.org/10.1021/acs.analchem.7b03144.
(b) X.P. Wang, Y. Zhang, et al., iSERS microscopy guided by wide field immunofluorescence: analysis of HER2 expression on normal and breast cancer FFPE tissue sections, Analyst 141 (2016) 5113–5119. Available from: https://doi.org/10.1039/C6AN00927A.
[87] Y.W. Wang, A. Khan, M. Som, D. Wang, Y. Chen, S.Y. Leigh, et al., Rapid ratiometric biomarker detection with topically applied SERS nanoparticles, Technol. (Singap. World Sci.) 2 (2) (2014) 118–132. Available from: https://doi.org/10.1142/S2339547814500125.
[88] S. Kang, Y.W. Wang, X. Xu, E. Navarro, K.M. Tichauer, J.T.C. Liu, Microscopic investigation of topically applied nanoparticles for molecular imaging of fresh tissue surfaces, J. Biophoton. 11 (4) (2018) e201700246. Available from: https://doi.org/10.1002/jbio.201700246.
[89] A. Oseledchyk, C. Andreou, M.A. Wall, M.F. Kircher, Folate-targeted surface-enhanced resonance Raman scattering nanoprobe ratiometry for detection of microscopic ovarian cancer, ACS Nano 11 (2) (2017) 1488–1497. Available from: https://doi.org/10.1021/acsnano.6b06796.
[90] Y.W. Wang, J.D. Doerksen, S. Kang, D. Walsh, Q. Yang, D. Hong, et al., Multiplexed molecular imaging of fresh tissue surfaces enabled by convection-enhanced topical staining with SERS-coded nanoparticles, Small 12 (40) (2016) 5612–5621. Available from: https://doi.org/10.1002/smll.201601829.
[91] S. Kang, Y. Wang, N.P. Reder, J.T.C. Liu, Multiplexed molecular imaging of biomarker-targeted SERS nanoparticles on fresh tissue specimens with channel-compressed spectrometry, PLoS One 11 (9) (2016) e0163473. Available from: https://doi.org/10.1371/journal.pone.0163473.
[92] Y.W. Wang, N.P. Reder, S. Kang, A.K. Glaser, Q. Yang, M.A. Wall, et al., Raman-encoded molecular imaging with topically applied SERS nanoparticles for intraoperative guidance of lumpectomy, Cancer Res. 77 (16) (2017) 4506–4516. Available from: https://doi.org/10.1158/0008-5472.CAN-17-0709.

CHAPTER 10

Surface-enhanced Raman spectroscopy for cancer characterization

Wen Ren and Joseph Irudayaraj
Department of Bioengineering, Cancer Center at Illinois, Nick Holonyak Micro and Nanotechnology Laboratory, University of Illinois at Urbana-Champaign, Urbana, IL, United States

10.1 Introduction

Surface-enhanced Raman scattering (SERS) has been demonstrated to be a powerful analytical tool exhibiting strong potential in biomedical diagnostics. Until now, various biomolecules from proteins to nucleic acid sequences and cells were tested with SERS [1−6]. Meanwhile, SERS imaging of cells, tissues, organs, and even whole bodies has been demonstrated exhibiting its potential in diagnosis [7−11]. SERS could offer extremely enhanced spectra with an average enhancement factor of around 10^6 [12], reaching as high as $10^{14}-10^{15}$ under favorable conditions [13]. Furthermore, the exquisite enhancement enables the collection of signals from biomolecules which usually gives a weak response with other analytical methods. SERS spectra could reveal molecular structural information of the targets for diagnosis and disease characterization. The bandwidth of SERS shifts is much narrower than that of fluorescent peaks, thus facilitating multiplex detection and multichannel imaging depending on the choice of Raman reporter molecules with unique feature fingerprints. SERS spectra could be excited by lasers at different wavelengths including the near-infrared (NIR) laser, thus SERS could be used for in vivo examination in the NIR window (from 650 to 1350 nm). SERS substrates for enhancement could be functionalized with different ligands: (1) conjugation of recognition ligands such as antibody or aptamer for specific detection and identification; (2) loading of drug molecules for drug delivery; (3) encapsulation of reactive oxygen species (ROS) for photodynamic therapy along with SERS detection. Furthermore, the designed structure of SERS substrates can exhibit photothermal response

Principles and Clinical Diagnostic Applications of Surface-Enhanced Raman Spectroscopy
DOI: https://doi.org/10.1016/B978-0-12-821121-2.00007-X

that can be utilized in photothermal therapy. SERS substrates could be constructed with gold, which is nontoxic to humans. The enhancement is located in the region near the SERS substrates, thus the influence from other components away from the substrates could be reduced when complex biological samples are tested. SERS spectroscopy could be integrated with other analytical methods and imaging strategies to enhance its diagnostic capability. Various SERS imaging methods have been developed to illustrate specific biomarkers, characterize cells, and analyze tissues, tumors, and organs, which would help in our understanding of diseases and for cancer staging. The development of hardware enables the SERS detection and imaging for bedside and intraoperative examinations as well.

Limitations of SERS are that it is highly dependent on the strong surface plasmonic resonance (SPR) or localized SPR (LSPR) of substrates, which would change dramatically with the morphology of the SERS substrates. Nanoscale structures termed as "hot spots" with extremely high SERS activity on SERS substrates are critical for the enhancement of SERS signals [14–17]. Thus, the intensity variation of SERS signals might influence detection and imaging. In addition, the detection based on SERS requires the substrates to enhance the Raman signal from targets or labeled SERS reporters. However, nonspecific binding of SERS substrate, especially in the detection with cells, tissues, and in vivo applications, would influence the distribution and labeling. Benefiting from the advances in nanotechnology and SERS, novel SERS substrates (detailed in Chapter 2, *Nanoplasmonic materials for surface-enhanced Raman scattering*) have emerged and some attempts were reported to overcome these limitations.

In this chapter, we focus on applications of SERS in cancer which has a ~39.3% incidence of life risk of developing cancer at any site per the 2012–2016 reports by National Institutes of Health [18], suggesting that almost every one of us might have a likelihood of the disease or has a cancer patient in our family. The World Health Organization indicated in 2018 that cancer is the second leading cause of death around the world [19]. The challenge is on the mechanism of how cancer occurs and develops based on biological clues, and the lack of effective therapy for most types of cancers which makes early diagnosis important and valuable. Therefore cancer-related research has been in the spotlights in recent decades resulting in an enormous number of publications. SERS as a powerful analytical tool is capable of recognizing different types of biomarkers to diagnose cancer ranging from proteins to genes and microRNA sequences. The high sensitivity due to strong SERS enhancement makes the detection of trace levels of biomarkers for early diagnosis of cancer. SERS imaging could illustrate the distribution of interesting biomolecules in cells or tissues to improve the understanding of cancers. Furthermore, although tumors look like normal tissues under white light, SERS imaging has the ability to delineate tumors from the nearby normal tissues to enhance resection in operations. In the following sections, applications of SERS in cancer characterization and therapy will be discussed.

10.2 SERS diagnosis of cancer biomarkers

For cancer diagnosis the determination of key biomarkers is a common route. Although there are different types of biomarkers for cancer, SERS enables various detection strategies based on the SERS spectra from target biomarkers themselves or from SERS probes modified with Raman reporters and recognition ligands such as antibody or aptamer to label target biomarkers or indicators. SERS could detect the biomarkers extracted from cancer cells, tissues, and blood or other biophysical samples. Based on the biomarkers on the surface of malignant cells labeled with SERS probes, cancer cells and tumors could also be recognized from the SERS signal.

Antibody is one of the most common recognition ligands capable of recognizing various targets from small molecules to proteins. Labeling of the targets with antibody-modified SERS probes should be the primary strategy for the detection of cancer biomarkers. A classic pattern of antibody-based recognition with SERS is a sandwich-structure that targets biomarkers captured by antibodies immobilized on substrates and labeled with antibody-modified SERS substrates, like the one reported by Grubisha et al. [20]. As initial experiments, the detection of prostate-specific antigen (PSA) was performed in buffer. The antibodies conjugated on gold chips captured the target PSA as the biomarker for prostate cancer and gold nanoparticles (GNPs) modified with antibodies and Raman reporters labeled the captured PSA. The SERS intensity from the Raman reporters was used to determine the concentration of target molecules of PSA. Wang et al. detected mucin protein MUC4 as the target for pancreatic adenocarcinoma in a sandwich pattern with SERS substrates of antibody labeled GNPs [21]. The detection method was extended for the detection of biomarkers in cell lysates and pooled serum. Furthermore, Lee et al. prepared hollow gold nanospheres modified with Raman reporters and antibodies as SERS probes and patterned gold microarray for improved reproducible immunoassay of cancer biomarkers of alpha-fetoprotein (AFP) and angiogenin [22]. To construct the gold microarray, titanium was first deposited on glass substrates to reduce the fluorescence from glass substrates. Further, the gold pattern was fabricated as hydrophilic while the other region is hydrophobic, thus aqua samples could be concentrated on the gold surface. This design resulted in an improved sensitivity and wider detectable dynamic range. On a gold array substrate, with different antibodies, multiplex detection of various biomarkers was demonstrated for the diagnosis of specific cancers [23]. Li et al. prepared gold triangle nanoarrays with a confined 3D plasmonic field to obtain stronger SERS enhancement [24]. Immunoassay was developed with a sandwich target-capture concept with the nanoarrays with improved sensitivity to detect cancer biomarkers of vascular endothelial growth factor in blood plasma. Instead of solid substrates with a large area, Liu et al. proposed a microprobe-based SERS immunoassay [25]. The antibody-modified microprobe was inserted into a cell to capture targets. After washing, the SERS probes were labeled on the targets captured on the microprobes for SERS detection. This strategy enabled detection with high sensitivity and the illustration of the spatial distribution of targets in a single cell.

Utilizing a sandwich pattern–based detection from solid substrates to solution, Chon et al. used antibody-modified hollow gold nanospheres and magnetic bead to detect lung cancer biomarker, carcinoembryonic antigen (CEA) [26]. With a magnetic separation-concentration step, the SERS response from Raman reporter conjugated to hollow gold nanosphere enabled improved detection of CEA. Then they extended the method for multiplex detection of biomarkers in buffer and serum [27,28]. Not only biomolecules, the magnetic NPs modified with antibodies could also concentrate cancer cells for SERS detection. Noh et al. fabricated magnetic SERS probes, which were antibody-modified SiO_2 coated AgNPs/magnetic NPs, to capture and separate bronchioalveolar stem cells (BASCs) [29]. Raman reporters linked to AgNPs gave SERS spectra for the identification of BASCs isolated from total lung cells extracted from lung tissues with around four to five fold magnetic enrichment. The same concept was applied for the detection of breast cancer cells (SKBR3) and floating leukemia cells (SP2/O) [30].

Beyond magnetic separation for the detection of cancer cells, other strategies were also applied with biomarker labeled by antibody-modified SERS probes. Conde et al. modified Cetuximab, antibody–drug conjugate, to GNPs which were linked with Raman reporters first and wrapped by polyethylene glycol (PEG) layer [31]. The PEG layer could improve the stability, biodistribution, and pharmacokinetic properties of the obtained nanoantennas [32], and the SERS fingerprint from 3,3′-diethylthiatricarbo-cyaniniodid enhanced by GNPs was used to indicate the presence of cancer cells. Similarly, silver nanoparticles (AgNPs) with antibodies were used for multiplex detection of cancer cells. MacLaughlin et al. used GNP probes with different antibodies and Raman reporters to identify Leukemia and Lymphoma cells in a triplex format [33]. This concept enabled a SERS-based flow cytometry to sort labeled cells. Instead of SERS probes integrating Raman reporters and antibodies on a single nanoparticle, in our past work, a strategy to bind multiple SERS beacons of SERS-active nanoparticles modified with Raman reporters to antibody-modified probes was shown as Fig. 10.1 to detect specific surface expressions of CD44 and CD24 in cancer cells [34]. The indicators were recognized with probes modified with antibodies, and DNA sequences conjugated on SERS beacons and linked the beacons to the probes through DNA hybridization to generate reversible nanoparticle networks that can be excited at specific wavelengths for feature signals. Furthermore, the obtained nanoparticle networks could be imaged by a complementary hyperspectral LSPR scattering with darkfield microscopy. SERS can also recognize the cells in clinical samples. Circulating tumor cells (CTCs) in blood are responsible for cancer metastasis, thus SERS detection of CTCs is attractive for cancer diagnosis and therapy evaluation. Wang et al. developed a SERS-based CTCs detection method in

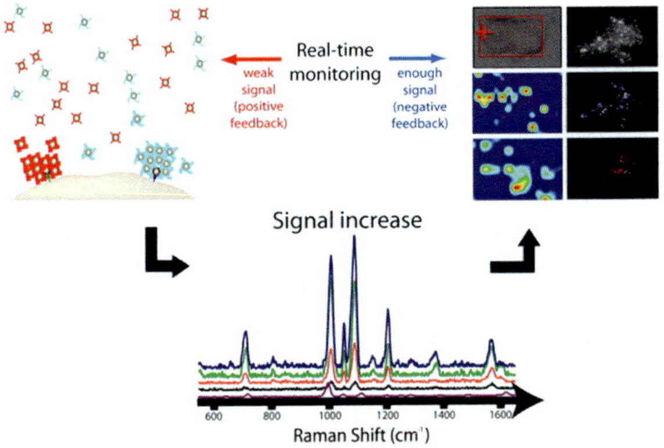

Figure 10.1 Antibody-based SERS probes for cancer cell surface marker detection and SERS imaging. *Reprinted with permission from K. Lee, V.P. Drachev, J. Irudayaraj, ACS Nano 5 (2011) 2109–2117. Copyright 2011 American Chemical Society.*

blood samples [35]. GNPs were synthesized as SERS substrates which were further modified with Raman reporters and epidermal growth factor (EGF). EGF could interact with the epidermal growth factor receptor (EGFR) overexpressed on squamous cell carcinoma of the head and neck (SCCHN) cell line. Therefore the obtained SERS substrate could specifically label the CTCs cells in blood samples for the detection. Assisted with a separation with magnetic beads, Sha et al. demonstrated a SERS-based detection of SKBR3 tumor cells [36]. The presence of the signature SERS spectra from GNP-based SERS tags confirms the existence of target SKBR3 cancer cells at a concentration of as low as 50 cells mL^{-1}.

For in vivo applications, antibody-modified SERS probes could be used to identify tumors, but more commonly used for imaging. Qian et al. prepared GNP-based SERS probes and performed in vivo detection of tumors in mice [37]. The GNPs were first functionalized with Raman reporter molecules then wrapped with thiol-terminated PEG which facilitated the antibody conjugation and improved the biocompatibility of the obtained SERS probes. With the NIR excitation, the SERS signals from probes in mice were obtained for the determination of tumors.

Beyond antibody, aptamer could also be used for specific recognition of particular targets, such as cancer biomarkers. Linked by the aptamer against Mucin-1, a protein biomarker for breast cancer expressed in patient tissue, and its complementary sequences, AgNPs and gold nanorods (GNRs) were assembled to be core-satellite structures by Feng et al. [38]. The presence of target Mucin-1 would dehybridize the aptamer-based double-stranded DNA linker between AgNPs and GNRs, thus a decrease in SERS intensity from the Raman reporters modified on AgNPs was observed for the determination of Mucin-1 due to the separation of AgNPs with Raman reporters away from GNRs. Aptamer was also used to detect cancer cells. Li et al. linked aptamers to Ag/Au shell/Au core nanoparticles to recognize cancer cells while the Raman reporters between the shell and core offered SERS signal and imaging [39].

10.3 SERS detection of nucleic acid sequence indicators in cancer

In addition to detecting protein biomarkers, several genes and microRNA implicated in cancer were evaluated by SERS with high sensitivity and in a multiplex format [40,41]. In these strategies, cancer-related genes sequences and microRNA sequences can be tested with SERS for cancer diagnosis. Multiplex SERS–based detection of alternative splice variants of breast cancer susceptibility gene 1 was proposed by hybridizing the target sequence with a capture sequence and a probing sequence concept [42]. Wang et al. reported SERS detection of erbB-2 gene encoding a transmembrane glycoprotein in the family of EGFR and Ki-67 gene encoding Ki-67 protein as a cell proliferation marker [43]. AgNPs modified with Raman label-terminated single-stranded DNA probe sequences in hairpin conformation were used as SERS probe. In hairpin conformation, the Raman label is near to the surface of AgNPs where the SERS activity is high, and the presence of target DNA sequences would disrupt the hairpin conformation and Raman label is away from the surface of AgNPs. Thus, the decrease of SERS signals could be used to determine the presence of target DNA. Similar probe sequences in hairpin conformation were also used for cancer-related microRNA detection [44]. To Ag-coated gold nanostars, Raman reporter–terminated probe sequences that were hybridized with placeholder DNA sequences were linked. The hybridized double-stranded DNA structure kept the Raman label away from nanostar resulting in a weak SERS signal. Target microRNA could hybridize the placeholder sequences and displace the probe sequences which generated hairpin conformation resulting in a stronger SERS signal. The detection protocol enabled multiplex detection of different microRNA targets. With nucleic acid sequences as indicators for cancer detection, an advantage is that the detection could be enhanced by nucleic acid–related reactions. For instance, Ye et al. introduced circular exponential amplification reaction (EXPAR) for the detection of microRNA in lung cancer cells [45]. The presence of target microRNA could trigger linear and circular exponential amplification with EXPAR, and the resulting single-stranded products could be linked to Raman reporter–terminated DNA probes conjugated on GNPs to generate SERS signal for the detection. This strategy could provide a 10^6-fold improvement in the sensitivity referring to that without amplification. Besides the presence of the cancer-related nucleic acid sequences, the DNA damage could be recognized with SERS as an indicator to differentiate cancer and normal cells. Panikkanvalappil et al. extracted DNA samples from cells with AgNPs to develop a SERS test [46]. The DNA conformational change induced by ROS was characterized with SERS fingerprints, and the difference in these fingerprints between human oral squamous carcinoma cancer cells and human keratinocytes normal cells showed a possible cancer diagnosis route based on the SERS spectra from the DNA samples.

10.4 SERS diagnosis based on other indicators

Narrow peak width in SERS spectra enables the observation of the peak shift due to the interaction between recognition ligands and cancer-related targets. Therefore some unique strategies could be applied for SERS detection of cancer-related targets. For instance, Guerrini et al. utilized the heterodimerization of c-Fos to c-Jun which is an oncoprotein related to the carcinogenic mechanism of several human cancers to establish a SERS-based c-Jun detection [47]. The interaction between c-Fos and target c-Jun induced the fingerprint change of the SERS spectra from c-Fos linked to AgNPs, which was then used to determine the concentration of target c-Jun.

The enriched information in the SERS spectra collected from the complex biophysical samples could reveal the difference in samples due to cancers, thus label-free diagnosis of cancer is possible. For instance, Lin et al. demonstrated the detailed difference in the SERS spectra of total serum proteins between gastric cancer patients and normal persons, which could be useful as a clinical tool for cancer screening [48]. The whole detection process was simple: total serum was collected, and albumin and globulin were separated from blood serum with membrane electrophoresis and redispersed in acetate solution. These proteins were then enhanced with silver nanoparticles for SERS spectra. Analyzed with principal component analysis (PCA), a clear difference in the SERS fingerprint could be seen to differentiate cancer samples from the normal. Later, this group further simplified the protocol and applied it for colorectal cancer, cervical cancer, and nasopharyngeal cancer detection [49–53]. In their work on cervical cancer detection, without electrophoresis the SERS spectra from cervical cancer patients demonstrated a clear difference from that of normal, especially in the ranges of 1310–1430 cm^{-1} and 1560–1700 cm^{-1} which were attributed to $\delta(CH2)$ from adenine, collagen, and phospholipids, $\delta(C=C)$ from phenylalanine, and $\nu(C=O)$ from amide, respectively. PCA and linear discriminant analysis (LDA) were used to develop a diagnostic with a sensitivity of 96.7% and specificity of 92% based on a test of 60 cervical cancer positive patients and 50 negative controls. Indirectly, SERS was used to evaluate target-capture efficacy of cervical cancer biomarkers in lateral flow sensors [54]. A similar strategy was also applied for the screening of prostate cancer by Li et al. [55]. It should be noted that multivariate analysis such as PCA and LDA is a powerful tool for the data analysis, especially in deconvolving the complex SERS

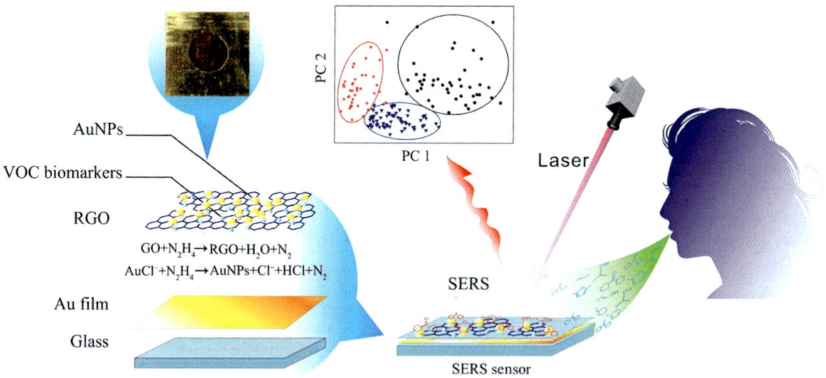

Figure 10.2 Breath analysis of VOCs based on SERS assisted with PCA for the diagnosis of early and advanced gastric cancer. *Reprinted with permission from Y. Chen, Y. Zhang, F. Pan, J. Liu, K. Wang, C. Zhang, et al., ACS Nano 10 (2016) 8169–8179. Copyright 2016 American Chemical Society.*

fingerprints of biomolecules of the biological samples, for cancer detection. Beyond blood samples, PCA method was used to analyze the SERS spectra of exosome isolated from H1299 and H522 (human lung carcinoma) cells for the diagnosis of lung cancer [56].

SERS was also used to detect volatile organic compounds (VOCs) in human breath, which are produced by cancer cells or tissues during metabolism or oxidative stress and could be used as indicators for cancer diagnosis [57–59]. Chen et al. synthesized reduced graphene oxide (rGO)-based SERS substrates to detect these VOCs as indicators for the diagnosis of early and advanced gastric cancer [60]. As shown in Fig. 10.2, VOCs from breath were captured by the rGO and the SERS signals from the captured VOCs were enhanced by GNPs and gold film. With the feature Raman bands, 14 VOC were tested as indicators. PCA analysis showed that the differences in these feature bands could be used for the determination of cancer with high specificity and sensitivity.

10.5 Multifunctional SERS substrates for diagnosis and therapy

Benefiting from the development of nanotechnology, nanostructures could not only provide SERS activity but also other functions. SERS substrates with suitable architecture could be used for drug delivery, photothermal therapy, and photodynamic therapy in cancer. For drug delivery, a common route is to construct nanostructures based on SERS-active materials and porous structure such as TiO_2 or mesoporous SiO_2 where the SERS signal and drugs or other molecules were assigned at different sections [61–63]. For instance, Wang et al. synthesized multifunctionalized nanostructures with AgNP linked with Raman reporter molecules, SiO_2 as first layer shell and mesoporous TiO_2 as the second layer shell to load cancer drugs and fluorescent beacons [62]. SiO_2 shell could improve the stability of SERS signals by locking in the Raman reporters to AgNP core and locking out the other molecules outside the SiO_2 layer. Mesoporous TiO_2 shell loaded cancer drug molecules for drug delivery and fluorescent molecules for multichannel detection. SERS-active material–based hollow nanostructures able to load drug molecules are another choice for drug delivery and SERS detection. Tian et al. synthesized hollow gold nanocages for doxorubicin (DOX) loading and feature SERS intensity was used to determine the DOX release from the nanocage [64]. With photothermal effect, the decrease of SERS signal indicated that the DOX loaded could be released in few minutes. Similarly, hollow gold shells with branches were prepared to integrate drug delivery, photothermal therapy, and SERS detection of tumors in mice [65]. Vesicles with hollow structures could also be used for drug delivery. Song et al. prepared GNP-containing vesicles for SERS detection and drug delivery [66]. The antibodies on the vesicles enabled the recognition of target cancer cells with SERS signals. Furthermore, the SERS signal was changed upon drug release due to the pH-induced vesicle destruction, making it possible to monitor the drug release with SERS.

Usually, nanostructures with strong absorbance in long wavelength range in their extinction spectra would potentially be used for photothermal therapy. Meanwhile, if the nanostructures are made by SERS-active materials then they could act as SERS substrates. Thus a series of substrates based on gold or silver in morphologies with absorption in the range of more than 650 nm in the extinction spectra were reported for SERS characterizations as well as photothermal therapy. In the list, there are GNPs [67], gold nanostars [68–70], GNRs [71], gold nanopopcorn [72], GNP aggregates [73], and even architectures composited with SERS-active nanoparticles [74]. One of the advantages of the combination of photothermal treatment and SERS investigation is that the cell death process could be revealed by the corresponding SERS fingerprints obtained in the process. Aioub et al. collected SERS spectra from HSC-3 cells enhanced with GNPs according to time-dependent cell death due to photothermal therapy [67]. The changes in SERS peak position and intensity in fingerprints were assigned to the change of proteins during the photothermal therapy. The increase in phenylalanine content and break of disulfide bonds was confirmed based on the observed time-dependent SERS change. Meanwhile, the change in SERS fingerprints demonstrated the influence of photothermal therapy due to the difference in excitation power and the concentration of GNP substrates. Because of the unique physical and chemical properties, graphene and graphene oxide have been widely used in the fabrication of various nanostructures for photothermal SERS substrates. For instance, Chen et al. constructed an rGO-based nanostructure for photothermal therapy and SERS recognition of A549, overexpressing EGFR lung cancer cells [75]. With rGO as substrates, carbon porous silica nanosheets were prepared where GNPs with rhodamine 6G (R6G) as Raman reporters were loaded and antibodies were immobilized to recognize target cancer cells. With antibody-based binding, obtained nanostructures are light-sensitivity for photothermal therapy while the GNP-R6G could yield SERS signature to present the existence of target cells. In this work the rGO acts as a good platform to integrate the functions of SERS enhancement, antibody-based recognition, and photothermal therapy, while exhibiting the advances of graphene-related materials in applications. Beyond graphene and graphene oxides, carbon nanotubes could also be used to construct SERS substrates with photothermal property. Beqa et al. simply attached gold nanostars to single-wall carbon nanotubes for SERS detection and photothermal therapy of human breast cancer cells [76]. For in vivo application, Song et al. assembled a carbon nanotube ring structure covered by GNPs, thus resulting in a SERS and photothermally active nanostructure for cancer detection and therapy [77]. They demonstrated detection and photothermal therapy with U87MG cancer cells and in vivo in mice. A further effort to combine SERS and photothermal therapy was to construct biodegradable plasmonic vesicles integrating SERS recognition, photothermal therapy, and drug delivery [74]. As shown in Fig. 10.3, the vesicles were assembled with polymers PEG and polylactide-modified GNRs which provide a photothermal response to enhance the SERS spectra of Raman reporters. The vesicles encapsulated DOX, an anticancer drug, thus could be combined with the photothermal response from GNRs for a higher death rate of cancer cells. Proteinase K was used to demonstrate the biodegradation of the vesicles. Antibodies against epithelial cell adhesion molecule (EpCAM) linked on

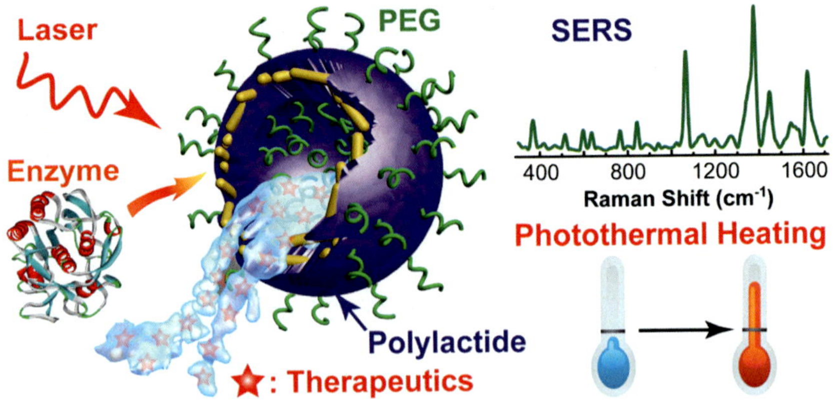

Figure 10.3 SERS substrates of GNR-based vesicles for detection, photothermal therapy and drug delivery. *Reprinted with permission from J. Song, L. Pu, J. Zhou, B. Duan, H. Duan, ACS Nano 7 (2013) 9947–9960. Copyright 2013 American Chemical Society.*

the vesicles enabled SERS-based cancer cell recognition. Beyond antibody-based recognition, the interaction between folic acid and folate receptor overexpressed on cancer cells was also utilized in the SERS detection combined with photothermal therapy of cancer cells. Boca-Farcau et al. modified silver nanotriangles with folic acid for cancer cell labeling and Raman reporters for SERS detection of human ovarian carcinoma (NIH: OVCAR-3) cells [78]. The extinction of silver nanotriangles in the range of around 700–1050 nm in the UV–vis spectrum offered the photothermal response with a continuous wave NIR laser.

Nanostructures with SERS activity could be used for photodynamic therapy due to the optical responses. For instance, Fales et al. fabricated gold nanostar–based nanostructures for SERS imaging and photodynamic therapy with BT-549 breast cancer cells [79]. On the as-prepared nanostars, protoporphyrin IX molecules were conjugated, to generate ROS for the treatment of cancers and some other diseases with suitable light excitation, and 3,3′-diethylthiadicarbocyanine iodide (DTDC) was modified as Raman reporter for the SERS imaging of cancer cells. Seo et al. prepared SiO_2-wrapped GNRs loaded with methylene blue for the integration of photothermal/photodynamic therapy and SERS detection [80]. GNRs provided photothermal and SERS activity while SiO_2 shell loaded methylene blue which is chosen as the photosensitizer. The SERS spectra from methylene blue upon the laser excitation revealed the molecular structure change of methylene blue in photodynamic therapy.

10.6 SERS imaging for cancer imaging and delineation

SERS imaging could not only provide the SERS spectra from interesting target or SERS probes but also illustrate the spatial distribution of these elements and more importantly could picture the tumors for cancer diagnosis and resection. SERS imaging could be used for imaging the component distribution on the cell membrane or in cells, and the obtained information could help in the detection and understanding of cancer. Antibody-modified SERS probes enable selective imaging of cells. For instance, Lee et al. constructed SERS substrates of Au/Ag core–shell nanostructures with Raman reporters between Au core and Ag shell and antibodies on the Ag shell for selective SERS image of HEK293 cells [81]. As shown in Fig. 10.4, with the monoclonal antibody against phospholipase Cγ1 (PLCγ1) biomarker overexpressed in some hyperproliferative tissues, such as HEK293 cell line, the SERS-active nanostructures could specifically bind to HEK293 surface where the PLCγ1 biomarkers are located. Thereby, the obtained SERS images show only the shape of cancer cells where the SERS intensity illustrated the distribution of the biomarkers, while normal cells did not provide positive response in the SERS images. The multiplex recognition capability of SERS-based identification would also improve SERS imaging in a multiplex-channel format [82]. Nima et al. modified four different Raman reporters to Ag-coated GNRs which were conjugated with four antibodies against four biomarkers on MCF7 cancer cells, thereby proposed a four-channel SERS imaging method [83]. Silver layers around the GNR further improved SERS activity of the substrates. Different antibodies along with

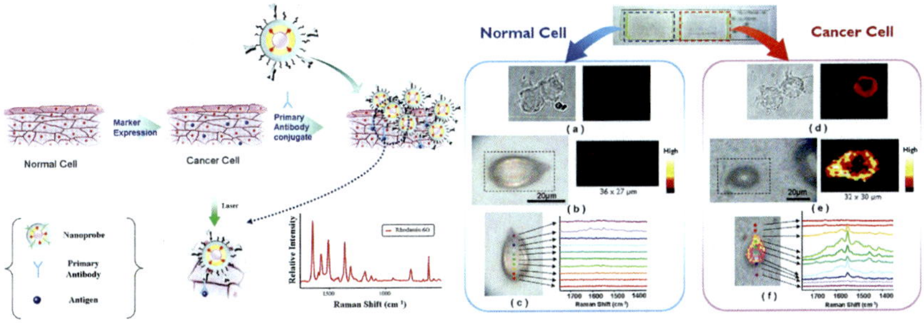

Figure 10.4 Schematic strategy of SERS imaging of cells with antibody-modified SERS probes and the SERS images normal cell and cancer cell. *Reprinted with permission from S. Lee, S. Kim, J. Choo, S.Y. Shin, Y.H. Lee, H.Y. Choi, et al., Anal. Chem. 79 (2007) 916–922. Copyright 2007 American Chemical Society.*

the corresponding Raman reporters enabled the simultaneous imaging of EpCAM, IGF-1 Receptor β, CD44, and Keratin18 biomarker on cancer cells. Requiring no additional enrichment, this imaging method could differentiate specific cancer cells from seven million blood cells.

Compared to cancer cells, SERS imaging of organs and tumors would be more valuable for cancer therapy. Under white light, tumors are hard to be differentiated from normal tissues. SERS imaging would improve the recognition of tumors and assist in the resection [84]. For instance, combined with fluorescent imaging, Mohs et al. systemically investigated the use of SERS imaging in tumor intraoperative detection [85]. Mammary tumors in nude mice were used as models to evaluate the combined imaging in operation. Karabeber et al. reported the brain tumor resection with SERS imaging obtained with a hand-held Raman scanner [86]. Although the experiments were conducted with fixed mice brains, it illustrated a possible route for SERS imaging—based tumor resection. With different Raman reporters, multichannel SERS imaging could be achieved in vivo [87]. SERS nanotags of antibody-modified nanoparticles with feature SERS fingerprints pictured the distribution of three different cancer biomarkers on MDA-MB-231 breast cancer cell surface [88]. By injecting the SERS probes into mouse models, the SERS spectra could be collected in vivo. Based on the SERS intensity in mice, it was noted that direct injection into tumors resulted in a higher concentration of SERS nanotags in tumors than that in livers, spleen or kidney, while intravenous injection would cause the accumulation of SERS nanotags in these organs than that in tumors [89,90]. SERS probes without antibody could also be used for in vivo SERS imaging. Zavaleta et al. prepared SiO_2-wrapped GNPs with different Raman reporters exploring the multichannel imaging in mice [91]. The injected SERS probes were accumulated in livers of mice naturally and SERS imaging was obtained. The narrow SERS peak width of different Raman reporters demonstrated imaging in the corresponding channels.

Advances in hardware for SERS provide more options for in vivo imaging. For instance, instead of microscopy, an endoscopy-based SERS imaging strategy was proposed by Zavaleta et al. [92]. Optical fiber—based endoscopic device was constructed to induce a 785-nm excitation to the imaging area where nanoparticles with different Raman reporters were used to develop SERS images. Human colon tissues and a porcine colon were used as model samples to validate the endoscopic strategy. Later, the same group improved the endoscopic device by incorporating a motor-driven scan mirror as shown in Fig. 10.5 for rapid scanning of luminal

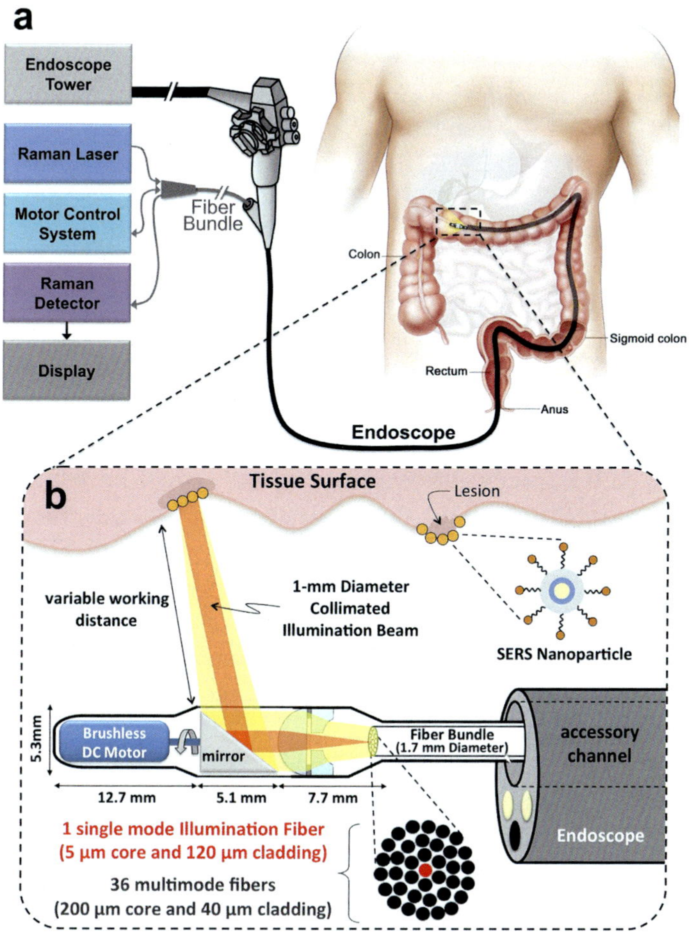

Figure 10.5 Scheme of device (a) for endoscopy-based system and its distal end for the rapid intraluminal SERS imaging and the for early cancer screening in colon. *Reprinted with permission from E. Garai, S. Sensarn, C.L. Zavaleta, N.O. Loewke, S. Rogalla, M.J. Mandella, et al., PLoS One 10 (2015). This is an open-access article distributed under the terms of the Creative Commons Attribution License, which permits unrestricted use, distribution, and reproduction in any medium, provided the original author and source are credited.*

surfaces of hollow organs to generate quantitative SERS images [93]. Although SERS substrates of nanoparticles were not applied in patients, in their clinical study, the overlay of the spectra from SERS substrates to the Raman spectra was collected from the colon wall of the patients to simulate the SERS signal for in vivo imaging.

During in vivo SERS imaging, nonspecific binding of SERS substrates to peritoneal or visceral surfaces is expected, thus even with the recognition ligands immobilized to SERS substrates, the specificity of the SERS detection could not be completely guaranteed with a potential risk of false positive in the corresponding cancer diagnosis. Blocking the surface of SERS substrates with protein or polymer molecules is commonly used to reduce nonspecific binding [94–97], however, due to the complexity of tissue samples anatomical crevices would still adhere to the blocked SERS substrates. Oseledchyk et al. proposed a ratiometric strategy with the antifolate receptor antibody-modified surface-enhanced resonance Raman scattering (SERRS) NPs and nontargeted SERRS-NPs for the SERS imaging of pancreatic tissues [98]. These two NPs were labeled with different Raman reporter molecules; thus with the corresponding feature fingerprint in SERS spectra, the existence of NPs could be simultaneously identified. Assisted with direct classical least-squares analysis, the obtained SERS spectra from each pixel from the tissues with both SERRS-NPs were processed and the ratio between two SERRS-NPs was calculated. The positive region of tumors could be recognized with a high ratio of antibody-modified SERRS-NPs to nontargeted SERRS-NPs. This strategy was used to detect microscopic residual tumors as small as 370 μm for ovarian tumor visualization.

Cancer diagnosis and the investigation of the cancer mechanism and therapy require numerous information. SERS spectroscopy, especially the label-free SERS with spectra from biomolecules in cancer cells and tissues, could reveal the change in molecular structure. A combination of SERS and other analytical techniques offering different information could be used to further enhance our understanding of cancer and confirm diagnosis. For instance, by integrating fluorescent dyes to the nanostructures of SERS substrates, fluorescent/SERS dual-model imaging is commonly used to locate the target biomarkers in cancer cells/tissues [99,100]. Furthermore, SERS-active materials, such as gold and silver, have been shown to possess optical properties that can be harnessed with other analytical techniques. For instance, Jokerst et al. combined photoacoustic and SERS imaging with GNRs for imaging tumors in mice [101]. The photoacoustic imaging could be enhanced by nanoparticles and was able to provide in vivo tomographic images with spatial resolution as high as 50 μm and penetration depth up to 5 cm, while SERS imaging could guide tumor resection. El-Said et al. prepared indium tin oxide glass slides covered with gold nanoflowers as SERS substrates and electrochemical electrode to monitor the response of cancer cells to chemotherapeutic agents with both SERS and cyclic voltammetry (CV) [102]. With the substrates, the observed SERS fingerprint changes were noted as the response in real-time from HepG2 cells to drugs. Microfluidic chip enables rapid detection in an automatic process, thus facilitating high throughput screening of a large number of samples. Lee et al. proposed GNP probe–based SERS detection of cancer biomarkers on a microfluidic chip [103]. The detection protocol enabled the rapid detection of AFP, a biomarker for hepatocellular carcinoma, in less than 60 min. Pallaoro et al. combined microfluidic chip with SERS-based labeling to establish a rapid detection protocol for cancer cells [104]. The microfluidic system was able to concentrate the cancer cells to the center of the channels in a signal file and SERS labels yielded well-defined spectra for the identification of the cells. The detailed review on microfluidic system can be found in Chapter 7, *Surface-enhanced Raman spectroscopy-based microdevices for in vitro diagnostics.*

10.7 Summary

We have attempted to provide numerous examples of SERS-related applications in cancer diagnosis. More importantly, common strategies for SERS-based disease characterization were highlighted based on clinical relevance. SERS substrates with recognition ligands and Raman reporters were found to provide enhanced signals for diagnosis and imaging. The SERS signals from label-free tests were usually combined with multivariate statistical data analysis for identification. SERS imaging was used to delineate tumor regions in normal tissues to help the resection in operational therapy. In summary, SERS has been found to be a promising analytical technique for disease characterization and diagnosis. Specific attempts to exploit this sensing modality in complex molecular diagnostics where chemical fingerprinting could be integrated with quantitative measurements can be expected.

References

[1] H. Hwang, S.-H. Kim, S.-M. Yang, Lab Chip 11 (2011) 87–92.
[2] H. Xu, E.J. Bjerneld, M. Käll, L. Börjesson, Phys. Rev. Lett. 83 (1999) 4357.
[3] Y. Wang, K. Lee, J. Irudayaraj, Chem. Commun. 46 (2010) 613–615.
[4] L. Sun, J. Irudayaraj, Biophys. J. 96 (2009) 4709–4716.
[5] K. Lee, J. Irudayaraj, J. Phys. Chem. C. 113 (2009) 5980–5983.

[6] C. Yu, E. Gestl, K. Eckert, D. Allara, J. Irudayaraj, Cancer Detect. Prev. 30 (2006) 515–522.
[7] J. Kneipp, H. Kneipp, B. Wittig, K. Kneipp, Nanomed. Nanotechnol. Biol. Med. 6 (2010) 214–226.
[8] A.F. Palonpon, J. Ando, H. Yamakoshi, K. Dodo, M. Sodeoka, S. Kawata, et al., Nat. Protoc. 8 (2013) 677–692.
[9] M. Schütz, D. Steinigeweg, M. Salehi, K. Kömpe, S. Schlücker, Chem. Commun. 47 (2011) 4216–4218.
[10] S.E. Bohndiek, A. Wagadarikar, C.L. Zavaleta, D. Van de Sompel, E. Garai, J.V. Jokerst, et al., Proc. Natl. Acad. Sci. U.S.A. 110 (2013) 12408–12413.
[11] Y. Wang, J.L. Seebald, D.P. Szeto, J. Irudayaraj, ACS Nano 4 (2010) 4039–4053.
[12] M. Fan, G.F. Andrade, A.G. Brolo, Anal. Chim. Acta 693 (2011) 7–25.
[13] S. Nie, S.R. Emory, Science 275 (1997) 1102–1106.
[14] K.A. Willets, R.P. Van Duyne, Annu. Rev. Phys. Chem. 58 (2007) 267–297.
[15] Z.Q. Tian, B. Ren, D.Y. Wu, J. Phys. Chem. B 106 (2002) 9463–9483.
[16] L. Ouyang, Y. Hu, L. Zhu, G.J. Cheng, J. Irudayaraj, Biosens. Bioelectron. 92 (2017) 755–762.
[17] K. Lee, J. Irudayaraj, Small 9 (2013) 1106–1115.
[18] Bethesda, M., National Cancer Institute, SEER Cancer Stat Facts: Cancer of Any Site, https://seer.cancer.gov/statfacts/html/all.html. 2019.
[19] World Health Organization, Cancer: Key facts, https://www.who.int/en/news-room/fact-sheets/detail/cancer. 2018.
[20] D.S. Grubisha, R.J. Lipert, H.-Y. Park, J. Driskell, M.D. Porter, Anal. Chem. 75 (2003) 5936–5943.
[21] G. Wang, R.J. Lipert, M. Jain, S. Kaur, S. Chakraboty, M.P. Torres, et al., Anal. Chem. 83 (2011) 2554–2561.
[22] M. Lee, S. Lee, J.-h Lee, H.-w Lim, G.H. Seong, E.K. Lee, et al., Biosens. Bioelectron. 26 (2011) 2135–2141.
[23] J.H. Granger, M.C. Granger, M.A. Firpo, S.J. Mulvihill, M.D. Porter, Analyst 138 (2013) 410–416.
[24] M. Li, S.K. Cushing, J. Zhang, S. Suri, R. Evans, W.P. Petros, et al., ACS Nano 7 (2013) 4967–4976.
[25] J. Liu, D. Yin, S. Wang, H.Y. Chen, Z. Liu, Angew. Chem. 128 (2016) 13409–13412.
[26] H. Chon, S. Lee, S.W. Son, C.H. Oh, J. Choo, Anal. Chem. 81 (2009) 3029–3034.
[27] H. Chon, S. Lee, S.-Y. Yoon, S.-I. Chang, D.W. Lim, J. Choo, Chem. Commun. 47 (2011) 12515–12517.
[28] Z. Cheng, N. Choi, R. Wang, S. Lee, K.C. Moon, S.-Y. Yoon, et al., ACS Nano 11 (2017) 4926–4933.
[29] M.S. Noh, B.-H. Jun, S. Kim, H. Kang, M.-A. Woo, A. Minai-Tehrani, et al., Biomaterials 30 (2009) 3915–3925.
[30] B.H. Jun, M.S. Noh, J. Kim, G. Kim, H. Kang, M.S. Kim, et al., Small 6 (2010) 119–125.
[31] J. Conde, C. Bao, D. Cui, P.V. Baptista, F. Tian, J. Controlled Release 183 (2014) 87–93.
[32] G.F. Paciotti, L. Myer, D. Weinreich, D. Goia, N. Pavel, R.E. McLaughlin, et al., Drug. Deliv. 11 (2004) 169–183.
[33] C.M. MacLaughlin, N. Mullaithilaga, G. Yang, S.Y. Ip, C. Wang, G.C. Walker, Langmuir 29 (2013) 1908–1919.
[34] K. Lee, V.P. Drachev, J. Irudayaraj, ACS Nano 5 (2011) 2109–2117.
[35] X. Wang, X. Qian, J.J. Beitler, Z.G. Chen, F.R. Khuri, M.M. Lewis, et al., Cancer Res. 71 (2011) 1526–1532.

[36] M.Y. Sha, H. Xu, M.J. Natan, R. Cromer, J. Am. Chem. Soc. 130 (2008) 17214–17215.
[37] X. Qian, X.-H. Peng, D.O. Ansari, Q. Yin-Goen, G.Z. Chen, D.M. Shin, et al., Nat. Biotechnol. 26 (2008) 83–90.
[38] J. Feng, X. Wu, W. Ma, H. Kuang, L. Xu, C. Xu, Chem. Commun. 51 (2015) 14761–14763.
[39] J. Li, Z. Zhu, B. Zhu, Y. Ma, B. Lin, R. Liu, et al., Anal. Chem. 88 (2016) 7828–7836.
[40] L. Sun, C. Yu, J. Irudayaraj, Anal. Chem. 79 (2007) 3981–3988.
[41] L. Sun, J. Irudayaraj, J. Phys. Chem. B 113 (2009) 14021–14025.
[42] L. Sun, C. Yu, J. Irudayaraj, Anal. Chem. 80 (2008) 3342–3349.
[43] H.-N. Wang, T. Vo-Dinh, Nanotechnology 20 (2009) 065101.
[44] H.-N. Wang, B.M. Crawford, A.M. Fales, M.L. Bowie, V.L. Seewaldt, T. Vo-Dinh, J. Phys. Chem. C. 120 (2016) 21047–21055.
[45] L.-P. Ye, J. Hu, L. Liang, C.-y Zhang, Chem. Commun. 50 (2014) 11883–11886.
[46] S.R. Panikkanvalappil, M.A. Mackey, M.A. El-Sayed, J. Am. Chem. Soc. 135 (2013) 4815–4821.
[47] L. Guerrini, E. Pazos, C. Penas, M.E. Vázquez, J.L. Mascareñas, R.A. Alvarez-Puebla, J. Am. Chem. Soc. 135 (2013) 10314–10317.
[48] J. Lin, R. Chen, S. Feng, J. Pan, Y. Li, G. Chen, et al., Nanomed. Nanotechnol. Biol. Med. 7 (2011) 655–663.
[49] D. Lin, S. Feng, J. Pan, Y. Chen, J. Lin, G. Chen, et al., Opt. Express 19 (2011) 13565–13577.
[50] S. Feng, D. Lin, J. Lin, B. Li, Z. Huang, G. Chen, et al., Analyst 138 (2013) 3967–3974.
[51] S. Feng, R. Chen, J. Lin, J. Pan, Y. Wu, Y. Li, et al., Biosens. Bioelectron. 26 (2011) 3167–3174.
[52] D. Lin, J. Pan, H. Huang, G. Chen, S. Qiu, H. Shi, et al., Sci. Rep. 4 (2014) 1–8.
[53] S. Feng, R. Chen, J. Lin, J. Pan, G. Chen, Y. Li, et al., Biosens. Bioelectron. 25 (2010) 2414–2419.
[54] W. Ren, S.I. Mohammed, S. Wereley, J. Irudayaraj, Anal. Chem. 91 (2019) 2876–2884.
[55] S. Li, Y. Zhang, J. Xu, L. Li, Q. Zeng, L. Lin, et al., Appl. Phys. Lett. 105 (2014) 091104.
[56] J. Park, M. Hwang, B. Choi, H. Jeong, J.-h Jung, H.K. Kim, et al., Anal. Chem. 89 (2017) 6695–6701.
[57] G. Konvalina, H. Haick, Acc. Chem. Res. 47 (2014) 66–76.
[58] G. Peng, M. Hakim, Y. Broza, S. Billan, R. Abdah-Bortnyak, A. Kuten, et al., Br. J. Cancer 103 (2010) 542–551.
[59] Z. Xu, Y. Broza, R. Ionsecu, U. Tisch, L. Ding, H. Liu, et al., Br. J. Cancer 108 (2013) 941–950.
[60] Y. Chen, Y. Zhang, F. Pan, J. Liu, K. Wang, C. Zhang, et al., ACS Nano 10 (2016) 8169–8179.
[61] D. Shao, X. Zhang, W. Liu, F. Zhang, X. Zheng, P. Qiao, et al., ACS Appl. Mater. Interfaces 8 (2016) 4303–4308.
[62] Y. Wang, L. Chen, P. Liu, Chem. Eur. J. 18 (2012) 5935–5943.
[63] S. Zong, Z. Wang, H. Chen, J. Yang, Y. Cui, Anal. Chem. 85 (2013) 2223–2230.
[64] L. Tian, N. Gandra, S. Singamaneni, ACS Nano 7 (2013) 4252–4260.
[65] J. Song, X. Yang, Z. Yang, L. Lin, Y. Liu, Z. Zhou, et al., ACS Nano 11 (2017) 6102–6113.
[66] J. Song, J. Zhou, H. Duan, J. Am. Chem. Soc. 134 (2012) 13458–13469.
[67] M. Aioub, M.A. El-Sayed, J. Am. Chem. Soc. 138 (2016) 1258–1264.

[68] Y. Liu, J.R. Ashton, E.J. Moding, H. Yuan, J.K. Register, A.M. Fales, et al., Theranostics 5 (2015) 946.
[69] Y. Gao, Y. Li, J. Chen, S. Zhu, X. Liu, L. Zhou, et al., Biomaterials 60 (2015) 31–41.
[70] Y. Liu, Z. Chang, H. Yuan, A.M. Fales, T. Vo-Dinh, Nanoscale 5 (2013) 12126–12131.
[71] C. Wang, J. Chen, T. Talavage, J. Irudayaraj, Angew. Chem. 121 (2009) 2797–2801.
[72] W. Lu, A.K. Singh, S.A. Khan, D. Senapati, H. Yu, P.C. Ray, J. Am. Chem. Soc. 132 (2010) 18103–18114.
[73] S. Jung, J. Nam, S. Hwang, J. Park, J. Hur, K. Im, et al., Anal. Chem. 85 (2013) 7674–7681.
[74] J. Song, L. Pu, J. Zhou, B. Duan, H. Duan, ACS Nano 7 (2013) 9947–9960.
[75] Y.W. Chen, T.Y. Liu, P.J. Chen, P.H. Chang, S.Y. Chen, Small 12 (2016) 1458–1468.
[76] L. Beqa, Z. Fan, A.K. Singh, D. Senapati, P.C. Ray, ACS Appl. Mater. Interfaces 3 (2011) 3316–3324.
[77] J. Song, F. Wang, X. Yang, B. Ning, M.G. Harp, S.H. Culp, et al., J. Am. Chem. Soc. 138 (2016) 7005–7015.
[78] S. Boca-Farcau, M. Potara, T. Simon, A. Juhem, P. Baldeck, S. Astilean, Mol. Pharm. 11 (2014) 391–399.
[79] A.M. Fales, H. Yuan, T. Vo-Dinh, Mol. Pharm. 10 (2013) 2291–2298.
[80] S.-H. Seo, B.-M. Kim, A. Joe, H.-W. Han, X. Chen, Z. Cheng, et al., Biomaterials 35 (2014) 3309–3318.
[81] S. Lee, S. Kim, J. Choo, S.Y. Shin, Y.H. Lee, H.Y. Choi, et al., Anal. Chem. 79 (2007) 916–922.
[82] K.K. Maiti, A. Samanta, M. Vendrell, K.-S. Soh, M. Olivo, Y.-T. Chang, Chem. Commun. 47 (2011) 3514–3516.
[83] Z.A. Nima, M. Mahmood, Y. Xu, T. Mustafa, F. Watanabe, D.A. Nedosekin, et al., Sci. Rep. 4 (2014) 4752.
[84] C. Andreou, V. Neuschmelting, D.-F. Tschaharganeh, C.-H. Huang, A. Oseledchyk, P. Iacono, et al., ACS Nano 10 (2016) 5015–5026.
[85] A.M. Mohs, M.C. Mancini, S. Singhal, J.M. Provenzale, B. Leyland-Jones, M.D. Wang, et al., Anal. Chem. 82 (2010) 9058–9065.
[86] H. Karabeber, R. Huang, P. Iacono, J.M. Samii, K. Pitter, E.C. Holland, et al., ACS Nano 8 (2014) 9755–9766.
[87] G. von Maltzahn, A. Centrone, J.H. Park, R. Ramanathan, M.J. Sailor, T.A. Hatton, et al., Adv. Mater. 21 (2009) 3175–3180.
[88] U. Dinish, G. Balasundaram, Y.-T. Chang, M. Olivo, Sci. Rep. 4 (2014) 4075.
[89] T. Lammers, P. Peschke, R. Kühnlein, V. Subr, K. Ulbrich, P. Huber, et al., Neoplasia (New York, NY.) 8 (2006) 788.
[90] H. Xie, B. Goins, A. Bao, Z.J. Wang, W.T. Phillips, Int. J. Nanomed. 7 (2012) 2227.
[91] C.L. Zavaleta, B.R. Smith, I. Walton, W. Doering, G. Davis, B. Shojaei, et al., Proc. Natl. Acad. Sci. U.S.A. 106 (2009) 13511–13516.
[92] C.L. Zavaleta, E. Garai, J.T. Liu, S. Sensarn, M.J. Mandella, D. Van de Sompel, et al., Proc. Natl. Acad. Sci. U.S.A. 110 (2013) E2288–E2297.
[93] E. Garai, S. Sensarn, C.L. Zavaleta, N.O. Loewke, S. Rogalla, M.J. Mandella, et al., PLoS One (2015) 10.
[94] X. Wu, L. Luo, S. Yang, X. Ma, Y. Li, C. Dong, et al., ACS Appl. Mater. Interfaces 7 (2015) 9965–9971.
[95] K.K. Maiti, U. Dinish, A. Samanta, M. Vendrell, K.-S. Soh, S.-J. Park, et al., Nano Today 7 (2012) 85–93.

[96] G.B. Braun, S.J. Lee, T. Laurence, N. Fera, L. Fabris, G.C. Bazan, et al., J. Phys. Chem. C. 113 (2009) 13622–13629.
[97] L. Jiang, J. Qian, F. Cai, S. He, Anal. Bioanal. Chem. 400 (2011) 2793.
[98] A. Oseledchyk, C. Andreou, M.A. Wall, M.F. Kircher, ACS Nano 11 (2017) 1488–1497.
[99] Z. Wang, S. Zong, J. Yang, J. Li, Y. Cui, Biosens. Bioelectron. 26 (2011) 2883–2889.
[100] S. Lee, H. Chon, S.-Y. Yoon, E.K. Lee, S.-I. Chang, D.W. Lim, et al., Nanoscale 4 (2012) 124–129.
[101] J.V. Jokerst, A.J. Cole, D. Van de Sompel, S.S. Gambhir, ACS Nano 6 (2012) 10366–10377.
[102] W.A. El-Said, T.-H. Kim, H. Kim, J.-W. Choi, Biosens. Bioelectron. 26 (2010) 1486–1492.
[103] M. Lee, K. Lee, K.H. Kim, K.W. Oh, J. Choo, Lab Chip 12 (2012) 3720–3727.
[104] A. Pallaoro, M.R. Hoonejani, G.B. Braun, C.D. Meinhart, M. Moskovits, ACS Nano 9 (2015) 4328–4336.

CHAPTER 11

Multivariate approaches for SERS data analysis in clinical applications

Duo Lin[1], Sufang Qiu[2], Yang Chen[3], Shangyuan Feng[1] and Haishan Zeng[4,5]
[1]Key Laboratory of OptoElectronic Science and Technology for Medicine, Ministry of Education, Fujian Provincial Key Laboratory for Photonics Technology, Fujian Normal University, Fuzhou, P.R. China
[2]Fujian Medical University Cancer Hospital, Fujian Cancer Hospital, Fuzhou, P.R. China
[3]Department of Laboratory Medicine, Fujian Medical University, Fuzhou, P.R. China
[4]Imaging Unit—Integrative Oncology Department, BC Cancer Research Centre, Vancouver, BC, Canada
[5]Department of Dermatology and Skin Science, Photomedicine Institute, University of British Columbia, Vancouver, BC, Canada

11.1 Introduction

As an ultrasensitive detection method, surface-enhanced Raman spectroscopy (SERS) technology is increasingly used in biology, medicine, and life science-related fields, while the corresponding data analysis procedure is very important for subsequent application for solving practical problems. This procedure determines whether the technique can be successfully implemented. The main content of this chapter is to explain the basic concepts and implementation processes of different multivariate algorithms and corresponding practical application cases. It is presented from two aspects: unsupervised analysis and supervised analysis. Starting from the label-free SERS detection data, a series of commonly used unsupervised and supervised analysis methods is introduced in general, followed by practical examples to demonstrate the use of specific multivariate approaches. At the end, the data of labeling SERS is specifically dealt with and the related treatment is supplemented with the original means in the case of the label-free SERS. In this way the contents of this chapter will cover a broad range of topics involved in the two mainstream directions of the SERS detection field.

Principles and Clinical Diagnostic Applications of Surface-Enhanced Raman Spectroscopy
DOI: https://doi.org/10.1016/B978-0-12-821121-2.00003-2

11.2 Data analysis for label-free surface-enhanced Raman spectroscopy measurements

11.2.1 Unsupervised data analysis and practical applications

Before starting the data analysis, it is common to preprocess the original Raman spectra. Because background autofluorescence, noises, and other factors will interfere with the useful information of the spectrum and affect the results, we need to do preprocessing of the Raman spectra such as noise reduction, baseline correction, and normalization [1–4]. The preprocessed data can then be used in subsequent multivariate data analysis. Unsupervised data analysis is based on unsupervised machine learning. It is a data analysis method that has no clear purpose, does not need to categorize the data in advance, and cannot quantify the effect. Through unsupervised data analysis, we can quickly classify a large amount of spectral data. We intend to introduce three commonly used unsupervised data analysis methods including principal component analysis (PCA), multivariate curve resolution (MCR), and cluster analysis (CA).

11.2.1.1 Principal components analysis

PCA is one of the commonly used dimensionality reduction methods, which aims to use the idea of dimensionality reduction to transform multiple indicators into a few comprehensive indicators. The PCA uses an orthogonal transformation to transform a set of variable data that may have a linear correlation into a set of new variables not linearly correlated. The transformed variables are called principal components (PCs). The first principal component (PC1) comprises the largest variance and the other PCs exhibit declining variance, meaning declining explanation of the original data so that the top PCs can be used to display the characteristics of the data with minimal loss of information.

The application of PCA in SERS has the following advantages:

1. Simplifying calculations. The data processed by PCA can keep the information contained in the Raman spectra as much as possible while maintaining the minimum number of variables, helping to simplify the calculation and interpretation of results.
2. Visualizing data using scatter plots. The main components (PC1, PC2, PC3, etc.) with the largest contribution can be selected as representatives for visualization. Two-dimensional scatter plots (PC1, PC2) and/or three-dimensional scatter plots (PC1, PC2, PC3) are often utilized.

3. Discovering hidden related variables. In the process of combining the variables of the Raman spectra to obtain the PCs, it may reveal that some original variables have similar contributions to the same PC; that is, there are correlations between these variables.

The processed spectral data are first imported into a typical software/algorithm, and the PCA is carried out to reduce the dimensions to obtain the PCs. Then the software will calculate the scores, load the PCs, draw the score plot and load the plot to visualize the data.

In clinical applications, the data obtained through SERS often need to be analyzed by PCA. It is worth noting that the data obtained after the dimensional reduction of the spectrum can also be used as new input data for analytical methods. SERS combined with PCA can be used to distinguish different types of samples, including cells [5–11], virus [7], bacteria [12–14], blood [15,16], cerebrospinal fluid [17], and urine [18].

In recent years, SERS technology has played a role in the rapid identification of pathogens and biomolecules. Lim et al. [7] proposed the detection of virus-infected cells based on SERS and PCA to identify new influenza viruses. Cells expressing A/WSN/33 H1N1 or A/California/04/2009 H1N1 virus envelope protein produced a distinct SERS signal. The min–max normalization method was adopted to eliminate the fluctuation of the Raman signal intensity during the SERS measurement process. To systematically capture key characteristic Raman shift information from mixed Raman spectral data to facilitate early detection of the influenza virus, PCA was applied to the Raman spectrum of uninfected and infected influenza virus cells as shown in Fig. 11.1. The obtained score plots showed that the Raman spectrum of cells infected with WSN strain was significantly different from the Raman spectrum of uninfected cells ($P < .05$). Similarly, the projected Raman spectrum of cells infected with the CAL strain was significantly different from the control case. Raman bands with high and low loading values in PC loading plots are the key to distinguish infected cells from uninfected cells. The same methods were also applied in the detection of urinary tract infections causative bacteria [13] and dengue diagnosis from clinical samples [15]. These studies show that the combination of SERS and PCA makes it possible to identify cells infected with different types of pathogenic microorganisms, providing a more accurate and sensitive diagnostic method for detecting pathogens.

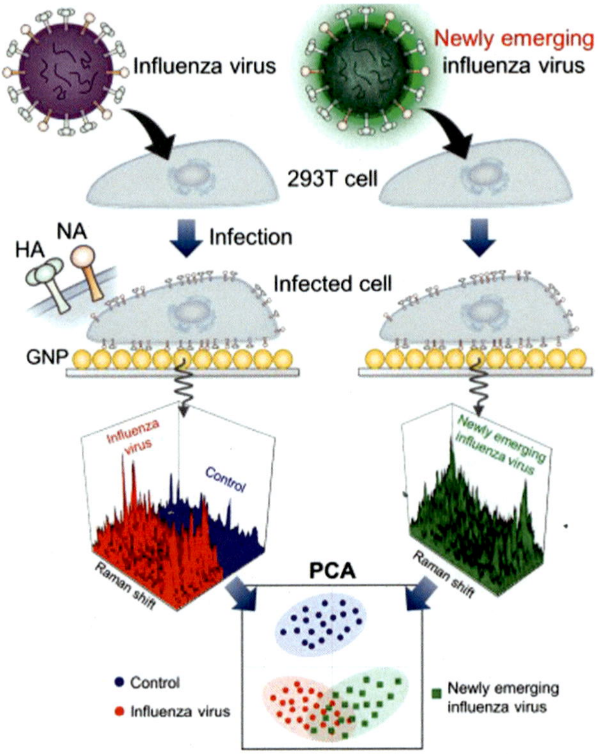

Figure 11.1 The key features of surface-enhanced Raman spectroscopy patterns via principal component analysis of Raman signals. *Reproduced from J.Y. Lim, J.S. Nam, H. Shin, J. Park, H.I. Song, M. Kang, et al., Identification of newly emerging influenza viruses by detecting the virally infected cells based on surface enhanced Raman spectroscopy and principal component analysis, Anal. Chem. 91 (9) (2019) 5677–5684, with permission from ACS Publications.*

11.2.1.2 Multivariate curve resolution

MCR is a statistical method of chemical dosimetry, which can extract various response curves of pure substances from the data containing many unknown mixtures, including spectrum, time curve, pH curve, etc., without the need to know the type and composition information of unknown substances. The curve resolution method mainly analyzes the matrix data and directly extracts the pure spectrum and concentration curve of each component from the data. It is based on the self-mode curve discrimination method (SMCR). The basic idea is to first obtain the pure spectrum of the pure species, and then use the least-squares method to obtain the concentration distribution curve to complete the qualitative and quantitative analysis. It is based on two assumptions: one is that the mixture spectrum is linearly additive and the other is that the spectral measurements and concentrations can only be positive. These two assumptions are valid for the data generated by general spectroscopy instruments, so they can be widely used in various chemical fields. Based on SMCR, many chemical experts have proposed a variety of new algorithms. Among these MCR methods, MCR-alternating least squares (MCR-ALS) method uses the results obtained by evolving factor analysis as iterative initial values and is often used in the resolution of chromatographic combined data in complex systems. Besides, it has also been widely used in other spectroscopy fields.

The algorithm of MCR-ALS follows the bilinear model where the matrix arranging with spectroscopic data (D) is decomposed into concentration matrix C (scores related to concentrations) and spectra matrix S^T (loadings related to pure substance spectra) as shown in Fig. 11.2 [19]. It is performed by an iterative algorithm which is carried out until the error (E) reaches a minimal value. The MCR-ALS model can then be expressed as:

$$D = CS^T + E$$

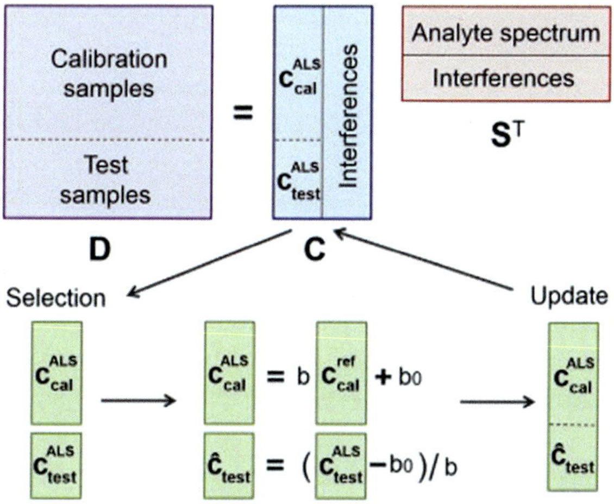

Figure 11.2 Description of the correlation constrained multivariate curve resolution–alternating least squares algorithm. *Reproduced from J.E.L. Villa, M.A.S. Afonso, D.P. Dos Santos, P.A. Mercadal, E.A. Coronado, R.J. Poppi, Colloidal gold clusters formation and chemometrics for direct SERS determination of bioanalytes in complex media, Spectrochim. Acta A 224 (2020) 117380, with permission from Elsevier.*

In the ideal case, only the number of components and suitable initial estimates for either concentration profiles or pure spectra are needed to perform the analysis. Additionally, for the decomposition of chemical spectra, nonnegative concentrations and spectra can be imposed as constraints to be applied to reduce the number of possible solutions for C and S^T (rotational ambiguity). Given the initial estimates by using simple-to-use interactive self-modeling mixture analysis with the iterative least squares–based algorithm, the quality of the final results is dependent on the initial estimates. An important advantage that might be achieved by the MCR-ALS-based methods is the so-called second-order advantage, which provides the capability to predict analyte concentrations even in the presence of unknown interference. The multivariate analysis focuses on extracting the spectrum of each component along with the concentration map with limited prior knowledge.

MCR techniques have, as the main goal, the species resolution and their identification from a mixture of overlapping multivariate signals. Particularly, MCR-ALS enables the recovering of pure species information with physicochemical meaning by using constraints during the iterative process. It has been used in vibrational microspectroscopic image analysis of biological samples, which are not simple and often contain various components to be simultaneously analyzed [20]. Spectra can be mathematically resolved, and the contribution of the components in the sample can be separated and quantified, even if they are not known or identified. Thus, the need for experimental procedures on the sample is minimized by utilizing a reduced number of standard and the analysis is highly simplified while maintaining the advantages of multivariate calibration models, as a result of its pseudo-univariate nature. This powerful technique can identify the chemical compounds and visualize their distribution across the sample to categorize chemically distinct areas from the full spectral profile [21]. It can be directly applied to different kinds of Raman spectroscopy images of any kind of biological samples and it covers all the important stages of the analysis in several steps [20]: (1) spectra preprocessing; (2) MCR-ALS analysis of the data, which can find the spectral signature (and the corresponding concentration profile) of each pure component, using only the image data (i.e., the spectra in each pixel of the image); (3) evaluation of the results in terms of chemistry by reference spectra matching; and (4) visualization and evaluation of the results in terms of anatomy using pure component maps and segmentation maps. Results are rapidly achieved and easy to interpret and evaluate both in terms of chemistry and biology.

MCR-ALS is a chemometric method with a highly efficient strategy that has been successfully employed to resolve overlapping spectroscopic bands, allowing the elucidation of the concentration profiles as well as the pure spectra of the components. It decomposes the spectral image stack into the corresponding spectrum of each major species and concentration maps in the samples. But, the results of MCR-ALS are not limited to the straightforward interpretation of pure component distribution maps and spectra. In contrast to PCA, MCR models the shapes of distribution maps and spectra according to natural chemical, spectroscopic, and mathematical properties of the image data, instead of imposing orthogonality among components. Besides, the maps provided by MCR-ALS can be used as starting information for subsequent cluster analsyis (CA). However, MCR can encounter difficulties for experiments with insufficiently sized data sets. It is a classical but evolving data analysis tool that is still in progress in terms of new applications [19,22,23]. Much more development can still be foreseen in different areas [24–26].

Villa et al. [22] proposed a label-free method for quantifying urinary adenosine using SERS and MCR-ALS. The MCR-ALS algorithm is shown in Fig. 11.3. The model provided the analyte concentration of the test sample calculated in true concentration unit. By plotting the MCR-ALS score of the analyte against the added concentration, thereby

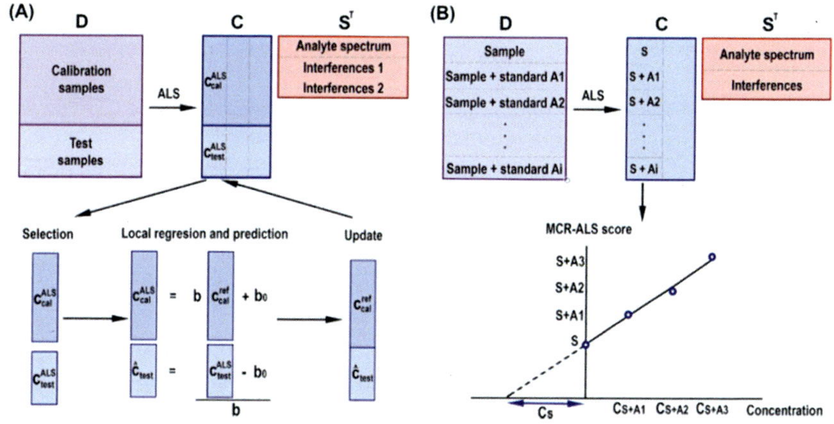

Figure 11.3 Description of (A) the correlation constrained multivariate curve resolution–alternating least squares method and (B) the multivariate curve resolution–alternating least squares standard addition method. *Reproduced from J.E.L. Villa, C. Pasquini, R.J. Poppi, Surface-enhanced Raman spectroscopy and MCR-ALS for the selective sensing of urinary adenosine on filter paper, Talanta 187 (2018) 99–105, with permission from Elsevier.*

extrapolating an added standard curve for calculating the analyte concentration in the sample, the SERS spectra from adenosine and the real human urine matrix were identified and recovered. This study verified that the SERS/MCR-ALS method has good selectivity, accuracy, and precision in determining physiologically relevant concentrations of urinary adenosine.

11.2.1.3 Cluster analysis

CA is a process of classifying data into different classes or clusters according to their characteristics, resulting in the great similarity between objects in the same cluster, and great dissimilarity between objects in different clusters. The larger the similarity is within the group and the larger the gap between the groups, the better the clustering effect. The CA, an exploratory analysis [27], was used as a multivariate data analysis technique for SERS, in which the abundant Raman spectra were divided into clusters according to their resemblance [28,29]. In the process of classification, there is no need to give a prior designation of classification standards. The analysis will calculate the distance between the sample and the cluster to determine the sample class. Hence, the CA can also be applied in the data preprocessing process. For multidimensional data with complex structures, the data can be aggregated through cluster distribution to standardize complex structure data. The CA is usually divided into four steps: (1) data preprocessing, (2) defining a distance function for measuring the similarity between data points, (3) clustering or grouping, and (4) evaluating output. There are two commonly used algorithms: hierarchical CA (HCA) and k-means CA (KMCA).

The HCA includes agglomerative HCA and divisive HCA. It can also be regarded as the bottom-up method and the top-down method. The basic idea of the bottom-up method is to treat each sample point as a cluster, calculate the distance between each cluster, aggregate the two nearest clusters into a new cluster, and repeat the above process until only one cluster left. The top-down rule is the opposite. At the beginning, all variables belong to a cluster, then the aliens are excluded according to the linkage, and finally, each variable becomes a cluster. Agglomerative HCA is often used. Because HCA does not specify the specific number of clusters, but only focuses on the distance between clusters, the final clustering effect obtained by selecting different distance indicators is also different. The results of HCA are displayed in a tree shape. Dendrogram can be visualized to simplify our understanding of the data.

HCA methods were successfully applied for SERS data analysis in pathogenic microorganism detection. Examples of these applications include detecting and identifying the foodborne pathogen [30] or influenza virus [31], which presents a clear characterization based on the barcode spectral data reduction. For a large number of variables contained in the original spectral data set, data preprocessing are carried out to reduce the redundancy of data before HCA analysis using methods, such as PCA and mean-centered linear discriminate analysis [32]. PCA is the common approach performed to reduce the dimensionality of complex SERS data and capture the subtle spectral differences between the similar spectra, especially in the case of high correlating Raman variables in the CA routine. The PCA results can then be used as the inputs in HCA modeling procedures, which construct the dendrogram to discriminate and categorize the samples. For example, HCA combined with PCA is capable of exploring spectral fingerprint of pathogens groups that have been used for discrimination of bacteria [33–36], mycoplasma [37,38], and virus [39,40].

KMCA is the most famous nonhierarchical CA. Due to its simplicity and efficiency, it is the most widely used among all clustering algorithms. In KMCA, one predivides the data into *k* groups, randomly selects *k* objects as the initial cluster center, calculates the distance between each object and each seed cluster center, and then assigns each object to the closest cluster center. The cluster centers and the objects assigned to them represent a cluster. For each sample assigned, the cluster center of the cluster is recalculated based on the existing objects in the cluster. This process is repeated until a certain termination condition is met. The termination condition may be that no (or minimum number of) objects are reassigned to different clusters, no (or minimum number of) cluster centers change again, or the squared error reaches a local minimum. KMCA was utilized for the analysis of the different structures present in the studied samples and the identification of the tumoral region. This method can be employed for the study of individual cells by SERS to identify the cellular spatial distribution and biochemical characteristics and give insights into cells that have similar or different compositions [41,42].

CA is an initial processing for spectral analysis and aims to obtain an insight into the data structure. Different cluster analyses could not only perform differently but also work on different principles. For a large number of variables, KMCA is simple with relatively short computation time than HCA, where the dendrogram could be so big that it becomes too burdensome to visualize and interpret. KMCA requires the user to specify the number (k) of clusters for the data set in advance. Different k number choices would result in different clusters. And the selection of proper initial clusters is also an important task.

11.2.2 Supervised data analysis and practical applications

Supervised data analysis is an analytical method based on supervised machine learning. First, a sample set with certain characteristics is used as a training set to build a mathematical model, and then the established model is used to predict the unknown sample. Supervised learning is a common technique for training neural networks and decision trees (DT). For neural networks, classification systems use the information to determine network errors, and then continuously adjust network parameters. For DTs, classification systems use it to determine which attributes provide the most information. Common supervised analysis methods include partial least square (PLS) method, linear discriminant analysis (LDA), support vector machine (SVM), classification and regression trees (CRT), and artificial neural network (ANN).

Before performing supervised data analysis, the Raman spectral data after simple preprocessing may be processed for dimension reduction using methods such as PCA. The PCA analysis output data is used as new input data for subsequent analysis.

11.2.2.1 Partial least square method

PLS is a regression model of multiple dependent variables to multiple independent variables. It can simultaneously implement regression modeling (multivariate linear regression analysis), simplifying data structure (PCA), and the correlation analysis between two groups of variables (canonical correlation analysis).

The construction method of the PLS model is as follows. There are q dependent variables and p independent variables. To study the statistical relationship between the dependent variable and the independent variable, n sample points are observed, thereby forming the data tables X and Y of the independent variable and the dependent variable. PLS extracts t and u from X and Y, respectively, where t and u should carry the variation information in their respective data tables as much as possible, and the correlation between t and u can reach the maximum. After the first component is extracted, the PLS performs X-to-t regression and Y-to-t regression. If the regression equation has reached a satisfactory accuracy, the algorithm is terminated; otherwise, the residual information after X is interpreted by t and the residual information after Y is interpreted by t is used for the second round of component extraction. Until a satisfactory accuracy is reached, a regression equation is finally generated. Then the prediction accuracy of the model can be evaluated by calculating the determination coefficient (R^2), root-mean-square error of calibration, root-mean-square error of prediction, and root means the square error of cross-validation.

Compared with traditional multiple linear regression models, the characteristics of PLS regression are:

1. Regression modeling can be performed under the condition of severe multiple correlations of independent variables.
2. Regression modeling can be performed when the number of sample points is less than the number of variables.
3. All independent variables will be included in the final model.
4. System information and noise (even some nonrandom noise) will be easily identified.
5. The regression coefficients for each independent variable will be easier to explain.

PLS is the most commonly used method for spectral multivariate quantitative correction. It has been widely used in the establishment of Raman spectral quantitative models and has almost become a general method for establishing linear quantitative correction models in spectral analysis. Recently, with the in-depth application of the PLS method in spectral analysis, the PLS method has also been used to solve qualitative analysis problems such as pattern recognition, quantitative correction model applicability judgment, and abnormal sample detection. Since PLS method extracts loadings and scores from both the spectral array and the concentration array, as well as overcomes the shortcomings of the PCA methods that do not use the concentration array, it can effectively reduce the dimension, eliminate the complex collinear relationship that may exist between the spectra, and thus achieves very satisfactory quantitative analysis results from cells [43], blood [44–46], urine [47], and follicular fluid [48]. PLS is a multivariate statistical analysis method widely applied to SERS spectrum processing. A combined method, PLS-DA, will be described in Section 11.2.2.2.

Using PCA and PLS, Raman spectral data can be converted into quantifiable and easy to visualize information. While removing noise and reducing human error, PCA extracts key information across the entire spectral range. Score plots for PCA show three different clusters. Each cluster indicates that they can be easily distinguished. PCA can effectively extract related SERS spectral changes with high specificity. The use of PLS will help to establish a standard calibration curve for detection in a variety of situations, including pure analytes, complex pure analytes, multiple analytes in artificial urine, and analytes in real patient urine samples. PCA recognizes the Raman spectrum corresponding to the target metabolite while eliminating possible interferences. PLS then converts all Raman spectral information in the entire spectral window into quantifiable data, generating a linear range of concentration changes. PLS analysis is also capable of monitoring small changes in metabolite concentrations, enabling rapid quantitative screening of key metabolites in patient samples. The combination of the SERS platform and metabolomics offers more advantages, including highly specific molecular fingerprints for clear metabolite identification, label-free capabilities, ultralow detection limits, simple sample handling, low sample size requirements, and noninvasive rapid quantification. This method can potentially be applied to screening for various diseases in many clinical settings and is particularly useful when only a small number of samples are available.

11.2.2.2 Linear discriminant analysis

LDA is a supervised classification technique based on uncorrelated linear discriminant functions, fulfilling an important condition: the canonical scores obtained are independent. LDA also allows visualizing how the discriminant functions lead to samples' grouping by plotting the individual canonical scores for the discriminant functions.

The number of variables is less than or equal to the number of samples and as a direct consequence, the LDA can be efficiently applied. Therefore, in some circumstances, when samples are very similar, or data sets are very large, some methods should be performed before LDA to simplify the complex data sets. The dimensionality reduction method, PCA, can determine the key variables in a multidimensional data set that best explains the differences in the observations. It becomes ideal to combine LDA with PCA, which often leads to better discrimination of the samples,

The following outlines the steps of a typical PCA-LDA analysis of Raman spectra:

1. PCA was performed to reduce the high dimension of variables into a few PCs while retaining the most diagnostically significant information for clinical classification.

2. Independent-sample *T*-test was used to search diagnostically significant PCs for each case.
3. These selected diagnostically significant PCs ($P < .05$) were input into the LDA model with the leave-one-out, cross-validation method for classification.

It should be noted that PLS can also be used as a dimension reduction technique similar to PCA, which shows better prediction results than that of PCA-LDA [49]. Recently, SERS technology with LDA-based spectral classification method has been widely applied for research in cancer tissues [50], cells [43,51–53], blood samples [54–60], and body fluids [61,62]. The capabilities of label-free SERS are demonstrated for intracellular analysis and its ability to provide a way of characterizing intracellular composition. The biocompatibility offered by plasmonic gold nanoparticles (AuNPs) is one of the major advantages of their use in intracellular SERS experiments. Employed as SERS probes, the extent to which normal cellular metabolism is affected by AuNPs internalization was studied with the combination of SERS and PCA-LDA by Taylor et al. [52]. With the same analysis strategy, Zhang et al. [53] detected and classified cancer-related exosomes and paved the way for new real-time diagnosis and classification of cancer by using exosome as a cancer marker. This powerful clinical tool for label-free, noninvasive disease detection, screening, and classification were also applied in type II diabetes [63], Huntington's disease [64], and Alzheimer's disease [65].

The SERS-based detection methods were also applied in distinguishing microorganism, coupled with multivariate methods to achieve visualization differentiation results. Stephen et al. [66] successfully distinguished a large number of very closely related Arthrobacter strains. Additionally, Dina et al. [67] used this method to achieve the diagnosis of invasive fungal infections. Three different clinical isolates of relevant filamentous fungal species were discriminated using fuzzy PCA (FPCA) in combination with LDA. FPCA was performed to solve the problems associated with PCA: the sensitivity to outliers, missing data, and poor linear correlation between variables. This method resulted in a better separation of data among three fungal isolates (*Aspergillus fumigatus* sensu stricto, cryptic *A. fumigatus* complex species, and *Rhizomucor pusillus*).

The exploratory work using SERS combined with an LDA-based classification method was developed for disease diagnosis with excellent diagnostic efficiency, which further supports this label-free blood SERS technique for practical clinical applications. It has been used in noninvasive detection in cancer, including nasopharyngeal cancer (NPC) [54,55], colorectal cancer [68], prostate cancer [69,70], lung cancer [71], cervical cancer [72], bladder cancer [73], and breast cancer [74]. A simple and label-free blood test for esophageal cancer detection was developed by Lin et al. [75]. A multivariate statistical analysis based on PCA/PLS-LDA can reveal hidden relationships between biochemical parameters and establish the relevant characteristics for classification and grouping. It was employed to incorporate the entire spectrum into analyses and automatically determine the most diagnostically significant features for improving the efficiency of clinical analysis and differentiation.

Feng et al. [55] applied PCA algorithm to simplify the complex spectral data of blood SERS. And the most diagnostically significant PCs identified by the *T*-test were used to generate the classification model (2D and 3D) for discriminating normal and NPC blood samples, as shown in Fig. 11.4. In the work of Lin et al. [57], the PCA-LDA method was employed to analyze and differentiate the SERS spectra from normal, T1 stage NPC and T2–T4 stage NPC. SERS technology combined with PCA-LDA algorithm was also applied for real blood circulating DNA detection for the first time by the same researchers (Fig. 11.5), and a diagnostic sensitivity of 83.3% and specificity of 82.5% were obtained for

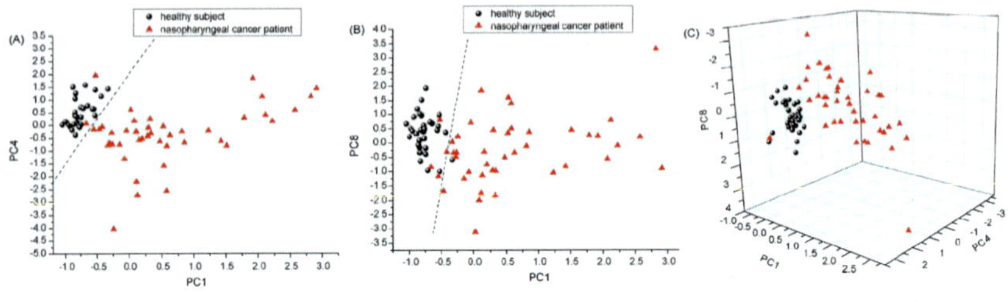

Figure 11.4 (A) Plots of the first principal component (PC1) versus the fourth principal component (PC4) for normal group versus nasopharyngeal cancer group. The dotted line (PC4 = 3.91PC1 + 1.89) as diagnostic algorithm separates the two groups very well. (B) Plot of the first principal component (PC1) versus the eighth principal component (PC8) for normal group versus nasopharyngeal cancer group. The dotted line (PC8 = 12.6PC1 + 4.8) as diagnostic algorithm separates the two groups very well. (C) A three-dimensional mapping of the principal component analysis result for the nasopharyngeal cancer group (*red triangle*) and the healthy volunteer group (*black circle*). *Reproduced from S. Feng, R. Chen, J. Lin, J. Pan, G. Chen, Y. Li, et al., Nasopharyngeal cancer detection based on blood plasma surface-enhanced Raman spectroscopy and multivariate analysis, Biosens. Bioelectron. 25 (11) (2010) 2414–2419, with permission from Elsevier.*

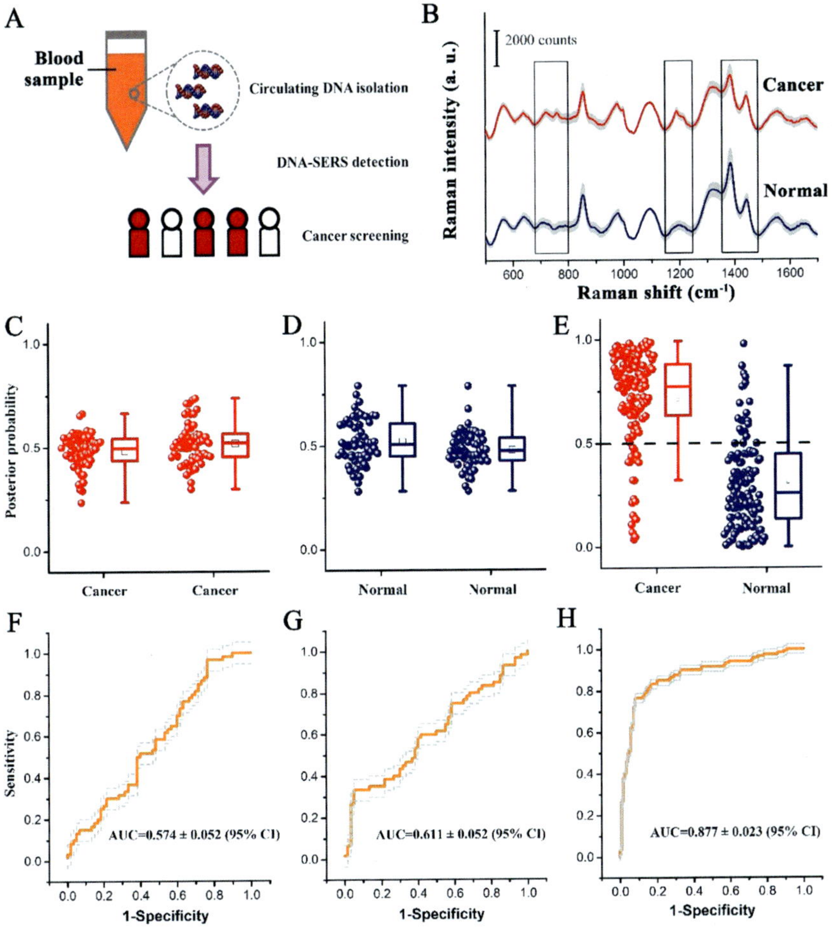

Figure 11.5 (A) Schematic of circulating DNA detection using surface-enhanced Raman spectroscopy (SERS) for cancer screening. (B) Mean SERS spectrum of circulating DNA from normal and cancer subjects. The shaded areas represent the standard deviations of the means. (C–E) Scatter plots of the posterior probabilities belonging to cancer and normal DNA samples for intra and inter comparison using the principal component analysis-linear discriminant analysis. (F–H) Relative receiver operating characteristic (ROC) curves of classification results for DNA sample classification. AUC: the integration areas under the ROC curves. *Reproduced from D. Lin, Q. Wu, S. Qiu, G. Chen, S. Feng, R. Chen, et al., Label-free liquid biopsy based on blood circulating DNA detection using SERS-based nanotechnology for nasopharyngeal cancer screening, Nanomed-Nanotechnol. 22 (2019) 102100, with permission from Elsevier.*

differentiating NPC from the normal group, demonstrating promising potential as alternative nanotechnology for NPC screening based on liquid biopsy [56].

11.2.2.3 Support vector machine

SVM is a powerful supervised learning algorithm for classifying complex groups based on the machine-learning approach [76]. The fundamental idea of SVM is that it looks for the optimal hyperplane that maximizes the margin of separation between the hyperplane and closest data points on either side of the hyperplane. As a classifier, SVM is capable of processing classification problems with nonlinear boundary by mapping sample data set into a higher-dimensional space. SVM has attracted great attention due to the ability to reveal nonlinear relationships and produce models that achieve better classification results than traditional methods.

The process of discrimination and diagnosis was divided into the following steps:

1. For SVM implementation, the already preprocessed data set was further normalized (to the [0, 1] range) to put all the variables on the same scale.
2. Kernel functions were used to make the project data to the feature space, including radial basis function (RBF), polynomial, linear, or sigmoid to obtain an SVM classifier with good classification ability. Gaussian RBF is the most frequently used one.
3. The optimal tuning parameters were found using a grid search and then used to calculate the sensitivity and specificity for the different comparisons. A penalty factor *C* was introduced to allow some training data to be misclassified to avoid the problem that there are countless separating hyperplanes, leading to the risk of overfitting once the spectra are mapped to the feature space [65]. The penalty factor *C* and the kernel-related parameter can be optimized by grid search.
4. The cross-validation was used to evaluate the SVM diagnostic algorithm. To value the efficiency of the model developed by SVM, the employed parameters include specificity, sensitivity, accuracy, Matthew correlation coefficient, and rigidity. The combination of SERS and SVM-based method has been successfully applied in cancer screening, disease prediction, gene selection, etc.

   Disease screening

It has been confirmed that serum SERS analysis combined with SVM could discriminate cancer patients from healthy volunteers [77–83]. SERS measurements were performed on serum protein samples from 104 liver cancer patients, 100 NPC patients, and 95 healthy volunteers by Yu et al. [49]. PLS was employed for feature analysis of the SERS spectra by extracting a set of components [latent variables (LV)]. With the number of components being compressed to 3, SVM was then used to form a diagnostic algorithm and classify multiple cancers simultaneously. Fig. 11.6A shows the classification results of the RBF kernel SVM model in the feature space. Circles represent the support vectors. Separating hyperplanes were created in the feature space to distinguish samples. Fig. 11.6B shows the results of classifying SERS spectra in the testing set using the diagnostic model as shown in Fig. 11.6A. Based on the PLS-SVM algorithm, high diagnostic sensitivities of 92.31% and 96% and specificities of 100% and 88%, respectively, were achieved for screening liver cancer patients and NPC patients simultaneously.

To test the capability of plasma SERS spectra for discriminating esophageal cancer from normal, multivariate statistical methods based on SVM and PCA-LDA were performed on the measured blood plasma SERS spectra by Lin et al. [75]. They found different types of kernel functions achieved different classification accuracies and the SVM based on the RBF kernel function achieved the best classification result.

Pathogens identification

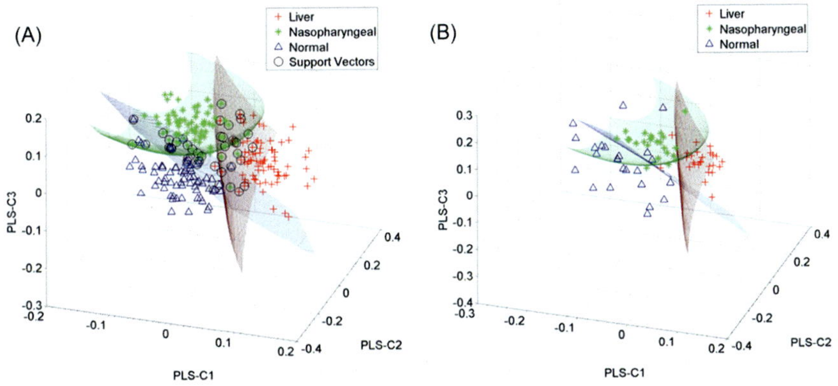

Figure 11.6 (A) Support vector machine classification results for the three groups of samples. (B) Prediction results of the testing set. *Reproduced from Y. Yu, Y. Lin, C. Xu, K. Lin, Q. Ye, X. Wang, et al., Label-free detection of nasopharyngeal and liver cancer using surface-enhanced Raman spectroscopy and partial lease squares combined with support vector machine, Biomed. Opt. Express 9 (12) (2018) 6053–6066, with permission from The Optical Society (OSA).*

Walter et al. [84] demonstrated that the validation of bacteria can be fulfilled by measuring ultrasonic busted bacteria using microfluidic lab-on-a-chip based SERS. Taking advantage of the SERS enhancement, the spectral recording time was reduced to 1 s and a database of 11200 spectra was established for a model *E. coli* system including nine different strains. The validation of the bacteria on the strain level was achieved employing SVM with accuracies of 92%.

Thrift et al. [85] presented the first SERS odor compass. Using a grid array of SERS sensors, machine-learning analysis enabled reliable identification of multiple odor sources arising from the diffusion of analytes from one or two localized sources, which was used to identify the location of an *E. coli* biofilm via its complex signature of volatile organic compounds. Convolutional neural network (CNN)- and SVM-based classifier models achieved over 90% accuracy for a multiple odor source problem. It opens a path toward an Internet of things approach to monitoring toxic gases and indoor pathogens.

Gene selection

Choi et al. [86] used SERS to identify oligonucleotides corresponding to neuraminidase (NA) stalk motifs, which are associated with enhanced influenza virulence. Multivariate classification method, SVM-DA, was able to distinguish between different NA stalk motifs linked to NA-enhanced influenza virus virulence with >95% sensitivity and specificity in both synthetic RNA sequences and the isolated viral RNA. This study demonstrated the feasibility of SERS for direct identification of influenza NA stalk mutations associated with virulence without sample amplification or labeling.

Overall, SVM is a method based on the mechanism of separating classes with a hyperplane between them, which has certain advantages over LDA in SERS spectral discrimination for disease diagnosis due to its hyperplane property [87]. Moreover, it is a very powerful statistical tool for differentiating between multiple classes and additionally it can reveal the SERS signals that have a strong impact on the classification.

11.2.2.4 Decision tree

DT is a prediction model that represents a mapping relationship between object attributes and object values. DT contains a root node, several internal nodes, and several leaf nodes. Each internal node represents a test on an attribute, each branch represents a test output, and each leaf node represents a category. The algorithm of CRT is a common method in the process of building the DT model. It achieves the prediction goal by constructing a binary tree, which is easy to understand, explain, and use. The DT prediction constructed by CRT is more accurate. The more complex the data and the more variables are, the more significant the superiority of CRT is.

Building a DT requires three steps: feature selection, tree generation, and pruning of the tree.

First, a feature is selected from the many features as the current node-splitting criterion. The square error minimization criterion is used for the regression tree and the Gini index minimization criterion is used for the classification tree for feature selection. CRT assumes that the DT is a binary tree and calculates the Gini index of the training data set at the existing feature pair nodes. Each feature A corresponds to all possible values B. According to the test of sample point A = B, the training data set is divided into two parts and the Gini index when A = B is calculated. Among all possible features A and all possible segmentation points B, the feature with the smallest Gini index and its corresponding possible segmentation points are selected. According to this optimal feature and optimal segmentation point, two child nodes are generated from the current node, and the training data set is distributed to the two child nodes according to the characteristics.

Then, according to the training data set, starting from the root node, DT recursively generates child nodes from top to bottom, constructs a binary tree, and stops until the data set is indivisible.

Finally, since DTs are prone to overfitting, pruning is generally required to reduce the size of the tree structure and prevent the overfitting. There are two types of pruning: prepruning and postpruning. Prepruning is to set an index during the growth of the tree and stop growing when this index is reached. In postpruning, the tree must first be fully grown until the leaf nodes are minimal, and then pruning is performed with the minimum loss function as the standard for pruning. Postpruning does not need to retain samples for cross-validation and can make full use of the entire training set information. However, the amount of postpruning calculation is large, and it is only suitable for small sample size.

CRT is easy to understand and implement and can make feasible and effective results for large data sources in a relatively short time. The credibility of the model is easy to determine. In recent years, it has also shown excellent suitability in researches based on SERS.

Cells

Ou et al. [88] used SERS to study the DNA damage of the NPC CNE2 cell line after irradiation. Based on different radiation doses and different cell incubation time after irradiation, SERS was used to monitor chemical changes in DNA. Multivariate statistical analysis was performed on the collected Raman spectra using PCA and random forest (RF) analysis methods. The obtained SERS spectra were divided into a test set and a validation set, and an RF model was established based on the CRT, and a 10-time cross-validation method was used to evaluate the prediction error of the model. The classification accuracy of the CRT model varied with the incubation time and radiation dose. The separation effect was better under X-ray irradiation at 10 Gy (24 h incubation) and 20 Gy (48 h incubation). The SERS spectra under these two conditions were significantly different. The results provided important information for SERS to study radiation-induced DNA damage at the molecular level.

Blood

Zhang et al. [89] used SERS to collect spectra of plasma from patients with amyotrophic lateral sclerosis and analyze metabolic characteristics to explore prognostic factors related to patient survival. PCA and DT were used to statistically analyze all SERS spectra and band intensities of the short-course and long-course groups. Based on the entire SERS spectrum and the intensity of each band, a DT classification model was constructed from the data. In the cross-validation process of the DT model, CRT algorithm was used to illustrate the importance of the variables. Variable importance was the contribution of each variable to model building. Through CRT screening, the 10 most important variables to distinguish between the short-duration group and the long-duration group were identified. The corresponding biological characteristics of the variables included four major categories of nucleic acids, amino acids, antioxidants, and glucose. Then the posterior probability distribution of the DT model constructed by the CRT algorithm was used to estimate the receiver operating characteristic curve. The results proved that SERS can be used to distinguish the duration of amyotrophic lateral sclerosis patients with high sensitivity and specificity. CRT analysis revealed the role of glucose metabolism, amino acid metabolism, nucleic acid metabolism, and antioxidant levels in disease progression, all of which are potential therapeutic targets for amyotrophic lateral sclerosis.

11.2.2.5 Artificial neural network

ANN is a mathematical model of distributed parallel information processing produced by biological neural networks. It simulates the processing of information by the human brain. ANN is a computing model formed by a large number of connected neurons. Each neuron represents a specific output function, called an activation function. The result of the linear calculation is transformed into a nonlinear function result of fitting probability theory through a function, and the transformed function is an activation function. The connection between every two neurons represents a weighted value for the signal passing through the connection, called the weight, and the connection weight between neurons reflects the strength of the connection between the units. By adjusting the interconnection relationship between a large number of internal nodes, the purpose of information processing is achieved. The output of the network depends on the structure of the network, how it is connected, weights, and activation functions.

There are three types of processing units in the network: input unit, output unit, and hidden unit. The input unit accepts external signals and data. Multiple layers of hidden units calculate and convert the data in the input layer, and then pass it to the subsequent output unit. The output unit realizes the output of the system processing results. The neural network needs a set of input data and output data pairs. After selecting the network model and the transfer and training functions, the ANN calculates the output results and corrects the weights according to the error between the actual output and the expected output. ANN can continuously adjust the neuron weights and thresholds until the output error of the network reaches the expected result.

The neural network system consists of a topology structure formed by a large number of neuron connections. It depends on the huge number of neurons and the connections between them to achieve complex information processing. It has obvious advantages in processing fuzzy data, random data, and nonlinear data. It is especially applicable to systems with large scale, complex structure, and ambiguous information. ANN has the following basic characteristics:

1. Parallel processing capability. ANN has the characteristics of parallel processing, which can greatly improve the speed of work and make fast judgments, decisions, and processing on many problems.
2. Nonlinear model. Each neuron of ANN accepts a large number of other neurons' inputs and generates output through a parallel network, which affects other neurons. These kinds of neurons restrict and interact with each other to achieve a nonlinear mapping from input state to output state space.

3. Associative memory function and good fault tolerance. ANN stores the processed data information in the weights between neurons through its unique network structure. It has an associative memory function. The distributed storage form makes the network very fault-tolerant and can be used from incomplete information and noise interference and can restore the original complete information.
4. Self-learning ability and adaptive ability. Based on the data provided, ANN obtains the weights and structure of the network through learning and training and presents a strong self-learning ability and adaptive ability to the environment.
5. Nonconvexity. The evolution direction of a system will depend on a certain state function under certain conditions. Nonconvexity means that this function has multiple extreme values in the ANN, so the system has multiple stable equilibrium states, which will lead to the diversity of system evolution.

In recent years, research work of ANN has been continuously deepened, and great progress has been made. It has successfully shown good intelligent characteristics in the fields of biology and medicine. Due to the complexity and unpredictability of the human body and diseases, the acquired Raman spectral data has a very complicated nonlinear relationship, which is suitable for application of ANNs.

Guselnikova et al. [90] established a method based on SERS combined with ANN to identify and quantify light-induced DNA damage. The main advantage of ANN is the ability to discover nonlinear relationships, which is more suitable for complex biological samples. Before the SERS spectra were input into the neural network, the spectral data was preprocessed using the developed automatic algorithm, which also eliminated the potential uncertainty caused by spectral background interference. A special ANN, called CNN, was used to process the obtained spectrum. The given spectrum was used as the input, and the corresponding damage degree was used as the output. The formula for the classification problem was established. The prediction ability of ANN was studied with the validation data set, and the ANN training results obtained are shown in the Fig. 11.7. Using the training part of the spectral data set as input to the ANN classifier, the neural network-based spectral evaluation results provide their categories. In Fig. 11.7A, the probability that each line corresponds to the input spectrum belonging to a given category is represented by different colors. In the training classification diagram, 0, 1, 2, and 3 were used to indicate not damaged, slightly damaged, moderately

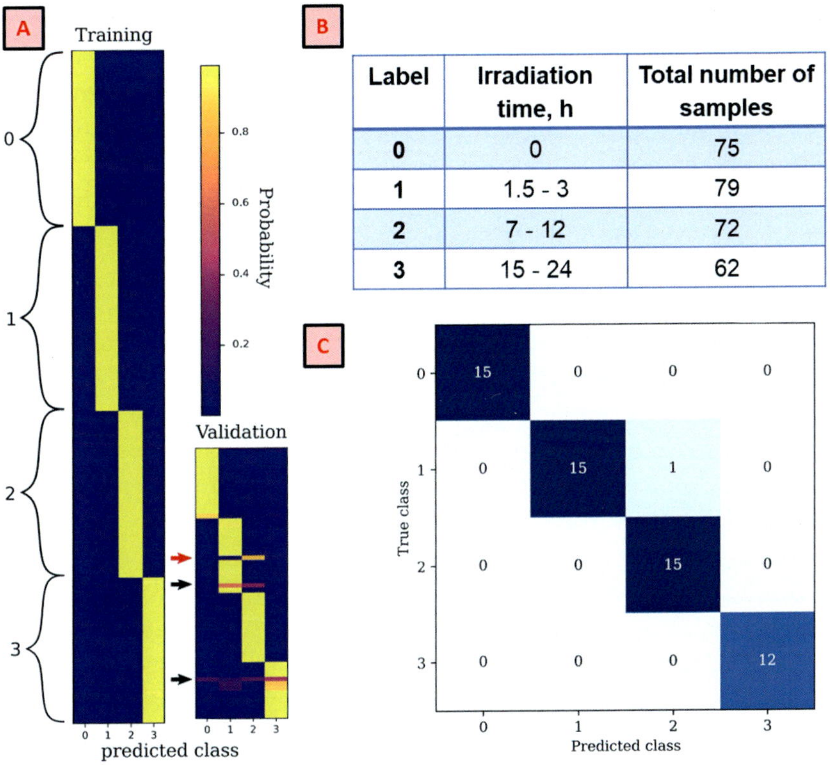

Label	Irradiation time, h	Total number of samples
0	0	75
1	1.5 - 3	79
2	7 - 12	72
3	15 - 24	62

Figure 11.7 Artificial neural network analysis of surface-enhanced Raman spectroscopy (SERS) spectra for recognition of photoinduced DNA damage. (A) Classification maps; (B) table of labels corresponding to the irradiation time; and (C) relation between true and predicted classes for tested SERS spectra of UV-light damaged oligonucleotide. *Reproduced from O. Guselnikova, A. Trelin, A. Skvortsova, P. Ulbrich, P. Postnikov, A. Pershina, et al., Label-free surface-enhanced Raman spectroscopy with artificial neural network technique for recognition photoinduced DNA damage, Biosens. Bioelectron. 145 (2019) 111718, with permission from Elsevier.*

damaged, and significantly damaged, respectively. In the validation the red arrow indicated that only one data was misclassified, and the black arrow highlights data that was classified correctly but with low confidence. Fig. 11.7B and C shows the classification of SERS spectral data and accurate estimation of DNA damage. The results show that the classification and corresponding graphs show that the prediction accuracy of this model is higher than 98%, and only one of the 58 verification spectra was misclassified. Using the proposed functional SERS/ANN method, it can not only classify samples but also provide ANN analysis feedback to link the spectral changes of the DNA structure with the chemical changes, or even identify extremely small DNA damage. This ANN-based DNA damage recognition method has the advantages of high sensitivity, convenient portability, short processing time, simple operation, low cost, easy automation, and high-throughput analysis.

Li et al. [60] explored the feasibility of using SERS in combination with PLS-SVM/DA/ANN to identify serum from patients with liver cancer, patients with cirrhosis and healthy people. The Raman spectra were obtained from 44 healthy people, 45 patients with liver cancer, 42 patients with liver cancer after treatment, and 45 patients with liver cirrhosis. The SERS spectra were analyzed using SVM, PLS-DA, and ANN to determine its diagnostic potential for liver disease. For the PLS-ANN model, PLS first extracted information from the entire spectral region and used the first two potential variables (LV) as the neurons in the input layer of the network. ANN was used to classify and output data by repeatedly weighting and passing input variables between neurons. Since ANN tends to overfit, it uses only one hidden layer to build a three-layer network consisting of an input layer and an output layer. Fig. 11.8 shows the complete ANN network. The input layer has two neurons corresponding to the first 2 LV components (I1 and I2), the hidden layer has different numbers of neurons for training (H1–H18), and the output layer has four neurons representing four classification groups (O1, O2, O3, and O4). The characteristics of the original SERS spectrum were retained by multivariate statistical analysis, and the SERS spectra of different groups were classified. The retained sample cross-validation method was used to check the performance of the model. For 176 samples, the diagnostic accuracy of PLS-SVM, PLS-DA, and PLS-ANN was 91.5%, 89.2%, and 90.3%, respectively. This preliminary study demonstrates that serum SERS can be used for liver cancer screening and is an analytical tool for the rapid prediction of liver disease.

11.3 Additional applications in labeling SERS measurements

11.3.1 Practical applications of unsupervised analysis

11.3.1.1 Cells

SERS technique is used for a stereoscopic description of the intrinsic chemical nature of cells and the precise localization of molecular composition at different cellular locations under highly confocal conditions. Chen

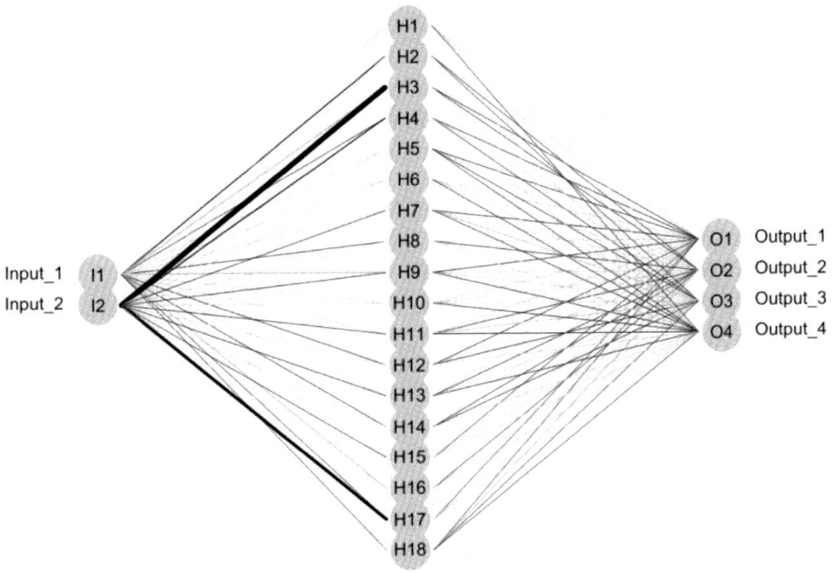

Figure 11.8 The PLS-artificial neural network (ANN) analysis of blood serum surface-enhanced Raman spectroscopy spectra was used to detection liver diseases. The ANN architecture is shown. *Reproduced from X. Li, T. Yang, S. Li, L. Jin, D. Wang, Guan D., et al., Noninvasive liver diseases detection based on serum surface enhanced Raman spectroscopy and statistical analysis, Opt. Express 23 (14) (2015) 18361–18372, with permission from The Optical Society (OSA).*

et al. [41] detected double organelles (nucleus and membrane) in single HeLa cells using labeled SERS probe and reported on the time-dependent changes of cell nuclei as well as membrane receptor proteins during apoptosis using multivariate methods. Raman spectra were clustered by KMCA and the Raman signals of two kinds of membrane-targeting SERS probes and one nuclear-targeting SERS probe from the same cell in three dimensions were obtained, achieving a new SERS strategy for triplex three-dimensional SERS imaging, allowing for both temporal (real-time) and spatial (multiple organelles and molecules in three-dimensional space) live-cell imaging.

Dugandžić et al. [91] presented a SERS-based strategy for the detection of early-stage atherosclerosis, which allowed the assessment of the vulnerability of plaques. The ability to detect macrophages as highly abundant cells in prone-to-rupture atherosclerotic plaques would enable detection and early diagnosis of the vulnerable plaques. So, they used mannose-modified SERS-active gold nanoparticles tagged with a suitable reporter for targeting macrophages and traced the uptake of the nanoparticles by macrophages by confocal Raman microspectroscopy in vitro. KMCA was applied to estimate the macrophages areas and detect the nanoparticles within the cells. They examined the possibility of the detection and characterization of vulnerable atherosclerotic plaques.

Similar methods have been developed for the detection of microscopic tumor cells by Potara and coworkers [92]. They also designed and assessed a carboplatin (CBP) nanotherapeutic delivery system, which allowed combinatorial functionalities of chemotherapy, pH sensing, and multimodal traceable properties inside live NIH: OVCAR-3 ovarian cancer cells, and generated a robust SERS traceable system by anchoring a pH-sensitive Raman reporter, 4-mercaptobenzoic acid (4-MBA) onto the surface of chitosan-coated silver nanotriangles (chit-AgNTs). CBP is then loaded to 4-MBA labeled chit-AgNTs (4-MBA-chit-AgNTs) core under alkaline conditions to endow this nanoplatform with chemotherapeutic abilities. KMCA was used to perform the simultaneously Raman mapping of the whole NIH: OVCAR-3 cell and accurately track the SERS spectra of the internalized CBP-4-MBA-chit-AgNTs in the combined Raman and SERS spectral microimaging. It demonstrated the ability of biosynthesized silver nanoparticles to act as contrast agents when tested in vitro.

11.3.1.2 Pathogens

Kearns et al. [93] developed a new type of biosensor using magnetic separation and SERS technology for the separation and detection of a variety of bacterial pathogens. Bacteria were first captured and isolated from the sample matrix using lectin-functionalized magnetic nanoparticles, and then SERS-active nanoparticles with strain-specific antibodies were used to specifically detect bacterial pathogens. *E. coli*, *Salmonella typhimurium*, and methicillin-resistant *Staphylococcus aureus* were isolated and detected at cell concentrations as low as 10 CFU mL^{-1}. In addition to single pathogen detection, SERS was also used to identify a mixture of three bacteria in the same sample matrix, and triple detection was confirmed by PCA. The PCA score plots show the relationship between the multiple spectra and each of the three single pathogen spectra and identify four unique groupings. PCA successfully identified and distinguished bacterial pathogens in the complex and was able to distinguish bacterial complexes from control samples, as shown in Fig. 11.9. In the case that the bacterial concentration is lower than the clinical diagnosis requirements, it proves that

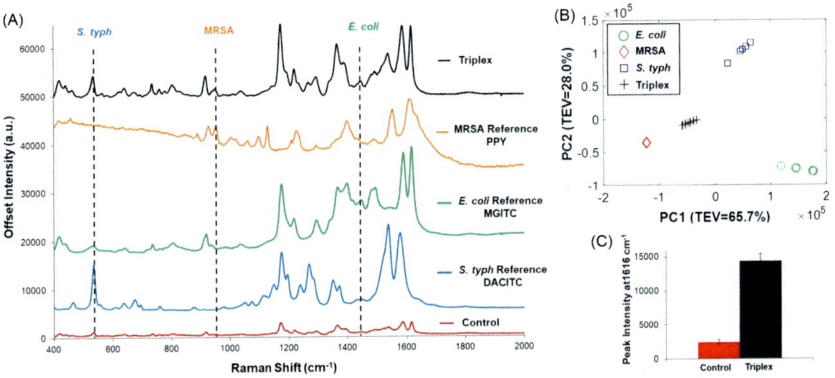

Figure 11.9 SERS spectra obtained from multiple pathogen detection using the surface-enhanced Raman spectroscopy (SERS) assay. (A) Stacked SERS spectra showing the raw spectra obtained from the detection of all three bacterial pathogens simultaneously using the detection assay and a control sample for comparison when no bacteria was present plus the SERS spectra obtained from the detection of each pathogen separately. Control spectrum is red. 7-dimethylamino-4-methylcoumarin-3-isothiocyanate spectrum (blue) represents *Salmonella typhimurium*; malachite green isothiocyanate spectrum (green) represents *Escherichia coli*. 4-(1H-pyrazol-4-yl)-pyridine spectrum (orange) represents methicillin resistant Staphylococcus aureus, and the black spectrum is the multiplex containing all three bacterial pathogens. The black dotted lines show peaks that are unique to each Raman reporter and hence used to identify the presence of the bacterial targets. The bacteria concentration used was 10^3 CFU mL^{-1}. (B) Principal component analysis scores plot showing the relationship between the multiplex spectra and each of the three single pathogen spectra. The black crosses represent the triplex spectra; green circles are *E. coli*; blue squares are *S. typh*; and red diamonds are methicillin resistant Staphylococcus aureus. (C) Comparative peak intensities at 1616 cm^{-1} for assay and control. SERS spectra were recorded using a 532 nm laser excitation with an accumulation time of 1 s. Peak intensities were obtained by scanning three replicate samples five times, and the error bars represent one standard deviation. *TEV*, Total explained variance. *Reproduced from H. Kearns, R. Goodacre, L.E. Jamieson, D. Graham, K. Faulds, SERS detection of multiple antimicrobial-resistant pathogens using nanosensors, Anal. Chem. 89 (23) (2017) 12666–12673, with permission from Analytical Chemistry.*

this multichannel biosensor can quickly and sensitively identify bacterial pathogens, which will provide opportunities to the progress for future medical equipment, clinical diagnosis, and biomedical applications.

Wang et al. [94] combined nanodielectric microfluidic devices with SERS technology to construct a new type of portable biosensor to facilitate the detection and characterization of *E. coli* O157: H7 at high sensitivity levels (single cells/mL). By deploying three different SERS-labeled molecular probes targeting different epitopes of the same pathogen,

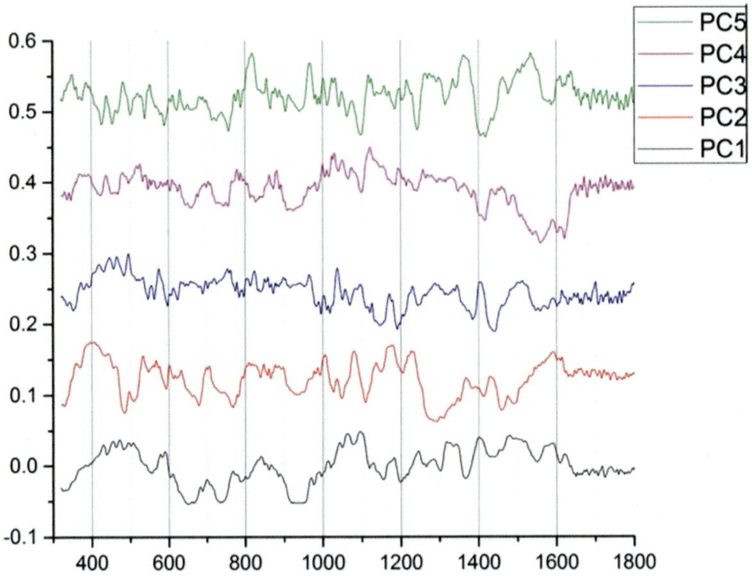

Figure 11.10 The first five principal component loadings of the principal component analysis performed on the surface-enhanced Raman spectroscopy spectra acquired from multiplex antibodies functionalized nanoprobes conjugating with *Escherichia coli O157: H7* bacteria sample. *Reproduced from C. Wang, F. Madiyar, C. Yu, J. Li, Detection of extremely low concentration waterborne pathogen using a multiplexing self-referencing SERS microfluidic biosensor, J. Biol. Eng. 11 (2017) 9, with permission from Journal of Biological Engineering.*

subspecies-specific detection of pathogen targets at the single-cell level was achieved. In the presence of nontarget interference, PCA still confirmed the positive recognition of the target, and the detection accuracy was above 95% (Fig. 11.10). It has the potential to become a powerful biosensor with high sensitivity and specificity, greatly reducing operating procedures, shortening detection time, and reducing costs. It has been used to detect very low levels of pathogens in the field.

11.3.1.3 Other subjects

It was reported that SERS detection of protein receptors and ligands can provide insight into the protein—ligand interactions, which play important roles in many biological processes. In work by Kong et al. [95], they made a stable, highly cost-effective and easy to fabricate, SERS-active substrate to develop a rapid and label-free immunoassay for clinical use. With this strategy, different antigen concentrations (10—1000 ng mL^{-1})

in urine samples were effectively identified using classification model generated by PCA algorithm.

Hornemann et al. [96] reported SERS hybrid probes generated by conjugating different reporter molecules to bovine serum albumin (BSA), with advantages of high biocompatibility, stabilization of the gold nanoparticles in the biological environment, stable reporter signals, and easy preparation compared to other SERS nanoprobes. They applied the BSA–reporter conjugate hybrid probes in 3T3 fibroblast cells to demonstrate SERS enhancement and changes in the chemical structures of the reporter and BSA. HCA was employed for imaging distribution of SERS hybrid nanoprobes in 3T3 fibroblast cells using the full spectral range (300–1700 cm^{-1}) of normalized spectra. It suggested that characteristic intensities in the conjugate spectra may vary depending on the cellular environment. By multivariate classification, the reporter was identified despite changes in one specific band, in addition to contributions from the cellular surroundings. It suggested more successful applications of the SERS hybrid probes with HCA in unordered high-density bioanalytical sensing.

11.3.2 Practical applications of supervised data analysis

Zhang et al. [97] used a microfluidic chip to effectively screen circulating tumor cells (CTCs) from blood, based on the size difference between CTCs and blood cells. SERS probes with orthogonal spectra were used to analyze cell membrane proteins and identify cancer subpopulations. Jurkat cells for the control group were treated with aptamer-free control probes. Three breast cancer cell lines, SKBR3, MCF7, and MDA-MB-231, were selected as models, and three different SERS aptamer nanovector (SAV) cocktails were injected into the chip. Three kinds of aptamers, which showed excellent specificity and affinity to human epidermal growth factor receptor 2 (HER2), epithelial cell adhesion molecule (EpCAM), and human epidermal growth factor receptor 1 (EGFR), respectively, were used to recognize the target biomarkers. These three cell lines showed very different expressions of EpCAM, HER2, and EGFR. Therefore, these SAVs showed different affinities for these three cell types. After full contact, the Raman spectrum of each cell was collected with SERS. PLS-DA used the basic principle of PCA, not only reduced the dimension of the spectral data from several hundred wavenumbers to several key components but also further rotated the components to maximize the

Table 11.1 Application of multivariate approaches for surface-enhanced Raman spectroscopy data analysis.

Detection target	Multivariate statistical approaches
Cell	PCA [5–11], MCR-ALS [20], CA [29,41,42,91,92,96], PLS [43], PLS-DA [97], PCA-LDA [43,51–53], SVM [42], PCA-CRT [88]
Blood	PCA [15,16], PLS [44–46], PCA-LDA [54–58,63–65,68–75], PLS-DA [60], SVM [75,78–80], PCA-SVM [77,83,87], PLS-SVM [49], PCA-CRT [89], PCA-ANN, PLS-ANN [60]
Urine	PCA [18,95], MCR-ALS [21–25], PCA-PLS [47]
Pathogens	PCA [12–14,93,94], CA [27], PCA-CA [30,32–40], PCA-LDA [66], PLS-LDA [67], PCA-SVM [84], SVM-ANN [85]
Tissue	PCA-LDA [50]
Cerebrospinal fluid	PCA [17]
Follicular fluid	PCA-PLS [48]
Pleural effusion	PCA-LDA [60]
Knee synovial fluid	PCA-LDA [62]
DNA/RNA	ANN [90], PCA-LDA [59], MCR-ALS [26], PLS-CA [31], SVM-DA [86]

ALS, Alternating least squares; *ANN*, artificial neural network; *CA*, cluster analysis; *CRT*, classification and regression trees; *DA*, discriminant analysis; *LDA*, linear discriminant analysis; *MCR*, multivariate curve resolution; *PCA*, principal component analysis; *PLS*, partial least square method; *SVM*, support vector machine.

separation of the groups. PLS-DA was able to accurately classify cells of different subtypes. According to the root-mean-square error cross-validation results, the first four LV were selected as the PCs to establish a diagnostic algorithm. Most of the samples were correctly arranged into their categories. The sensitivity to SKBR3, MCF7, MDA-MB-231, and Jurkat were 87.5%, 90.91%, 93.33%, and 96.77%, and the specificities were 95.75%, 97.85%, 97.92%, and 97.90%, respectively. Combined with the PLS-DA classification algorithm, cells of different breast cancer subtypes can be reliably classified, with high sensitivity and selectivity.

11.4 Concluding remarks

In the past decades, SERS technology has been broadly applied and developed in many fields. Methodology based on SERS is constantly updated, and the corresponding data analysis methods have also entered various biomedical applications. In this chapter, we reviewed the commonly used unsupervised and supervised methods to promote the overall goal of optimized analysis of SERS data (summarized in Table 11.1). Hope this will spark more discussions on topics that as yet unexplored, but worthy of investigation and study.

References

[1] A. Rinnan, F. van den Berg, S.B. Engelsen, Review of the most common pre-processing techniques for near-infrared spectra, TrAC, Trends Anal. Chem. 28 (10) (2009) 1201–1222.

[2] T. Bocklitz, A. Walter, K. Hartmann, P. Rosch, J. Popp, How to pre-process Raman spectra for reliable and stable models? Anal. Chim. Acta 704 (1-2) (2011) 47–56.

[3] S.A. Schonbichler, L.K. Bittner, A.K. Weiss, U.J. Griesser, J.D. Pallua, C.W. Huck, Comparison of NIR chemical imaging with conventional NIR, Raman and ATR-IR spectroscopy for quantification of furosemide crystal polymorphs in ternary powder mixtures, Eur. J. Pharm. Biopharm. 84 (3) (2013) 616–625.

[4] J. Zhao, H. Lui, D.I. Mclean, H. Zeng, Automated autofluorescence background subtraction algorithm for biomedical Raman spectroscopy, Appl. Spectrosc. 61 (11) (2007) 1225–1232.

[5] A.B. Veloso, J.P.F. Longo, L.A. Muehlmann, B.F. Tollstadius, P.E.N. Souza, R.B. Azevedo, et al., SERS investigation of cancer cells treated with pdt: quantification of cell survival and follow-up, Sci. Rep. 7 (1) (2017) 7175.

[6] Y.J. Zhang, Q.Y. Zeng, L.F. Li, M.N. Qi, Q.C. Qi, S.X. Li, et al., Label-free rapid identification of tumor cells and blood cells with silver film SERS substrate, Opt. Express 26 (25) (2018) 33044–33056.

[7] J.Y. Lim, J.S. Nam, H. Shin, J. Park, H.I. Song, M. Kang, et al., Identification of newly emerging influenza viruses by detecting the virally infected cells based on surface enhanced Raman spectroscopy and principal component analysis, Anal. Chem. 91 (9) (2019) 5677–5684.

[8] O. Liang, P. Wang, M. Xia, C. Augello, F. Yang, G. Niu, et al., Label-free distinction between p53 + / + and p53 −/− colon cancer cells using a graphene based SERS platform, Biosens. Bioelectron. 118 (2018) 108–114.

[9] A. Kaminska, T. Szymborski, E. Witkowska, E. Kijenska-Gawronska, W. Swieszkowski, K. Nicinski, et al., Detection of circulating tumor cells using membrane-based SERS platform: a new diagnostic approach for 'liquid biopsy', Nanomaterials (Basel) 9 (3) (2019).

[10] J. Zhang, X. Ma, Z. Wang, Real-time and in-situ monitoring of Abrin induced cell apoptosis by using SERS spectroscopy, Talanta 195 (2019) 8–16.

[11] I. Aguilar-Hernandez, D.L. Cardenas-Chavez, T. Lopez-Luke, A. Garcia-Garcia, M. Herrera-Dominguez, E. Pisano, et al., Discrimination of radiosensitive and radioresistant murine lymphoma cells by Raman spectroscopy and SERS, Biomed. Opt. Express 11 (1) (2020) 388–405.

[12] J. Sundaram, B. Park, Y. Kwon, K.C. Lawrence, Surface enhanced Raman scattering (SERS) with biopolymer encapsulated silver nanosubstrates for rapid detection of foodborne pathogens, Int. J. Food Microbiol. 167 (1) (2013) 67–73.

[13] N.E. Mircescu, H. Zhou, N. Leopold, V. Chis, N.P. Ivleva, R. Niessner, et al., Towards a receptor-free immobilization and SERS detection of urinary tract infections causative pathogens, Anal. Bioanal. Chem. 406 (13) (2014) 3051–3058.

[14] E. Avci, N.S. Kaya, G. Ucankus, M. Culha, Discrimination of urinary tract infection pathogens by means of their growth profiles using surface enhanced Raman scattering, Anal. Bioanal. Chem. 407 (27) (2015) 8233–8241.
[15] S.K. Gahlaut, D. Savargaonkar, C. Sharan, S. Yadav, P. Mishra, J.P. Singh, SERS platform for dengue diagnosis from clinical samples employing a hand held Raman spectrometer, Anal. Chem. 92 (3) (2020) 2527–2534.
[16] J. Lin, R. Chen, S. Feng, J. Pan, Y. Li, G. Chen, et al., A novel blood plasma analysis technique combining membrane electrophoresis with silver nanoparticle-based SERS spectroscopy for potential applications in noninvasive cancer detection, Nanomed-Nanotechnol 7 (5) (2011) 655–663.
[17] A. Kaminska, E. Witkowska, A. Kowalska, A. Skoczynska, I. Gawryszewska, E. Guziewicz, et al., Highly efficient SERS-based detection of cerebrospinal fluid neopterin as a diagnostic marker of bacterial infection, Anal. Bioanal. Chem. 408 (16) (2016) 4319–4327.
[18] S. Lee, J.-M. Namgoong, H. Yu, M. Jue, G. Kim, S. Jeon, et al., Diagnosis in a preclinical model of bladder pain syndrome using a Au/ZnO nanorod-based SERS substrate, Nanomaterials. 9 (2) (2019) 224.
[19] J.E.L. Villa, M.A.S. Afonso, D.P. Dos Santos, P.A. Mercadal, E.A. Coronado, R.J. Poppi, Colloidal gold clusters formation and chemometrics for direct SERS determination of bioanalytes in complex media, Spectrochim. Acta A 224 (2020) 117380.
[20] J. Felten, H. Hall, J. Jaumot, R. Tauler, A. de Juan, A. Gorzsas, Vibrational spectroscopic image analysis of biological material using multivariate curve resolution-alternating least squares (MCR-ALS), Nat. Protoc. 10 (2) (2015) 217–240.
[21] D. Zhang, P. Wang, M.N. Slipchenko, J.X. Cheng, Fast vibrational imaging of single cells and tissues by stimulated Raman scattering microscopy, Acc. Chem. Res. 47 (8) (2014) 2282–2290.
[22] J.E.L. Villa, C. Pasquini, R.J. Poppi, Surface-enhanced Raman spectroscopy and MCR-ALS for the selective sensing of urinary adenosine on filter paper, Talanta 187 (2018) 99–105.
[23] J.E. Villa, R.J. Poppi, A portable SERS method for the determination of uric acid using a paper-based substrate and multivariate curve resolution, Analyst 141 (6) (2016) 1966–1972.
[24] M.B. Mamian-Lopez, R.J. Poppi, Quantification of moxifloxacin in urine using surface-enhanced Raman spectroscopy (SERS) and multivariate curve resolution on a nanostructured gold surface, Anal. Bioanal. Chem. 405 (24) (2013) 7671–7677.
[25] M.B. Mamian-Lopez, R.J. Poppi, Standard addition method applied to the urinary quantification of nicotine in the presence of cotinine and anabasine using surface enhanced Raman spectroscopy and multivariate curve resolution, Anal. Chim. Acta 760 (2013) 53–59.
[26] M.B. Mamian-Lopez, P. Corio, M.L. Temperini, Cooperative hydrogen-bonding of the adenine-thymine pair as a strategy for lowering the limit of detection of thymine by surface-enhanced Raman spectroscopy, Analyst 141 (11) (2016) 3428–3436.
[27] R. Prucek, V. Ranc, L. Kvitek, A. Panacek, R. Zboril, M. Kolar, Reproducible discrimination between gram-positive and gram-negative bacteria using surface enhanced Raman spectroscopy with infrared excitation, Analyst 137 (12) (2012) 2866–2870.
[28] M.Z. Pacia, K. Czamara, M. Zebala, E. Kus, S. Chlopicki, A. Kaczor, Rapid diagnostics of liver steatosis by Raman spectroscopy via fiber optic probe: a pilot study, Analyst 143 (19) (2018) 4723–4731.
[29] S.F. El-Mashtoly, D. Petersen, H.K. Yosef, A. Mosig, A. Reinacher-Schick, C. Kotting, et al., Label-free imaging of drug distribution and metabolism in colon cancer cells by Raman microscopy, Analyst 139 (5) (2014) 1155–1161.

[30] C. Wei, M. Li, X. Zhao, Surface-enhanced Raman scattering (SERS) with silver nano substrates synthesized by microwave for rapid detection of foodborne pathogens, Front. Microbiol. 9 (2018) 2857.
[31] P. Negri, R.A. Dluhy, Detection of genetic markers related to high pathogenicity in influenza by SERS, Analyst 138 (17) (2013) 4877–4884.
[32] S. Liu, H. Li, M.M. Hassan, J. Zhu, A. Wang, Q. Ouyang, et al., Amplification of Raman spectra by gold nanorods combined with chemometrics for rapid classification of four Pseudomonas, Int. J. Food Microbiol. 304 (2019) 58–67.
[33] M. Chisanga, D. Linton, H. Muhamadali, D.I. Ellis, R.L. Kimber, A. Mironov, et al., Rapid differentiation of Campylobacter jejuni cell wall mutants using Raman spectroscopy, SERS and mass spectrometry combined with chemometrics, Analyst 145 (4) (2020) 1236–1249.
[34] W. Liao, Q. Lin, S. Xie, Y. He, Y. Tian, Y. Duan, A novel strategy for rapid detection of bacteria in water by the combination of three-dimensional surface-enhanced Raman scattering (3D SERS) and laser induced breakdown spectroscopy (LIBS), Anal. Chim. Acta 1043 (2018) 64–71.
[35] X. Wu, Y.W. Huang, B. Park, R.A. Tripp, Y. Zhao, Differentiation and classification of bacteria using vancomycin functionalized silver nanorods array based surface-enhanced Raman spectroscopy and chemometric analysis, Talanta 139 (2015) 96–103.
[36] O.E. Rivera-Betancourt, R. Karls, B. Grosse-Siestrup, S. Helms, F. Quinn, R.A. Dluhy, Identification of mycobacteria based on spectroscopic analyses of mycolic acid profiles, Analyst 138 (22) (2013) 6774–6785.
[37] S.L. Hennigan, J.D. Driskell, R.A. Dluhy, Y. Zhao, R.A. Tripp, K.B. Waites, et al., Detection of Mycoplasma pneumoniae in simulated and true clinical throat swab specimens by nanorod array-surface-enhanced Raman spectroscopy, PLoS One 5 (10) (2010) e13633.
[38] O.E. Rivera-Betancourt, E.S. Sheppard, D.C. Krause, R.A. Dluhy, Layer-by-layer polyelectrolyte encapsulation of Mycoplasma pneumoniae for enhanced Raman detection, Analyst 139 (17) (2014) 4287–4295.
[39] V. Hoang, R.A. Tripp, P. Rota, R.A. Dluhy, Identification of individual genotypes of measles virus using surface enhanced Raman spectroscopy, Analyst 135 (12) (2010) 3103–3109.
[40] P. Negri, G. Chen, A. Kage, A. Nitsche, D. Naumann, B. Xu, et al., Direct optical detection of viral nucleoprotein binding to an anti-influenza aptamer, Anal. Chem. 84 (13) (2012) 5501–5508.
[41] Y. Chen, X. Bai, L. Su, Z. Du, A. Shen, A. Materny, et al., Combined labelled and label-free SERS probes for triplex three-dimensional cellular imaging, Sci. Rep. 6 (2016) 19173.
[42] M. Hassoun, J. Rüger, T. Kirchberger-Tolstik, I.W. Schie, T. Henkel, K. Weber, et al., A droplet-based microfluidic chip as a platform for leukemia cell lysate identification using surface-enhanced Raman scattering, Anal. Bioanal. Chem. 410 (3) (2018) 999–1006.
[43] Y. Yu, J. Lin, D. Lin, S. Feng, W. Chen, Z. Huang, et al., Leukemia cells detection based on electroporation assisted surface-enhanced Raman scattering, Biomed. Opt. Express 8 (9) (2017) 4108–4121.
[44] J. Kuligowski, M.R. El-Zahry, Á. Sánchez-Illana, G. Quintás, M. Vento, B. Lendl, Surface enhanced Raman spectroscopic direct determination of low molecular weight biothiols in umbilical cord whole blood, Analyst 141 (7) (2016) 2165–2174.
[45] S. Fornasaro, A. Bonifacio, E. Marangon, M. Buzzo, G. Toffoli, T. Rindzevicius, et al., Label-free quantification of anticancer drug imatinib in human plasma with surface enhanced Raman spectroscopy, Anal. Chem. 90 (21) (2018) 12670–12677.

[46] D.R. Parachalil, D. Commerford, F. Bonnier, I. Chourpa, J. McIntyre, H.J. Byrne, Raman spectroscopy as a potential tool for label free therapeutic drug monitoring in human serum: the case of busulfan and methotrexate, Analyst 144 (17) (2019) 5207–5214.
[47] Y.-C. Kao, X. Han, Y.H. Lee, H.K. Lee, G.C. Phan-Quang, C.L. Lay, et al., Multiplex surface-enhanced Raman scattering identification and quantification of urine metabolites in patient samples within 30 min, ACS Nano 14 (2) (2020) 2542–2552.
[48] A. Momenpour, P.D.A. Lima, Y.A. Chen, C.R. Tzeng, B.K. Tsang, H. Anis, Surface-enhanced Raman scattering for the detection of polycystic ovary syndrome, Biomed. Opt. Express 9 (2) (2018) 801–817.
[49] Y. Yu, Y. Lin, C. Xu, K. Lin, Q. Ye, X. Wang, et al., Label-free detection of nasopharyngeal and liver cancer using surface-enhanced Raman spectroscopy and partial lease squares combined with support vector machine, Biomed. Opt. Express 9 (12) (2018) 6053–6066.
[50] S. Feng, J. Lin, Z. Huang, G. Chen, W. Chen, Y. Wang, et al., Esophageal cancer detection based on tissue surface-enhanced Raman spectroscopy and multivariate analysis, Appl. Phys. Lett. 102 (2013) 043702.
[51] S. Feng, Z. Li, G. Chen, D. Lin, S. Huang, Z. Huang, et al., Ultrasound-mediated method for rapid delivery of nano-particles into cells for intracellular surface-enhanced Raman spectroscopy and cancer cell screening, Nanotechnology 26 (6) (2015) 065101.
[52] J. Taylor, J. Milton, M. Willett, J. Wingfield, S. Mahajan, What do we actually see in intracellular SERS? Investigating nanosensor-induced variation, Faraday Discuss. 205 (2017) 409–428.
[53] P. Zhang, L. Wang, Y. Fang, D. Zheng, T. Lin, H. Wang, Label-free exosomal detection and classification in rapid discriminating different cancer types based on specific Raman phenotypes and multivariate statistical analysis, Molecules (Basel, Switz.) 24 (16) (2019) 2947.
[54] D. Lin, G. Chen, S. Feng, J. Pan, J. Lin, Z. Huang, et al., Development of a rapid macro-Raman spectroscopy system for nasopharyngeal cancer detection based on surface-enhanced Raman spectroscopy, Appl. Phys. Lett. 106 (2015) 013701.
[55] S. Feng, R. Chen, J. Lin, J. Pan, G. Chen, Y. Li, et al., Nasopharyngeal cancer detection based on blood plasma surface-enhanced Raman spectroscopy and multivariate analysis, Biosens. Bioelectron. 25 (11) (2010) 2414–2419.
[56] D. Lin, Q. Wu, S. Qiu, G. Chen, S. Feng, R. Chen, et al., Label-free liquid biopsy based on blood circulating DNA detection using SERS-based nanotechnology for nasopharyngeal cancer screening, Nanomed. Nanotechnol. Biol. Med. 22 (2019) 102100.
[57] D. Lin, J. Pan, H. Huang, G. Chen, S. Qiu, H. Shi, et al., Label-free blood plasma test based on surface-enhanced Raman scattering for tumor stages detection in nasopharyngeal cancer, Sci. Rep. 4 (2014) 4751.
[58] Y. Lu, Y. Lin, Z. Zheng, X. Tang, J. Lin, X. Liu, et al., Label free hepatitis B detection based on serum derivative surface enhanced Raman spectroscopy combined with multivariate analysis, Biomed. Opt. Express 9 (10) (2018) 4755–4766.
[59] Y. Chen, G. Chen, X. Zheng, C. He, S. Feng, Y. Chen, et al., Discrimination of gastric cancer from normal by serum RNA based on surface-enhanced Raman spectroscopy (SERS) and multivariate analysis, Med. Phys. 39 (9) (2012) 5664–5668.
[60] X. Li, T. Yang, S. Li, L. Jin, D. Wang, D. Guan, et al., Noninvasive liver diseases detection based on serum surface enhanced Raman spectroscopy and statistical analysis, Opt. Express 23 (14) (2015) 18361–18372.

[61] X. Shao, J. Pan, Y. Wang, Y. Zhu, F. Xu, X. Shangguan, et al., Evaluation of expressed prostatic secretion and serum using surface-enhanced Raman spectroscopy for the noninvasive detection of prostate cancer, a preliminary study, Nanomed. Nanotechnol. Biol. Med. 13 (3) (2017) 1051–1059.
[62] C.D. Bocsa, V. Moisoiu, A. Stefancu, L.F. Leopold, N. Leopold, D. Fodor, Knee osteoarthritis grading by resonant Raman and surface-enhanced Raman scattering (SERS) analysis of synovial fluid, Nanomed. Nanotechnol. Biol. Med. 20 (2019) 102012.
[63] J. Lin, Z. Huang, S. Feng, J. Lin, N. Liu, J. Wang, et al., Label-free optical detection of type II diabetes based on surface-enhanced Raman spectroscopy and multivariate analysis, J. Raman Spectrosc. 45 (2014) 884–889.
[64] A. Huefner, W.L. Kuan, S.L. Mason, S. Mahajan, R.A. Barker, Serum Raman spectroscopy as a diagnostic tool in patients with Huntington's disease, Chem. Sci. 11 (2) (2020) 525–533.
[65] C. Carlomagno, M. Cabinio, S. Picciolini, A. Gualerzi, F. Baglio, M. Bedoni, SERS-based biosensor for Alzheimer disease evaluation through the fast analysis of human serum, J. Biophotonics 13 (3) (2020) e201960033.
[66] K.E. Stephen, D. Homrighausen, G. DePalma, C.H. Nakatsu, J. Irudayaraj, Surface enhanced Raman spectroscopy (SERS) for the discrimination of Arthrobacter strains based on variations in cell surface composition, Analyst 137 (18) (2012) 4280–4286.
[67] N.E. Dina, A.M.R. Gherman, V. Chiş, C. Sârbu, A. Wieser, D. Bauer, et al., Characterization of clinically relevant fungi via SERS fingerprinting assisted by novel chemometric models, Anal. Chem. 90 (4) (2018) 2484–2492.
[68] D. Lin, H. Huang, S. Qiu, S. Feng, G. Chen, R. Chen, Diagnostic potential of polarized surface enhanced Raman spectroscopy technology for colorectal cancer detection, Opt. Express 24 (3) (2016) 2222–2234.
[69] A. Stefancu, V. Moisoiu, R. Couti, I. Andras, R. Rahota, D. Crisan, et al., Combining SERS analysis of serum with PSA levels for improving the detection of prostate cancer, Nanomedicine 13 (19) (2018) 2455–2467.
[70] J. Pan, X. Shao, Y. Zhu, B. Dong, Y. Wang, X. Kang, et al., Surface-enhanced Raman spectroscopy before radical prostatectomy predicts biochemical recurrence better than CAPRA-S, Int. J. Nanomed. 14 (2019) 431–440.
[71] K. Zhang, X. Liu, B. Man, C. Yang, C. Zhang, M. Liu, et al., Label-free and stable serum analysis based on Ag-NPs/PSi surface-enhanced Raman scattering for noninvasive lung cancer detection, Biomed. Opt. Express 9 (9) (2018) 4345–4358.
[72] S. Feng, D. Lin, J. Lin, B. Li, Z. Huang, G. Chen, et al., Blood plasma surface-enhanced Raman spectroscopy for non-invasive optical detection of cervical cancer, Analyst 138 (14) (2013) 3967–3974.
[73] S. Li, L. Li, Q. Zeng, Y. Zhang, Z. Guo, Z. Liu, et al., Characterization and noninvasive diagnosis of bladder cancer with serum surface enhanced Raman spectroscopy and genetic algorithms, Sci. Rep. 5 (2015) 9582.
[74] D. Lin, Y. Wang, T. Wang, Y. Zhu, X. Lin, Y. Lin, et al., Metabolite profiling of human blood by surface-enhanced Raman spectroscopy for surgery assessment and tumor screening in breast cancer, Anal. Bioanal. Chem. 412 (7) (2020) 1611–1618.
[75] D. Lin, S. Feng, H. Huang, W. Chen, H. Shi, N. Liu, et al., Label-free detection of blood plasma using silver nanoparticle based surface-enhanced Raman spectroscopy for esophageal cancer screening, J. Biomed. Nanotechnol. 10 (3) (2014) 478–484.
[76] M. Sattlecker, C. Bessant, J. Smith, N. Stone, Investigation of support vector machines and Raman spectroscopy for lymph node diagnostics, Analyst 135 (5) (2010) 895–901.
[77] Y. Hong, Y. Li, L. Huang, W. He, S. Wang, C. Wang, et al., Label-free diagnosis for colorectal cancer through coffee ring-assisted surface-enhanced Raman spectroscopy on blood serum, J. Biophotonics (2020) e201960176.

[78] B. Yan, B. Li, Z. Wen, X. Luo, L. Xue, L. Li, Label-free blood serum detection by using surface-enhanced Raman spectroscopy and support vector machine for the preoperative diagnosis of parotid gland tumors, BMC Cancer 15 (2015) 650.
[79] L. Zhou, Y. Liu, F. Wang, Z. Jia, J. Zhou, T. Jiang, et al., Classification analyses for prostate cancer, benign prostate hyperplasia and healthy subjects by SERS-based immunoassay of multiple tumour markers, Talanta 188 (2018) 238–244.
[80] M. Paraskevaidi, K.M. Ashton, H.F. Stringfellow, N.J. Wood, P.J. Keating, A.W. Rowbottom, et al., Raman spectroscopic techniques to detect ovarian cancer biomarkers in blood plasma, Talanta 189 (2018) 281–288.
[81] W. Kim, J.C. Lee, J.H. Shin, K.H. Jin, H.K. Park, S. Choi, Instrument-free synthesizable fabrication of label-free optical biosensing paper strips for the early detection of infectious keratoconjunctivitides, Anal. Chem. 88 (10) (2016) 5531–5537.
[82] W. Kim, S.H. Lee, S.H. Kim, J.C. Lee, S.W. Moon, J.S. Yu, et al., Highly reproducible Au-decorated ZnO nanorod array on a graphite sensor for classification of human aqueous humors, ACS Appl. Mater. Inter. 9 (7) (2017) 5891–5899.
[83] R. Botta, P. Chindaudom, P. Eiamchai, M. Horprathum, S. Limwichean, C. Chananonnawathorn, et al., Tuberculosis determination using SERS and chemometric methods, Tuberculosis 108 (2018) 195–200.
[84] A. Walter, A. März, W. Schumacher, P. Rösch, J. Popp, Towards a fast, high specific and reliable discrimination of bacteria on strain level by means of SERS in a microfluidic device, Lab Chip 11 (6) (2011) 1013–1021.
[85] W.J. Thrift, A. Cabuslay, A.B. Laird, S. Ranjbar, A.I. Hochbaum, R. Ragan, Surface-enhanced Raman scattering-based odor compass: locating multiple chemical sources and pathogens, ACS Sens. 4 (9) (2019) 2311–2319.
[86] J. Choi, S.J. Martin, R.A. Tripp, S.M. Tompkins, R.A. Dluhy, Detection of neuraminidase stalk motifs associated with enhanced N1 subtype influenza A virulence via Raman spectroscopy, Analyst 140 (22) (2015) 7748–7760.
[87] S.X. Li, Q.Y. Zeng, L.F. Li, Y.J. Zhang, M.M. Wan, Z.M. Liu, et al., Study of support vector machine and serum surface-enhanced Raman spectroscopy for noninvasive esophageal cancer detection, J. Biomed. Opt. 18 (2) (2013) 27008.
[88] L. Ou, Y. Chen, Y. Su, C. Zou, Z. Chen, Detection of genomic DNA damage from radiated nasopharyngeal carcinoma cells using surface-enhanced Raman spectroscopy (SERS), Appl. Spectrosc. 70 (11) (2016) 1821–1830.
[89] Q.J. Zhang, Y. Chen, X.H. Zou, W. Hu, X.L. Lin, S.Y. Feng, et al., Prognostic analysis of amyotrophic lateral sclerosis based on clinical features and plasma surface-enhanced Raman spectroscopy, J. Biophotonics 12 (8) (2019) e201900012.
[90] O. Guselnikova, A. Trelin, A. Skvortsova, P. Ulbrich, P. Postnikov, A. Pershina, et al., Label-free surface-enhanced Raman spectroscopy with artificial neural network technique for recognition photoinduced DNA damage, Biosens. Bioelectron. 145 (2019) 111718.
[91] V. Dugandžić, D. Drikermann, O. Ryabchykov, A. Undisz, I. Vilotijević, S. Lorkowski, et al., Surface enhanced Raman spectroscopy-detection of the uptake of mannose-modified nanoparticles by macrophages in vitro: a model for detection of vulnerable atherosclerotic plaques, J. Biophotonics 11 (12) (2018) e201800013.
[92] M. Potara, T. Nagy-Simon, A.M. Craciun, S. Suarasan, E. Licarete, F. Imre-Lucaci, et al., Carboplatin-loaded, Raman-encoded, chitosan-coated silver nanotriangles as multimodal traceable nanotherapeutic delivery systems and pH reporters inside human ovarian cancer cells, ACS Appl. Mater. Inter. 9 (38) (2017) 32565–32576.
[93] H. Kearns, R. Goodacre, L.E. Jamieson, D. Graham, K. Faulds, SERS detection of multiple antimicrobial-resistant pathogens using nanosensors, Anal. Chem. 89 (23) (2017) 12666–12673.

[94] C. Wang, F. Madiyar, C. Yu, J. Li, Detection of extremely low concentration water-borne pathogen using a multiplexing self-referencing SERS microfluidic biosensor, J. Biol. Eng. 11 (2017) 9.
[95] K.V. Kong, W.K. Leong, Z. Lam, T. Gong, D. Goh, W.K. Lau, et al., A rapid and label-free SERS detection method for biomarkers in clinical biofluids, Small 10 (24) (2014) 5030−5034.
[96] A. Hornemann, D. Drescher, S. Flemig, J. Kneipp, Intracellular SERS hybrid probes using BSA-reporter conjugates, Anal. Bioanal. Chem. 405 (19) (2013) 6209−6222.
[97] Y. Zhang, Z. Wang, L. Wu, S. Zong, B. Yun, Y. Cui, Combining multiplex SERS nanovectors and multivariate analysis for in situ profiling of circulating tumor cell phenotype using a microfluidic chip, Small 14 (20) (2018) e1704433.

Index

Note: Page numbers followed by "*f*" and "*t*" refer to figures and tables, respectively.

C

D

E

M

T

U

Printed in the United States
by Baker & Taylor Publisher Services